# PHYTOTHERAPEUTIC WISDOM OF INDIAN RURALS ABORIGINALS

## OTHER BOOKS

| Title | Author |
| --- | --- |
| A Manual of Ethnobotany 2nd Ed. | *Jain, S.K.* |
| Betelvine Cultivation & Management of Diseases | *Chaurasia, J.P.* |
| Bibliography of Indian Ethnobotany, 2002 | *Jain, S.K.* |
| Biodiversity & Sustainable Utilization of Biological Resource | *Sahu, T.R. et al.* |
| Biodiversity and Vegetation of Pachmarhi Hills | *Singh, V.P. et al.* |
| Biodiversity of Ranthambhore Tiger Reserve, Rajasthan | *Singh, V. et al.* |
| Dictionary of Indian Alchemy | *Saroya, A.P. Singh* |
| Ethnic Aphrodisiac Plants | *Sood, S.K. et al.* |
| Ethnic Indian Plants in Cure Diabetes | *Sood, S.K. et al.* |
| Ethnobotany and Medicinal Plants of India & Nepal Vol. 1-2 | *Singh, V. et al.* |
| Ethnobotany & Medicinal Plants of Indian Subcontinent | *Maheshwari, J.K.* |
| Ethnobotany of Rajasthan (India) | *Singh, V. et al.* |
| Flora of Andhra Pradesh Vol. 1-5 | *Pullaiah, T. et al.* |
| Floristic Biodiversity of Barda Hills and its Surrounding | *Nagar, P.S.* |
| Flowering Plants of Thrissur Forest | *Sasidharan, N. et al* |
| Flowers and Vegetables of India | *Sud, R.K. et al.* |
| Glossary of Medicinal Plants used in Ayurveda | *Saroya, A.P. Singh* |
| Handbook of Indian Medicinal Plants | *Joshi, M.C.* |
| Identification of Common Indian Medicinal Plants | *Naik, V.N.* |
| Indian Folk Medicines & Other Plant-based Products | *Singh, V.* |
| Leafy and Edible Plants of North-East India | *Kumar, S.* |
| Monograph on Indian *Leucas* R.Br. (Dronapushpi) Lamiaceae | *Singh, V.* |
| Monograph on Indian Subtribe Cassiinae (Caesalpiniaceae) | *Singh, V.* |
| The Economic Plants of North East India | *Kumar, S.* |
| The Flora of Jabalpur | *Oommachan M. et al.* |
| The Poisonous Plants of Bombay | *Kirtikar, K.R.* |
| Wetland Phytodiversity: A complete guide to Indian Helobieae | *Guha, Ratna et al.* |

# PHYTOTHERAPEUTIC WISDOM OF INDIAN RURALS ABORIGINALS

*Editor*

**V. SINGH**

**SCIENTIFIC PUBLISHERS (INDIA)**

P.O. BOX 91 JODHPUR

*Published by:*
Pawan Kumar
Scientific Publishers (India)
5-A, New Pali Road,
P.O. Box 91
Jodhpur - 342 001 (India)
E-mail: info@scientificpub.com
www.scientificpub.com



ISBN: 978-81-7233-533-5

Reprint from: *J. Econ. Taxon. Bot.*
Vol. 32 (Suppl.) 2008

Cover design: Tanay Sharma

Lasertype set: Rajesh Ojha
Printed in India

# EDITORIAL

The existence of man and animals on this planet has been directly related with vital role played by plant kingdom in sustaining life in symbiotic manner. The forest resources have direct or indirect impact on the life of local populations, tribals, forest dwellers etc. The socio-economic systems, customs, cultures and life patterns of these groups are closely related with forests. They use an enormous range of wild plants and have developed a unique understanding of forest resources and pass on these traditions and knowledge by word of mouth from one generation to other generation. The inventorisation and documentation of forest resources and tribals traditional know-how are important and crucial fields of research and development for attaining the theme of resource management, conservation of biodiversity and socio-economic development of the region. As such, the scope of ethnobotany has now greatly enlarged, both in terms of its theoretical contributions to an understanding of plant-human relationship, as well as for practical application and utilization of biological knowledge of tribals and their systems, particularly regarding medicines. During recent years, ethnobotany has emerged as a new branch of study documenting the age old wisdom, knowledge and experiences of the ethnic societies about the utility of diverse plant species and their miraculous properties as food, medicine and other miscellaneous uses, not yet known to the modern World.

Further, ethnobotanical studies in recent years have also received much attention due to their wide local acceptability and clues for new or less known medicinal plants. Today, it is widely accepted as a science of human interaction with plants and their ecosystems. The fundamental structure of ethnobotanical research is to examine the dynamic relationship between human population, cultural values and plants.

In India, the importance of traditional aboriginal medicines in the treatment of various human and cattle ailments is well known from older days. The epic Ramayana tells us about "Sanjivini Booti", a life saving drug. Our country has kept alive a number of traditional systems of medicine, like Ayurveda, Homeopathy, Sidha, Yoga, Unani, Tibatan etc. The ancient works like Charaka Samhita and Sushruta Samhita describe uses of more than 1200 plant-based drugs along with their action. In ancient literature it is also clearly mentioned that every plant on this earth is useful for human beings, animals and also for other plants. As such, organized ethnobotanical research is needed in this field. Not even a single family should be neglected during documentation of traditional knowledge since some information may be associated with that family. It is possible because traditional knowledge of a family or society may be unique based on beliefs, norms and culture. Hence, careful documentation and scientific analysis for further transmission are very important.

Unfortunately, the native people in all parts of World are loosing their culture and traditions, resulting lacuna in identification of economically important plants. Due to many fold urbanization, scientific break through and fast changing scenario of rural and tribal societies, not only biodiversity rich areas are shrinking but also the folk wisdom based on them. The tribal ways of life are rapidly breaking off under the impact of improved communication facilities and still more under various tribal welfare schemes of governments. Today, the impressed and urgent task of ethnobotanists is to record posterity whatever is valuable in the traditions of tribes, their way of life, their knowledge of plant etc, before all these disappear for ever. It has become more necessary for the countries like India because in recent times Ayurveda and Chinese systems of medicine have hit global market becoming a part of alternative system of medicine in many western and developed countries at one hand and India has been one of the major exporter of herbal products as well as raw drugs on the other hand.

Tribals constitute a notable segment of Indian population, representing about 8% of the total population. Therefore, immediate need of today is to highlight the useful plants to bring the uses to limelight for sustainable utilization of resources. The efficient utilization of available resources to an optimum level is supposed to bring prosperity in the country, since rich and diversified flora and tribal societies of India are most valuable storehouses of medicinal plants and medication. The medicines having roots in tribal knowledge are comparatively safe and environment friendly than synthetic ones. However, attempts are needed to validate ethnobotanical claims through phyto-chemistry, pharmacognosy, pharmacology, toxicology, screening and clinical tests.

In the end I wish to call ethnobotanists to document the ethnic traditions and knowledge regarding plants, especially medicinal ones and mode of medication at a faster rate, since culture has been the victim of civilization. Further, to achieve the goal of conservation, both bioresources and ethnic wisdom, they should make aware and encourage the traditional users of plants regarding their valuable know-how and its contributions to the human race worldwide. It may lead to encourage the younger generation to continue the use of traditional knowledge of their ancestors. Ethnobotanists should also act like a bridge in transmitting the ethnic knowledge of one group or society to other during their field work.

**V. Singh**
Chief Editor
Journal Economic & Taxonomic Botany

# CONTENTS

# A MIXTURE OF ALKALOIDS, FLAVONOIDS AND TERPENES/STEROLS WITH ANTIMYCOBACTERIAL ACTIVITY FROM METHANOL EXTRACT OF *ARTEMISIA ANNUA* L. (COMPOSITAE) (WHOLE PLANT)

A. UBA*, K. IBRAHIM**, E.B. AGBO* & A.A. MAKINDE***

**Biological Sciences Programme (Microbiology Unit), Abubakar Tafawa Balewa University, P. M. B. 0248, Bauchi, Nigeria*

***Department of Pharmaceutical Microbiology and Biotechnology, National Institute of Pharmaceutical Research and Development (NIPRD), P. M. B. 21, Garki, Abuja, Nigeria*

****Dermatophilosis and Diagnostics Unit, National Veterinary Research Institute (NVRI), Vom, P. M. B. 001, Jos, Plateau state, Nigeria*

*e-mail : ubaauwalu2000@yahoo.co.uk*

## INTRODUCTION

The World over reports indicated that the annual incidence of tuberculosis is increasing (American Thoracic Society, 1992; WHO, 1997; Lall & Meyer, 1999). Individuals infected with HIV become very susceptible to tuberculosis and develop the disease before other manifestations of AIDS become apparent (Grange & Davey, 1990). The HIV-tuberculosis co-infection has caused an impact on tuberculosis epidemiology all over the World in that the efficacies of the therapeutic schemes traditionally prescribed in the treatment of tuberculosis, such as isoniazid, rifampicin and pyrazinamide, have decreased due to the appearance of multidrug-resistant (MDR) *Mycobacterium tuberculosis* strains (Pietro *et al.,* 2000). It is estimated that between the years 2000 and 2020 nearly one billion people will be newly infected, 200 million will develop TB and 35 million will die from the disease (WHO, 2000). In 1993, the World Health Organization announced this chronic disease to be a global emergency (Raviglione *et al.,* 1995). Drug resistance by *Mycobacterium tuberculosis* and other atypical Mycobacteria pose serious problems to local and global TB control programmes. The present scenario, therefore, makes the search for novel antibacterial agents active against *Mycobacterium tuberculosis* and other atypical mycobacteria urgent.

*Artemisia* species are used for medicinal purposes and *Artemisia annua* is reported in the Chinese pharmacopoeia as a remedy for various fevers (Mueller *et al.*, 2000). The aerial parts of the plant contain 0.63-0.70 artemisinin per dry weight, and approximately 40% of this artemisinin can be extracted by simple tea preparation methods. *Artemisia* species are also used traditionally for the treatment of chest related diseases such as fever, cough and blood in the sputum (Ba-Ninge, 2000; personal communication). This indicates that species of the plant might contain substances active against multidrug resistant tubercle bacilli. This paper, therefore, reports on the antimycobacterial activity and phytochemical screening of extracts of *Artemisia annua*.

## MATERIALS AND METHODS

### Plant material

The plant sample was obtained from Maradi in the Republic of Niger. Preliminary identification of the plant was done at the herbarium of the Biological Sciences Programme, Abubakar Tafawa Balewa University, Bauchi, Nigeria where voucher specimen was deposited (AG 011). The dried plant sample was pounded into fine powder using a pistol and mortar. The powdered sample was transported to the Natural Products Research Laboratory, School of Biological Sciences, University of Wales Swansea, UK where the extracts were prepared and the plant finally confirmed.

### Test organisms

Four species of *Mycobacterium* were used for the study, namely : *Mycobacterium tuberculosis, M. bovis, M. avium-complex* and *M. smegmatis* ATCC 607. The first three species were isolated from sputum samples obtained from patients at Tuberculosis and Leprosy Hospital Bayara in Bauchi. Isolation was done according to standard procedures on Glycerol-Egg and Lowenstein-Jensen (L-J) media (Baker *et al.*, 1998). *Mycobacterium smegmatis* ATCC 607 was obtained from Microbiology Department of the National Institute for Pharmaceutical Research and Development Idu-Abuja.

### Preparation of plant extracts

The plant sample was subjected to sequential solvent extraction. This was done according to the methodology described by Newton *et al.* (2002). The plant material (1 kg) was soaked in 2.5 litre of hexane and allowed to stand for 72 hours with vigorous shaking after every 24 hours. At the end of extraction, the extract was filtered through Whatman No. 1 filter paper (England). Subsequently, the residue was extracted with chloroform, acetone, methanol and water in that order. All extracts except the aqueous were concentrated to dryness at <40°C using a rotary evaporator (Janke & Kunkee, Labortechnik). The aqueous extract was allowed to stand at room temperature under a fume cupboard until most of the water was evaporated. The resulted slurry was then concentrated to dryness at reduced pressure. The dried extracts were assessed for *in vitro* antimycobacterial activities.

## Determination of antimycobacterial activity

Each extract was screened for antimycobacterial activity using the Absolute Concentration Method (MIC Method) described by Canetti *et al.* (1969) and Vareldzis *et al.* (1993). An inoculum weighing 2 milligram (wet weight) from a solid growth was taken from an L-J culture slope within 2 weeks of the culture initially becoming positive for growth. The inoculum was suspended into 0.4 ml of sterile distilled water in a 7 ml screw capped bottle. A loopful of the suspension was spread on the surface of each L-J slope containing the plant extract. Slopes were incubated at 37°C and a preliminary reading made at 2 weeks. A final reading was made at 4 weeks. The plant extracts were separately incorporated into 250 ml batches of L-J medium before distribution into slants to obtain final concentrations of 0.1, 0.2, 1.0, 2.0, 5.0 and 10.0 mg/ml. The minimum inhibitory concentration (MIC) of each plant extract was taken as the lowest concentration that showed inhibitory activity. Control slants (medium without plant extracts) were also inoculated. The following criteria were used for the interpretation of results : Twenty or more colonies on slants with plant extract, the strain was considered resistant and less than 20 colonies, the strain was considered sensitive.

## Isolation of active fraction from *Artemisia annua* crude methanolic extract

The crude methanolic extract of *Artemisia annua* was subjected to solvent/solvent separation (Figure 1). Active fraction was further subjected to column chromatography eluting with hexane, ethyl acetate, acetone and methanol. Fractions were monitored by TLC (hexane : ethylactate : acetone) [15 : 3 : 2] using iodoplatinate as spray reagent and ultra violet light (T2202, SIGMA Chemical Company).

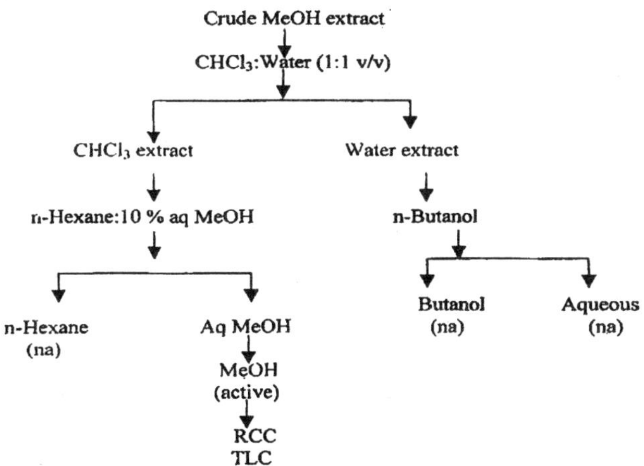


Symbols : na = not active, MeOH = methanol, $CHCl_3$ = chloroform, At = *Artemisia,* RCC = repeated column chromatography, TLC = thin layer chromatography.

Figure 1. Fractionation of methanol extract of *Artemisia annua* (Mahasneh & El-Oqlah, 1999).

## Identification of Major Chemical Groups in the Isolated Active Fractions

The identification of major chemical groups in the isolated active extract was carried out by TLC on pre-coated silica (silica gel $G_{F254}$) TLC plates (Merck, Germany). For terpenes and sterols, hexane: ethyl acetate (1 : 1) was used as a mobile phase and Liebermann-Burchard as reagent, a range of colours were produced after heating sprayed plates for 10 min at 100°C (Tona *et al.,* 1998). Alkaloids were detected by acid/base extraction procedure for alkaloids and analyzed by TLC in chloroform/ methanol/ammonia solution 25% 8 : 2 : 0.5 as solvent system. Spots were detected after spraying with Dragendorff's reagent.

For the detection of flavonoids, TLC was developed in n-butanol/acetic acid/ water 4:1:5. Spots were visualized with 1% aluminium chloride solution in methanol under ultra violet light (Harbone, 1974). Aqueous extracts were used for the detection of tannins with 1% gelatin solution & saponins by froth test (Tanira *et al.,* 1994).

Flavanols and flavanol-3-glycosides were detected with 0.5 g of metallic zinc and 10 drops of conc. sulphuric acid ($H_2SO_4$). Reducing compounds were detected by Benedict's test and, Cardiac glycosides detected by dissolving 0.5 g of sample in 2 ml of chloroform and a few drops of conc. $H_2SO_4$ added (Harbone, 1974).

## RESULTS

The minimum inhibitory concentrations of the crude extracts of *A. annua* (whole plant) obtained by sequential solvent extraction are shown in Table 1. Only the methanol extract inhibited all the four test organisms with MICs of 2.5, 2.0, 2.0 and 2.5 mg/ml against *M. smegmatis* ATCC 602, *M. tuberculosis, M. bovis* and *M. avium-complex* respectively. Chloroform and acetone extracts exhibited varying degrees of activity with MICs ranging between 4.0 mg/ml and 5.0 mg/ml. The hexane and water extracts did not show activity on any of the test organisms.

Only the methanol crude extract exhibited activity against all the test organisms and was, therefore, subjected to solvent/ solvent separation, TLC and column chromatography (Fig. 1). The isolated active fraction had MICs of 1.0, 0.5, 2.0 and 1.0 mg/ml against *M. smegmatis, M. tuberculosis, M. bovis* and *M. avium-complex* respectively. The major chemical groups detected in the fraction were alkaloids, flavonoids and terpenes/sterols.

Table 1. Minimum inhibitory concentrations of extracts of *A. annua* obtained by sequential solvent extraction (mg/ml).

| Test Orgs | Extract | | | | |
|---|---|---|---|---|---|
| | Hex | | Acet | Met | Wat |
| *M. smegmatis* | - | - | 5.0 | 2.5 | - |
| *M. tuberculosis* | - | 4.0 | - | 2.0 | - |
| *M. bovis* | - | 4.5 | - | 2.0 | - |
| *M. avium-complex* | - | - | - | 2.5 | - |

Symbols : Hex = Hexane, Chlo = Chloroform, Acet = Acetone, Met = Methanol, Wat = Water, - = No Activity.

## DISCUSSION

Among the extracts obtained by sequential solvent extraction, the methanol extract was found to be the most active against the test organisms (Table 1). The MIC values ranged between 2.0 mg/ml and 2.5 mg/ml. The activity of the methanol extract may be due to the fact that methanol being a polar solvent extract have a wide range of bioactive plant compounds such as terpenoids (Taylor *et al.,* 1996), flavones (Sato *et al.,* 1996), phenones (Peres *et al.,* 1997) and polyphenols (Vijaya *et al.,* 1995). Alkaloids, flavonoids and terpenes/ sterols were detected in the active fraction isolated from *A. annua* in the present study. This might explain the activity exhibited by the methanol extract against the test organisms in this study.

Although water is a polar solvent and most traditional healers recommend it as extractant in preparing herbal remedies, the aqueous extract in this study did not show activity on any of the test organisms. The possible explanation might be that methanol might have removed most of the active components before water was used as reported by Cowan (1999).

The active fraction from the methanolic crude extract of *A. annua* yielded a mixture of alkaloids, flavonoids and terpenes/sterols. The *in vitro* inhibitory activity noticed against the test organisms in the present study is in conformity with previous studies on the bioactivity of the three isolated plant compounds. Terpenes were reported to possess activity against bacteria and fungi (Goshal *et al.,* 1996; Barre *et al.,* 1997; Amaral *et al.,* 1998). This mode of action of terpenes involves membrane disruption by lipophilic compounds (Mendoza *et al.,* 1997). This might explain the *in vitro* inhibitory activity exhibited by the fraction isolated from the methanol extract of *A. annua* whole plant considering the fact that the test organisms have high lipid content in their cell walls. Flavonoids have also been reported to have *in vitro* antimicrobial activity against a wide range or microorganisms because lipophilic flavonoids disrupt microbial membranes (Tsuchiya *et al.,* 1994).

This study has established the pharmacological basis of the use of *A. annua* (whole plant) in the treatment of TB and/or its symptoms and, therefore, forms the basis for further research.

## ACKNOWLEDGEMENTS

We are thankful to Mal. Mahmoud Y.Y. of Microbiology Laboratory, Specialist Hospital, Bauchi and Mr. Emmanuel Mshelia of NIPRD, Abuja for their technical assistance. We are also thankful to all the staff of Bayara Tuberculosis and Leprosy Laboratory for their assistance in collecting sputum samples. The main author is thankful to the Association of Commonwealth Universities (ACU) for the fellowship grant to the University of Wales Swansea, UK and Mal. Bouboua Louali for providing the plant sample.

## REFERENCES

AMARAL, J.A., A. EKINS, S.R. RICHARDS & R. KNOWLES 1998. Effect of selected monoterpenes on methane oxidation, denitrification, and aerobic metabolism by bacteria in pure culture. *Applied and Environmental Microbiology* 64 : 520-525.

AMERICAN THORACIC SOCIETY 1992. Control of Tuberculosis in the United States. *American Review of Respiratory Diseases* 146 : 1623-1633.

BAKER, F.J., R.E. SILVERTON & C.J. PALLISTER 1998. *Introduction to Medical Laboratory Technology*, pp. 285-286. 7th ed. Butterworth Heinemann, Oxford.

BARRE, J.T., B.F. BOWDEN, J.C. COLL, J. JESUS, V.E. FUENTEN, G.C. JANAIRO & C.Y. RAGASA 1997. A bioactive triterpene from *Lantana camara. Phytochemistry* 45 : 321-324.

CANETTI, G., W. FOX, A. KHOMENKO, D.A. MITCHISON, N. RIST & N.A. SMELEV 1969. Advances in techniques of testing mycobacterial drug sensitivity and the use of sensitivity tests in tuberculosis control programs. *Bulletin of World Health Organisation* 41 : 21-43.

COWAN, M.M. 1999. Plant products as antimicrobial agents. *Clinical Microbiology Review* 12 (4) : 564-582.

GOSHAL, S., B.N. KRISHNA PRASAD & V. LAKSHMI 1996. Antiamoebic activity of *Piper longum* fruits against *Entamoeba histolytica in vitro* and *in vivo. Journal of Ethnopharmacology* 50 : 167-170.

GRANGE, J.M. & R.W. DAVEY 1990. Defection of antituberculosis activity in plant extracts. *Journal of Applied Bacteriology* 68 : 587-591.

HARBONE, J.B. 1974. *Phytochemical Methods : A Guide to Techniques in Plant Analysis*, pp. 89-131 and 279. Chapman and Hall, London.

LALL, N. & J.J.M. MEYER 1999. *In vitro* inhibition of drug resistant and drug sensitive strains of *Mycobacterium tuberculosis* by ethnobotanically selected South African plants. *Journal of Ethnopharmacology* 66 : 347-354.

MENDOZA, L., M. WILKENS & A. URZUA 1997. Antimicrobial study of the resinous exudates and of diterpenoids and flavonoids isolated from some Chilean *Pseudognaphalium* (Asteraceae). *Journal of Ethnopharmacology* 58 : 85-88.

MUELLER, M.S., I.B. KARHAGOMBA, H.M. HIRT & E. WEMAKOR 2000. The potential of *Artemisia annua* L. as a locally produced remedy for malaria in the tropics : agricultural, chemical and clinical aspects. *Journal of Ethnopharmacology* 73 : 487-493.

NEWTON, S.M., C. LAU, S.S. GURCHA, G.S. BESRA & C.W. WRIGHT 2002. The evaluation of forty-three plant species for *in vitro* antimycobacterial activities : isolation of active constituents from *Psoralea corylifolia* and *Sanguinaria canadensis. Journal of Ethnopharmacology* 79 (1) : 57-67.

PERES, M.T.L.P., F.D. MONACHE, A.B. CRUZ, M.G. PIZZOLATTI & R.A. YUNES 1997. Chemical composition and antimicrobial activity of *Croton urucurana* Baillon *(Euphorbiaceae). Journal of Ethnopharmacology* 56 : 223-226.

PIETRO, R.C., S. KASHIMA, D.N. SATO, A.H. JANUARIO & S.C. FRANCA 2000. *In vitro* antimycobacterial activities of *Physalis angulata* L. *Phytochemistry* 7 (4) : 335-338.

RAVIGLIONE, M.C., D.E.J. SNIDER & A. KOCHI 1995. Global epidemiology of tuberculosis : morbidity and mortality of a worldwide epidemic. *Journal of American Medical Association* 273 : 220-226.

SATO, M., S. FUJIWARA, H. TSUCHIYA, T. FUJII, M. LINUMA, H. TOSA & Y. OHKAWA 1996. Flavones with antibacterial activity against cariogenic bacteria. *Journal of Ethnopharmacology* 54 : 171-176.

TANIRA, M.O.M., A.K. BASHIR, R. DIB, C.S. GODWIN, I.A. WASFI & N.R. BANNA 1994. Antimicrobial and phytochemical screening of medicinal plants of the United Arab Emirates. *Journal of Ethnopharmacology* 41 : 201-205.

TAYLOR, R.S.L., F. EDEL, N.P. MANANDHAR & G.H.N. TOWERS 1996. Antimicrobial activities of Southern Nepalese medicinal plants. *Journal of Ethnopharmacology* 50 : 97-102.

TONA, L., K. KAMBU, N. NGIMBI, K. CIMANGA & A.J. VLIETINCK 1998. Antiamoebic and phytochemical screening of some Congolese medicinal plants. *Journal of Ethnopharmacology* 61 : 57-65.

TSUCHIYA, H., M. SATO, M. LINUMA, J. YOKOYAMA, M. OHYAMA, T. TANAKA, I. TAKASE & I. NAMIKAWA 1994. Inhibition of the growth of cariogenic bacteria *in vitro* by the plant flavonones. *Experimentia* 50 : 846-849.

VARELDZIS, P.B., J. GROSSET, I. DE KANTO, J. CROFTON, A. LASZLO, M. FELTEN, M.C. RAVIGLIONE & A. KOCHI 1993. *Laboratory evaluation of drug resistant tuberculosis,* pp. 1-12. WHO/TB/93.171, Geneva.

VIJAYA, K., S. ANANTHA & R. NALIMI 1995. Antibacterial effect of theaflavin, polyphenon 60 *(Camellia sinensis)* and *Euphorbia hirta* on *Shigella* spp. - a cell culture study. *Journal of Ethnopharmacology* 49 : 115-118.

WHO 1997. *Antituberculosis Drug Resistance Surveillance.* WHO Global tuberculosis Programme, Geneva.

———— 2000. *WHO Tuberculosis Fact Sheet No. 104.* http://who.int/inf-fs/en/fact 104.html.

# AN ETHNOBOTANICAL STUDY OF HALAM TRIBE OF DHALAI DISTRICT, TRIPURA, INDIA

R. HALAM, REEMA SAHA & B.K. DATTA

*Plant Taxonomy and Palynology Laboratory, Department of Life Sciences, Tripura University, Suryamaninagar, Tripura-West 799 130, India*

**ABSTRACT**

The present paper deals with the traditional uses of medicinal plants by the Halam tribal community inhabiting the forest outskirts of Dhalai district of Tripura. The particulars of the plant parts used, local names, mode of preparation and administration are given. In all 31 plant species belonging to 28 families used in primary health care are mentioned.

## INTRODUCTION

The north-east India is inhibited by a number of tribes of various ethnic groups, each with unique cultural heritage. This region is regarded as one of the world's 25 mega diversity hot spots. These ethnic groups have rich knowledge about medicinal plants. Their habitats in remote forest areas without modern medical facilities have compelled them to depend upon plants for their primary health care and by doing so generation after generation they have preserved rich lore of medicinal plants.

Tripura is situated between 22°56' and 24°32' N latitude and between 90°09' and 92°20' E longitude. It is bounded on the North-West, South and South-East by Bangladesh, where as in the East it has a common boundary with Assam and Mizoram.

The Halams are an important tribe inhabiting in North Tripura. They belong to the Kuki-Chin tribe of Tibeto-burma ethnic group of Sino-Tibetan family. The name of Halam is given by the Rajas of Tripura which comprises of several sub-tribes such as Ranglong, Chorai, Kaipang, Bong, Manc-hiphang, Molsom, Saimar, Dap, Longkai, Sakachap, Thtngachap, Hrangkhal, Rupini, Koloi, Korbong, Khulum etc.

Considering the significance of the study, a good number of researchers are now engaged in this field in north-east India and consequently a number of publications are available (Deb, 1968;

Borthakur, 1976; Debbarma, 1976; Majumder *et al.,* 1978; Rao & Jamir, 1982 etc). Though the State is rich in ethnobotanically important plant resources, little work have been done in this regard. The present paper deals with 31 ethnobotanically important plants used by the people of Halam community.

## METHODOLOGY

The present work is based on information collected from many herbal practioners and elders of the villages having the knowledge of herbal medicines. Field trips were conducted in entire Dhalai district, Tripura during 2003-2004, covering all seasons and entire Halam community. The local name, methods of preparation of drug and ailments treated were especially and carefully recorded from the tribals. Plant specimens collected during the survey were identified with the help of standard treaties and deposited in the herbarium, Department of Life Science, Tripura University, Suryamaninagar, 799 130, India.

## OBSERVATIONS

The plants are enumerated alphabetically with botanical names, family name in parenthesis and local name along with their ethnomedicinal uses.

**Allium ascalonicum** L. (Liliaceae)

*Local name* : Rumpurun.

The underground stem is crushed to make a paste which is mixed with water and is boiled for few minutes. The lukewarm mixture is used as a remedy for throat pain.

**Aloe vera** (L.) Burm. f. (Liliaceae)

*Local name :* Borelingte.

The juice extracted from the crushed leaves is given to ladies once daily for seven days suffering from white discharge.

**Alpinia nigra** (Gaertn.) Burtt (Zingiberaceae)

*Local name :* Mitdinglei.

The rhizome of the plant is crushed and applied on the forehead of the person having high temperature during fever.

**Andrographis paniculata** (Burm. f.) Wall. ex Nees (Acanthaceae)

*Local name :* Akhurlie.

Leaves are used for treatment of malaria. 5-6 matured leaves are taken from the fresh plants and are crushed into a paste. The paste is then boiled with 1 glass of water for few minutes. 1-2 teaspoon of the mixture are given to the patient.

**Cardiospermum helicacabum** L. (Sapindaceae)

*Local name :* Bakhate.

Mature leaves (4- 5) from the plant are mixed with 50 grams pepper and ginger each and crushed to make a paste which is applied on the forehead for quick relief from pain.

**Centella asiatica** (L.) Urban (Umbelliferae)

*Local name :* Ankhnaleng.

All the parts of the plant except flowers and roots are used for curing jaundice. It can be eaten directly (raw) or is crushed and the extract is taken.

**Crinum asiaticum** L. (Amaryllidaceae)

*Local name :* Amphut.

The plant is used as a remedy for chest pain due to cough. The underground stem is crushed, fried in mustard oil and is used for massaging the chest.

**Cuscuta reflexa** Roxb. (Cuscutaceae)

*Local name :* Ruiengte.

40-50 gms stem of the plant is crushed and extracted is filtered and given to person suffering from jaundice.

**Dicliptera bupleuroides** Nees (Acanthaceae)

*Local name :* Anrakhak.

The young 2-3 leaves are ground and the extract is given to the person suffering from dysentery for 3 days.

**Eupatorium odoratum** L. (Asteraceae)

*Local name :* Khorcheikpa.

The leaf juice 2-3 drops are applied on the wound or cut portion of the body for quick coagulation of blood.

**Euphorbia hirta** L. (Euphorbiaceae)

*Local name :* Mitnalei.

The leaf juice, 2-3 drops, is applied to the eyes in redness or burning.

**Flacourtia jangomas** (Lour.) Raeusch. (Flacourtiaceae)

*Local name :* Takchroi.

Capsule is made from leaf-paste and given to the person suffering from urine problem.

**Gomphrena globosa** L. (Amaranthaceae)

*Local name :* Parbosente.

The liquid leaf extract (2-3 drops) is applied on the wound or cut area for quick relief from pain.

**Hydrolea zeylanica** (L.) Vahl (Hydrophyllaceae)

*Local name :* Lunalei.

The leaves are ground and the paste is applied on the forehead for quick relief from pain.

**Jatropha curcas** L. (Euphorbiaceae)

*Local name :* Barathing.

The leaves are burnt, crushed into paste and applied on the swollen portion of the body due to injury.

**Kalanchoe pinnata** (Lam.) Pres. (Crassulaceae)

*Local name :* Patharkuchi.

Tender leaves (7 or 8 in numbers) are crushed with some drops of water, extract is filtered and given to person having stone in the stomach.

**Mirabilis jalapa** L. (Nyctaginaceae)

*Local name :* Parngaite.

The paste of the leaves is applied on the forehead for immediate relief from pain or migrain.

**Morus australis** Poir. (Moraceae)

*Local name :* Thurseamte.

The young leaf is crushed, boiled with water and given to the person for quick digestion.

**Parabaena sagittata** Miers. (Menispermaceae)

*Local name :* Ruirangte.

The leaves are used for treatment of fracture. 5-7 leaves are crushed and tightened over the fracture with the help of thin cloth for a week or longer.

**Persicaria barbata** (L.) Hara (Polygonaceae)

*Local name :* Tuichikte.

The leaf extract is given to a person in order to prevent Ascaris attack for 5 days.

**Punica granatum** L. (Punicaceae)

*Local name :* Dalim.

The tender leaves (5 leaves for adults, 2 for children) are eaten raw for quick relief from dysentery.

**Ricinus communis** L. (Euphorbiaceae)

*Local name :* Muchi.

The leaves are used for the treatment of arthritis. 2- 3 young leaves are crushed and tightened over the knee with the help of thin cloth for a week or longer.

**Sansevieria roxburghiana** Schult. & Schult. f. (Liliaceae)

*Local name :* Boreacheilpa.

The leaf-paste is applied on snake bite for quick relief from pain.

**Scoparia dulcis** L. (Scrophulariaceae)

*Local name :* Khongbaicheik.

Person suffering from indigestion takes the juice of the young leaves for 3 days.

**Senna alata** (L.) Roxb. (Caesalpiniaceae)

*Local name :* Saquaklei.

The young leaf paste is applied on the affected area of ringworm for seven days.

**Sida acuta** Burm. f. (Malvaceae)

*Local name :* Rumlumphe.

The leaves or stem (tender) are crushed to make into a paste, which is applied on boils and abscesses for quick and effective relief.

**Solanum nigrum** L. (Solanaceae)

*Local name :* Rummunta.

The roots alone are powdered and given orally for swellings due to liver disorders.

**Spilanthus paniculata** Wall. ex DC. (Asteraceae)

*Local name :* Uisnai.

The leaf-paste is applied on the affected area for quick relief from toothache.

**Tamarindus indica** L. (Caesalpiniaceae)

*Local name :* Tentary.

The one year old fruit extract of this plant is given orally to get relief from constipation.

**Torenia flava** Buch.-Ham. ex Benth. (Scrophulariaceae)

*Local name :* Pardumte.

The stem and leaves are crushed to make a paste which is given orally to cure indigestion and stomach pain.

**Zingiber zerumbet** (L.) Rosc. ex Smith (Zingiberaceae)

*Local name :* Rumithing.

The rhizomes mixed with pepper (50 gm) are crushed to make a paste which is then made into tablets and given 3 times a day to a person suffering from gastric.

## DISCUSSIONS

As Tripura is predominantly a rural area, it bears a great resource of ethno-botany and wide scope of research. Moreover, the plants used by the local tribes needs to be exposed properly. This detail study has been undertaken as an initiative approach to document the valuable information gathered from some tribal and non tribal medicine man of the Halam community and also from few old villagers having adequate knowledge about the use of the plants.

The mentioned plants are either used individually or sometimes in combination with many other different plants or plant parts. The herbal remedies mentioned in the present paper work against white discharge in women, swelling tooth and gum, bone fracture, fever, stomach-ache, eye inf ection, snake-bite, cough, head-ache, boils, dysentery, liver disorders, diabetes etc.

## REFERENCES

BORTHAKUR, S.K. 1976. Some known medicinal uses of plants among tribals of Mikir hills. *Bull. Bot. Surv. India* 10 : 166-171.

DEB, D.B. 1968. Medicinal plants of Tripura State. *Indian For.* 98 : 753-765.

DEBBARMA, M. 1976. 'Chuak'- A drink for Tribals in Tripura. *Folklore* 17 : 347-351.

MAJUMDER, R., K.C. TIWARI, S. BHATTA-CHARJEE & A.R. NAIR 1978. Some folklore medicines from Assam and Meghalaya. *Quart. J. Crude Drug Res.* 16 : 4-10.

RAO, R.R. & N.S. JAMIR 1982. Ethnobotanical studies in Nagaland-II : 54 medicinal plants used by Nagas. *J. Econ. Taxon. Bot.* 3 : 11-17.

# AN ETHNOBOTANICAL STUDY OF SAMBHARPAT OF WEST NIMAR DISTRICT OF MADHYA PRADESH

S.K. MAHAJAN*

*Retd. Professor, Botany Department, Govt. P.G. College, Khargone - 451 001, M.P., India*
**Mailing address: 31, Jain Mandir Path, Khargone - 451 001, M.P., India*

**ABSTRACT**

The present communication deals with the ethnobotanical study of Sambharpat and its neighbourhood areas (27°45' N longitude, 75°30' E latitude and 250 m above m.s.l.). The study revealed in all 16 taxa belonging to 16 genera and 12 angiospermic families that are used by the tribal people to cure various human diseases.

## INTRODUCTION

Ethnobotany or Medico-ethnobotany deals with the relationship between human societies and plants. Sambharpat is situated 25 km away from Khargone, the district headquarter of West Nimar, M.P. and it is a part of Central India. More than 30% of the total population of this area belongs to tribal community, viz. Barela, Tadwi, Bhilala, Bhil and Banjara.

A review of literature on ethnobotany indicates that although a lot of work has been done in India in general and M.P. in particular (Jain, 1963, 1965; Jain & Tarafdar, 1968; Maheshwari *et al.,* 1981; Maheshwari, 1990 and Oommachan & Masih, 1991), but it has been mainly limited to Mandla, Bastar, Balaghat and Jabalpur districts only. As regards the Southern M.P., a little work on ethnobotany has been done (Mahajan, 1990; 1995; Mahajan & Patel, 2003; Mahajan *et al.,* 2003). Earlier, the flora of West Nimar district was studied by Shastri (1977) and Mahajan (1987). Keeping this in view, the present investigation was undertaken to fill the gap.

## MATERIALS AND METHODS

The present work was done during 2002-03, and for this field work was done. The plants used by the tribal people to cure various diseases were collected with the consultation of old experienced tribal persons and tribal medicine-man. The herbarium sheets were prepared and identification was done following the standard literature (Cooke, 1901-08; Solanki, 1984; Verma *et al.,* 1993; Mudgal *et al.,* 1997; Singh *et al.,* 2001; Khanna *et al.,* 2001). The plants collected during the survey are arranged in alphabetical order with their scientific names, synonyms (if any), local names, plant parts used and names of families to which they belong. The herbarium sheets are deposited in the Botany Department, Govt. P.G. College, Khargone for future record.

## ENUMERATION

**Abrus precatorius** L. (Fabaceae)

*Local name :* Ratti

*Uses :* The root-extract is used for rheumatism and leaf-juice is given as blood purifier.

**Achyranthes aspera** L. (Amaranthaceae)

*Local name :* Hathijhara

*Uses:* Root-extract is given to cure pneumonia.

**Ageratum conyzoides** L. (Asteraceae)

*Local name :* Osadi

*Uses :* Leaves and fruits are used to check bleeding.

**Aloe vera** (L.) Burm. f. (Liliaceae)

*Local name :* Guarpatha

*Uses :* The juice of fleshy leaves is used in case of enlarged liver. Its juice when mixed with plaster is also used as repellant for white ants.

**Argemone maxicana** L. (Papaveraceae)

*Local name :* Satynasi

*Uses :* Its juice is used as diuretic.

**Aristolochia bracteolata** Retz. (Aristolochiaceae)

*Local name :* Girdhan

*Uses :* The decoction of dry leaves is administered twice daily in killing intestinal worms. Its leaves contain aristolochic acid.

**Barleria prionitis** L. (Acanthaceae)

*Local name :* Katasla

*Uses :* Leaves are chewed and swallowed in case of toothache.

**Cardiospermum halicacabum** L. (Sapindaceae)

*Local name :* Kanphuti

*Uses :* The roots, leaves and seeds are used in ear trouble.

**Cyperus rotundus** L. (Cyperaceae)

*Local name :* Nagar motha

*Uses :* The decoction of roots is given thrice in case of diarrhoea.

**Eclipta prostrata** (L.) L. (Asteraceae)

*Local name :* Kesari

*Uses :* Leaf-juice is used to cure sores on shoulders caused by carrying heavy loads and swellings on the ears of cattle.

**Leucas cephalotes** (Roth.) Spreng. (Labiatae)

*Local name :* Gattatumba

*Uses :* Flowers are used for cough and cold.

**Malvastrum coromandelianum** (L.) Garcke (Malvaceae)

*Local name :* Khareti

*Uses :* The leaves are used in diabetes and its bark is administered to check excess urination.

**Peristrophe paniculata** (Forssk.) Brummitt (Acanthaceae)

*Local name :* Atrilal

*Uses :* Roots are used in snake-bite.

**Trianthema portulacastrum** L. (Aizoaceae)

*Local name :* Gotlai

*Uses :* Leaf-juice is used to cure one-sided headache. It is diuretic and also used in rheumatism.

**Tridax procumbens** L. (Asteraceae)

*Local name :* Ghav-patti

*Uses :* Leaves are applied on cuts and wounds.

**Vitex negundo** L. (Verbenaceae)

*Local name :* Nirgur

*Uses :* Leaves are used in rheumatism and liver trouble. Roots are anthelmintic and flowers are used in diarrhoea.

## ACKNOWLEDGEMENTS

The author is thankful to the Principal, Govt. P.G. College, Khargone for facilities and encouragements.

## REFERENCES

COOKE, T. 1901-08. *Flora of the Presidency of Bombay,* Vol. I-II. London.

JAIN, S.K. 1963. Plants used in medicine by tribals of Madhya Pradesh. *Bull. Reg. Res. Lab.* Jammu 1 : 126-128.

———— 1965. Medicinal plantlore of the tribals of Bastar. *Econ. Bot.* 19 : 236-250.

———— & C.R. TARAFDAR 1968. Native plant remedies for snake bite among Adivasis of Central India. *Ind. Med. J.* 57(12) : 307-309.

KHANNA, K.K., A. KUMAR, R.D. RIXIT & N.P. SINGH 2001. *Supplement to the flora of Madhya Pradesh.* BSI, Calcutta.

MAHAJAN, S.K. 1987. Further additions to the flora of Khargone, Madhya Pradesh. *Proc. 74th Ind. Sci. Cong.* Part III (Bot. Sect.) Abst. pp. 106.

———— 1990. A preliminary ethnobotanical survey in the Sirwel hills of western M.P. *Natl. Seminar on Ethnobotany of Tribals of M.P.* (sponsored by MAPCOST) organised by S.G.S. Govt. P.G. College, Sidhi, M.P. Abst. No. 14

———— 1995. A preliminary ethnobotanical survey in the Bijagarh hills, M.P. *Seminar on Recent Researches on Science and Technology,* organised by Karimganj College Science Forum (KCSF), Karimganj under the ageis of Assam University, Silchar. Abst. pp. 4.

———— & PUSHPA PATEL 2003. Antiseptic plants of Khargone and its neighbourhood areas of Madhya Pradesh. Janarthanam, M. K. & D. Narasimhan (Eds.) *Plant Diversity, Human Welfare And Conservation,* pp. 227-229.

————, ————, P.P. SINGH & T. SAPRU 2003. A preliminary survey of the Anthelmintic plants used by the tribes of West Nimar District, Madhya Pradesh. *Ibid.* 289-292.

MAHESHWARI, J.K. 1990 The ethonobotany of tribals of Madhya Pradesh : An overview. Keynote Address, *Nat. Seminar on Ethnobotany on tribals of M.P.* (Sponsored by MAPCOST) and organised by Department of Botany, S.G.S. Govt. P.G. College, Sidhi, Abst. 01.

————, K.K. SINGH & S. SAHA 1981. *The Ethnobotany of Tharus of Kheri district, U.P.* E.B.I.S., Lucknow.

MUDGAL, V., K.K. KHANNA & P.K. HAJRA 1997. *Flora of Madhya Pradesh,* Vol. 2. BSI, Calcutta.

OOMMACHAN, M. & S.K. MASIH 1991. Ethnobotanical and conservational aspects of medicinal plants of Madhya Pradesh. *Indian J. Applied and Pure Biol.* 6 (1) : 39-44.

SHASTRI, M.B. 1977. *Flora of Khargone, Madhya Pradesh. Recent Trends and Contacts between Cytogenetics, Embryology and Morphology,* pp. 445-454.

SINGH, N.P., K.K. KHANNA, V. MUDGAL & R.D. DIXIT 2001. *Flora of Madhya Pradesh,* Vol. 3. BSI, Calcutta.

SOLANKI, C.M. 1984. *Flora of Indore.* Ph.D. Thesis, Devi Ahilya Vishwavidyalaya, Indore.

VERMA, D.M., N.P. BALAKRISHNAN & R.D. DIXIT 1993. *Flora of Madhya Pradesh,* Vol. 1. BSI, Calcutta.

# ASSESSMENT OF THE ANTIMICROBIAL AND PHYTOCHEMICAL PROPERTIES OF THE STEM BARK EXTRACT OF *DISTHEMONANTUS BENTAMIANUS* L. (AFRICAN SATIN WOOD) ON MICROORGANISMS ISOLATED FROM SEMEN

K.A. YONGABI, E.B. AGBO* & F. TAHIR*

*FMEnv/ZERI Centre, Abubakar Tafawa Balewa University*

**Biological Sciences Programme, Abubakar Tafawa Balewa University, PMB 0248, Bauchi, Nigeria.*

*Email : yongabika@yahoo.com*

## ABSTRACT

The assessment of the efficacy of the stem bark extracts of *Disthemonantus bentamianus* L., a component of a traditional preparation for infertility galenical, was carried out on microorganisms isolated from infected semen using the agar well diffusion method. The solvent polarity technique involving petroleum ether, Toluene, acetone, methanol and water was employed to extract the plant materials. The extracts were tested against seven organisms, viz. *Staphylococcus aureus, Escherichia coli, Klebsiella aerogenes, Pseudomonas aeruginosa, Enterobacter aerogenes, Proteus mirabilis* and *Candida albicans.* Only the acetone and methanol fractions of the extract showed appreciable inhibition on all the seven test microorganisms, with the acetone fraction exhibiting a higher inhibition on *Staphylococcus aureus.* No significant difference (P0.05) was observed between the activities of the methanol and acetone fractions using an independent T-test, while an F-test analysis revealed a significant difference (P0.05) in activities of the fractions between the various test microorganisms. Minimum inhibitory concentration (MIC) of the acetone fraction on all the test organisms ranged from 100-200 mg/ml except for *Enterobacter aerogenes* and *Proteus mirabilis* with 200-300 mg/ml. Preliminary phytochemical screening of the active fractions tested positive for alkaloids. The study justifies the use of this plant for treatment of most microbial infections.

## INTRODUCTION

Infectious diseases are responsible for one-quarter of all the death in the World, second only to cardiovascular diseases (WHO, 1999). They are associated with the new strains of bacteria and viruses emerged within the past twenty-five years, which are also highly resistant to drugs and antibiotics. Microorganisms affect different parts of the body such as gastrointestinal tract, ear, skin, urinary tract and even the reproductive organs. Their involvement in the seminal tract/ reproductive organs could result in several problems including infertility

It has been estimated that 8% of couples worldwide experience some form of infertility problem during their reproductive lives (WHO, 2000). Microbes have been implicated as partly responsible for this problem particularly *Staphylococcus aureus* and *Neisseria gonorrhoeae* (WHO, 2000). It is also becoming widely recognised that designing evermore-deadly drugs to kill bacteria is no longer a viable option, both financially and clinically. Apart from rapid resistance within short period after introduction, drugs that are toxic to bacteria are often toxic to human beings as well and the side effects may become much worse than the cure. Antimicrobials continue to be documented among plants that are used in traditional medical preparations. Researchers at Lucknow, India reported the antimicrobial activity of seedlings of plants grown on cow dung against eight bacteria and three fungal strains (Cho, 2000). There are many tried and tested antimicrobial agents from plants that improve the body's resistance to infection within traditional medical systems (Ahmad, 1998; Sharma & Chandra, 1998) which have been the subject of hundreds of Western scientific studies. These include ginseng, Astralgulus thistle, Echinacea and garlic as well as many others.

The plant under investigation (*Disthemonantus bentamianus*) belongs to the family Caesalpiniaceae and has been used in a traditional preparations to treat and manage infertility related problems in some parts of South West Cameroon. It has been commonly called "Quinine tree", because of its bitter taste. The tree is common in the tropical forests of Africa particularly in the western part of Cameroon as well as eastern part of Nigeria as reported by Hopkins & Stanfied (1966). Despite its use in traditional medicine, there has been little documented work on the plant especially on its antimicrobial activity. The objective of this study was to screen the extracts of this plant for antimicrobial and phytochemical properties using microorganisms isolated from infected semen as indicators.

## MATERIALS AND METHODS

### Test organisms

The test organisms obtained were *Staphylococcus aureus, Escherichia coli, Klebsiella aerogenes, Proteus mirabilis, Pseudomonas aeruginosa, Enterobacter aerogenes* and *Candida albicans*. These organisms were isolated from semen of patients who presented with infertility at various clinics in Bauchi metropolis. The method of semen collection and isolation of organisms was carried out using standard procedures specified by WHO (2000).

### Collection of plant samples (*Disthemonantus bentamianus* bark; Caesalpiniaceae)

Stem bark samples of the plant was collected and information on the plant was obtained by interviewing one of the traditional healers based in Kumba and Bamenda in the South and North Western parts of Cameroon. Samples collected and authenticated by experts at the Forest department (ONADEFO) in Kumba, the South West province of Cameroon and voucher herbarium specimens were deposited at the Biological Sciences Department of Abubakar Tafawa Balewa University, Bauchi, Nigeria.

## PREPARATION OF PLANT EXTRACTS

The fresh stem bark sample of the plant was dried under the shade at room temperature at the FMEnv/ZERI centre's laboratory. The sample was pounded into powder using a clean pestle and mortar.

The cold extraction procedure as described by Pavia & Kritz (1976) was employed. One hundred grams (100 g) of the pulverized fine poweder was soaked in 500 mls of the solvents. The various solvents employed were used in order to their increasing polarity, starting with the least polar. Each of the solvents was allowed to stand for 72 hours, after which it was filtered in order to get the fraction of that extract, then allowed to evaporate before treating with the next solvent. The order of treatment was as follows : Petroleum ether, Toluene, Acetone, Methanol and then water. The filtrate was evaporated to dryness in a fume cupboard at 25°C in order to remove all traces of the extracting solvent.

### Screening of extracts for antimicrobial activity

The Agar well diffusion method of the British Pharmacopoeia (1998) was employed. The assay media used were Nutrient agar (NA) for bacteria and potato dextrose agar (PDA) for fungi. From each of the extract, 0.7 grains were weighed using a solar weighing balance and dissolved in one ml of distilled water to give a final concentration of 700 mg/ml. The NA and PDA plates were seeded with viable cells of the isolates. Five wells were cut using a sterile 6 mm cork borer, thereafter each of the three wells was filled with only one fraction of extract while the fourth and fifth wells were filled with sterile distilled water and the extracting solvent respectively as controls. The plates were allowed to diffuse for two hours at room temperature before incubation at 37°C for 24 hours. The zone of inhibition produced by each fraction was measured to the nearest millimeter using a ruler.

### Determination of Minimum Inhibitory Concentration (MIC)

The Agar dilution method of Reeves *et al.* (1978) was employed. Prepared Nutrient agar plates containing varying concentrations of acetone and methanol fractions of the plant extract ranging from 700 mg/ml to 50 mg/ml in duplicate was carried out and plates allowed to solidify. The agar plates were spot inoculated with 0.01 ml of a $10^{-3}$ dilution of a 24 hours culture of the test organisms to a density of approximately $10^{6}$ cfu/ml using a calibrated micropipette and incubated at 37°C for 24 hour. A corresponding viable count of each volume or inoculums used was determined by plate count method. Following incubation the lowest concentration with inhibition was taken as minimum inhibitory concentration for the organism. Control agar plates with no extracts but with organisms inoculated were also set up for comparison.

**Preliminary phytochemical screening of extracts of *Disthmonantus bentamianus***

The tests for alkaloids, saponins, tannins and balsams (resins) were carried out on all the extracts to ascertain the class of active ingredient, following standard methods as reported by Brain & Turner (1973) and Evans (1989). The Dragendorff's and Mayer's tests were used to detect alkaloids, Froth test for saponins and Ferric chloride test for tannins and balsams (Evans, 1989).

## ANALYSIS OF RESULT

An independent T-test was used to compare the antimicrobial activities of the acetone to the methanol extracts while an F-test analysis was used to compare the efficacies of the extracts on all the various test isolates (Gay, 1992).

## RESULTS AND DISCUSSION

The results of the antimicrobial activity of five fractions of the extract of the plant on isolates of microorganisms are shown in Table 1. The results indicated that the petroleum ether, toluene and water extracts had no antimicrobial activity. Although the Acetone and Methanol extracts inhibited all the test organisms, there was no significant difference (P-0.05) between the activities of the two fractions on the test microorganisms.

There was a significant difference (P-0.05) on the antimicrobial activities of the fractions on the various test organisms. The inhibitory activity observed with acetone and methanol fraction against the test organisms justifies the use or the plant for the treatment of problems associated with infertility due to microbial infection. This finding also supports reports elsewhere (Dalziel, 1937; Ketharo & Bouquet, 1950; Bosisio *et al.,* 1997 and Tahir, 2001) that a number of African indigenous plants do possess antimicrobial activity.

The lack of activity for the petroleum ether extract could be attributed to its non-polar nature. Petroleum ether normally extracts fatty substances and conversely the antimicrobial activities of the acetone and methanol extracts suggest that the active ingredients are polar.

The lack of significant difference between the antimicrobial activities of acetone and methanol on the test organisms suggest that it is the same active ingredient in both fractions. It is possible that the active ingredient was not completely soluble in acetone, hence, the activity noticed with the methanol extract.

Table 1. Antimicrobial activity of fractions of extracts of the bark of *Disthemonantus bentamianus* L. (African satin wood) on Semen microbial isolates.

| **Fraction of extracts (700 mg/ml)** | **Zone of inhibition (mm) 24 hrs.** | | | | | | |
|---|---|---|---|---|---|---|---|
| | ***Staphylococcus aureus*** | ***Escherichia coli*** | ***Klebsiella aerogenes*** | ***Pseudomonas aeruginosa*** | ***Enterobacter aerogenes*** | ***Proteus mirabilis*** | ***Candida albicans*** |
| Petroleum ether | 0.0 | 0.0 | 0.0 | 0.0 | 0.0 | 0.0 | 0.0 |
| Toluene | 0.0 | 0.0 | 0.0 | 0.0 | 0.0 | 0.0 | 0.0 |
| Acetone | 25.0 | 14.8 | 19.0 | 14.3 | 12.0 | 12.0 | 16.0 |
| Methanol | 14.0 | 11.5 | 15.0 | 12.5 | 7.0 | 7.0 | 17.0 |
| Water | 0.0 | 0.0 | 0.0 | 0.0 | 0.0 | 0.0 | 0.0 |
| Solvent controls | 0.0 | 0.0 | 0.0 | 0.0 | 0.0 | 0.0 | 0.0 |

Symbols : 0.0 = No potency observed. (Average results of Duplicate Analysis).

Table 2. Minimum inhibitory concentration of acetone extracts of *Disthemonantus bentamianus* L. on Semen Microbial Isolates.

| **Fraction of extracts (700 mg/ml)** | ***Staphylococcus aureus*** | ***Escherichia coli*** | ***Klebsiella aerogenes*** | ***Pseudomonas aeruginosa*** | ***Enterobacter aerogenes*** | ***Proteus mirabilis*** | ***Candida albicans*** |
|---|---|---|---|---|---|---|---|
| 700 | NG | NG | NG | NG | NG | NG | NG |
| 600 | NG | NG | NG | NG | NG | NG | NG |
| 500 | NG | NG | NG | NG | NG | NG | NG |
| 400 | NG | NG | NG | NG | NG | NG | NG |
| 300 | NG | NG | NG | NG | G | G | NG |
| 200 | G | G | G | G | G | G | G |
| 100 | G | G | G | G | G | G | G |
| 50 | G | G | G | G | G | G | G |

Symbols : NG = No growth, G = Growth.

Table 3. Preliminary phytochemical screening of the bark extracts of *Disthemonantus bentamianus* L. (African satin wood).

| **Group of Metabolite** | **Petroleum ether extract** | **Toluene extract** | **Acetone extract** | **Methanol extract** | **Water extract** |
|---|---|---|---|---|---|
| Alkaloids | — | — | — | — | — |
| Tannins | — | — | — | — | — |

| | | | | | |
|---|---|---|---|---|---|
| Saponins | — | — | — | — | — |
| Balsams | — | — | — | — | — |

Symbols : + = Positive (present), - = Negative (absent), pH of the extracts = 7.0.

The results of the Minimum Inhibitory Concentration (MIC) of acetone extracts on the organisms are presented in Table 2. It was observed that the MIC of the acetone fraction on the organisms was within the range of 100-200 mg/ml except *Enterobacter aerogenes* and *Proteus mirabilis* at 200-300 mg/ml. The higher MIC values for these two organisms suggest that they are slightly less sensitive to the extract than the other organisms. The different responses exhibited may be associated to the fact that the active ingredient was not in high enough concentration to effect inhibition as previously noted by Garreft & Brown (1964) and Tahir (2001).

The results of the preliminary phytochemical screening of bark extract of *Disthemonantus bentamianus* are possibly alkaloids. This finding supports observations that compounds (phytochemical) like alkaloids, tannins, terpenes and unsaturated sterols etc do possess antimicrobial activity. For instance, the inhibitory effect of extracts from pawpaw (*Carica papaya*) has been attributed to the presence of a number of phytochemical agents, including alkaloids (Hashem *et al.,* 1980). Generally, the observation in this study attests to reports of Cho (2000) that antimicrobials continue to be documented among plants that are used in traditional medical systems, the World over.

The study, thus, confirms suggestion that *D. bentamianus* indeed possesses some antimicrobial activity. Its activity against organisms from infected semen also reinforce suggestions for its use in the treatment of fertility related problems. It is recommended that further studies be carried out on the plant to characterize the active ingredients for refined pharmaceutical applications.

## ACKNOWLEDGEMENTS

The authors gratefully acknowledge the funding from Research Grant Committee of Abubakar Tafawa Balewa University, Bauchi, Nigeria and Mr. Thomas Anchang and John Chechi of Kumba, Cameroon for helping in the identification of the plant.

## REFERENCES

AHMAD, I.M. 1998. Screening of some Indian medicinal plants for their antimicrobial properties. *Journal Ethnopharmacology* 62 (2).

BOSISIO, E., D. MASCETTI, L. VEROTTA, F. ZANI, P. MAZZA & M. TALBOT 1997. *Guiera senegalensis* J.F. Gmelin (Combretaceae) : Biological activities & chemical investigation. *Phytomedicine* 3 (4) : 339-348.

BRAIN, K.R. & T.D. TURNER 1973. *The Practical Evaluation of Pharmaceuticals.* Wright-Scientenchnica, Bristol. 1st ed., pp. 144, 152-153.

BRITISH PHARMACOPOEIA 1998. Printed by the majesty's stationary office at the University press, Cambridge, pp. 1102-1105. Updated version.

CHO, H.J. 2000. Traditional Medicine, Professional Monopoly and Structural Interest : A Korean Case. *Soc. Sc. Med.* 50 : 123-135.

DALZIEL, J.M. 1937. The useful plants of West Tropical Africa. Crown Agents.

EVANS, W.C. 1989. *Trease and Evan Pharmacognosy,* 13th ed., pp. 386-540. Bailliere Tindall, London.

GARREFT, E.R. & M.R.W. BROWN 1964. Kinetics and Mechanics of action of antibiotics on micro-organisms. *Journal Pharmaceutical Science* 53 : 179-181.

GAY, L.R. 1992. *Educational Research : Competencies for analysis and application,* 4th Ed., pp. 20-27. MacMillan Published Company.

HASHEM, F.M., M.Y. HAGGAG & A.M.A. GALEL 1980. A phytochemical study of *Carica papaya* L. grown in Egypt. *Egypt Journal Pharmaceutical Sciences* 21 (3-4) : 199-214.

HOPKINS & STANFIELD 1966. *Savanna trees of Nigeria. A field key.* Ibadan University Press.

KERHARO, J. & A. BOUQUET 1950. Plants medicinals Ectoxiques de la cote-D'voire Haute Volta, Vogot fere. Paris.

PAVIA & KRITZ LAMPMAN 1976. *Introductory organic laboratory techniques : A temporary Approach,* pp. 27-405. Saundors Company, Sanada, WB.

REEVES, D.S., I.D. PHILIPS & R. WIZE 1978. *Laboratory methods in Antimicrobial chemotherapy : Agar Dilution methods.* 1st ed. Churchlivingstone, England.

SHARMA, D.C. & U. CHANDRA 1998. Prophylactic uses of some medicinal plants in Bastar District in Madhya Pradesh. *Ancient Science of Life* 17 (4) : 284-289.

TAHIR, F. 2001. Screening and characterization of the antimicrobial components of some medicinal plants used in Bauchi. Ph.D. Dissertation, ATBU, Bauchi.

WHO 2000. Laboratory manual for the standardized investigation, diagnosis and management of infertile male. Cambridge.

# COLLECTION AND CONSERVATION OF TAMARIND (*TAMARINDUS INDICA* L.) - A MULTIPURPOSE TREE FOR LIVELIHOOD OF RURAL POOR

N. SUNIL, G. SUBBI REDDY*, N. SIVARAJ & S. SRIDHAR

*National Bureau of Plant Genetic Resources, Regional Station, Rajendranagar, Hyderabad 500 030, Andhra Pradesh*
**Acharya N. G. Ranga Agricultural University, Agricultural Research Station, Anantapur 515 001, Andhra Pradesh*

## ABSTRACT

Tamarind (*Tamarindus indica* L.) is an important tree in the rural areas not only for economic and environmental reasons but also for safeguarding the livelihood systems of the rural poor. The exploitation of the tamarind tree with its myriad uses has been limited mainly due to the non-availability of the elite high yielding genotypes for mass propagation. In this regard, a collaborative effort was made to collect the Tamarind diversity from parts of Andhra Pradesh by Regional Station of National Bureau of Plant Genetic Resources (NBPGR) with Acharya NG Ranga Agricultural University (ANGRAU). The survey was undertaken in parts of East Godavari and Visakhapatnam districts of Andhra Pradesh to identify trees exhibiting good variability in yield, bearing habit, size of the fruit, pulp content and taste etc for collection and conservation of representative diversity. Some areas in East Godavari, viz. Kakavada, are very famous for producing the quality Tamarind. In interior villages like Lankapakala and Velagapalli in East Godavari and Darakonda in Vishakhapatnam, large Tamarind population was observed which formed the base of the livelihood of rural population. The mission resulted in the collection of 44 Tamarind germplasm accessions. The collected scion material of Tamarind germplasm had been grafted on to the rootstocks by wedge grafting and the seed material has been deposited in the NGB for maintenance, evaluation, conservation and utilization of the germplasm.

## INTRODUCTION

Tamarind (*Tamarindus indica* L.) is an important tree in the rural areas not only for economic and environmental reasons but also for safeguarding the livelihood systems of the rural poor. It belongs to the family Caesalpiniaceae and has its origin in Africa, now widely distributed in tropical and sub-tropical regions of Asia and Africa with a good concentration in India. It is a large evergreen, drought resistant tree. Almost all the plant parts are economically important. Most economic part, the fruit pulp, is an important ingredient in the Indian cuisine as a condiment and dried pulp is used for polishing brass and copper wares. Leaves, flowers and fruits are used in dyeing industry as auxiliaries; flowers and leaves are also used for culinary purposes. The seed of Tamarind is used in the preparation of jelly powder. It also has many medicinal properties, besides utility as a timber and fuel. The soil binding ability of the Tamarind tree supplemented with its capacity to fix nitrogen, makes it a suitable species in the marginal and disturbed soils in the wild, agency areas and wastelands.

Tamarind, a slow-growing, long-lived tree reaches under favorable conditions a height of about 80-100 ft (24-30 m) and may attain a spread of 40 ft (12 m) with a trunk circumference of 25 ft (7.5 m). Foliage is composed of pinnate leaves, 3 to 6 inches (7.5-15 cm) in length with oblong leaflets. It is an evergreen tree with a leafing period in March-April. It flowers in May-June. Flowers borne in small racemes are 5-petalled, yellow with orange or red streaks. The flower-buds are pink due to the outer colour of the 4 sepals which are shed when the flower opens.

The flattish, bean like, irregularly curved and bulged pods are borne in the new branches and usually vary from 2 to 7 inches long and from 3/4 to 1 and 1/4 inches (2-3.2 cm) in diameter. The pods are brown or greyish-brown externally and at first are tender-skinned with green, highly acid flesh and soft, whitish, under-developed seeds. At maturity the skin becomes brittle and shell like and the pulp inside turns juicy and brown or reddish-brown. The pulp dehydrates naturally to a sticky paste with strands of fiber extending lengthwise from the stalk. Each pod may contain 1 to 12 fully formed hard glossy-brown seeds, squarish in form and each enclosed in a parchment like membrane.

## SURVEY AREA

The survey was focussed in the agency areas of Rampachodavaram and Chintapalli regions of East Godavari and Vishakhapatnam districts respectively. The area surveyed was mostly of high altitude area with height ranging from 500-3000 ft with undulating landscape. The important soil groups of the region surveyed are alluviums and red soils with clay base. The area receives a rainfall of 800-1100 mm mostly from the South-west monsoon. The maximum and minimum air temperatures during the southwest monsoon period in the region range from 32°C to 36°C and 23°C to 27°C respectively. Low temperatures of upto 5-6°C are also recorded during the winter season.

## MATERIALS AND METHODS

The survey was undertaken in parts of East Godavari and Visakhapatnam districts of Andhra Pradesh (Table 1) to identify trees exhibiting good variability in yield, bearing habit, size of the fruit, pulp content and taste etc for collection and conservation of representative diversity of the surveyed area. The required root-stock material was procured from central nursery under the conservator of forests, Rajahmundry. Biased and selective sampling strategy was followed as trees, which exhibited unique characters, were considered for collection from the existing populations in the village common areas. Semi- hardwood budwood material, which is reported to be successful in grafting, was collected. The sphagnum moss soaked in water was used to maintain the scion budwood material collected in good condition for grafting onto the rootstock on the same day. Each accession was replicated thrice while grafting by selecting the best scion budwood material collected. The wedge grafting was carried out with skilled assistance of HRS, Pandirimamidi research station of ANGRAU. A duplicate set of the entire germplasm is being maintained at ARS, Anantapur station. The fruits of the accessions, wherever available (37 accessions), were collected and seed extracted for *ex-situ* conservation in the gene banks.

Table 1. Areas surveyed for Tamarind germplasm collection in Andhra Pradesh.

| District | Mandal | Village |
|---|---|---|
| East Godavari | Rampacho-davaram | Anampalli |
| | | Kakavada |
| | | Lankapakala |
| | | Vusirijonnala |
| | | B. velamalakota |
| | | Vootla |
| | Maredumilli | Maredumilli |
| | | Jayavalasa |
| | | Rindikuravalasa |
| | Y. Ramavaram | Panasalapalem |
| | | Devaramadugula |
| | Addateegala | Nimmalapalem |
| | Devipatnam | Pamugandi |
| | | Velagapalli |
| Vishakhapatnam | G. K Veedi | R.V Nagar |
| | | Gudem colony |
| | | Lankapakala |
| | | Darakonda |
| | | Kongapakala |
| | Chintapalli | Bailakinchangi |
| | | Sarapalli |
| | | Choutupalli |

## RESULTS AND DISCUSSION

The survey aimed at collecting the representative diversity of the area explored, resulted in the acquisition of 44 diverse germplasm accessions of Tamarind. Some areas in East Godavari district, viz. Kakavada, are very famous for the production of quality Tamarind. In villages like Lankapakala and Velgapalli in East Godavari and Darakonda in Vishakhapatnam district, which are situated in the interior agency areas, a large natural population of the Tamarind trees numbering from 500-1000 were noted. From such populations, germplasm exhibiting good diversity in yield, bearing habit, size of the fruit, pulp content and taste etc was sampled. Based on time of flowering, Tamarind has been classified as early, mid and late groups. All the Tamarind germplasm accessions collected belong to mid to late groups and these are reported to be high yielding genotypes (Usha & Bhupinder, 1996). Priority was given for the collection of high yielding germplasm lines, followed by other important traits referred above. The yield reported for the accessions ranged between 200-1000 kgs/ tree, in comparison with reported high yields of 250 kgs/tree. It could be due to less stress during the flowering period in these high rainfall areas as inferred by Usha & Singh (1996). Various types of Tamarind genotypes, having different fruit shapes, viz. straight and bulged, straight and flattened, curved and bulged and curved and straight were collected as reported by Shivanandam & Thimmaraju (1988). For selecting high pulp content, fruit size rather than fruit shape was considered as suggested by Karale *et al.* (1999). Some of the diverse types collected during the survey are listed in table 2.

Table 2. Diversity in the Tamarind germplasm collected.

| S.N | Character | Collector no. |
|---|---|---|
| 1 | Early maturing types | IC-345504, IC-345525, IC-345529 |
| 2 | Late maturing type | IC-345524 |
| 3 | Heavy bearers | IC-345511 IC-345512, IC-345518, IC-345523, IC-345528, IC-345503, IC-345536, IC-345538, IC-345539, IC-345543. |
| 4 | Cluster bearing | IC-345502. |
| 5 | Regular bearers | IC-345497, IC-345498, IC-345518 |
| 6 | Long straight pods | IC-345500, IC-345503, IC-345528 |
| 7 | Small size pods | IC-345531, SSS-45. |
| 8 | White shell types | IC-345521 |
| 9 | High pulp content | IC-345536, IC-345539 |
| 10 | Bitter taste | IC-345542 |
| 11 | Sweet taste | IC-345527, IC-345524, IC-345531, IC-345550. |

The accessions collected from Velaga0 palli (IC-345536 with a very high yield) and Pamugandi (IC-345531 small fruits, sweet taste and high yield) in East Godavari district and accession numbers IC-345539, IC-345540 collected from Gudem colony and IC-345550, a very sweet tasting Tamarind collected from Bailankinchangi from Vishakhapatnam district are some of the prominent Tamarind accessions.

## CONSERVATION AND UTILISATION OF THE GERMPLASM

The collected bud-wood Tamarind germplasm material has been wedge grafted on to the rootstocks both at Horticultural Research Station (HRS), Pandirimamidi and Agricultural Research Station (ARS), Ananatapur. The established material is being conserved at the above stations of ANGRAU, which have a mandate of collection, evaluation and conservation of Tamarind germplasm. The seed material of the above germplasm has been sent along with the passport data to NBPGR, New Delhi for conservation in the gene bank and for assigning IC numbers. A set of the above germplasm (seed material) is also being conserved at NBPGR, RS, Hyderabad. The IC numbers received will be intimated to the above stations for assigning the same to the collected germplasm. The collected germplasm material needs to be characterized and evaluated to capitalize on the Tamarind germplasm diversity collected.

**Tamarind tree plantations-economics and its role in sustaining the livelihood system**

India is one of the foremost exporters of Tamarind in the World. In India, Tamarind is collected and marketed mainly by rural communities and forms a significant contribution to the family income of small and marginal farmers. Research done at Periakulam Research Station on the economics of Tamarind plantation shows that there is no income from the plantations for the first three years. However, inter-cropping with the annual crops would compensate for the labour and other costs involved in the initial establishment of the plantation, which worked out to Rs. 3,500 for 3 years. The trees yield upto 40 Kg of fruits per tree from the $4^{th}$ year onwards which translates into a net profit of Rs. 14,000 acre. A ten-year old tree, when the bearing is maximum, yields about 100 Kg of fruit with a profit of Rs. 38,000 acre. In tribal villages of South Bihar and South West Bengal, where there is a natural Tamarind population, the annual quantity and earnings from non-wood forest products including Tamarind per family is reported to be substantial (220 Kg of Tamarind earning Rs. 514) (Rao & Singh, 1996).

In the areas surveyed, which were identified from the local sources, the Tamarind formed the base of their livelihood apart from other fruit crops like jackfruit. In most of the villages, the harvesting was done on community basis. The returns from the sale of the produce were apportioned based on their contribution made during harvest time. The Tamarind was being sold after processing to Girijan Cooperative Cooperation (GCC). The Tamarind is being used only for culinary purpose apart from its use as fuelwood. In such areas where there is an abundance of natural Tamarind population and where the tamarind forms integral part of the livelihood systems of the rural people, alternative uses for effective utilization of the material with a bit of government support would go a long way in improving the living standards of the people.

## CONCLUSION

The diverse Tamarind germplasm may be tapped more effectively after characterisation and evaluation of the economic components like the pulp for sugar content. Such accessions which are sweet can be used in the preparation of sauce and shakes as in countries like Brazil, Mexico etc (Mortan, 1987). Seeds have to be evaluated for the polyose, a pectin which can be used in the preparation of jams, jellies and marmalades and also as a stabilizer in ice cream preparation. The seeds are also known to yield an amber oil which can be used as an illuminant and in varnish in painting dolls and idols. The leaves and flowers need to be analyzed for the dye which is used as a mordant in dyeing woollen and silk clothes, and the potential of Tamarind in lac industry as a host to lac insect (*Kerria lacca*) under Indian conditions also need to be studied. Thus, new avenues are to be explored in rural sector for value added products of Tamarind so as to improve the rural economy. The diversity of the germplasm collected will be a key aid in realizing this goal.

## REFERENCES

KARALE, A.R., A.P. WAGH, B.G. PAWAR & T.A. MORE 1999. Association of fruit characters in tamarind. *Journal Maharashtra Agriculture University* 24 (3) : 319-320.

MORTAN, J. 1987. Tamarind. In Mortan, J.F. & F. Miami (Eds.) *Fruits of warm climates,* pp. 115-121.

RAO, A.R. & B.P. SINGH 1996. Non-wood forest products contribution in tribal economy-A case study in South Bihar and South West Bengal. *Indian Forester* 122 (4) : 337-342.

SHIVANANDHAM, V.R. & K.R. THIMMARAJU 1988. Correlation between some fruit characters of four tamarind (*Tamarindus indica* L.) types. *Mysore Journal Agriculture Science* 22 : 229-231.

USHA, K. & B. SINGH 1996. Selection of Tamarind tree types for higher yield. *Crop Improvement* 23 (2) : 199-202.

# DENTAL CARE IN HERBAL WAY : SOME ETHNOBOTANICAL FINDINGS

ENAKSHI RAY CHAUDHURI, H.A. MOLLA & S.K. BASU

*Industrial Section, Indian Museum, Botanical Survey of India, Calcutta*

**ABSTRACT**

The paper deals with 119 medicinal plant species distributed under 106 genera and 57 families used by the tribal for different gum and tooth diseases. The species are arranged alphabetically.

## INTRODUCTION

A good gum with strong teeth makes a person healthy, wealthy and beautiful. Septic teeth, dental caries and pyorrhoea alveolaris are common causes of dyspepsia and serious ill health. The causes of dental caries are still debated. It is predisposed to by some underlying systematic factors and by deficiency of calcification of the teeth. To overcome the latter, various drugs containing Calcium and Vitamin D are now being prescribed. The fluridonation of drinking water probably helps to resist dental caries.

The commonest cause of gum and teeth diseases is lack of oral hygiene. The soft, pulpy, sweet and farinaceous foods around the teeth and gum encourage the growth of some acid producing microorganism. These acid producing microorganism are responsible for dental caries. Hence, adequate cleaning of teeth after talking every food is necessary.

The present paper deals with the plants used by the aboriginals against toothache and gum diseases. Keeping in view the importance and richness of the traditional healing methods a documentary work was carried out to gather information about various methods of treatment of toothache.

The traditional healing knowledge is inherited and transmitted by word of mouth from one generation to another generation. Under the duress of rapid urbanization and acculturation of indigenous people, the ethnobotanical knowledge is liable to be lost for ever. Therefore, documentation of this knowledge is the necessity of today. The information about some plants marked with asterisk (*) are new to the knowledge, collected by Sri H. A. Molla from different tribal localities of Jalpaiguri during the ethnobotanical tours. The voucher specimens are deposited in the Economic Botany Section, Central Botanical Laboratory, Botanical Survey of India, Calcutta.

In total, 120 plant species have been listed alphabetically with correct scientific name, family, local names/common names, uses of plant parts and related diseases.

## ENUMERATION

| Scientific Name/ Family | Common Name | Parts used, Disease & Remedy |
|---|---|---|
| *Acacia catechu* (L. f.) Willd. Mimosaceae | Khair(H), (Khaderi(Bo), Khadira(S), Kuth (B), Karangali (Tam), Khadiramu (Tel) | Bark is burnt and the ash is applied over the paining tooth for curing toothache. (Bora, 1999) |
| *Acacia nilotica* (L.) Willd. ex Del. ssp. *indica* (Bth.) Brenan Mimosaceae | Babla (B), Kikar (H), Babhula (Bo), Karuvelum (Tam), Nallatumma(Tel) | Leaf juice is applied to cure mouth sore. (Sur *et al.*, 1980). The tender branches are used as tooth brushes (Shah *et al.*, 1981) |
| *Achillea millefolium* L. Asteraceae | Arrowroot (Eng), Gandana (H), Rojmari (Bo), Momadruchopandiga (Kash) | Fresh herb is used to cure toothache (Kapali, 1990). Leaves are very effective against toothache and gum swelling (Kaul, 1990). The powdered rhizome is made into a paste and applied into the tooth cavity (Ganai & Nawchoo, 2003) |
| *Achyranthes aspera* L. Amaranthaceae | Latjira (H.), Apang(B), Kutri (P), Apamarga (S), Nayurivi(Tam), Uttareni (Tel) | Root pieces are chewed in toothache (Singh, 1999). Roots are used for cleaning teeth (Sen, 1990) |
| **Achyranthes bidentata* Blume Amaranthaceae | Ultakanta (N) | Pasted root is applied on the gum in infection and caries by the Nepalese. |
| *Aconitum heterophyllum* Wall. ex Royle Ranunculaceae | Patris (Kash) | The rhizome is powdered and mixed with mustard oil and the paste is then applied in to the tooth cavity (Ganai & Nawchoo, 2003). |
| *Acorus calamus* L. Araceae | Bach (H & B), Vaj (Bo), Vashambu (Tam), Vadaja (Tel), Bhadra (S) | Rhizome is used to cure toothache (Pal *et al.*, 1999). |
| *Aegle marmelos* (L.) Corr. Rutaceae | Bel (H, B & Bo), Bilva (S), Villuvam (Tam), Maredu (Tel) | Young branches are cut into pieces and used is brushing teeth (Das, 1997). |
| *Albizia lebbeck* (L.) Benth. Mimosaceae | Siris (H, B & Bo), Vagai (Tam), Dirasana (Tel), Shirisha (S) | Decoction of leaves and flowers is used for gargling in weak and spongy gums and chronic pharyngitis (Das, 1997). |
| *Allium sativum* L. Liliaceae | Garlic (Eng) | The paste of the bulb is applied on the gums and cavity of the infected teeth (Ganai & Nawchoo, 2003). |

| | | |
|---|---|---|
| *Alstonia scholaris* (L.) R. Br.<br>Apocynaceae | Chatium (H), Chhatim (B), Sapta (S), Edakulapala (Tel), Pala (Tam & Mal) | Leaf extract is used in toothache (Singh, 1999). |
| *Argemone mexicana* L.<br>Papaveraceae | Shialkanta (H & B), Srigala Kantaka (S), Bramadandu (Tam), Brahmadandi (Tel), Ponnummattam (Mal) | The latex is used in toothache (Singh, 1999). |
| *Asphodelus tenuifolius* Cav.<br>Liliaceae | Bokat (H), Bringharbij (Pu), Dungro (Guj) | Seed are kept in red hot earthen pot to give out fumes which are inhaled through mouth for carious teeth (Singh, 1999). |
| *Azadirachta indica* A. Juss.<br>Meliaceae | Nim (B, Bo & H), Nimba (S), Vepu (Tam, Tel, Mal) | Young stem is commonly used as tooth brush. Leaf extract is used to manufacture toothpaste and in other dental ailments such as bleeding/swelling of gums and pyorrhoea. (Nargas & Trivedi, 1999). |
| *Bambusa arundinacea* (Retz.) Roxb.<br>Poaceae | Bans (B & H), Mandgay (Bo), Vansh (S), Mungli (Tam & Mal), Bongu veduru (Tel) | Stem is used as toothbrush (Singh, 1999). |
| *Barleria cristata* L.<br>Acanthaceae | Jhinti (S & B), Tadrelu (P), Udamulli (Tam), Kodikannu (Tel), Jhinli (Assam) | The paste of the root is applied in the tooth pain (Chaudhuri *et al.*, 1975). |
| *Barleria prionitis* L.<br>Acanthaceae | Katsareya (H), Kantajati (B), Vajra-danti (Bo), Skemmuli (Tam & Mal), Mullugoranta (Tel), Karanta (S) | Leaves used in toothache (Pal *et al.*, 1999). |
| *Barleria strigosa* Willd. emend. Malhotra & Moorthy<br>Acanthaceae | Dassi (B), Wahiti (Bo), Nilakusuma (S), Nili (Tam), Nilambaramu (Tel), Nilakurunni (Mal) | Leaf-juice is applied in skin disease and as a gargle for mouth wash in gum pain (Tribedi *et al.*, 1982). |
| *Berberis lycium* Royle<br>Berberidaceae | Kashmol (H), Darhald (Bo) | It is used to cure toothache (Pal *et al.*, 1999). |
| *Bergia suffruticosa* (Delile) Fenzl<br>Elatinaceae | Karbuja, Rohwan (Rajathan) | Stems are used as tooth brushes (Shah *et al.*, 1981). |
| *Brassica rapa* L. ssp. *campestris* (L.) Clapham<br>Brassicaceae | Tilgogul (B) | Oil extracted from the seeds is mixed with salt and sodium bicarbonate and then applied into the tooth cavity to relieve pain (Ganai & Nawchoo, 2003). |
| *Butea monosperma* (Lam.) Taub.<br>Fabaceae | Palasha (S), Palas (B & Marathi), Dkak (H), Moduga (Tel), Parasa (Tam) | Pasted bark and leaves are given in toothache (Banerjee, 1999). |
| *Cajanus cajan* (L.) Millsp.<br>Fabaceae | Arhar (H, B & P), Adhaki (S), Tuver (Bo), Thovary (Tam), Kandulu (Tel), Thuvara (Mal) | Tender leaves of the plant are chewed in case of gum infection. (Soren *et al.*, 1999). |

| | | |
|---|---|---|
| *Calotropis gigantea* (L.) R. Br.<br>Asclepiadaceae | Ak (H), Akanda (B & Bo), Arkkam (Tam), Jilledu (Tel), Erikku (Mal) | Latex is used in toothache (Banerjee, 1999) and decay of teeth (Jain *et al.,* 1973). |
| *Cannabis sativa* L.<br>Cannabinaceae | Bang (Local Name), Bhang, Charas, Ganja (H, B & Bo) | The roots of the plant are chewed under the infected teeth to relieve pain (Ganai & Nawchoo, 2003). |
| *Capparis decidua* (Forssk.) Edgew.<br>Capparaceae | Karen (H), Karira (S), Sengun (Tam), Delha (P), Kari (Bo), Kariramu (Tel) | The decoction of the twigs with Brassica oil is used externally during toothache (Saxena *et al.,* 1983). |
| *Catharanthus roseus* (L.) G. Don<br>Apocynaceae | Nayantara (Beng), Ainskati (Oriya), Rattanjot (Punj), Sadaphul (Marathi), Bill ganeru (Tel) | The root is used to cure toothache (Sur *et al.,* 1987). |
| *Cinnamomum bejolghota* (Buch.-Ham.) Sweet<br>Lauraceae | Tezpat (B), Patichanda (Assam), Barasingoli (Nepal), Phalgoli (Kumaon) | Oil extracted from the bark is used in toothache (Islam, 2000). |
| *Clerodendrum viscosum* Vent.<br>Verbenaceae | Bhat (H), Ghentu (B), Vami (Santal), Kulamarscal (Kol) | Leaf extract is used to check bleeding of teeth (Singh, 1999). Fresh shoot is chewed with the paining tooth for treatment of toothache (Bora, 1999). |
| *Cocos nucifera* L.<br>Arecaceae | Nariyal (H), Dab, Narikel (B), Narel (Bo), Narikela (S), Tenkai (Tam), Narikelamu (Tel), Narikelam (Mal) | Gargling with root decoction relieves toothache (Rosakutly *et al.,* 1999). |
| *Commiphora wightii* (Arn.) Bhandari<br>Burseraceae | Guggulu (Sans), Guggul (H, Beng, Guj), Guggule (Mar & Kan), Guggul (Tel) | Lotion made of resin of the plant is applied to the toothache, week spongy gums pyorrhoea, tonsilitis and pharyngitis (Das, 1997). |
| *Cordia dichotoma* Forst. f.<br>Ehretiaceae | Lasora (H), Buhul, Bohodari (B), Gondan (Bo), Naruvili (Tam), Chinna nakkeru (Tel) | Decoction of bark strengthens teeth and gums (Kothari & Londhe, 1999). |
| *Dalbergia lanceolaria* L. f.<br>Fabaceae | Takoli (Hind), Chakemdia (Beng), Nat valanga (Tam) | The branches are used as tooth brush (Chaudhuri *et al.,* 1975). |
| *Datura metal* L.<br>Solanaceae | Dhatura (B), Dhatura (Bo), Sadahdhatura (H), Dhustura (S), Vellummattai (Tam) | Fruit is smoked for toothache (Siwakoti & Siwakoti, 1999). |
| *Datura stramonium* L.<br>Solanaceae | Datur (H) | Seeds, either crushed into a paste applied to the infected teeth or seeds mixed with butter and burnt and the smoke is inhaled into the mouth. After keeping the smoke in the mouth for one or two munutes, the sputum is thrown out (Ganai & Nawchoo, 2003). |

| | | |
|---|---|---|
| *Daucus carota* L. Apiaceae | Gajar (H, B, P), Skikla-mulam (S), Hayjara Kelangu (H) | Leaves are chewed to relieve toothache (Kumar & Naqshi, 1990). |
| *Delphinium denudatum* Wall. ex Hk. f. & Th. Ranunculaceae | Nirbisi (H), Judwar (P), Niruvisha (S) | The dried roots are chewed as stimulant and valued in toothache (Paliwal & Bedoni, 1990). |
| *Delphinium roylei* (Munz) Webb. Ranunculaceae | Judwar (H) | Small pieces of root put inside tooth cavities for pain relief (Kaul *et al.,* 1990). |
| *Desmodium triflorum* (L.) DC. Fabaceae | Kodalia (B), Kudaliya (H), Jangli methi (Bo), Sirapulladi (Tam), Munta mandu (Tel) | Crushed leaves applied locally for toothache (Ramchandram & Nair, 1981). |
| *Dicoma tomentosa* Cass. Asteraceae | Cholohamacharo (Guj), Navanayichopala (Marathi) | Roots are used in toothache (Singh & Pandey, 1980). |
| *Ehretia aspara* Roxb. Ehretiaceae | Tellajuvi (Tel), Chamror (P), Kupa (Marathi) | The Munda women blacken their teeth with the plant as a remedy of toothache (Soren *et al.,* 1999). |
| *Elephantopus scaber* L. Asteraceae | Gobhi (H), Gojialata (B), Hastipata (Bo), Anashovadi (Tam & Mal), Gojihva (S) | Roots chewed or paste applied to relieve toothache (Soren *et al.,* 1999; Siwakoti & Siwakoti, 1999). The root of this plant and root of *Zanthoxylum armatum* DC. are crushed and applied to kill worms in teeth (Jain *et al.,* 1973). |
| *Erycibe paniculata* Roxb. Convolvulaceae | Unamkodi (Tam), Puttapalatiga (Tel), Irimpiyatale (Mel), Kari (Santh) | The influsion of herbs is used as gargle in gum pain (Tribedi *et al.,* 1982). |
| *Eucalyptus globulus* Labill. Myrtaceae | Karpura marum (Mar) | Twigs used in toothache (Banerjee, 1999). |
| *Evolvulus nummularius* (L.) L. Convolvulaceae | Bichhamotia (Oriya) | The plant paste is applied on the affected tooth to cure toothache (Sur *et al.,* 1987). |
| *Ferula assafoetida* L. Apiaceae | Hing (B & H), Hingara (Bo), Hingu (S), Kayam (Mar) | The gum resin is an effective remedy in toothache when it is kept inside the cavity of the aching tooth (Goel & Aswal, 1990). |
| *Ficus benghalensis* L. Moraceae | Bor (H), Bar, Bat (B), Vad (Bo), Bor (P), Vada (Marathi), Pudavam (Tam), Peddamatti (Tel) | The latex collected through incision made in the bark is used in toothache (Paliwal & Benodi, 1990). Decoction of stem bark is used as gurgle in bleeding and spongy gums and as lotion in chronic irritable ulcers (Das, 1997). |
| *Ficus racemosa* L. Moraceae | Gular (H), Jagyadumar (B), Umbar (Bo), Atti (Tam, Tel & Mal) | Decoction of stem bark is useful for spongy gums (Das, 1997). |

| | | |
|---|---|---|
| *Flacourtia jangomas* (Lour.) Raeusch. Flacourtiaceae | Talispatri (H & N), Paniyala (B), Jaggam (Bo), Talisha (S) | Bark extract is given in bleeding gums and toothache (Islam, 2000). |
| *Fragaria nilgerrensis* Schlecht. ex Gay Rosaceae | Nilgiri straw berry (Eng) | The root is chewed for treatment of toothache and tooth decay (Megoneitso *et al.*, 1983). |
| *Gaultheria fragrantissima* Wall. Ericaceae | Machino (Nep), Kalomba (Lepcha) | A flavouring agent in tooth paste and tooth powders (Kumar *et al.*, 1999). |
| *Glochidion multiloculare* Voigt ex Muell-Arg. Euphorbiaceae | Tilbario (Mech) | Stem is used as tooth brush. |
| *Glycomis mauritiana* (Lamk.) Taraka Rutaceae | Ban nebu (H) | The twigs are used as tooth brush and said to be useful for toothache (Singh, 1999). |
| *Hedychium spicatum* Buch.-Ham. ex Smith Zingiberaceae | Sitruti (H), Kapurakachari (Marathi), Gandhashati (B), Sutti (Bo), Simaikkichilik, Kilhangu (Tam) | Stem is chewed to remove the bad smell from their teeth & mouth (Arya & Prakash, 1999). |
| *Hyoscyamus niger* L. Solanaceae | Bazar Bhag, Khurasani-ajvayan (H), Khorasani ajowan (B) | The smoke of the seeds on burning in a cup containing a coating of butter or ghee is inhaled into the mouth. After one or two minutes, the sputum is thrown out (Ganai & Nawchoo, 2003). |
| *Hyptis suaveolens* (L.) Poit. Lamiaceae | Gangatulsi (Oriya), Bilati tubi (B), Wilayati tubi (H) | Plant yields an essential oil which is applied as a remedy for toothache. The stem used as tooth brush (Das *et al.*, 1983). |
| *Indigofera caerulea* Roxb. Fabaceae | Nil (Merwara), Surmainil (Hindi), Karu nili (Tel), Kut avert (Tam), Kalaklitaka (Sans) | Stem branches are used as tooth brushes (Shah *et al.*, 1981). |
| *Jatropha curcas* L. Euphorbiaceae | Bagbherenda (H & B), Karanevranda (S), Mogalieranda (Bo), Kattava nakku (Mal), Katiya mudamu (Tel) | The juice of the plant is very effective for teeth or gum infection. The twig can be used as tooth brush during tooth complaint (Soren *et al.*, 1999). |
| **Jatropha gossypifolia* L. Euphorbiaceae | Samri (H) | Stem is used as tooth brush for recovery from gum infection by the Santals. |
| *Juglans regia* L. Juglandaceae | Akbort (H & B), Akroda (Bo & Marathi), Akrotuu (Tam & Tel), Akschota (S) | Root, bark and twigs for teeth and gums disease and for mouth cleaning and teeth. (Singh, 1999; Paliwal & Bedoni, 1990). |
| *Lannea coromandelica* (Houtt.) Merrill Anacardiaceae | Jhingan (Hind), Jiyal (Beng), Mol (Masthi), Mavedi (Guj), Appriyada (Tel) | Twigs chewed in toothache (Banerjee, 1999). |

| | | |
|---|---|---|
| *Lannea grandis* (Dennst.) Engl. Anacardiaceae | Jiyol (B), Moja (Bo), Jingan (H), Kalmina (Garhwal), Oli (Mal), Kimlu (P), Odi (Tam), Gumpenal (Tel) | Decoction of bark is used in apathy of mouth and toothache or swelling (Islam, 2000). |
| *Lawsonia inermis* L. Lythraceae | Hena, Mehndi (H), Mehedi (B) | Leaf-powder purchased from market is mixed with water and the paste is then applied on the cheeks corresponding to the infected ailing teeth (Ganai & Nawchoo, 2003). |
| *Leucas zeylanica* (L.) R. Br. Lamiaceae | Gattatumba (Sing) | The fresh leaves and young shoots are chewed for toothache (Magoneitso *et al.,* 1983). |
| **Lygodium flexuosum* (L.) Sw. Lygodiaceae | Binna (Raj) | Decoction of the root is used for gargling in toothache by Rajbansi. |
| *Madhuca longifolia* (Koenig) Macbride Sapotaceae | Mahuva (R), Mohwa (Bo), Mohua (M), Madhuka (S), Mippa (Mal) | Decoction of leaves or stem bark is used as gargle in bleeding and spongy gum. Twig is used as tooth brush (Das, 1997). |
| *Mangifera indica* L. Anacardiaceae | Am (B), Am, Amb (H & Bo), Mamaram (Tam), Amran (Mal), Amramu (Tel) | Twigs are commonly used as tooth-brush (Singh, 1999). |
| *Maytenus senegalensis* (Lamk.) Exell Celastraceae | Bharati, Yenkhad (Mar.) | Bark-decoction is used for tooth complaints (Kothari & Londhe, 1999). |
| **Mimosa pudica* L. Mimosaceae | Lajjabati (B), Lajja (O) | Decoction of the root is used for gargling in toothache and gum infection by Oraon.<br>The root is tied in neck with the belief that it will expel the worms from teeth. |
| *Mimusops elengi* L. Sapotaceae | Bakul (H & B), Borsali (Bo), Bakula (S), Bakulam (Mal), Malsari (P), Magilam (Tam), Vakulamu (Tel) | Extract of leaves and stem bark is made as gargle in weak and spongy gums, pyorrhoea, stomatitis and ulcerated throat. Twigs are used as tooth-brush in toothache (Das, 1997).<br>The bark of the tree is boiled in water and used as a mouth wash repeatedly to cure toothache(Sur, 1987). |
| *Morinda angustifolia* Roxb. Rubiaceae | Asho, Darhharidra (B), Asu-goch (Assam), Dieng - norg (Khasi), Chhennong (Garo), Kawnpel (Lushai) | Bark is used for toothache (Islam, 2000). |
| *Mukia maderaspatana* (L.) Roem. Cucurbitaceae | Bilari (Hind), Musa musukkai (Tam), Kutaru budama (Tel) | Roots is used in the treatment of toothache (Khanna *et al.,* 1994). |
| **Murraya paniculata* (L.) Jack. Rutaceae | Kamini (B), Chulajuti (Bo), Pandry (Kan), Marchula (H), Simaikkonji (Tam), Nagagolunga (Tel) | Hot water extract of leaves is used as toothwash for toothache (Rahman, 1999). Mech and Rajbansi of Jalpaiguri district use the decoction of leaves for gargling in toothache and caries. |

| | | |
|---|---|---|
| *Myrica esculenta* Buch.-Ham. ex D. Don Myricaceae | Kaiphal (H & B), Kaya phala (Man), Kariphal (G), Kaiduryanu (Tel), Marudam (Tam), Kirishivani (Kan), Marutal (Nal) | Bark with vinegar is given in gum and toothache (Islam, 2000). |
| **Neolamarckia cadamba* (Roxb.) Bosser Rubiaceae | Kadam (H) | The stem is used as tooth brush for one month in gum infection and caries. |
| *Nepeta cataria* L. Lamiaceae | Catmint (Eng) | Leaves chewed to relieve toothache (Sharma & Rana, 1999). |
| *Nicotiana tabacum* L. Solanaceae | Tambaku (H & Bo), Tamak (B), Tamaku (P), Tamakhu (S), Pugaiyilay (Tam), Pogaku (Tel), Pokala (Mal) | Leaves are chewed in case of spongy gums and toothache (Malhotra & Moorty, 1973). |
| *Origanum vulgare* L. Lamiaceae | Sathra (H), Mirganijosha (P & Urdu), Mridubaru-vanu (Tel) | The paste of the plant is applied for toothache (Kumar *et al.,* 1990). |
| *Oroxylum indicum* (L.) Vent. Bignoniaceae | Arlu (H), Sona (B), Tetu (Bo), Shyonakia (S), Vangam (Tam), Moppavepa (Tel), Aralu (Mal), Tatmorang (P) | The extract of the fruit is applied as a paint over teeth to prolong its life (Sharma & Rana, 1999). |
| *Paederia scandens* (Lour.) Merrill Rubiaceae | Gandhali (H), Gandhabhadulia (B), Savirela (Tel), Prasaram (Bo), Talanili (Mal), Hiravel (Marathi) | The extract of the fruit is applied as a pain killer over teeth to prolong its life (Sharma & Rana, 1999). Pasted leaves are used as tooth paste in toothache and gum infection by some tribal in Jalpaiguri district. |
| *Phoenix sylvestris* (L.) Roxb. Arecaceae | Khejur (H) | The mid rib of the leaf is used as tooth brush in the different villages of the 24 Paraganas, West Bengal. |
| *Phyllanthus fraternus* Webs. Euphorbiaceae | Jaramla (Hind), Bhuivali (B), Nile usirika (Tel), Nela nelli (Kan) | Chewing the leaves strengthen the teeth (Rosakutty *et al.,* 1999). |
| *Phyllanthus reticulata* Poir. Euphorbiaceae | Panjuli (H), Panjull (B), Datwan (Guj), Pavana (M), Nallapali, Palisar (Tel), Abiranji (Tam), Anamsha (Kan), Kilanchi (Mal) | Juice of the leaves is applied to bleeding gums (Das *et al.,* 1983). |
| *Physalis minima* L. Solanaceae | Tulatipati (H), Ban tepariya (B), Thanmari (Bo), Habbikaknaj (P), Kupanti (Tel) | The fruits are dried and powdered and rolled in a dry leaf and smoked like a cigar to relieve toothache (Sharma & Rana, 1999). |
| *Piper mullesua* D. Don Piperaceae | Thippalli (Kerala) | Stem crushed and applied for toothaches (Kumar *et al.,* 1999). |
| *Pogostemon benghalensis* (Burm. f.) Kuntze Lamiaceae | Gondrikulu (Oriya), Kusurijang (Tel), Lujra (Garhwal) | The young shoot ground with water is used to cure sores of mouth and tongue (Siwakoti & Siwakoti, 2000). |

| | | |
|---|---|---|
| *Pongamia pinnata* (L.) Pierre Fabaceae | Karanja (S & H), Dahur karanja (B), Karanj (Bo & P), Punnu (Mal), Pungu (Tam), Kranjuga (Tel) | The heated bark compressed over to cure tooth pain (Sur *et al.,* 1987). |
| *Potentilla fulgens* Wall. ex Hook. Rosaceae | Bhuitara (Beng), Lyngianbru (Khasi) | Local people use fresh roots for the treatment of stomachache and toothache (Arya & Prakash, 1999). |
| *Potentilla lineata* Trevir. ex Rich. Rosaceae | Rattonjot (Panjabi) | The powdered dried leaves and roots are used for washing the teeth (Paliwal & Bedoni, 1990). |
| *Punica granatum* L. Punicaceae | Anar (H) | The stem twigs are boiled in water and the water extract is used for mouth washing (Ganai & Nawchoo, 2003). |
| *Quirivelia frutescens* (L.) M. R. & S. M. Almeida Apocynaceae | Kalidudhi (H), Dudhi, Shyamlata (B), Savira (S), Illukkatti (Tel), Udargodi (Tam), Palvalli (Mal) | Leaf extraction is used in stomach pain. Stem extraction is used in toothache and gum swellings (Siwakoti & Siwakoti, 1999). |
| *Rheum australe* D. Don Polygonaceae | Revand Chuni (Hindi) | Plant is used to make tooth powder (Singh, 1999). |
| *Salvadora oleoides* Decne. Salvadoraceae | Bahapilu (H), Pilu (S), Kankhina (Bo), Pil (P), Kalwa (Tam) | Roots are used in toothache. (Singh & Pandey, 1980). |
| *Schrebera swietenioides* Roxb. Oleaceae | Ghantaparul (B), Mokagantha (Bo), Moka (H), Ghantapatali (S), Magalingum (Tam), Magalinga (Tel) | The leaf juice is applied to relieve toothache and sometimes painless removal of teeth (Shah *et al.,* 1981). |
| **Scoparia dulcis* L. Scrophulariaceae | Tina (Mech) | Decoction of the leaves is used for gargling in toothache by Mech. |
| *Sesamum orientale* L. Pedaliaceae | Til (Hindi & Mar), Tal (Guj), Nuvhulu (Tel), Ellu (Tam & Kan), Klasa (Oriya) | About 3 gms of seeds is chewed for strengthening the teeth (Saxena *et al.,* 1990). |
| *Shorea robusta* Gaertn. f. Dipterocarpaceae | Sal (H, P, B, Bo), Sala (S), Kungiliyam (Tam), Sarijakamu (Tel), Mulappumarutu (Mal) | Twigs are favourite tooth brushes (Banerjee, 1999). |
| *Sida acuta* Burm. f. Malvaceae | Bonmethi (B), Bariara (H), Jaglimethi (Bo), Vattatiruppi (Tam), Visaboddi (Tel), Malatanni (Mal) | Stem is used as tooth brush (Sur *et al.,* 1992). |
| *Smilax lanceaefolia* Roxb. Liliaceae | Chobehini (H), Guteashukchina (B) | The root is used in toothache and gum pain (Tribedi *et al.,* 1982). |
| *Smilax perfoliata* Lour. Liliaceae | Ram dataun (H) | Twigs are used as tooth brush in pyorrhoea (Singh, 1999). |

| | | |
|---|---|---|
| *Smilax zeylanica* L. Liliaceae | Kumarika (B), Langli auskleala (Bo), Guti (Marathi), Malaitha marai (Tam), Kondatamara (Tel), Kaltamara (Mal) | Stems are used as tooth brush to cure gum infection (Chaudhuri *et al.,* 1982). |
| *Solanum melongena* L. Solanaceae | Begun (B), Baigun (H), Vankaya (Tel), Kaththiri (Tam), Valuthina (Mal) | Berries vapours cure toothache (Kothari & Londhe, 1999). |
| *Solanum myriacanthum* Dunal Solanaceae | | Fruits are smoked for relieving toothache (Siwakoti & Siwakoti, 1999). Ripe and unripe fruit is boiled and cut into two helves and the halves kept over decayed teeth for half an hour. Fruits are crushed and boiled in a pot covered with banana leaf; in a small hole a hollow bamboo is fitted and the vapour is allowed to go to the mouth and teeth (Borthakur, 1976). |
| *Solanum viarum* Dunal Solanaceae | | Smoke of burning seeds is brought in contact with the teeth for expulsion of worms (Chaudhuri *et al.,* 1982). |
| *Solanum virginianum* L. Solanaceae | Kantakari (S & B), Kancliari (P), Kathali (H), Bhuringri (Bo), Kandargatheiri (Tam), Challamulaga (Tel), Kantankattiri (Mal) | Fumigation with the vapour of the burning seeds is very useful to relieve toothache. Juice of berries given in toothache (Rosakutty *et al.,* 1999). |
| *Spilanthes paniculata* Wall. ex DC. Asteraceae | Pirajha (Assam), Pipulka (Mar), Maratimogga (Tel), Kuppa mangel (Mal) | Leaf paste is applied over tooth for relieving toothache (Bora, 1999). |
| *Synedrella nodiflora* (L.) Gaertn. Asteraceae | Mudiyendrapacha (Mal) | The root is made into a paste with water and applied on gums (Molla & Roy, 1984). |
| *Syzygium cumini* (L.) Skeels Myrtaceae | Jambu (S & Bo), Jamun (H), Jam (B), Naval (Mel), Naredam (Tam), Jambuvu (Tel) | Leaf is used to treat blisters in mouth (Khanna *et al.,* 1994). |
| *Tectona grandis* L. f. Verbenaceae | Sagwan (H, P & Bo), Saka (S), Segun (B), Tekku (Tam), Tekka (Mal), Tega (Kan), Peddateku (Tel), Ching Jagu (Assam) | The juice when applied cures mouth sores and syphilis (Sur *et al.,* 1987). |
| *Telosma pallida* (Roxb.) Craib Asclepiadaceae | Surkila (Kumaon) | Root is said to be used by natives for making tooth brushes (Chaudhuri *et al.,* 1982). |
| *Terminalia chebula* Retz. Combretaceae | Haritaki (S & B), Harir (H), Hirda (Bo), Kadukkai (Tam), Karitaki (Tel), Katukka (Mal), Hilikha (Assam) | Decoction of the fruit with salt is applied to cure toothache (Sur *et al.,* 1990). |

| | | |
|---|---|---|
| *Thymus linearis* Benth. Lamiaceae | Banajwain (H), Marizha, Masho (P), Hasha (Oriya) | The paste of leaves and floral tops is applied in toothache (Paliwal & Bedoni, 1990). |
| *Toona hexandra* (Wall. ex Roxb.) Roem. Meliaceae | Tun (H & B), Tuni (Bo), Tunumaram (Tam), Khusing (P), Nandichetal | The pounded stem-bark is given to cure toothache (Siwakoti & Siwakoti, 2000). |
| *Urtica dioica* L. Urticaceae | Soi (H), Bichu (H & P) | The roots are chewed to relieve pain. (Ganai & Nawchoo, 2003). |
| *Vitex negundo* L. Verbenaceae | Nisinda (B), Nirgundi (S, Bo & B), Nirgandi (H), Vennochi (Tam & Mal), Nallavavilli (Tel), Niragundi (Kam) | The water obtained after boiling leaves in water is said to produce antiseptic effect, kills germs in teeth and relieves pain due to swelling of gums by washing the mouth with it (Tarafder, 1978). |
| *Wedelia chinensis* (Osbeck) Merrill Asteraceae | Bhanra (H), Bhimrakesraj (Beng), Pivalamaka (Mar), Bhangaro (Guj), Aswagandhi (Mal), Kalsarjii (Kan) | Yellow flower bearing inflorescence is rubbed on teeth to cure toothache (Rahman, 1999). |
| *Xylocarpus granatus* Koen. Meliaceae | Pussur (H), Dhundul (B), Kandalangay (Tam) | The seed-paste is applied in toothache (Saxena & Vyas, 1975). |
| *Zanthoxylum acanthopodium* DC. Rutaceae | Tejphal, Darmar (H), Tambul (Beng) | Small branches of the tree are used as tooth brushes (Paliwal & Bedoni, 1990). |
| *Zanthoxylum armatum* DC. Rutaceae | Tejbal (H & P), Tumburu (S), Nepalidpamia (B), Timru (Kumaon), Tumburudu (Kan), Kababe (Urdu) | Stem is used as tooth sticks and also remedy for toothache (Kapur, 1993). |

## ACKNOWLEDGEMENTS

Authors are grateful to Dr. H.S. Debnath, Scientist-C, Industrial Section, Indian Museum, Botanical Survey of India for his valuable suggestions in preparation of this manuscript.

## REFERENCES

ARYA, K.R. & VED PRAKASH 1999. Ethnobotanical study of a remote tribal area of Almora District : A Survey Report Part I. *J. Econ. Taxon. Bot.* 23 (2) : 247-252.

BANERJEE, ARCHANA 1999a. Ethnobotany of a few plant species in the eroded soil of Birbhum, West Bengal. *J. Econ. Taxon. Bot.* 23 (2) : 527-530.

——————— 1999b. Ethnobotany of some trees in the Santhal village of Birbhum, West Bengal. *J. Econ. Taxon. Bot.* 23 (2) : 531-534.

BORA, P.J. 1999. A study on ethnomedicinal uses of plants among the Bodo tribe of Sonitpur District, Assam, India. *J. Econ. Taxon. Bot.* 23 (2) : 604-608.

BORTHAKUR, S.K. 1976. Less known medicinal uses of plants among the tribes of Kabri-Anglong (Mikir Hills), Assam. *Bull. Bot. Surv. India* 18 (1-4) : 166-171.

CHAUDHURI, H.N. RAI, D.C. PAL & C.R. TARAFDER 1975. Less known uses of some plants from the tribal areas of Orissa. *Bull. Bot. Surv. India* 17 (1-4) : 132-136.

————, D.K. BANERJEE & A. GUHA 1977. Ethnobotanical uses of Herberia. *Bull. Bot. Surv. India* 9 (1-4) : 256-261.

————, H.A. MOLLA, D.C. PAL & B. ROY 1982. Plants used in traditional medicine by some tribes of Jalpaiguri Dist., West Bengal. *Bull. Bot. Surv. India* 24 (1-4) : 87-90.

DAR, G.H., VIRJEE, P. KACHROO & G.M. BUTH 1984. Ethnobotany of Kashmir-I, Sindh Valley. *J. Econ. Taxon. Bot.* 5 (3) : 668-675.

DAS, S.N. 1997. A study of Ethnobotany of Karauli and Sawai Madhopur District, Rajasthan. *J. Econ. Taxon. Bot.* 21(3): 587-605.

————, K.P. JANARDHANAN & S.C. ROY 1983. Some observations on the Ethnobotany of the tribes of Totopara and adjoining areas in Jalpaiguri District, West Bengal. *J. Econ. Taxon. Bot.* 14 (2) : 453-474.

GANAI, KHURSHEED AHMAD & INSHAD A. NAWCHOO 2003. Traditional treatment of toothache by the Gujjar and Bakevwal tribes of Kashmir in India. *J. Econ. Taxon. Bot.* 27 (1) : 105-107.

GOEL, A.K & B.S. ASWAL 1990. Less known uses of some plants in indigenous Folklore from Northern India. *J. Econ. Taxon. Bot.* 14 (1) : 185-188.

ISLAM, M. 2000. Ethnobotany of bark of certain plants of North East India. *J. Econ. Taxon. Bot.* 24 (2) : 419-432.

JAIN, S.K., D.K. BANERJEE & D.C. PAL 1973. Medicinal plants among certain Adibasi in India. *Bull. Bot. Surv. India* 15 (1 & 2) : 85-91.

KAPALI, B.K. 1990. Ethnobotanical investigation in Lahul (Himachal Pradesh). *J. Econ. Taxon. Bot.* 14 (1) : 49-55.

KAPUR, S.K. 1993. Ethnomedico plants of Kangra valley (Himachal Pradesh). *J. Econ. Taxon. Bot.* 17 (2) : 395-408.

KAUL, M.K. *et al.,* 1990. Ethnobotanical studies in North-West and Trans Himalaya II. Approaches to the study of Ethnobotany towards the human welfare in remote North West and Trans Himalayas. *J. Econ. Taxon. Bot.* 14 (2) : 271-285.

KHANNA, K.K., V. MUDGAL, G. SUKLA & S.K. SRIVASTAVA 1994. Unreported Ethnomedicinal uses of plants as aphrodisiac from the folklore of Uttar Pradesh plains, India. *Bull. Bot. Surv. India* 36 (1-4) : 91-94.

KOTHARI, M.J. & A.N. LONDHE 1999. Ethnobotany in human health care of Chikhaldara Amarawati District in Maharashtra State, India. *J. Econ. Taxon. Bot.* 23 (2) : 273-281.

KUMAR, G.M. & NAQSHI 1990. Ethnobotany of Jammu-II Banihal. *J. Econ. Taxon. Bot.* 14 (1) : 67-74.

KUMAR, K. KISHOR, N. SASIDHARAN & K. SWARUPANDAN 1999. Ethnobotanical studies on the Hill tribes in the shola forests of high ranges, Kerala, South India. *J. Econ. Taxon. Bot.* 23 (2) : 451-466.

MAGONEITSO *et al.,* 1983. Ethnobotanical studies in Nagaland - 4 : Sixty two medicinal plants used by the Agami Nagas. *J. Econ. Taxon. Bot.* 4 :167-172.

MALHOTRA, S.K. & S. MOORTY 1973. Some useful and medicinal plants of Chandrapur District. *Bull. Bot. Surv. India* 15 (1 & 2) : 13-21.

MOLLA, H.A. & B. ROY 1984. Folklore about some medicinal plants from the tribal areas of Jalpaiguri Dist., West Bengal. *Bull. Bot. Surv. India* 26 (3 & 4) : 160-163.

NARGAS, JYOTI & P.C. TRIVEDI 1999. Traditional medicinal importance of *Azadirachta indica* Juss. in India. *J. Econ. Taxon. Bot.* 23 (2) : 369-375.

PAL, D.C., A. GUHA & R. SEN 1999. Medicinal plants and plant products used in children diseases among aboriginals in India. *J. Econ. Taxon. Bot.* 23 (1) : 121-136.

PALIWAL, G.S. & A.K. BEDONI 1990. Ethnobotany of the Hill Tribes of Uttarkashi : Medicinal Plants. *J. Econ. Taxon. Bot.* 14(2):421-422.

PANDA, P.C. & P. DAS 1999. Medicinal plantlore of the tribals of Baliguda Sub-division, Phulbani Dist., Orissa. *J. Econ. Taxon. Bot.* 23 (2) : 515-521.

RAHMAN, M.A. 1999. Ethnomedico-botanical knowledge among tribals of Bangladesh. *J. Econ. Taxon. Bot.* 23 (1) : 89-93.

RAMCHANDRAM, V.S. & N.C. NAIR 1981. Ethnobotanical observations on rurals of Tamilnadu (India). *J. Econ. Taxon. Bot.* 2 : 183-190.

ROSAKUTTY, P.J., A. STELLAROSLIN & V. GNACIMATH 1999. Some traditional Folklore Medicinal plants of Kannya Kumari District, Tamilnadu. *J. Econ. Taxon. Bot.* 23 (2) : 369-375.

SAXENA, A.P. & K.M. VYAS 1983. Ethnobotany of Dhasan Valley. *J. Econ. Taxon. Bot.* 4 : 121-128.

SAXENA, H.O. & P.K. DUTTA 1975. Studies of Ethnobotany of Orissa. *Bull. Bot. Surv. India* 17 (1-4) : 124-131.

SAXENA, SUNEEL KUMAR & JAI PRAKASH TRIPATHI 1990. Ethnobotany of Bundelkhand-II. Folklore therapy through herbs among inopulent parishioners and of aboriginal tribes. *J. Econ. Taxon. Bot.* 14 (2) : 263-270.

SEN, R. *et al.,* 1990. Ethnobotanical uses of Herberia–IX. *J. Econ. Taxon. Bot.* 14 (2) : 340-355.

SHAH, G.L., A.R. MENON & G.V. GOPAL 1981. An account of the Ethnobotany of Saurashtra in Gujarat State (India). *J. Econ. Taxon. Bot.* 2 : 173-182.

SHARMA B.D. & J.C. RANA 1999. Traditional medicinal uses of plants of Himachal Hills. *J. Econ. Taxon. Bot.* 23 (1) : 173-176.

SHARMA, U.K. 1999. Folk and herbal medicine among Nepalese of Assam. *J. Econ. Taxon. Bot.* 23 (2) : 599-603.

SINGH, AJAI KUMAR 1999. A contribution to the ethnobotany of sub Himalayan region of Eastern Uttar Pradesh. *J. Econ. Taxon. Bot.* 23 (1) : 237-246.

SINGH, S. GOPAL 1999. Ethnobotanical study of useful plants of Kulu District in North India. *J. Econ. Taxon. Bot.* 23 (1) : 185-198.

SINGH, V. & R.P. PANDEY 1980. Medicinal plantlore of the tribals of Eastern Rajasthan (India). *J. Econ. Taxon. Bot.* 1 : 137-147.

SIWAKOTI, M. & S. SIWAKOTI 1999. Ethnomedicinal uses of plants among the Satar tribe of Nepal. *J. Econ. Taxon. Bot.* 23 (1) : 99-108.

——— & ——— 2000. Ethnomedicinal uses of plants among the Satar tribe of Nepal. *J. Econ. Taxon. Bot.* 24 (2) : 323-333.

SOREN, A.M., R. SEN & D.C. PAL 1999. A contribution to the Ethnobotany of Bankura District, West Bengal. *J. Econ. Taxon. Bot.* 23 (2) : 545-555.

SUR, P.R., RATNA SEN, A.C. HALDER & S. BANDYOPADHYAY 1987. Observation on the Ethnobotany of Malda-West Dinajpur District, West Bengal-I. *J. Econ Taxon. Bot.* 10 (2) : 395-401.

———, ———, ——— & ——— 1990. Observation on the Ethnobotany of Malda-West Dinajpur District, West Bengal-II. *J. Econ Taxon. Bot.* 14 (2) : 453-459.

———, ———, ——— & ——— 1992. Ethnomedicine in then Ajodhya Hills Region of Purulia District, West Bengal, India. *J. Econ. Taxon. Bot.* Addl. Ser. 10 : 333-337.

TARAFDAR, C.R. 1978. Some interesting new uses of the well known plant *Vitex negundo* Linn. (Nisinda). *Bull. Bot. Surv. India* 20 (1-4) : 176-177.

TRIBEDI, G.N., R.N. KAYAL, RAI, H.N. CHAUDHURI 1982. Some medicinal plants of Mayurbhanj (Orissa). *Bull. Bot. Surv. India* 24 (1-4) :117-120.

# DYE YIELDING PLANTS OF ASSAM AND ITS NEIGHBOURING AREAS

M. ISLAM & F. HASIN

*Department of Life Sciences, Dibrugarh University, Dibrugarh- 786 004 (Assam)*

## ABSTRACT

A comprehensive study on dye yielding plants of Assam and its neighbouring areas revealed that 62 species belonging to 52 genera and 34 families are used for different colours by the natives. The dicotyledonous species are found to be dominant over all others in yielding dyes.

## INTRODUCTION

Dyes are used for imparting colour to a wide variety of materials, as textile, leather, food, drugs, paints, varnishes, paper, printing, cosmetics etc and these are obtained from natural resources or prepared chemically. Natural dyes are obtained from roots, barks, flowers, fruits and wood of several plants since the beginning of civilization.

Several works have been done from time to time in India, as in 1881 Leotard drew up a memorandum of vegetable dyes of India, McCann (1883) and Fraser (1896) reported the occurrence of dyes and tans from Bengal and Burma respectively, Watt (1883) made an account of Indian dyes exhibited in the International Exhibition held at Calcutta, Burkill (1908) gave an idea in connection with the use of dyes from flowers in India and Dastur (1951) gave an account of certain useful plants. Recently, Manju Nath *et al.* (1948), Sastri *et al.* (1950-72) and others compiled update information in this regard. It was noted that all these studies refer chiefly to textile industry, but very little work has been done on non-toxic dyes worthy of use in food-stuff. Dyes obtained from *Bixa orellana, Carthamus tinctorius, Curcuma longa, Daucus carota, Garcinia xanthochymus, Hibiscus rosa-sinensis, Mallotus philippensis* etc. are non-toxic and are used for colouring food stuff.

Assam, the easternmost State of India, lying between the latitude 22°19' and 28°16' north and 84°42'and 96°30' east longitude is bounded by Bangladesh in the west, Meghalaya in the south, Nagaland in the south-east and Arunachal and Bhutan in the north. It is one of the monsoon belts of India with favourable climatic conditions. Physical, climatic and geographical characteristics of the region have been described elsewhere (Islam, 1984, 86) and the same are not reiterated here. The aim of the present investigation has been to enumerate dye yielding plants occurring in different localities of plains and hilly tracts of Assam and neighbouring areas.

## MATERIALS AND METHODS

During several ethnobotanical tours the dyeing properties of different parts of the plants have been recorded through personal observations or data collected from different users of tribal and non-tribal areas or from literature. Source of information has been indicated in the Table 1.

## ENUMERATION

Table 1. List of dye yielding plants of Assam and its neighbouring areas.

| Botanical names (local names, if any) | Families | Parts used | Colour produced | Source of information |
|---|---|---|---|---|
| *Acacia catechu* (L. f.) Willd. | Mimosaceae | Wood | Brown | Local people, Watt (1883) |
| *A. farnesiana* (L.) Willd. | Mimosaceae | Leaves, bark and fruit | Black | Leotard (1881), Watt (1883) |
| *Aegle marmelos* (L.) Corr. (Bel) | Rutaceae | Bark and fruit rind | Dull brown or yellow | -do- |
| *Albizia lebbeck* (L.) Benth. | Mimosaceae | Bark | Brown | Local people |
| *Annona reticulata* L. | Annonaceae | Leaves | Black | -do- |
| *Artocarpus heterophyllus* Lamk. (Kathal) | Moraceae | Bark and wood | Yellow | Watt (1883) |
| *A. lakoocha* Roxb. | Moraceae | Root | Yellow | Watt (1883), local people |
| *Baccaurea ramiflora* Lour. | Euphorbiaceae | Leaves | Green | Leotard (1881) |
| *Basella alba* L. | Basellaceae | Stem | Pink | Local people |
| *Bauhinia purpurea* L. | Caesalpiniaceae | Bark | Red, black | Local people, Watt (1883) |
| *B. variegata* L. | Caesalpiniaceae | Flowers | Yellow | Local people |
| *Bixa orellana* L. | Bixaceae | Fruit-pulp, seed, bark | Orange | Leotard (1881), Watt (1883) |
| *Bombax ceiba* L. | Bombacaceae | Petals | Brown | Local people |
| *Butea monosperma* (Lamk.) Taub. | Fabaceae | Petals, bark | Yellow | Leotard (1881), Watt (1883) |
| *Caesalpinia bonduc* (L.) Roxb. | Caesalpiniaceae | Wood and fruit | Red | Local people, Watt (1883) |
| *Carthamus oxycantha* Bieb. | Asteraceae | Flowers | Yellow | Local people |

| | | | | |
|---|---|---|---|---|
| *Cassia fistula* L. (Sonaru) | Caesalpiniaceae | Bark | Red | Leotard (1881), McCann (1883) |
| *Cestrum diurnum* L. | Solanaceae | Fruits | Blue, black | -do- |
| *Clitoria ternatea* L. (Aparajita) | Fabaceae | Flowers | Blue | -do- |
| *Curcuma longa* L. (Halodhi) | Zingiberaceae | Rhizome | Yellow | Local people, Leotard (1881) |
| *C. zedoaria* Rosc. (Bon-halodhi) | Zingiberaceae | Rhizome | Red | Watt (1883), McCann (1883) |
| *Cuscuta reflexa* Roxb. | Cuscutaceae | Whole plant | Yellow | Local people |
| *Diospyros peregrina* (Gaertn. f.) Gurke | Ebenaceae | Fruits | Brown | Local people, McCann (1883) |
| *Erythrina variegata* L. | Fabaceae | Flowers | Red | Local people, Watt (1883) |
| *Ficus racemosa* L. | Moraceae | Bark | Black | Local people |
| *Garcinia xanthochymus* Hk. f. ex T. Andr. | Clusiaceae | Bark, fruit (unripe) | Yellow | -do- |
| *Hibiscus rosa-sinensis* L. | Malvaceae | Flowers | Red, purple | -do- |
| *H. sabdariffa* L. | Malvaceae | Petals | Yellow | -do- |
| *Hyptis suaveolens* (L.) Poit. | Lamiaceae | Leaves | Blue | -do- |
| *Indigofera linifolia* (L.) Retz. | Fabaceae | Leaves | Blue | -do- |
| *I. tinctoria* L. | Fabaceae | Leaves | Blue | Leotard (1881) |
| *Jatropha curcas* L. | Euphorbiaceae | Bark, fruit-juice | Black | Local people, Watt (1883) |
| *Lannea coromandelica* (Houtt.) Merr. | Anacardiaceae | Bark | Brown | Local people, Sastri *et al.* (1950-72) |
| *Lawsonia inermis* L. | Lythraceae | Leaves | Red or reddish | Local people, Watt (1883) |
| *Leea asiatica* (L.) Ridsd. | Leeaceae | Root | Brown | Local people |
| *Mallotus philippensis* (Lam.) Muell.-Arg. | Euphorbiaceae | Fruit | Red, orange | Local people, Leotard (1881), Watt (1883) |
| *Mangifera indica* L. | Anacardiaceae | Leaves, bark | Green, yellow | Leotard (1881), McCann (1883) |
| Michelia champaca L. | Magnoliaceae | Flowers | Yellow | Local people, Watt (1883) |
| *Mimusops elengi* L. (Bakul) | Sapotaceae | Bark | Yellow | McCann (1883), Watt (1883) |
| *Morinda angustifolia* Roxb. | Rubiaceae | Wood, bark and root | Yellow, red and scarlet | Local people, Leotard (1881), Watt (1883) |
| *Nyctanthes arbor-tristis* L. | Oleaceae | Petals | Orange | Leotard (1881) |
| *Oroxylum indicum* (L.) Vent. | Bignoniaceae | Bark | Black | Local people |

| | | | | |
|---|---|---|---|---|
| *Opuntia elatior* Mill. | Cactaceae | Fruits | Red | Local people (infrequently used) |
| *Psidium guajava* L. (Madhuri) | Myrtaceae | Bark | Black | Local people, Watt (1883) |
| *Rubia cordifolia* L. | Rubiaceae | Roots | Red | Local people, Leotard (1881) |
| *Sarcochlamys pulcherrima* Gaud. | Urticaceae | Leaves | Dark or Maddar brown | Local people (infrequently used), McCann (1883) |
| *Semecarpus anacardium* L. f. | Anacardiaceae | Fruit | Brown | -do- |
| *Senna tora* (L.) Roxb. (Medelua) | Caesalpiniaceae | Seed | Yellow | Watt (1883), McCann (1883) |
| *Shorea robusta* Roxb. ex Gaertn. f. | Dipterocarpaceae | Bark | Red, black | Local people, McCann, Watt (1883) |
| *Syzygium cumini* (L.) Skeels (Kalajamu) | Myrtaceae | Bark, fruits | Black, brown | Local people, Leotard (1881), Watt (1883) |
| *Tabernaemontana divaricata* (L.) R. Br. | Apocynaceae | Fruit-pulp | Red | Burkill (1908) |
| *Tagetes erecta* L. (Genda) | Asteraceae | Petals | Yellow | Burkill (1908), Watt (1883) |
| *T. patula* L. | Asteraceae | Petals | Yellow | Watt (1881) |
| *Tamarindus indica* L. | Caesalpiniaceae | Leaves | Yellow | Watt (1883) |
| *Tectona grandis* L. f. | Verbenaceae | Leaves | Red | Leotard (1881), McCann (1883) |
| *Terminalia bellirica* (Gaertn.) Roxb. | Combretaceae | Fruits | Brown | Local people |
| *T. chebula* Retz. | Combretaceae | Bark, fruits | Black | Local people, Watt (1883) |
| *T. cuneata* Roth | Combretaceae | Bark | Brown | -do- |
| *Thespesia populnea* (L.) Soland. ex Corr. | Malvaceae | Flowers | Yellow | Watt (1883) |
| *Toona hexandra* (Wall. ex Roxb.) Roem. | Meliaceae | Flowers | Red | Leotard (1881), Watt (1883) |
| *Trewia polycarpa* Benth. | Euphorbiaceae | Bark | Black | Local people |
| *Ziziphus mauritiana* Lamk. | Rhamnaceae | Bark | Black | Local people, Watt (1883) |

## DISCUSSION

62 species belonging to 52 genera and 34 families have been recorded in the present work and it has been observed that out of these 62 species, 60 species alone belong to dicotyledonous group and other two species are monocotyledonous. Thus, dicotyledonous plants yielding dye are

dominant over all other groups. And amongst the dicotyledons, the family Caesalpiniaceae is found to be dominant with 6 species, followed by Fabaceae (5), Euphorbiaceae (4), Anacardiaceae, Asteraceae, Combretaceae, Mimosaceae, Moraceae and Malvaceae with 3 species each and all other families are with 2 or 1 species.

The species commonly used by both tribal and non-tribal people of the region are : *Curcuma longa, Garcinia xanthochymus Hibiscus rosa-sinensis, Lawsonia, inermis, Morinda angustifolia, Mallotus philippensis, Syzygium cumini* and *Terminalia chebula.* Curcuma longa, a monocotyledonous species, is very commonly used in dyeing food-stuff. Due to increasing demand of non toxic natural products in industry as well as in domestic aspects, it is desirable to acquaint the people with dye yielding plants of the State and its neighbouring areas, since the synthetic dyes in many cases are less permanent/ persistent and more harmful particularly in handling and when used in food stuff. In many advanced countries, the use of synthetic dyes in food-stuffs has been banned and huge amounts are being spent on research in search of natural dyes. Therefore, it is worthwhile to survey and assess the dye yielding plants of the country.

## REFERENCES

BURKILL, I.H. 1908. Dyes from flowers. *Agric. Ledg.* 2 : 7-29.

DASTUR, J.F. 1951. Useful Plants of India and Pakistan. Bombay.

FRASER, J.D. 1896. *Account of dyes and dyeing in Burma.* Govt. Press, Calcutta.

LEOTARD, L. 1881. Memorandum on dyes on Indian growth and production. Govt. Press, Calcutta.

ISLAM, M. 1984. Certain fibre yielding plants of North-eastern region, India. *J. Econ. Taxon. Bot.* 5 : 767-783.

———— 1986. Certain poisonous plants of North-eastern region, India. *J. Econ. Taxon. Bot.* 8 : 51-63.

MCCANN, H.W. 1883. *Report on the dyes and tans in Bengal.* Govt. Press, Calcutta.

MANJUNATH, B.L. & B.N. SASTRI 1948. The *Wealth of India, Raw Materials,* Vol. I. CSIR, New Delhi.

SASTRI, B.N. *et al.* (ed.). 1950-72. *The Wealth of India, Raw Materials,* Vol. II-IX. CSIR, New Delhi.

TROUP, R.S. 1907. *Indian Forest Utilisation.* Govt. of India, Calcutta.

WATT, G. 1883. *Economic Products of India,* Govt. Press, Calcutta.

# ETHNOBOTANICAL STUDIES IN KEDARNATH WILDLIFE SANCTUARY, CENTRAL HIMALAYAS, INDIA

D.P. SEMWAL & P. PARDHA SARADHI

*Department of Environmental Biology, University of Delhi, Delhi-110 007, India*
*E-mail : dinusem@rediffmail.com*

## ABSTRACT

An ethnobotanical survey along with botanical collections was made in selected villages of the Kedarnath Wild Life Sanctuary, Central Himalayas, India. The information regarding use of plant species for food, medicine, dye, fuel and construction materials, as well as for miscellaneous uses was collected and classified systematically. A total of 86 ethnobotanically important plant species were screened from the present study area, which belong to 74 genera and 54 families. Of the 48 food plants, 21 (43.7%) species provide edible fruits, 12 (25%) edible leaves, 8 (17.6%) edible flower/flower-buds, 3 (6.25%) provide all aerial parts, 1 species each (2.08%) provides edible roots, rhizome, seed and bark. Further, about 19 species have found multiple uses in the area. It is suggested that conservation measures are essentially required for sustainable use of these valuable resource.

## INTRODUCTION

Ethnobotany has emerged as a new branch of study documenting the age old wisdom and knowledge of the ethnic societies about the utility of diverse plant species and their miraculous properties as food, medicine and other miscellaneous uses, not yet known to the modem World. The cultivation of world's food crops have centered around 103 odd plant species, although three main cereal crops, viz. wheat, rice and maize have contributed to the larger share (Allen & Allen, 1990). India is the home to a great variety of ethnobotanically important plant species. Himalayas is one of the hot spots of floral and faunal diversity. It supports about 18,440 species of plants, of which 25.3 per cent are endemic to Himalaya (Singh & Hajra, 1997; Samant *et al.*, 1998). The rich plant diversity of the Indian Himalaya is utilized by the native communities in various forms including medicine and food. In past several years, some important contributions have been made on ethnobotanical, edible (food) and medicinal plants of Central Himalayas (Gaur, 1997, 1999; Gaur *et al.*, 1983; Naithani, 1973, 1984; Jain & Saklani 1991; Pundeer & Singh, 2002; Singh & Pundeer 2004 etc). The present study area has unique physiognomic, climatic and topographic conditions, but the Central Himalayas has remained neglected in spite of being ethnobotanically rich. Therefore, the ethnobotanical information was collected in the study area, including ethnomedicinal uses.

The local "Vaidya" (local medical practitioner) are usually not willing to disclose their knowledge about the uses of different plant species ("Jari Buti" or medicinal plants). The reasons behind this secrecy have been accounted to (i) improper use of the medicine, (ii) loss of plant species due to over exploitation and (iii) fear of loosing their status in the local community.

Many wild food plants have already been reported from Central Himalaya (Arora & Pandey, 1996; Gaur, 1999; Gaur *et al.*, 1983; Naithani, 1973; Jain, 1991; Semwal & Gaur, 1981 etc). Eastern Himalaya is one of the ethnobotanically important hot spots in India, with many species and their uses similar to Central Himalaya (Sundriyal & Sundriyal, 2000, 2001, 2004). This communication focusses on valuable information about botanical richness, distribution and availability of ethnobotanical plants used by the local people of the Kedarnath Sanctuary in Garhwal Himalaya.

## STUDY AREA AND CLIMATE

The study area covers 10 villages in and around the Kedarnath Sanctuary (Okhimath Block) of Rudraprayag district of Uttaranchal, Central Himalaya. The Sanctuary, famous for Musk Deer, is situated in the north-eastern part of Garhwal Himalaya. Beside the grandeur of Himalayan wilderness, the area encloses many important shrines notably Ansuya devi, Kalimath, Madhyamaheshwar, Rudranath, Trijuginarayan, Tungnath and Kedarnath - almost on its northern boundary. The major forest types of the study area are broad leaves/oak, conifer/pine and alpine scrub forests. Geologically the whole region comprises of five well defined physiographic belts, each being a distinct geological unit. These belts from south to north are identified as: Tarai and Bhabhar, The Siwaliks, The Lesser Himalaya, The Great Himalaya and the Tethys and Tibetan Himalaya. The soils of this region vary according to aspect, altitude and climate. On the whole, soils of the Himalaya are young and thin, generally the hill soil type (Raychaudhary, 1963) which is dark brown to brown at surface and brown to yellowish brown in the subsoil in nature and endodynomorphic. The lower part of the study area experiences montane climate and high altitude area possesses alpine climate which is greatly influenced by the adjacent Greater Himalayan range. The area receives 300 cm annual precipitation and the rainy months (June-August) contribute approximately 60%. The relative humidity varies from 40% to 80% annually. There is moderate to heavy snowfall during December- February in low altitude area. The mean maximum temperature varies between 2.5°C to 12°C.

Economy of local people depends mainly upon agriculture and also to some extent on tourism for six months i.e. May-

October. Because of the limited opportunities for different type of jobs, people often migrate from the villages to the metro cities.

## METHODOLOGY

The vegetation of different parts of study area was thoroughly scanned by repeated visits in different seasons of the year during 2002-2004, covering all ecological habitats represented in area. Plant specimens were collected and preserved following methods suggested by Jain & Rao (1977). The specimens were identified using Flora of Chamoli (Naithani, 1984) and Flora of the District Garhwal (Gaur, 1999).

The authors interviewed 110 knowledgeable people in 10 villages situated inside and near the Sanctuary. The informers included "Vaidyas" (local medical practitioner), knowledgeable old men/ women, medicine men and shepherd boys who were fully aware about their forest wealth. During consultation, information about important ethno-medicinal plants, plants used for fuel, dye and construction etc by the natives was recorded. Information was also gathered on plant parts used and the vernacular names of useful plants. Identified specimens of the plants and their parts were preserved in the herbarium of the Department of Environmental Biology, University of Delhi.

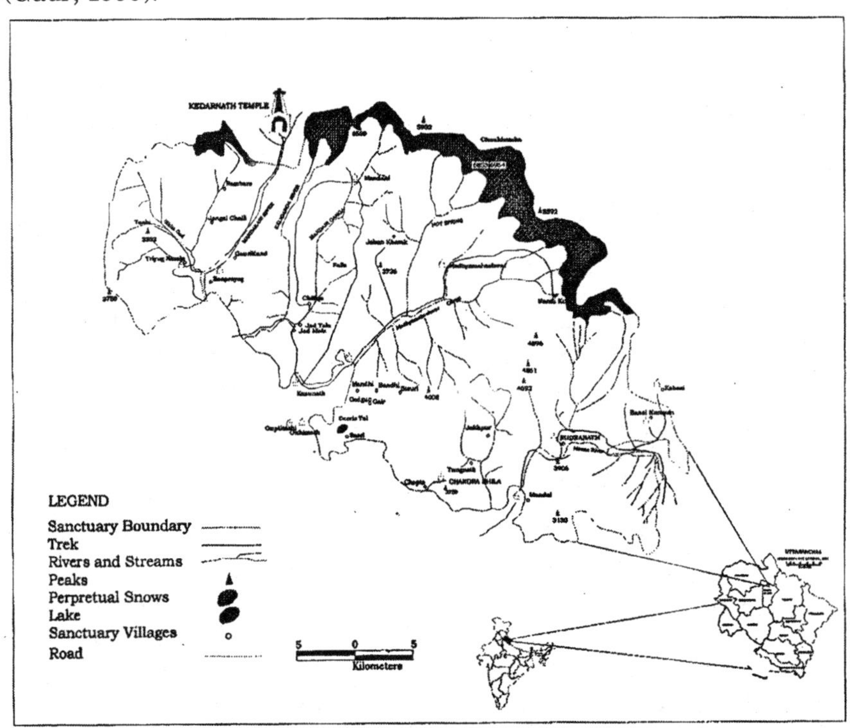


Map 1. Kedarnath Wildlife Sanctuary, Uttranchal, Central Himalaya

## ENUMERATION

Table 1. The plants used as food in Kedarnath Sanctuary, Central Himalayas, India.

| Family and species names | Local name | Parts used | Use/Application | Availability season |
|---|---|---|---|---|
| ACANTHACEAE | | | | |
| *Justicia adhatoda* L. | Basinga | Flower buds | Cooked | Feb-May |
| ANACARDIACEAE | | | | |
| *Rhus parviflora* Roxb. | Tungla | Fruit | Eaten fresh | Sep-Nov |
| APIACEAE | | | | |
| *Angelica glauca* Edgew. | Choru | Roots | As condiment | Jul-Aug |
| *Centella asiatica* (L.) Urban | Brahmi | Leaves | Eaten fresh | Aug-Nov |
| APOCYNACEAE | | | | |
| *Carissa opaca* Stapf ex Haines | Karaunda | Fruit | Eaten fresh | Mar-May |
| ASTERACEAE | | | | |
| *Senecio alatus* Wall. ex DC. | Ghuni- dhool | Leaves | Crushed extract | Aug-Sep |
| BERBERIDACEAE | | | | |
| *Berberis asiatica* Roxb. ex DC. | Kirmor | Fruit | Eaten fresh | May-Jun |
| *Berberis lycium* Royle | Daruhaldi | Fruit | Eaten fresh | Jun-Aug |
| CAESALPINIACEAE | | | | |
| *Bauhinia variegata* L. | Kuiral | Flower buds | Cooked | Mar-Apr |
| *Tamarindus indica* L. | Imli | Fruit | Eaten fresh | Apr-May |
| CAPRIFOLIACEAE | | | | |
| *Viburnum continifolium* D. Don | Bansura | Fruit | Eaten fresh | Sep-Oct |
| CRASSULACEAE | | | | |
| *Sedum imbricatum* (Edgew.) Walp. | Sedum | Leaves | Cooked | Aug-Sep |
| CHENOPODIACEAE | | | | |
| *Chenopodium album* L. | Bhyatu | Leaves | Cooked | Jun-Oct |
| CORNACEAE | | | | |
| *Benthamidia capitata* (Wall.) Hara | Bhamora | Fruit | Eaten fresh | Sep-Oct |
| DIOSCOREACEAE | | | | |
| *Dioscorea belophylla* (Prain) Voigt ex Haines | Tairu | Tuber | Cooked | Jul-Sep |
| ELAEAGNACEAE | | | | |
| *Elaeagnus parviflora* Wall. ex Royal | Giwain | Fruit | Eaten fresh | Apr-Jun |
| ERICACEAE | | | | |
| *Rhododendron arboreum* Smith | Burans | Flower | As juice | Mar-Apr |
| *Rhododendron barbatum* Wall. ex G. Don | Chimula | Flower | Eaten fresh | Mar-Apr |
| *Rhododendron campanulatum* D. Don | Simru | Flower | Eaten fresh | Mar-Apr |
| *Rhododendron hypenanthum* Balf. f. | Kodya | Leaves | As tea | Apr-May |

| | | | | |
|---|---|---|---|---|
| HELVELLACEAE | | | | |
| *Morchella esculenta* L. | Chyun | Aerial part | Cooked | Jul-Aug |
| JUGLANDACEAE | | | | |
| *Juglans regia* L. | Akhrot | Fruit | Seed | Sep-Oct |
| LAMIACEAE | | | | |
| *Colebrookia oppositifolia* Sm. | Binda | Fruit | Eaten fresh | May-Jun |
| LAURACEAE | | | | |
| *Cinnamomum tamala* (Buch.-Ham.) Nees | Tejpata | Leaves | Spices | Feb-Jun |
| LILIACEAE | | | | |
| *Allium consanguineum* Kunth | Pharan | Leaves | Spices | Sep-Oct |
| *Allium wallichii* Kunth | Jambo | Leaves | Flavours | Aug-Sep |
| LYTHRACEAE | | | | |
| *Woodfordia fruticosa* (L.) Kurz | Dhaula | Flower bud | Eaten fresh | Mar-May |
| MORACEAE | | | | |
| *Ficus oligodon* Miq. | Timla | Fruit | Eaten fresh | Apr-Jun |
| *Ficus palmata* Forssk. | Bedu | Fruit | Eaten mature | Apr-Jun |
| *Ficus semicordata* Buch.-Ham. ex J. E. Sm. | Khaina | Fruit | Eaten mature | Apr-Jun |
| *Ficus subincisa* Buch.-Ham. ex J. E. Sm. | Chhanchri | Fruit | Eaten fresh | Apr-Jun |
| MYRICACEAE | | | | |
| *Myrica esculenta* Buch.-Ham. ex D. Don | Kaphal | Fruit | Eaten fresh | Apr-Jun |
| PINACEAE | | | | |
| *Pinus roxburghii* Sarg. | Chir | Seed | Mature eaten | Feb-Apr |
| PODOPHYLLACEAE | | | | |
| *Podophyllum hexandrum* Royle | Bankakri | Fruit | Mature fruit | Jun-Jul |
| POLYPODIACEAE | | | | |
| *Dryopteris chinensis* Christ. | Kuthra | Leaves | Cooked | Mar-Jun |
| POLYGONACEAE | | | | |
| *Oxyria digyna* (L.) Hill | Kalmora | Leaves | Eaten fresh | Jul-Aug |
| *Polygonum rumicifolium* Royle ex Bab. | Kanthala | Leaves | Cooked | Jun-Jul |
| *Rheum australe* D. Don | Archa | Flower buds | Eaten fresh | Jun-Jul |
| *Rheum moorcroftianum* Royle | Dolu archa | Flower buds | Eaten fresh | Jun-Jul |
| *Rumex hastatus* D. Don | Almoru | Leaves | Eaten fresh | Apr-May |
| RANUNCULACEAE | | | | |
| *Caltha palustris* Linn. | Kushrya | Leaves | Cooked | Jun-Jul |
| ROSACEAE | | | | |
| *Fragraria deltoniana* Gay | Bhui kaphal | Fruit | Eaten fresh | Sep-Oct |
| *Pyracantha crenulata* (Don) Roem. | Ghingaru | Fruit | Mature fruit | May-Jul |
| *Pyrus pashia* Buch.-Ham. ex D. Don | Melu | Fruit | Mature fruit | Mar- Aug |

| | | | | |
|---|---|---|---|---|
| *Rubus ellipticus* Sm. | Hisole | Fruit | Eaten fresh | Apr-Jun |
| *Rubus niveus* Sm. | Kali hisole | Fruit | Eaten fresh | May-Aug |
| TAXACEAE | | | | |
| *Taxus wallichiana* Zucc. | Thunyor | Bark | As tea | Sep-Oct |
| URTICACEAE | | | | |
| *Urtica dioica* L. | Kandali | Fresh leaves | Cooked | Jun-Aug |

Table 2. The plants used as dye in Kedarnath Sanctuary, Central Himalayas, India.

| **Family and species names** | **Local name** | **Parts used** | **Use/Application** | **Availability season** |
|---|---|---|---|---|
| BERBERIDACEAE | | | | |
| *Berberis asiatica* Roxb. ex DC. | Kirmor | Root | It dyes yellow colour | Round the year |
| EUPHORBIACEAE | | | | |
| *Mallotus philippensis* (Lam.) Muell.-Arg. | Ruina | Fruit | It dyes red colour | March-June |

Table 3. Some important medicinal plants used in Kedarnath Sanctuary, Central Himalayas, India.

| **Family and species names** | **Local name** | **Parts used** | **Use for / against** |
|---|---|---|---|
| APIACEAE | | | |
| *Angelica archangelica* L. | Gandhrayan | Roots | Urinary disorder |
| ASTERACEAE | | | |
| *Anaphalis cuneifolia* Hook. f. | Buglu | Leaves | Cuts (wounds) |
| *Saussurea obvallata* (DC.) Edgew. | Brahmkamal | Leaves | Burns (wounds) |
| BERBERIDACEAE | | | |
| *Berberis asiatica* Roxb. ex DC. | Kirmor | Roots | Eye diseases |
| *Berberis lycium* Royle | Kilmor | Roots | Eye diseases |
| BETULACEAE | | | |
| *Alnus nepalensis* D. Don | Utees | Roots | Diarrhoea |
| BORAGINACEAE | | | |
| *Arnebia benthamii* (Wall. ex G. Don) Johnston | Ratanjot | Roots | Cuts (wounds) |
| EQUISETACEAE | | | |
| *Equisetum arvense* L. | Sarsyot | Leaves | Leaf-paste is used to cure teeth |
| ERICACEAE | | | |
| *Rhododendron arboreum* Sm. | Burans | Flower | Headache, fever |
| ORCHIDACEAE | | | |
| *Dactylorhiza hatagirea* (D. Don) Soo | Hatajari | Tubers | Diarrhoea, fever |
| PODOPHYLLACEAE | | | |
| *Podophyllum hexandrum* Royle | Bankakri | Roots | Skin diseases, cancer |

| | | | |
|---|---|---|---|
| POLYGONACEAE | | | |
| *Rheum australe* D. Don | Archa | Rhizome | Asthma, fever |
| *Rheum moorcroftianum* Royle | Dolu archa | Rhizome | Dysentery, internal pain |
| RANUNCULACEAE | | | |
| *Aconitum falconeri* Stapf | Bish | Tubers | Cough, snake-bite |
| *Aconitum heterophyllum* Wall. ex Royle | Ateesh | Tubers | Diarrhoea, tonic |
| ROSACEAE | | | |
| *Potentilla fulgens* Wall. ex Hook. | Bajradanti | Leaves | Pyorrhoea disease |
| RUTACEAE | | | |
| *Xanthoxylum armatum* DC. | Timru | Bark | Teeth cleaning |
| SCROPHULARIACEAE | | | |
| *Picrorhiza scrophulariflora* Pennell | Kutki | Roots | Liver diseases |
| TAXACEAE | | | |
| *Taxus wallichiana* Zucc. | Thyunor | Bark | Used as anti cancer drug |
| VALERIANACEAE | | | |
| *Nardostachys grandiflora* DC. | Massi | Roots | Epilepsy, jaundice |

Table 4. Plants used for house construction, fuel, fodder, agricultural implements etc in Kedarnath Sanctuary, Central Himalayas, India.

| **Scientific name** | **Local name** | **Family name** | **Uses** |
|---|---|---|---|
| *Acer acuminatum* Wall. ex D. Don | Kanchula | Aceraceae | As timber for house construction and also as fuel |
| *Alnus nepalensis* D. Don | Utees | Betulaceae | As fuel |
| *Cupresuss torulosa* D. Don | Surai | Cupressaceae | As resin, house construction and as fuel |
| *Lyonia ovalifolia* (Wall.) Drude | Anyar | Ericaceae | Only for fuel |
| *Ougeinia oojeinensis* (Roxb.) Hochr. | Sandan | Fabaceae | For making parothi/pariya and other household items |
| *Quercus dilatata* Lindl. ex Royle | Moru | Fagaceae | Agricultural tools particularly plough |
| *Quercus glauca* Thunb. | Hareenj | Fagaceae | Agricultural tools, fuel and leaves for fodder |
| *Quercus leucotrichophora* A. Camus | Banj | Fagaceae | For agricultural tools, fuel and leaves as fodder |
| *Quercus semecarpifolia* J. E. Sm. | Kharsu | Fagaceae | Mainly agricultural tools, fuel and as fodder |
| *Aesculus indica* (Wall. ex Camb.) Hook. f. | Pangar | Hippocastanaceae | For fuel and as fodder |
| *Toona hexandra* (Wall. ex Roxb.) Roem. | Toon | Meliaceae | Timber for house construction |
| *Albizia lebbeck* (L.) Willd. | Siris | Mimosaceae | Fuel and as fodder |
| *Fraxinus micrantha* Lingelschein | Angu | Oleaceae | As fuel |

| | | | |
|---|---|---|---|
| *Orchis chusua* D. Don | Chusa | Orchidaceae | Flowers for ornamental purpose |
| *Pinus roxburghii* Sarg. | Chir | Pinaceae | As resin, timber and also for fuel |
| *Pinus wallichiana* Jacks | Kail | Pinaceae | Wood is used as resin, timber and also for fuel purposes |
| *Cedrus deodara* (Royle ex D. Don) G. Don | Deodar | Pinaceae | As timber, resin and also for fuel |
| *Cynodon dactylon* (L.) Pers. | Doob | Poaceae | For religious ceremonies |
| *Danthonia cachemyriana* Jaub. | Tham | Poaceae | For making huts |
| *Arundinaria falcata* Nees | Ringal | Poaceae | Making basket and as climber support |
| *Thamnocalamus spathiflora* (Trin.) Munro | Dev ringal | Poaceae | Sticks are used for making basket |
| *Pyrus pashia* Buch.-Ham. ex D. Don | Melu | Rosaceae | Wood is used for house construction and also as agricultural tools |
| *Pyracantha crenulata* (D. Don) Roem. | Ghingaru | Rosaceae | As agricultural tools |
| *Madhuca longifolia* (Koen.) Macbr. | Mahwa | Sapotaceae | Used as fuel |
| *Symplocos paniculata* (Thunb.) Miq. | Lodh | Symplocaceae | As fuel |
| *Grewia optiva* Drumm. | Bhimal | Tiliaceae | As fiber, fuel and fodder |
| *Celtis australis* L. | Khareek | Ulmaceae | House construction and as fodder |

## RESULTS AND DISCUSSION

The study revealed that only six species listed in Table 1 are used for food in all the selected villages, viz. *Angelica glauca* Edgew., *Allium consanguineum* Kunth., *Allium wallichii* Kunth., *Chenopodium album* L., *Cinnamomum tamala* (Buch.-Ham.) Nees and *Rhododendron arboreum* Sm.

Plant species used for fuel purposes in most of the villages included *Alnus nepalensis* D. Don, *Lyonia ovalifolia* (Wall.) Drude, *Quercus leucotrichophora* A. Camus etc (Table 4). In general, local people use fallen wood as fuel, cut branches of tree for fodder as well as for fuel. For specific occasions or ceremonies, villagers sought permission from Forest Department officials to cut one or two trees for the same purpose, especially *Quercus leucotrichophora* A. Camus. and *Quercus glauca* Thunb.

Two species, *Berberis asiatica* Roxb. ex DC. and *Mallotus philippensis* (Lam.) Muell.-Arg. used as dye plants were found in most of the villages under investigation (Table 2). Several species are being used for more than one purpose, however, such multipurpose plants are under serious threat as compared to other plant species. For example, *Rhododendron arboreum* Smith is used as food (fresh flowers are eaten), as juice, as wood and as ornamental in the region. It is found that the *Rhododendron* species are loosing their habitat in and around the villages.

Plant species used for medicinal purposes in most of the villages included *Anaphalis cuneifolia* Hook. f., *Berberis asiatica* Roxb. ex DC., *Rhododendron arboreum* Sm. and *Xanthoxylum armatum* DC. etc. (Table 3).

Wild plants that play a very important role in economic upliftment of local people include *Allium consanguineum* Kunth, *Allium wallichii* Kunth, *Angelica glauca* Edgew., *Chenopodium album* L., *Cinnamomum tamala* (Buch.-Ham.) Nees, *Myrica esculenta* Buch.-Ham. ex D. Don, *Dryopteris chinensis* Christ., *Rhododendron arboreum* Sm., *Rhododendron barbatum* Wall. ex G. Don which are sold in markets and local village or towns.

The major crops and fruit plants species of the villages include paddy *(Oryza sativa* L.), wheat *(Triticum aestivum* L.), finger millet *(Eleusine coracana* L.), barley *(Hordeum vulgare* L.), maize *(Zea mays* L.), Potato *(Solanum tuberosum* L.), tor *(Cajanus cajan* (L.) Millsp.), bhatt *(Glycine max* L.), seb *(Pyrus malus* L.), nimbu *(Citrus medica* L.) and nashpati *(Pyrus communis* L.) etc.

The compiled information was compared with the earlier studies by Gaur (1999), Gaur *et al.* (1983), Naithani (1984), Jain (1991), Jain & Saklani (1991), Pundeer & Singh (2002) and Singh & Pundeer (2004) and it was found that the local names of the different species are identical in surrounding districts or regions, excepting a few species.

Out of the 86 species, 48 (55.81%) are used for food, 20 (23.25%) for medicine, 16 (18.6%) for fuel and two (2.3%) for dye. Eight species (9.3%) are used as construction material and 19 species have found multiple uses. *Pinus roxburghii* Sarg. is the only species which has found wide use for food, medicine, fuel, construction material and other miscellaneous uses.

Various parts of the plants are used by the villagers for food; of the 48 food plants, 21 (43.7%) provide edible fruits, 12 (25%) leaves, 8 (16.6%) flower/flower buds and two (4.16%) all aerial parts. One species each (2.08%) provides edible roots, rhizome, seed and bark. It was noticed that few decades back the fresh leaves of *Urtica dioica* L., locally known as Kandali (Bichhoo ghas), were being used by local people for food in huge amounts, but now a days only one or two local vaidyas (local medical practioner) are using it as medicinal and for "tantra-mantra" purposes.

This study will be helpful in data base generation of the traditional uses of plants by local people. Unfortunately, the native people of all parts of the World are loosing their culture and traditions, resulting in difficulties in identification of economically important wild plant wealth. As per our experiences, not even a single tribal family from the forest should be neglected, since some information may be associated with that. It has become rather more necessary in present days in view of rapid socio-economic and socio-cultural changes in the society.

## ACKNOWLEDGEMENTS

DPS is grateful to CSIR and Department of Science & Technology, Govt. of India for the award of Young Scientist project. We thank to Dr. D.S. Rawat, Dr. B.P. Nautiyal, University of Garhwal,

Srinagar for identification of specimens. The help provided by the local people, medicine men and Vaidyas during field visits is thankfully acknowledged.

## REFERENCES

ALLEN, R.P. & C.P. ALLEN 1990. How many plants feed world?. *Conservation Biology* 4 : 365-374.

ARORA, R.K. & A. PANDEY 1996. *Wild edible plants of India: Conservation and use.* Indian Council of Agricultural Research, National Bureau of Plant Genetic Resources, New Delhi.

GAUR, R.D. 1977. Wild edible fruits of Garhwal Hills. *The Himalaya* 1 : 66-70.

———— 1999. *Flora of the District Garhwal, North West Himalaya with Ethnobotanical notes.* Transmedia publication, Srinagar, Garhwal, Uttaranchal.

————, J.K. SEMWAL & J.K. TIWARI 1983. A survey of high altitude medicinal plants of Garhwal Himalayas. *Bull. Med. Ethnobot. Res.* 4 : 102-116.

JAIN, S.K. 1991. Dictionary of Indian Folk Medicine and Ethnobotany. Deep Publication, New Delhi.

———— & A. SAKLANI 1991. Observations on ethnobotany of the tons valley region of Uttarakashi district of North-West Himalayas. *Mount. Res. Dev.* 11 (2) : 177-183.

———— & R.R. RAO 1977. *Handbook of Field & Herbarium Methods.* Goyal Offsets, Delhi.

NAITHANI, B.D. 1973. Medicinal Plants of Western Garhwal. *Khadi Gramoudhyog* 19 : 269-278.

———— 1984. *Flora of Chamoli,* Vol. I & II. Botanical Survey of India, Dehra Dun.

PUNDEER, Y.P. & DHAN SINGH 2002. Ethnobotanical wild food plants of Jaunsar-Bawar (Western Himalaya), Uttaranchal. *Indian Forester* 128 (5) : 571-582.

RAYCHAUDHARY, S.P. 1963. *Indian Soil - Their classification, occurrence and properties.* Land Resources of India, Planning Commission, Government of India, New Delhi.

SAMANT, S.S., U. DHAR & L.M.S. PALNI 1998. Medicinal Plants of Indian Himalaya : Diversity, Distribution and Potential values. Gyanodaya Prakashan, Nainital.

SEMWAL, J.K. & R.D. GAUR 1981. The alpine flora of Tungnath Garhwal Himalaya. *Journ. Bombay Nat. Hist. Soc.* 78 : 498-512.

SINGH, D.K. & P.K. HAJRA 1997. Floristic diversity. In Gujral (Ed.) *Biodiversity Status in the Himalaya,* pp. 23-38. British Council, New Delhi.

SINGH, D. & Y.P.S. PUNDEER 2004. Wild medicinal plants of Jaunsar-Bawar (Western Himalayas) Uttaranchal-I. *Indian Forester* 130 : 1259-1271.

SUNDRIYAL, M. & R.C. SUNDRIYAL 2000. Potential of wild edible plants in the Sikkim Himalaya : conservation concerns. *Journal Non-Timber Forest Product* 7 (3/4) : 253-262.

———— & ———— 2001. Wild edible plants of the Sikkim Himalaya : Nutritive values of selected species. *Econ. Bot.* 55 (3) : 377-390.

———— & ———— 2004. Wild edible plants of the Sikkim Himalaya : Marketing, value addition and implications for management. *Econ. Bot.* 58 (2) : 300- 315.

# ETHNOBOTANICAL STUDIES IN SANGLA VALLEY, DISTRICT KINNAUR IN HIMACHAL HIMALAYAS

VIRAJ MAN & BHUPENDER DUTT

*Department of Forest Products, Dr. Y. S. Parmar University of Horticulture and Forestry, Nauni-173 230, Solan (Himachal Pradesh) India*

**ABSTRACT**

An ethnobotanical study has been carried out in the Sangla valley of tribal district Kinnaur in Himachal Pradesh. The information on traditional practices of 46 species of medicinal and aromatic plants belonging to 40 genera from 28 families were collected. The plants of ethnobotanical importance are arranged alphabetically and highlighted with their botanical name, family, Hindi name, local names, parts used and their ethnobotanical uses.

## INTRODUCTION

The traditional ethnobotanical studies in recent years have received much attention due to their wide local acceptability and clues for new or lesser-known medicinal plant (Tripathi, 2000). Today, it is widely accepted as a science of human interaction with plants and their ecosystems. The fundamental structure of ethnobotanical research is to examine the dynamic relationships between human population, cultural values and plants. The epic Ramayana tell us about "Sanjivani Booti", a life saving drug. In India, the importance of traditional and folk medicine in the treatment of various human ailments is well known from olden days and our country has kept alive a number of traditional systems of medicines, like Ayurveda, Sidha, Yoga, Unani, Tibetian etc, the oldest being the Ayurveda. The ancient works like Charaka Samhita and Sushruta Samhita, describe uses of more than 1200 plant based drugs along with their action and specific therapeutic applications (Rawat & Chowdhury, 1988).

Western Himalayas being an emporium of medicinal and aromatic plants holds great references to ethnobotanical studies. In recent years some valued ethnobotanical works have also been done on Himachal Pradesh (Gupta, 1964; Chauhan, 1974; Uniyal & Chauhan, 1971, 1979; Arora *et al.*, 1980; Sharma & Rana, 1999; Singh, 1999; Natarajan *et al.*, 2000;

Singh & Kaushal, 2000 and Singh & Chauhan, 2005). Generally the tribal people have close relationship with nature, therefore, they have very good knowledge of traditional medicinal practices. Keeping it in view, a detailed ethnobotanical study was conducted in Sangla valley during year 2003-2005.

## MATERIALS AND METHODS

Sangla valley (also known as Baspa valley) is one of the most beautiful valleys of the tribal district Kinnaur in Himachal Pradesh. It starts from Karcham (1800 m above msl) at the confluence of Baspa river with Satluj. The valley opens beyond Sangla village and is extended upto Chhitkul village (3450 m above msl). During old days goods were supplied from Chhitkul to the Indo-Tibetan border, which is just 35 kms away. The valley supports rich and diverse flora. The information on important medicinal and aromatic plants generally used by local people was collected through interviews and discussions with elderly knowledgeable people, herbal healers and local Vaids. The voucher specimens of ethnobotanical importance were collected and identified with the help of authentic literature on Himachal Pradesh (Chowdhery & Wadhwa, 1984; Nair, 1977; Collect, 1902). Voucher specimens are deposited in the Herbarium of Dr. Y.S. Parmar University of Horticulture and Forestry, Nauni, Solan (Himachal Pradesh).

## ENUMERATION OF SPECIES

Traditional knowledge of any society presents a unique picture on the basis of the belief, norms and culture of the society to which they belong. Hence, careful documentation and scientific analysis for further transmission is very important. The ethnobotanical plants have been arranged alphabetically with their botanical name, family, local names, parts used and ethnomedical uses.

**Aconitum heterophyllum** Wall. ex Royle (Ranunculaceae)

*Hindi name :* Atish; *Local name:* Aatish.

*Uses :* The root-powder is given twice daily with cold water for 3-5 days for the treatment of diarrhoea. In case of fever and vomiting, especially in infants, the root-powder is given with mother's milk 2-3 times a day for 5 days.

**Aconitum violaceum** Jacq. ex Stapf (Ranunculaceae)

*Hindi name :* Mitha-patish; *Local name :* Oonyalkas.

*Uses :* The dried roots are boiled in water and used in small doses to cure cough of children.

**Allium carolinianum** DC. (Liliaceae)

*Hindi name :* Jungli-pyaz; *Local name :* Chamcham.

*Uses :* Leaves and bulbs are used in indigestion, cold and also as spices and condiments. The leaves are frequently used as vegetable.

**Anaphalis triplinervis** (Sims) Clarke (Asteraceae)

*Local name :* Bacha-chi.

*Uses :* Powdered roots are given in cold, stomach disorders and also to heal scars. Leaves are dried, rubbed to obtain wool, which is used by the shepherd and Gujjars to ignite fire with stone.

**Angelica glauca** Edgew. (Apiaceae)

*Hindi name :* Chora; *Local name :* Sapal.

*Uses:* Roots are used as spice and condiments.

**Arisaema flavum** (Forsskal) Schott (Araceae)

*Hindi name :* Saafki-makki; *Local name :* Jamush.

*Uses :* Tubers are used in preparation of medicine for snake-bite. The fruit is covered with flour and is taken internally to cure for stomach diseases. The fruit is used locally for preparation of wine.

**Arnebia benthami** (Wall. ex G. Don) Johnston (Boraginaceae)

*Hindi name :* Ratanjot; *Local name :* Khome.

*Uses :* Used in fever and also in dyeing and hair oils. Paste of roots made in water is applied to wounds and burns.

**Berberis asiatica** Roxb. ex DC. (Berberidaceae)

*Hindi name :* Kashmal; *Local name :* Khapacho.

*Uses :* Ripe fruits are edible and are given as a mild laxative to children. Young shoots are generally used to drive away the evil spirits (ghost) from the houses. During "Khaipa" (the festival of ghosts) held during November-December, the young shoots are kept at four corners of the house and also at each door of the house with the belief that evil spirits (ghosts) can't enter the house.

**Bergenia ciliata** (Haw.) Sternb. (Saxifragaceae)

*Hindi name :* Phashanbed; *Local name :* Laoo.

*Uses :* Rhizomes are used to cure fever and swollen joints. Leaves are kept on head under the cap for relieving fever. It is considered to be an important drug for dissolving kidney and bladder stones.

**Bergenia stracheyi** (Hook. f. & Thoms.) Engl. (Saxifragaceae)

*Hindi name :* Phashnbed; *Local name:* Laoo.

*Uses :* Leaves and rhizomes are used in indigestion, fever, burn and in cleaning teeth. Rhizome-paste is given internally to cure jaundice.

**Betula utilis** D. Don (Betulaceae)

*Hindi name :* Bhojpatra; *Local name :* Bhuj, Bhooj.

*Uses :* Papery bark is used in houses for roofing and as an aromatic during 'Yajana' ceremonies like marriage. The fungal growth on bark is used against muscular pain and swellings.

**Cannabis sativa** Linn. (Cannabaceae)

*Hindi name :* Bhang; *Local name :* Kasa, Bhang.

*Uses :* Leaf-paste is applied on swellings developed by the sting of honey bee. Flower tops and dried leaves are used for making Bhang. Seeds are roasted and eaten as culinary by the local people and fibres are used for making ropes.

**Caragana brevispina** Rolye ex Benth. (Papilionaceae)

*Local name :* Shuroo.

*Uses :* Representatives of diety offer the flowers to the local people on local festival "Bishu" (Basant Panchmi) during 13th/14th April every year. Then these flowers are kept at the corner of the room alongwith each type of local cereals and dry fruits for one day. Then, they pray for the better crop in coming season. These dry fruits are exchanged among the neighbours and relatives.

**Cassiope fastigiata** (Wall.) D. Don (Ericaceae)

*Local name :* Tolashang.

*Uses :* The plant is mainly used in 'Yajana' ceremonies as an incence. Leaves are used for making namkeen tea. The aromatic leafy twigs are ground into a paste and applied on burns.

**Dactylorhiza hatagirea** (D. Don) Soo (Orchidaceae)

*Hindi name :* Hathapanja, Salampanja; *Local name :* Salampanja, Hathpanja.

*Uses :* Fresh or dried tubers are eaten raw in summers to check the bleeding from nose and to increase sexual power. The tubers are also used in diarrhoea and dysentry. Paste is applied on bone fractures.

**Delphinium brunonianum** Royle (Ranunculaceae)

*Local name :* Loskar.

*Uses :* Leaves and flowers are crushed and boiled in water and the extract obtained is used to cure dysentery and fever.

**Delphinium denudatum** Wall. ex Hook. f. & Thoms. (Ranunculaceae)

*Local name :* Loskar.

*Uses :* Seeds are used for the treatment of skin eruptions. Roots are pressed between teeth to get relief from toothache.

**Dioscorea deltoidea** Wall. ex Griseb. (Dioscoreaceae)

*Hindi name :* Shingli-mingle; *Local name :* Shingli-mingli.

*Uses :* The crushed rhizomes are used for washing wool and hairs. Powder of rhizomes is given in gastric complaints.

**Fagopyrum esculentum** Moench (Polygonaceae)

*Hindi name :* Ogla, Kuttu; *Local name :* Ogla, Olgo.

*Uses :* The seeds are ground into flour and used as food by the natives and also to fill the stomach of goats during "Khaning" (an annual function) to make a special preparation. Juice of plant is useful in urinary disorders.

**Ferula jaeschkeana** Vatke (Apiaceae)

*Hindi name :* Jungli-Hing; *Local name :* Kaith, Kaidmoh.

*Uses :* Mushroom is used as vegetables and considered as a digestive and stimulative.

**Gentiana moorcroftiana** Wall ex. G. Don (Gentianaceae)

*Local name :* Tikta.

*Uses :* Flowers and leaves are ground into a paste and applied externally for curing backache and headache.

**Heracleum lanatum** Michx. (Apiaceae)

*Hindi name :* Patrala; *Local name :* Poral.

*Uses :* Flowers and leaves are made into a paste and applied on forehead to cure headache.

**Hippophae salicifolia** D. Don (Elaeagnaceae)

*Hindi name :* Seabuck-thorn; *Local name :* Surcha.

*Uses :* Juice prepared from the berries is used as a substitute of tomato and sauce. Juice is also given to eradicate poison taken incidentally by livestock.

**Juglans regia** Linn. (Juglandaceae)

*Hindi name :* Akhrot; *Local name :* Kaa.

*Uses :* Nuts are edible and also used for extraction of oil. Oil is considered as digestive, carminative and stimulative. Bark and leaf are used as dye and also for cleaning teeth.

**Juniperus recurva** Buch.-Ham. ex D. Don (Cupressaceae)

*Hindi name :* Indian Juniper, Duhp; *Local name :* Jaleum, Theleru.

*Uses :* Needles are used as incense during 'Yajana' ceremonies and also as a substitute of agarbatti.

**Jurinea dolomiaea** Bioss. (Asteraceae)

*Hindi name:* Jari-dhop; *Local name:* Googul.

*Uses :* The roots are aromatic and used as incense. The bruised roots are applied to skin eruptions.

**Mentha longifolia** (Linn.) Hudson (Lamiaceae)

*Hindi name :* Jungle-pudina; *Local name :* Jungli-judina, Pudina.

*Uses :* The leaves and shoots are taken in the form of chutney and considered to be carminative, antiseptic and stimulant. It is also useful in stomach disorders and vomiting.

**Oxyria digyna** (Linn.) Hill (Polygonaceae)

*Local name :* Shup-Chi.

*Uses :* Leaves are made in to pickles and are considered as useful in stomach disorders.

**Picrorhiza kurrooa** Royle ex Benth. (Scrophulariaceae)

*Hindi name :* Kurru; *Local name :* Karru, Kauri.

*Uses :* Rhizomes and stolons are used in cold and cough. Paste of roots is used in stomach problems, and powder is used as a pain-killer.

**Pinus gerardiana** Wall. ex Lambert (Pinaceae)

*Hindi name :* Chigoza; *Local name :* Ree, Chilgoza.

*Uses :* Kernels are eaten as raw or roasted. Seeds are valued as carminative, stimulative and expectorant. Wood splinters are locally used as torch.

**Podophyllum hexandrum** Royle (Podophyllaceae)

*Hindi name :* Bankakri; *Local name:* Bankakri.

*Uses :* Paste made from root-powder is applied externally on forehead as a remedy against headache. Fruit-paste with honey is used in abdomen pain.

**Polygonum plebeium** R. Br. (Polygonaceae)

*Local name :* Piplichi.

*Uses :* Young leaves are used as chutney by the local people and considered as digestive and useful in pneumonia.

**Prunus armeniaca** Linn. (Rosaceae)

*Hindi name :* Chulli; *Local name :* Chul.

*Uses :* Unripe fruits are eaten raw and kernels are used for making chutney. Ripe fruits are eaten raw, while kernels are used for extraction of edible oil. The warm oil is applied on joints to get relief from rheumatic pain.

**Prunus persica** (L.) Batsch (Rosaceae)

*Hindi name :* Jungli Aadu, Bemi; *Local name :* Bemi, Rag.

*Uses :* Ripe fruits are eaten raw. Oil extracted from kernels is considered to be highly medicinal and edible. Oil is applied on joints to get relief from rheumatic pain.

**Ranunculus laetus** Wall. ex D. Don (Ranunculaceae)

*Local name :* Chip-chi.

*Uses :* Paste made from the leaves is used for curing hydrosel, gas troubles and joint pains.

**Rheum australe** D. Don (Polygonaceae)

*Hindi name :* Revandchini, Chukri; *Local name :* Archa.

*Uses :* Root-paste with water is commonly used to cure swellings developed by fracture. Powdered roots are used for cleaning teeth.

**Rhododendron anthopogon** D. Don (Ericaceae)

*Hindi name :* Talispatra; *Local name :* Mutashang.

*Uses :* Young shoots are used as incense during 'Yajana' ceremonies like "Griha Pravesh" (First time entry into newly constructed house) and marriage.

**Rorippa officinale** R. Br. (Brassicaceae)

*Local name :* Bolku.

*Uses :* Aerial parts are used as vegetable and considered as digestive.

**Saussurea costus** (Falc.) Lipsch. (Asteraceae)

*Hindi name :* Kuth; *Local name :* Kuth.

*Uses :* Roots are used for indigestion, nausea, stomach-pain and also given to cattle for promoting easy delivery.

**Saussurea graminifolia** Wall. ex DC. (Asteraceae)

*Local name :* Khasbal.

*Uses :* The woolly tomentum of aerial parts of plant is used to stop bleeding from fresh cuts and also to heal it in quick time.

**Saussurea obvallata** (DC.) Edgew. (Asteraceae)

*Hindi name :* Bhram-kamal; *Local name :* Dongar.

*Uses :* The whole plant is offered to the diety (through malli) on local fair 'Fulaich' (Festival of flowers); if flowers are eaten by the malli then it is considered that the diety is happy and it is a good sign of whole year for the local people.

**Sedum ewersii** Ledeb. (Crassulaceae)

*Local name :* Tindi.

*Uses :* Leaves and flowers are pounded into a paste and applied on forehead to treat headache.

**Skimmia laureola** (DC.) Sieb. & Zucc. ex Walp. (Rutaceae)

*Hindi name :* Kasturi-pata; *Local name :* Shashuri.

*Uses :* Twigs and leaves of the plant are offered to deity (through malli) by the local people on local festival "fagul" during the month of February or March each year; if the leaves offered are odd in number then it is considered good to fulfil the wishes of people, but if leaves are even in number, then it is a bad sign.

**Taxus wallichiana** Zucc. (Taxaceae)

*Hindi name :* Himalayan Yew; *Local name :* Van-Chiapati, Yamdhal.

*Uses :* Bark and leaves of tree are used for making namkeen tea by the local people.

**Thymus linearis** Benth. (Lamiaceae)

*Hindi name :* Banajwain; *Local name :* Banajwain.

*Uses :* Infusion of leaves with 'Jaggary' is taken to cure stomach complaints, cough and cold.

**Verbascum thapsus** Linn. (Scrophulariaceae)

*Local name :* Bhotiya-chi.

*Uses :* The plant is used to drive out ghost spirits from the houses. Leaves and flowers are sun dried, powered and taken orally to treat vomiting.

## ACKNOWLEDGEMENTS

Authors are very grateful to local tribes and Amchis who have willingly shared their knowledge on the local medicinal flora.

## REFERENCES

ARORA, R.K., M.L. MAHESHWARI, K.P.S. CHANDEL & R. GUPTA 1980. Mano (*Inula racemosa*): little known aromatic plant of Lahaul Valley, India. *Economic Botany* 34 : 175-180.

CHAUHAN, N.S. 1974. *Ainslaea aptera* DC. A hitherto unknown herb used in acute gastritis. *Sachitra Ayurveda* Feb. : 479-499.

CHOWDHERY, H.J. & B.M. WADHWA 1984. *Flora of Himachal Pradesh,* 3 Vols. Botanical Survey of India, Howrah.

COLLECT, H. 1902. *Flora Simlensis.* Thacker, Sprink & Co., London.

GUPTA, R. 1964. Survey record of medicinal and aromatic plants of Chamba forest division, Himachal Pradesh. *Indian Forester* 90 : 54-468.

NAIR, N.C. 1977. *Flora of Bashahr Himalaya.* International Bioscience Publisher, Hissar & Madras.

NATARAJAN, B., B.S. PAULSEN & V. KORN-ELIUSSEN 2000. An ethno-pharmacological study from Kullu district, Himachal Pradesh, India. *Pharmaceutical Bulletin* 3 (2) : 129-138.

RAWAT, M.S. & S. CHOWDHURY 1988. *Ethno-medico-botany of Arunachal Pradesh : Nishi and Apatani tribes.* Bishen Singh Mahendra Pal Singh, Dehra Dun.

SHARMA, B.D. & J.C. RANA 1999. Traditional uses of plants of Himachal Pradesh hills. *J. Econ. Taxon. Bot.* 23 (1) : 173-176.

SINGH, G.S. 1999. Ethno-botanical study of useful plants of Kullu district in North-western Himalayas, India. *J. Econ. Taxon. Bot.* 23 (1) : 185-198.

SINGH, K.K. & K. KAUSHAL 2000. Observation on ethno-veterinary medicine among the Gaddi tribes of Kangra valley, Himachal Pradesh. *Ethnobotany* 12 : 42-44.

SINGH, V. & N.S. CHAUHAN 2005. Traditional practices of herbal medicines in the Lahaul Valley, Himachal Himalayas. *Indian Jour. Taditional Knowledge* 4 (2) : 208-220.

TRIPATHI, Y.C. 2000. Ethnobotanical treasure of tribal Rajasthan. *Journ. Non-Timber Forest Products* 7 (1-2) : 77-84.

UNIYAL, M.R. & N.S. CHAUHAN 1971. Medicinal plants of Uhal Valley in Kangra Forest Division, H.P. *Journ. Research Indian Medicine* 6 (3) : 287-299.

——————— & ——————— 1979. A preliminary study of the indigenous drugs used at the Tibetian medicinal centre, Dharamshala (HP). *Nagurjun* 22 (8) : 190-193.

# ETHNOBOTANICAL STUDIES OF SRIKHOLA, LEPCHAJAGAT, TONGLU AND PANKHABARI AREAS OF DARJEELING DISTRICT, WEST BENGAL

RAMA PRASAD BHATTACHARYA & CHANDRAKALA RAI*

*Deptt. of Botany, Bidhannagar College, EB-2, Sector-1, Salt Lake, Kolkata-700 064, West Bengal*
**C/o Youn Rai, Assam Lingyey, East Sikkim (Via Ranipool), Gangtok-737 135*

## ABSTRACT

An account on the ethnobotanical uses of 36 plant species belong to 26 families and 36 genera by the tribals (Lepcha, Bhutia, Nepali) of Darjeeling district have been discussed in this paper.

## INTRODUCTION

Darjeeling, the northernmost frontier district of West Bengal, is located in the Eastern Himalayas between 27°13'05" and 26°27'10" North latitude and 85°53'00" and 87°55'30" East longitude. It is bounded on the west by Nepal, on the north by Sikkim, on the east by Bhutan and Bangladesh and on the south by Jalpaiguri district of West Bengal and Purnia district of Bihar. The district belongs to the lower Himalayan zone and consists of long tortuous ranges stretching from north to south throughout its length. Geologically, it is a sort of neutral country and in the hills the greater portion of the underlying rock consists of Sikkim gneiss. The vegetation is mainly of two types - tropical forest up to an altitude of about 1800 m dominated by *Shorea robusta* and temperate forest from about 1800-2700 m dominated by *Quercus*, *Rhododendron* and coniferous plant species.

The total population of the district is 8,50,000 (1991 census).The majority of the hill people belong to various Nepalese castes of Mongolian group and also include various ethnic communities like Lepcha, Bhutia, Tibetan etc. The acculturation among the people is noticeable but still their original cultural traditions, customs in respect of functions and festivities have not been changed. The acculturation in the material culture is much more among the workers of the various tea gardens. In this paper, the ethnobotanical uses of various plants of Srikhola, Lepchajagat, Tonglu and Pankhabari areas of Darjeeling district have been given.

Ethnobotanical studies of some areas of Darjeeling have been carried out by Rai, P.C. *et al.* (1998), Yonzone *et al.* (1981), Biswas (1956, 1966), Bhujel *et al.* (1984), Yonzone *et al.* (1984), Lama (1989), Nandi (1997), Chhetri (2004) etc. But, they have not touched the above mentioned area. So the survey and the ethnobotanical studies have been taken up.

## METHODOLOGY

Field studies were conducted following Jain's method (1965). The areas were visited season wise to collect the plant materials and ethnobotanical information accompanied with knowledgeable persons. The weekly markets were also visited to get the local plant products. Professional herbalists or Vaidyas, medical practitioners were approached and interviewed from time to time to examine the information. The plant specimens collected or supplied by the tribals were identified in the Central National Herbarium, Botanical Survey of India, Kolkata (CAL) and voucher specimens were deposited in the herbarium of Darjeeling Govt. College for future reference.

## OBSERVATIONS

The plant species have been arranged in alphabetical order by their botanical names, followed by their family name in parenthesis, local names in italics, short description, locality, collection number, ecological status, phenology and ethnobotanical uses.

Abbreviations used : *Fl. & Fr.*= Flowering and Fruiting; N=Nepali, L=Lepcha.

**Aconogonum molle** (D. Don) H. Hara (Polygonaceae); *Thotne* (N)

Scandant shrub; stipule ochreate; inflorescence racemose; flowers white.

Tonglu, *CR*-24, Common.

*Fl. & Fr.:* June-September.

Young peeled internodes of the stem are used to prepare pickles. Terminal leaves are eaten as vegetable with fish.

**Acorus calamus** L. (Acoraceae); *Bojo* (N)

Perennial herb; leaves distichous; inflorescence spadix; flowers bisexual, trimerous.

Pankhabari, *CR*-34, Rare.

*Fl. & Fr.:* July-September.

Rhizome gives a tonic of bitter taste used against high fever, bronchitis, lung-trouble, rheumatism, skin diseases. The dried ones are chewed to remove cough.

**Ageratum conyzoides** L. (Asteraceae); *Ilamae* (N)

Annual herb; leaves ovate, serrate; flowers epigynous.

Srikhola, *CR*-12, Common.

*Fl. & Fr.:* November-April.

Extract of young terminal leaves is applied to cuts, wounds and sores, considered as antiseptic and antitetanic. Root extract is often taken orally to prevent gall-stone formation. Decoction of leaves is also used to check high fever.

**Artemisia indica** Willd. (Asteraceae); *Titepati* (N)

Shrub with offensive scent; leaves pinnatipartite; stamens syngenecious.

Pankhabari, *CR*-35, Rare.

*Fl. & Fr.:* October- December.

The juice obtained from the leaves and stems is applied on cuts and bruises. It is also used to make a plug on to the nostril to stop nose bleeding. The plant extract is applied to the head to prevent convulsion. The terminal leaves are taken to reduce the hypertension.

**Begonia inflata** C.B. Clarke (Begoniaceae); *Magarkachae* ( N )

Succulent herb; leaves stipulate; flowers in cyme; stamens many.

Lepchajagat, *CR*-8, Common.

*Fl. & Fr.:* April-May.

Young shoots are eaten as chutney.

**Bergenia ciliata** (Haw.) Sternb. (Saxifragaceae); *Pakhenbet* (N)

Perennial herb; leaves ovate-rounded; flowers white, pink or purple.

Pankhabari, *CR*-37, Rare.

*Fl. & Fr.:* January-February.

The fresh or dried rhizome, sometimes fresh leaves, are used as paste or powder on cuts and bruises. It is also used in body pain after making mixture with *Astilbe* & *Fraxinus*. It is effective against diarrhoea, cough, fever and also as an antivomiting agent. Root extract is taken against food poisoning and to relieve labour pain.

**Bidens pilosa** L. (Asteraceae); *Kuro* (N)

Annual herb; leaves trifoliate; flowers in capitulum inflorescence.

Pankhabari, *CR*-36, Common.

*Fl. & Fr.:* November-March.

Dried fruits are boiled which give a kind of flavour and colour and often drunk like tea.

**Castanopsis hystrix** A.DC. (Fagaceae); *Katush* (N)

Tree; leaves elliptic; inflorescence spike.

Lepchajagat, *CR*-7, Common.

*Fl. & Fr.:* May- July.

Seeds are edible.

**Centella asiatica** (L.) Urban (Apiaceae ); *Golpata, Dhungri Jhar* (N)

Annual herb; leaves simple, reniform; inflorescence umbel.

Pankhabari, *CR*-33, Common.

*Fl. & Fr.:* June- September.

Leaves and young shoot are taken fresh and crushed and eaten to reduce blood pressure, throat complaints and in dysentery The leaves are taken to improve memory and are also useful in skin diseases .The leaf juice removes cataract after prolonged use.

**Clematis tongluensis** (Bruhl) Tamura (Ranunculaceae); *Pinasay lahara* (N)

Twiner; leaflets ovate; inflorescence solitary.

Tonglu, *CR*-22, Rare.

*Fl. & Fr.:* July-August.

Cleaned roots are crushed, wrapped in cloths, held tightly and inhaled to cure sinusitis and nose blockage.

**Dichroa febrifuga** Lour. (Hydrangeaceae); *Basak* (N)

Terrestrial shrub; leaves simple, lanceolate; flowers in panicle inflorescence.

Lepchajagat, *CR*-40, Common.

*Fl. & Fr.:* May-July.

Decoction made from the stem and root is used as febrifuge. Root and tender leaves are also used as febrifuge. Root extract is taken orally against dysentery.

**Drymaria villosa** Cham. & Schlecht. (Caryophyllaceae); *Abijalo* (N)

Sub-erect herb; leaves opposite, ovate - cordate; stamens -5.

Lepchajagat, *CR*-1, Rare.

*Fl. & Fr.:* June-August.

Leaves mixed with *Fragaria indica* are taken against jaundice. The plant is used to cure hay fever, asthma, cold, sinusitis, diphtheria, pneumonia.

**Edgeworthia gardneri** Meissn. (Thymelaeaceae); *Kagate, Argayle, Algeri* (N)

Shrub; leaves elliptic -lanceolate; flowers golden-yellow.

Srikhola, *CR*-19, Common.

*Fl. & Fr.:* March-April.

Bark is used in making rope and paper.

**Eupatorium adenophorum** Spreng. (Asteraceae); *Kalighar* (N)

Erect herb; leaves opposite, ovate; inflorescence capitulum.

Srikhola, *CR*-14, Common.

*Fl. & Fr.:* March-May.

The leaves and tender stem are taken fresh and crushed and the juice is applied on cuts and boils. In larger wounds the squeezed remains of the plant is placed over the wound to stop bleeding and protect the wound from infection.

**Fragaria vesca** L. (Rosaceae); *Bhui asilo* (N)

Perennial herb; leaves stipulate; flowers white.

Tonglu, *CR*-21, Common.

*Fl. & Fr.:* March-July.

Ripe fruits are eaten. They are also used to prepare local wine after fermentation with the help of "marcha" (yeast).

**Gentiana quadrifaria** Bl. (Gentianaceae); *Khumbu* (L)

Erect herb; leaves sessile, lanceolate; flowers bluish.

Srikhola, *CR*-16, Rare.

*Fl. & Fr.:* March-May.

Young shoots are used to prepare chutney.

**Heracleum wallichii** DC. (Apiaceae); *Chingping* (N)

Perennial shrub; leaves compound; flowers greenish

Tonglu, *CR*-23, Rare.

*Fl. & Fr.:* July-October.

Dried flowers and fruits are chewed to cure body ache and influenza. Fruits are also used to prepare chutney.

**Ipomoea purpurea** (L.) Roth (Convolvulaceae); *Jalap* (N)

Climbing herb; leaves simple, trilobed; flowers violet.

Pankhabari, *CR*-28, Common.

*Fl. & Fr.:* July-August.

The jalap is made from the tubers which is used as laxative.

**Mussaenda frondosa** L. (Rubiaceae); *Sittalu* (N)

Shrub; leaves stipulate; flowers yellowish-white in terminal cyme.

Pankhabari, *CR*-39, Frequent.

*Fl. & Fr.:* July-August.

The decoction of the root cures pneumonia, throat disturbances. The flowers are externally applied to clean foul ulcers and to cure skin eruptions.

**Nasturtium officinale** R. Br. (Brassicaceae); *Simrayo* (N)

Perennial herb; leaves alternate; inflorescence racemose.

Srikhola, *CR*-18, Common.

*Fl. & Fr.:* May-June.

The leafy stem is eaten as green spinach. The plant is effective against jaundice, cardiac diseases, low blood pressure, gout and tuberculosis.

**Ophiopogon intermedius** D. Don (Liliaceae); *Nakima* (N)

Scapigerous herb; leaves tufted, narrowly linear; flowers campanulate

Lepchajagat, *CR*-37, Rare.

*Fl. & Fr.:* May- August.

Inflorescence is used as vegetable.

**Oxalis corniculata** L. (Oxalidaceae); *Chariamilo* (N)

Herb; leaves compound; flowers yellow.

Lepchajagat, *CR*-5, Common.

*Fl. & Fr.:* May- July.

Leaves are used as refrigerant and stomachic. The fresh juice of leaves is used in dysentery, treatment of boils and also as febrifuge. Leaves also improve appetite and digestion of dyspeptic patients.

**Physalis minima** L. (Solanaceae); *Jangali fakfakay* (N)

Perennial shrub; leaves simple, cordate; inflorescence solitary flower.

Pankhabari, *CR*-32, Rare.

*Fl. & Fr.:* July- December.

The dried fruits are mashed and rolled inside the leaf to make cigar like structure and smoked to relieve toothache. They are also used to cure wounds.

**Phytolacca acinosa** Roxb. (Phytolaccaceae); *Jaringo* (N)

Terrestrial herb; leaves alternate, lanceolate; inflorescence racemose.

Lepchajagat, *CR*-2, Rare.

*Fl. & Fr.:* May-June.

Young shoots are eaten as vegetable which cures body ache and rheumatism.

**Pieris ovalifolia** D. Don (Ericaceae); *Angeri* (N)

Tree; leaves tristichous, ovate-elliptic; flowers white.

Lepchajagat, *CR*-3, Rare.

*Fl. & Fr.:* June-July.

Extract of young shoots cures skin diseases.

**Plantago major** L. (Plantaginaceae); *Kaneyjhar, Musay kanay, Nasho jhar* (N)

Herb; leaves simple; spike on long scape.

Srikhola, *CR*-10, Common.

*Fl. & Fr.:* June-August.

Leaves are crushed and paste is applied on major cuts. Decoction of the whole plant is used as an antipyretic agent.

**Polygala arvensis** Willd. (Polygalaceae); *Marcha* (N)

Terrestrial shrub; leaves tristichous; inflorescence racemose.

Tonglu, *CR*-25, Rare.

*Fl. & Fr.:* May-June.

Roots are used to prepare starter culture of yeast called "Marcha". It is used to ferment cereals and to prepare local beer.

**Rhododendron arboreum** Sm. (Ericaceae); *Rato guras* (N)

Tree; leaves oblong-lanceolate; flowers blood red

Srikhola, *CR*-13, Rare.

*Fl. & Fr.:* February-May.

Young leaves are used to poison fish. The red flowers are used in dysentery accompanied by gastrointestinal bleeding and against throat infections.

**Rubia manjith** Roxb. ex Fleming (Rubiaceae); *Majito* (N)

Scrambling herb; stems tetragonous; leaves ovate-cordate; flowers reddish-brown.

Tonglu, *CR*-24, Common.

*Fl. & Fr.:* July- September.

Dye is obtained from roots and stem. Yellowish red dye obtained from fruits is commonly used as cosmetics and as colouring ink. Root decoction is admini-stered in paralysis, jaundice, menstrual disorders, chest troubles and fever.

**Rubus ellipticus** Sm. (Rosaceae); *Aselụ* (N)

Prickly shrub; leaves stipulate; flowers white.

Srikhola, *CR*-17, Common.

*Fl. & Fr.:* February- May.

Ripe fruits are eaten. The tender stem is peeled fresh and eaten in colic pain and as anti-helminthic.

**Rumex nepalensis** Spreng. (Polygonaceae); *Halhale* (N)

Perennial herb; leaves simple, stipulate; flowers green.

Srikhola, *CR*-11, Common.

*Fl. & Fr.:* June-July.

The extract of roots, dried or fresh, used in liver disorders. Paste made from grinded root is applied in skin diseases and in loss of hairs. Young shoots are eaten as vegetable.

**Solanum khasianum** C.B. Clarke (Solanaceae); *Rhambhenri kara* (N)

Spiny shrub; leaves subopposite; inflorescence cymose.

Pankhabari , *CR*-32

*Fl. & Fr.:* June-August, Rare.

The seeds are burnt and the smoke is inhaled to relieve tooth pain.

**Swertia chirayita** (Roxb.) Karsten (Gentianaceae); *Chirato* (N)

Annual herb; leaves opposite; flowers in cyme.

Tonglu, *CR*-29, Rare.

*Fl. & Fr.:* August-September.

The entire plant is a favourite remedy in intermittent fever. It is used as a tonic, laxative and when taken with sandal wood paste, it stops internal haemorrhage of stomach.

**Thysanolaena latifolia** (Roxb. ex Hornem.) Honda (Poaceae); *Kucho, Omliso* (N)

Large grass; leaves flat; panicle large.

Srikhola, *CR*-19, Common.

*Fl. & Fr.:* December-February.

It is used for making brooms. A decoction of the root is used to induce abortion and to treat mumps, boils and abscesses.

**Urtica ardens** Link (Urticaceae); *Gharia sisnu* (N)

Perennial herb with stinging hairs; leaves ovate-cordate; flowers unisexual.

Tonglu, *CR*-26, Common.

*Fl. & Fr.:* August- September.

Young leaves and flowers are boiled with spices to make soup. Fibres obtained from stem are used to make cloth for traditional garments. Young leaves and flowers are taken to reduce high blood pressure.

**Viburnum erubescens** Wall. (Caprifoliaceae); *Asare* (N)

Perennial shrub; leaves simple; inflorescence subterminal corymbose; flowers epigynous.

Tonglu, *CR*-24, Common.

*Fl. & Fr.:* May-June.

Ripe fruits are edible.

## CONCLUSION

An account of the ethnobotanical uses of 36 plants has been presented. Hilly areas of Darjeeling harbour a rich variety of ethnically valuable plants. During present study, 21 medicinal, 10 edible and 5 species for other ethnic uses have been recorded. However, indiscriminate collection of these plants from nature by local people and external agencies have posed threats to the useful plants of this ecological zone. The detailed studies of the ethnic uses of plant species as native food, material culture, medicine will unveil the glimpse on their culture, which is to be documented for IPR purpose because the culture is the victim of present acculturation.

## ACKNOWLEDGEMENTS

The authors are grateful to the Director, Botanical Survey of India, Kolkata for affording the herbarium and library facilities and Mr. K.K. Tamang, Ex-Head, Deptt. of Botany, Darjeeling Govt. College for information and helpful suggestions. We thank Dr. D.C. Pal, Ex-Scientist, Botanical Survey of India, Kolkata for critically reading the manuscript and offering constructive suggestions.

## REFERENCES

ANONYMOUS 1980. West Bengal District Gazetteers, Darjeeling. Govt. of West Bengal.

BHUJEL, R.B., K.K. TAMANG & G.S. YONZONE 1984. Edible wild plants of Darjeeling district. *J. Bengal Nat. Hist. Soc.* 3 : 76-83.

BISWAS, K. 1956. *Common medicinal plants of Darjeeling and Sikkim Himalayas.* Superintendent, Govt. Printing West Bengal Govt. Press, Alipore, West Bengal.

———— 1966. *Plants of Darjeeling & Sikkim Himalayas,* Vol. 1. Superintendent, Govt. Printing West Bengal Govt. Press, Alipore, West Bengal.

CHHETRI, D.R. 2004. Medicinal plants used as antipyretic agents by the traditional healers of Darjeeling Himalayas. *Indian J. Traditional Knowledge* 3 (3) : 271-275.

JAIN, S.K. 1965. Medicinal plant-lore of the tribals of Bastar. *Econ. Bot.* 19 : 236-250.

LAMA, P.C. 1989. A preliminary report on the ethnobotanical importance of the Sukia Pokhari region of Darjeeling Himalayas. *J. Bengal Nat. Hist. Soc.* 8 (1) : 56.

NANDI, R.P. 1997. *Little known medicinal plants of Darjeeling Himalayan hills.* Indian Council of Alternative Medicines, Calcutta.

RAI, L.K. & E. SHARMA 1994. *Medicinal plants of Sikkim Himalayas-status, uses and potential.* Bishen Singh Mahendra Pal Singh, Dehra Dun.

RAI, P.C., A. SARKAR, R.B. BHUJEL & A.P. DAS 1998. Ethnobotanical studies in some fringe areas of Sikkim & Darjeeling Himalayas. *J. Hill Research* 11 (1) : 12-21.

RAI, S.K. & R.B. BHUJEL 1999. Notes on some less-known ethnobotanical plants from Darjeeling Himalayas. *J. Hill Research* 12 (2) : 160.

———— & ———— 2002. Ethnic uses of some monocotyledonous plants in the Darjeeling Himalayan region. In Das, A.P. (Ed.) *Perspectives of Plant Biodiversity.* Bishen Singh Mahendra Pal Singh. Dehra Dun.

YONZONE, G.S., D.K.N. YONZONE & K.K. TAMANG 1984. Medicinal plants of Darjeeling district. *J. Econ. Taxon. Bot.* 5(3): 605-616.

YONZONE, R. & S. MANDAL 1982. A preliminary report on the ethnobotanical survey of Kalimpong in Darjeeling district, West Bengal. *J. Bengal Nat. Hist. Soc.* (n.s.) 1 (1) : 96-99.

————, ———— & S. CHANDA 1981. A contribution to the ethnobotany of Darjeeling hills. *Trans. Bose Res. Inst.* 44 (3) : 74-81.

# ETHNOBOTANICAL STUDY OF KALAHANDI DISTRICT, ORISSA

A.M. SAREN, A.C. HALDER & P.R. SUR

*Central Botanical Laboratory, Botanical Survey of India, P.O. Botanic Garden, Howrah-711 103*

**ABSTRACT**

The paper deals ethnobotanically important 52 plant species under 42 genera of Kalahandi district, Orissa, India. The knowledge and wisdom of different tribals, like Kondh, Oraon and Santhal, have been summarised. The valid scientific names, vernacular names, plant parts used and their ethno-economic uses are documented in the present communication.

## INTRODUCTION

Kalahandi district is located between 19°12' and 20°27' North latitude and 82°312' and 83°48' East longitude. The total forest area of the district is 1413.69 sq km, i.e. only 4.61 per cent of the total area. Kalahandi hits the headlines in newspapers for its acute drought. The repeated drought situation has broken the economic backbone of cultivators since last three decades. A long history of drought covering more than a century in Kalahandi has occurred. Kalahandi, which was previously known as "Mahakantara" means Great Forest in ancient India, is now not more than a wasteland. The people who were once self sufficient and self contented due to abundant forest products and ample harvest, are now lacking even the fire wood and facing great food problems. The land of Kalahandi which was also famous for bumper paddy cultivation is now depending upon external helps.

Kalahandi, also known as "Karunda Mandala", is the treasure of precious stones like Karandan (Manik), Garnet (Red stones), Berz, Neelam (Blue stone) and Alexandra. The historical significance of naming Kalahandi as Karunda Mandala is based on the availability of Karundam in this region. The total population of Kalahandi district is 11,30,903; out of this, 192366 are SC and 3,26,648 ST. The main tribe is Kondh.

The climate of Kalahandi district is extreme type. There are three distinct seasons, like summer ,rainy and winter. The summer season is prolonging. The highest temperature in summer ranges from 48° to 50°C. The rain fall is average 655.5 mm in rainy season. Due to independent and combined effect of climate, topography, soil and the passed treatment, the forests of Kalahandi district vary from tropical semi-evergreen to dry deciduous scrub forest. They, however, belong to moist tropical forest and dry tropical forest. One of the distinctive feature of these types is the occurrence of both Sal and Teak, which is known as transition belt or Ecoton.

## MATERIAL AND METHODS

Three field tours were undertaken for covering different tribal populated areas of Madanpur, Biswanathpur, Kesinga, Bhabanipatna, Junagarh, Thuamulrampur, Jacom, Jurakhaman, Bhatang poddor, Rampur, Bazargarh etc. Plants of ethno-economic importance were collected and ethnobotanical information was recorded from sources such as Vedyas, Hakims and local tribals who are considered to have good knowledge of the plants of that region. Repeated queries were made to get the information confirmed. The voucher specimens have been deposited in Ethnobotanical herbarium, Central Botanical Laboratory, Botanical Survey of India, Howrah.

## OBSERVATION AND DISCUSSION

The plants used by the tribal people are summarised in alphabetical order by their botanical names together with their respective families, vernacular names (V. n) and collection numbers (No). This is followed by a brief note on uses and method of using the drugs as revealed by the informants during field trips. Under the vernacular names, the names of tribes are abbreviated like K (Kondh), Or (Oriya). About 52 plant species under 46 genera are discussed; out of these, 29 species are medicinal, 10 provide wild food and 15 are used for other purposes.

**Acacia leucophloea** (Roxb.) Willd. (Mimosaceae)

V.n: Gondo-goira - K, Goira - Or.

*Loc.:* Junagarh, Bh-Patna, Kalahandi (No. 18210).

*Use :* Seeds are used as bed-bug killer. Wood is used as fuel.

**Achyranthes aspera** Linn. (Amaranthaceae)

V.n: Apamara - K, Or.

*Loc.:* Gunpur,Kalahandi (No.18255).

*Use :* In one-side headache a portion of the root is applied with castor oil for quick relief.

**Ailanthus excelsa** Roxb. (Simaroubaceae)

V.n: Mahanim - K.

*Loc.:* Kinkhola, Bh-patna (No. 18330).

*Use :* Pasted root is used for driven out field rats from the paddy fields and food stores.

**Asparagus racemosus** Willd. (Liliaceae)

V.n: Subundri - K, Debarun - Or.

*Loc.:* Kinkhola, Bh-patna (No. 18272, 18256).

*Use :* The drink prepared from the roots is very useful for cooling stomach. Normally it can be taken for healthy stomach. 2 inch of root and 12 pieces of black pepper (Gol morich) are pasted together and a drink is prepared. Two doses for three days are prescribed for the treatment of gonorrhoea.

**Bauhinia vahlii** Wt. & Arn. (Caesalpiniaceae)

V.n: Sealimar - K, Chearo-chal - Or.

*Loc.:* Jogespodor, Bh-patna (No. 18247).

*Use :* Leaves are used for preparing plates and rain coats called "Hatur".

**Bauhinia variegata** Linn. (Caesalpiniaceae)

V.n: Kandul, Bhumij - K.

*Loc.:* Thuamal Rampur, Kalahandi (No. 18243).

*Use :* Seeds are consumed as "dal' and boiled seed juice is very effective in jaundice.

**Cajanus cajan** (Linn.) Millsp. (Fabaceae)

V.n: Kandul - K.

*Loc.:* Gunpur, Kalahandi (No. 18243).

*Use :* Mature seeds are taken as "Dal", boiled seed juice is very effective in jaundice.

**Calotropis gigantea** (Linn.) R. Br. (Asclepiadaceae)

V.n: Arak mara - K, Arak - Or.

*Loc.:* Gunpur, Kalahandi (No. 18231).

*Use :* Latex of the leaf is applied on the decayed teeth for uprooting the teeth easily. Caution should be taken that the latex may not touch other teeth, otherwise would be adversely effected.

**Cannabis sativa** Linn. (Cannabaceae)

V.n: Ganja, Bhang - Or.

*Loc.:* Kathgora, Gunpur, Kalahandi (No. 18251).

*Use:* Leaves are smoked like tobacco. "Bhang and "Siddhi", consisting of dried leaves and fruiting shoots are used for intoxicating liquor.

Native doctors prescribe it in bowel complaints and as a nervous stimulant.

**Cascabella thevetia** (L.) Lippold (Apocynaceae)

V.n: Bija-mara - K.

*Loc.:* Postigudaw, Kalahandi (No. 18336).

*Use :* Extract of stem juice is useful in diabetes.

**Cassia fistula** Linn. (Caesalpiniaceae)

V.n: Funnel mara - K, Sonorigoch - Or.

*Loc.:* Gunpur,Kalahandi (No. 18220).

*Use :* It is believed by the Kond people that when a dead body taken out for funeral if touches this plant, the dead man's spirit may come back in the same village and would make horror by throwing stones in the night time in that village.

**Catharanthus roseus** (Linn.) G. Don (Apocynaceae)

V.n: Boa - Or, K.

*Loc.:* Kinkhola, Kalahandi, Bh-patna (No. 18325).

*Use :* Leaf-juice is very effective in any insect sting (wasp, bee etc). Village Vaidya worship the idol of Ravana made with flowers of this plant.

**Centella asiatica** (Linn.) Urban (Apiaceae)

V.n: Thalkuni - Or.

*Loc.:* Sursanai, Gunpur (No. 18262).

*Use :* Leaf-paste is given in common amoebiasis.

**Cleistanthus collinus** (Roxb.) Benth. ex Hook. f. (Euphorbiaceae)

V.n: Karla - K.

*Loc.:* Jacom, Kalahandi (No. 18268).

*Use :* Leaves are largely used as insect repellent in paddy fields; also used for thatching huts, cottage etc. Stem is very durable and resistant to white ants.

**Clerodendrum viscosum** Vent. (Verbenaceae)

V.n: Ping mara - K, Vondari goch - Or.

*Loc.:* Gunpur, Kalahandi (No. 18321).

*Use :* Flowers are used for worshipping village god and in the marriage ceremony.

**Combretum album** Pers. (Combretaceae)

V.n: Atan - K.

*Loc.:* Gokolama, Gunpur, Kalahandi (No. 18310).

*Use :* The juice oozing out during burning of its wood in the non-burning side, is useful in curing eczema, ring worm etc.

**Cryptolepis buchanani** Roem. & Schult. (Periplocaceae)

V.n: Dudellaha - K.

*Loc.:* Gunpur, Kalahandi (No. 18224).

*Use :* For unhealthy babies, this creeper is tied in their hands by making "balas", which results in good health.

**Desmodium pulchellum** (Linn.) Benth. (Fabaceae)

V.n: Ummah - Or.

*Loc.:* Gunpur, Kalahandi (No. 18254, 18302).

*Use :* A portion of the root, 1-2 inch, and a black pepper are pasted together and given orally in snake-bite and the leaf-paste is applied in the affected portion of the body. If the twigs of this plant are kept under mattresses, the bed bugs are driven out.

**Dendrocalamus strictus** (Roxb.) Nees (Poaceae)

V.n: Bans - K, Or.

*Loc.:* Bhatangpodor, Bh-patna, Kalahandi (No. 18316).

*Use :* It is largely used for making huts and for paper making in paper industry.

**Dendrophthoe falcata** (Linn. f.) Etting. (Loranthaceae)

V.n: Modong - Or, K.

*Loc.:* Churadangor hills, Junagarh, Kalahandi (No. 18323).

*Use :* It is believed that plant would protect the house from evil eye and evil spirit if the twigs are tied in any place of the house.

**Dioscorea bulbifera** Linn. (Dioscoreaceae)

V.n: Mundi konda, Ranu konda - K.

*Loc.:* Gunpur, Kalahandi (No. 18218).

*Use :* Root is taken as vegetable.

**Diospyros melanoxylon** Roxb. var. **tupru** (Buch.-Ham.) Singh (Ebenaceae)

V.n: Kendu govh - Or, Kendu mara - K.

*Loc.:* Gunpur, Kalahandi (No. 18217).

*Use :* Bark of this plant with that of *Phyllanthus emblica* L. (Ahallya goch) is made into a paste and then a 'sarbat' is prepared. This 'sarbat' is hotted on tawai and then collected in a bottle and given in blood dysentery 2-teaspoonful in empty stomach for 2-3 days.

**Elephantopus scaber** Linn. (Asteraceae)

V.n: Smja-chere - K.

*Loc.:* Jacom, Kalahandi (No. 18282).

*Use :* Pasted root is applied on fore-head to relieve the pain of hemicrania quickly.

**Euphorbia hirta** Linn. (Euphorbiaceae)

V.n: Kherui mara - K, Or.

*Loc.:* Kinkhola, Bh-patna, Kalahandi (No. 18332).

*Use :* Plant decoction is given for curing bronchial affections and asthma.

**Ficus racemosa** Linn. (Moraceae)

V.n: Toin mara - K, Dumber - Or.

*Loc.:* Gokolama, Gunpur (No. 18223).

*Use :* Ripe fruits are eaten only in summer, not in rainy season.

**Ficus rumphii** Blume (Moraceae)

V.n: Pepul - Or.

*Loc.:* Gunpur, Kalahandi (No. 18265).

*Use :* Ripe fruits are edible.

**Flacourtia indica** (Burm. f.) Merr. (Flacourtiaceae)

V.n: Gutur - K, Kote kul - Or.

*Loc.:* Kathghora, Gunpur (No. 18248).

*Use :* Ripe fruits are eaten during May-June.

**Gardenia latifolia** Soland. (Rubiaceae)

V.n: Garura goch - Or, Gasrdu-mara - K.

*Loc.:* Gunpur, Kalahandi (No. 18261).

*Use :* Tablets are prepared from the pasted fruits and given to cure amoebiasis, 1-2 tablets for 4 days in empty stomach. Ripe fruits are edible.

**Hemidesmus indicus** (L.) R. Br. (Periplocaceae)

V.n: Dudellaha - K, Dud mal - Or.

*Loc.:* Postiguda, Gunpur, Kalahandi (No. 18224).

*Use :* For unhealthy babies this creeper is tied in the form of a "Bala" in their hands, for improving child's health.

**Holarrhena pubescens** (Buch.-Ham.) Wall. ex G. Don (Apocynaceae)

V.n: Kurchi - Or.

*Loc.:* Jugsaipatna, Kalahandi (No. 18317).

*Use :* Bark-paste is given in dysentery. The seed cotton is used for stuffing small pillows which are effective in 'sannipatic' diseases (cold, cough, headache, tonsil etc) if the victim person uses the pillow during sleeping time.

**Hygrophila schulli** (Buch.-Ham.) M. R. & S. M. Almeida (Acanthaceae)

V.n: Ekakheri - K.

*Loc.:* Gunpur, Kalahandi (No. 18263).

*Use :* The whole plant is cooked and the juice extracted is given to anaemic patients.

**Imperata cylindrica** (Linn.) Raeuschel. (Poaceae)

V.n: Dab-khar - Or.

*Loc.:* Gunpur, Kalahandi (Orissa No.18257).

*Use :* 7 pieces of root and 7 pieces of Kendu leaf are pasted together and made into three parts; one part is prescribed in the morning, 2nd part in the evening (mixing with little water) and third part in the next morning to cures blood in urine.

**Indigofera cassioides** Rottl. ex DC. (Fabaceae)

V.n: Girliphul - Or, Girel pui - K.

*Loc.:* Postiguda, Gunpur, Kalahandi (No. 18227).

*Use :* Flowers are taken as a tasteful vegetable.

**Lagenaria siceraria** (Molina) Standl. (Cucurbitaceae)

V.n: Onka - K.

*Loc.:* Gokolama, Gunpur, Kalahandi (No. 18297).

*Use :* Dried fruits are used as indigenous spoon and for worshipping village god.

**Madhuca longifolia** (Koen.) Macbr. var. **latifolia** (Roxb.) Chev. (Sapotaceae)

V.n: Pui - K, Mahua - Or.

*Loc.:* Gunpur, Kalahandi (No. 18232).

*Use :* Flowers are used in preparation of country liquor. The seeds called "Toora", are used for preparing vegetable oil (Vanaspati).

**Meyna spinosa** Roxb. ex Link (Rubiaceae)

V.n: Moyuna-konta - Or, K.

*Loc.:* Kathgora, Gunpur, Kalahandi (No. 18250).

*Use :* Dried fruits are grind to make powder which is used as indigenous spice. Strong fancing are made by this prickly shrub around the paddy fields.

**Morus alba** L. (Moraceae)

V.n: Tut phal goch, Hitmara - K.

*Loc.:* Postiguda, Kalahandi (No. 18226).

*Use :* Ripe fruits are eaten.

**Ocimum americanum** Linn. (Lamiaceae)

V.n: Gharadhana - Or.

*Loc.:* Postiguda, Kalahandi (No. 18269, 18206).

*Use :* A drink is prepared with the seed paste and sugar and given in diarrhoea for immediate effect. Leaf-juice mixed with water is used for washing the sores of cattle.

**Persicaria glabra** (Willd.) Gomez (Polygonaceae)

V.n: Bihangi - Or.

*Loc.:* Bhatangpodor, Bh-patna (No. 18311).

*Use :* The whole plant is used to stupefy fishes in stagnant water ponds. Leaf-juice is prescribed in colic.

**Pongamia pinnata** (Linn.) Pierre (Fabaceae)

V.n: Karanja - K.

*Loc.:* Postiguda, Kalahandi (No. 18283).

*Use :* Seed oil is good for oiling cart excels and wheels.

**Pterocarpus marsupium** Roxb. (Fabaceae)

V.n: Bija-mara - K.

*Loc.:* Postguda, Kalahandi (No. 18236).

*Use :* Stem-juice is given to cure diabetes.

**Ricinus communis** Linn. (Euphorbiaceae)

V.n: Kitteyamara - K, Jara-mara - Or.

*Loc.:* Thuamulrampur, Bh-patna, Kalahandi (No. 18224, 18240).

*Use :* Oil extracted from seeds is very effective for healthy hairs. Three pieces of three years old root and root of *Achyranthes aspera* (Apamara-Or)-2 pieces are tied together with a thread around the ear on the one sided pain for immediate relief.

**Senna occidentalis** (Linn.) Link (Caesalpiniaceae)

V.n: Kalkasunda - K, Or.

*Loc.:* Gunpur, Kalahandi (No. 18242).

*Use :* Seeds serve a good fodder for sheep.

**Senna tora** (Linn.) Roxb. (Caesalpiniaceae)

V.n: Chakundadal - K.

*Loc.:* Gunpur, Kalahandi (No. 18216).

*Use :* Young leaves are eaten as vegetable.

**Setaria italica** (Linn.) P. Beauv. (Poaceae)

V.n: Mondia - Or, K.

*Loc.:* Gokolama, Gunpur (No. 18239).

*Use :* Seeds of 'mondia' are made into powder and a drink is prepared with water and sugar. The drink is prescribed to the patient of jaundice, one glass each day in empty stomach for 7 days. It is more effective if added a few drops of lemon juice.

**Shorea robusta** Gaertn. f. (Dipterocarpaceae)

V.n: Sorgi, Sal - K.

*Loc.:* Gokolama, Gunpur (No. 18229).

*Use :* Burnt ash of the wood is used as detergent powder for cleaning clothes and washing utensils.

**Sida acuta** Burm. f. (Malvaceae)

V.n: Jitkabara - K.

*Loc.:* Thuamulrampur, Kalahandi (No. 18241).

*Use :* Brooms are prepared from the mature plants.

**Smilax zeylanica** Linn. (Liliaceae)

V.n: Mutri - K, Muter - Or.

*Loc.:* Postiguda, Kalahandi (No. 18230).

*Use :* Bread of flowers or juice prepared by burning flowers with the wrapper of the leaves is given to the child who passes urine in the bed while sleeping.

**Solanum myriacanthum** (Dunal) C. B. Clarke (Solanaceae)

V.n: Kanta-vegri - K.

*Loc.:* Mathurapur, Bh-patna (No. 18202).

*Use :* Flowers are pasted with honey and given in whopping cough to children and the pasted seed are given to the adults.

**Solanum torvum** Swartz (Solanaceae)

V.n: Tutuguna - Or, Tutguna polson - K.

*Loc.:* Postiguda, Kalahandi (No. 18235).

*Use :* Fruits are taken as 'chatni' with tamarind or mango.

**Solanum virginianum** L. (Solanaceae)

V.n: Churna-vegri - Or.

*Loc.:* Bazargarh, Kalahandi (No. 18209).

*Use :* 7 pieces of roots tied with a lace made of 7 pieces of threads and hanged on the neck of person affected by 'Pali fever' (constant irregular fever).

**Thysanolaena maxima** (Roxb.) O. Ktze. (Poaceae)

V.n: Beddum - K.

*Loc.:* Kathghara, Gunpur (No. 18325).

*Use :* Brooms are prepared from the grass (Phul-jharu).

**Tridax procumbens** Linn. (Asteraceae)

V.n: Visalwakarani - Or, K.

*Loc.:* Bazargarh, Kalahandi (No. 18204).

*Use :* Leaf-paste is applied on fresh cuts on the body to stop bleeding immediately.

**Woodfordia fruticosa** (Linn.) Kuntze (Lythraceae)

V.n: Dhatum phula - Or, Droi - K.

*Loc.:* Bhatangpodor, Bh-patna, Kalahandi (No. 18305).

*Use :* Nectar of the flowers is very sweet and taken as instant drink in early morning by the children and local people and also country liquor is prepared from its flowers.

## ACKNOWLEDGEMENTS

We are grateful to Director, Botanical Survey of India, Kolkata for providing facilities and encouragement. Thanks are also due to the Divisional Forest Officer, Kalahandi Division, Orissa and tribal people of Kalahandi district for their help and co-operation during this work.

# ETHNOMEDICINAL PLANTS USED AS ANTIDIABETIC

P.R. SUR, A.M. SAREN & A.C. HALDER

*Central Botanical Laboratory, Botanical Survey of India, P.O. Botanic Garden, Howrah-711 103*

## ABSTRACT

This paper deals with 24 ethnobotanical plants, which are used by villagers and tribal people of different regions of India as antidiabetic. The plant species are arranged in alphabetical order followed by vernacular names, distribution and their uses.

## INTRODUCTION

Diabetes is an old enemy, being mentioned in ancient Indian, Greek and Roman medicinal texts. About 2500 years ago, ayurvedic authorities Sushruta and Charaka noted that its victims frequently pass large amounts of sugar-laden urine which attract ants. Hence, they called the disease *madhu meha* (Honey urine), a term still commonly used. The modern name was coined by the second century physician Aretaeus from the Greek word 'diabetes', which means siphon, because he observed that the disease act like a siphon on its victims, melting down flesh and limbs in urine.

Now a days each and every family has at least one member with diabetes. Diabetes has affected about 3% of the Indian total population. From ancient time, the Indians are using the plants for curing or controlling the diabetes disease. The plants, which are used by tribals or village people as antidiabetic, are discussed below.

In the following account, the plant species are arranged alphabetically by their generic names, then the specific epithet, followed by families, vernacular names, distribution and medicinal uses among the tribal people of different areas. Pharmacological and clinical trials will help confirmation of the efficacy of the reported herbs.

## ENUMERATION

Abbreviation used :- V.n.: Vernacular name; H: Hindi name; B: Bengali name; S: Sanskrit name; E: English name; Sant = Santal; O = Oraon; Ori = Oriya; N = Naga.

**Alpinia galanga** (Linn.) Swartz (Zingiberaceae)

*V.n.:* Barakulanjar (H, B), Sugandhavacha (S), Kangu (Sant)

*Distribution:* Wild in Western Ghats; cultivated throughout India.

*Uses:* The aromatic rhizome is recommended in diabetes.

**Anacardium occidentalis** Linn. (Anacardiaceae)

*V.n.:* Kaju (H), Hajli Badam (B), Kajutaka (S), Geru-beeja (O)

*Distribution:* Naturalized and cultivated in the coastal districts of India.

*Uses:* Infusion of the bark is recommended in the treatment of diabetes insipidus.

**Borassus flabellifer** Linn. (Arecaceae)

*V.n.:* Tar (H), Tal (B, S), Tali (Sant)

*Distribution:* More or less all over India in the dry regions.

*Uses:* The fresh sweetish juice obtained from cuts made in the flowering stalk is fermented and given in diabetes.

**Butea monosperma** (Lam.) Taub. (Fabaceae)

*V.n.:* Dhak (H), Palas (B), Palasha (S), Khakro (O), Marup (Sant)

*Distribution:* Throughout greater parts of India, upto 1400 m height.

*Uses:* Decoction of leaves is given in diabetes. Flower decoction is also given to treat diabetes.

**Cassia fistula** Linn. (Caesalpiniaceae)

*V.n.:* Amaltas (H), Sundali (B), Suvarnaka (S), Bahava (O)

*Distribution:* Throughout India.

*Uses:* The pulp of the fruit (decoction) is given as a confection in diabetes. Seed powder is also given for treatment of diabetes.

**Ceiba pentandra** (Linn.) Gaertn. (Bombacaceae)

*V.n:* Safed Semal (H), Schwet Simul (B), Sveta Salmali (S)

*Distribution:* Widely distributed in the forest in the hotter parts of Western and Southern India.

*Uses:* The root juice is highly valued as a cure for diabetes.

**Celastrus paniculata** Willd. (Celastraceae)

*V.n.:* Malkagni (H, B), Kanguni (S), Peng (Ori)

*Distribution:* Throughout hilly regions of India.

*Uses:* The oil from seeds is given one teaspoonful in the morning to diabetic patient.

**Chromolaena odorata** (Linn.) King & Robinson (Asteraceae)

*V.n.:* Bhabri, Ganduri, Kotrojara (Sant)

*Distribution:* West Bengal, Assam.

*Uses:* Decoction of flowers used in diabetes.

**Coccinia grandis** (Linn.) Voigt (Cucurbitaceae)

*V.n:* Kaduri (H), Telakucha (B), Bimba (S), Galedu (Sant)

*Distribution:* Throughout India.

*Uses:* The juice of root and leaf is given to treat diabetes.

**Curcuma longa** L. (Zingiberaceae)

*V.n.:* Haldi, Halud (H,B), Haridra (S), Termeric (E)

*Distribution:* Cultivated throughout India.

*Uses:* The juice of rhizome with that of Amla (*Phyllanthus emblica* L.) is prescribed for diabetes.

**Ficus benghalensis** Linn. (Moraceae)

*V.n.:* Bor (H), Bat (B), Vata (S)

*Distribution:* Planted throughout India.

*Uses:* Infusion of stem bark, one part in ten part of water, is regarded as a remedy for diabetes.

**Ficus racemosa** Linn. (Moraceae)

*V.n.:* Gular (H), Jagya-dumur (B), Udumbara (S)

*Distribution:* Throughout India.

*Uses:* Infusion of stem bark is used for curing diabetes. Ripe fruits are used for treatment of diabetes.

**Gymnema sylvestre** (Retz.) R. Br. ex Schult. (Asclepiadaceae)

*V.n.:* Merasingi (H), Gurmari, Meshasingi (B), Meshashringi (S), Podapatri (O)

*Distribution:* Almost throughout India.

*Uses:* The leaf juice is given 2-3 teaspoonful daily in the morning to diabetic patient or 10 mg powder of dried leaves is given once in a day in the morning.

**Helicteres isora** Linn. (Sterculiaceae)

*V.n.:* Marorpshaḷi (H), Atmora (B), Mriga-singha (S)

*Distribution:* More or less throughout India.

*Uses:* A decoction of the bark or its juice is considered useful in diabetes, as it reduce the quantity of sugar.

**Mangifera indica** Linn. (Anacardiaceae)

*V.n.:* Amb (H), Aam (R), Amra (S)

*Distribution:* Cultivated in the tropics.

*Uses:* A powder made of the dried tender leaves is beneficially given in diabetes. The seed powder is also given for curing diabetes.

**Musa paradisiaca** Linn. (Musaceae)

*V.n.:* Kela (H), Kala (R), Kadali (S)

*Distribution:* Cultivated throughout India, wild in the north-eastern States.

*Uses:* The cooked flowers are given in diabetes.

**Parthenium hysterophorus** Linn. (Asteraceae)

*V.n.:* Santamaria, Ragweed, White top, Congress grass (E), Bish-goch (H, B)

*Distribution:* More or less all over India.

*Uses:* Tablets made from pasted leaves are given to diabetic patients, 2 tablets twice daily in the morning and night with water before meals.

**Pongamia pinnata** (Linn.) Pierre (Fabaceae)

*V.n.:* Karanja (H, S), Dahur-karanja (B)

*Distribution:* All over India.

*Uses:* The decoction of flowers is useful in diabetes.

**Senna auriculata** (Linn.) Roxb. (Caesalpiniaceae)

*V.n.:* Tarwar (H, B), Anwal (Sant)

*Distribution:* Wild in the dry regions of Madhya Pradesh, W. Peninsula and Rajasthan; cultivated in other parts of India.

*Uses:* The decoction of bark is given in diabetes.

**Senna sophera** (Linn.) Roxb. (Caesalpiniaceae)

*V.n.:* Kasunda (H), Kalkasunda (B), Kasamarda (S), Rawasan (Sant)

*Distribution:* Throughout India.

*Uses:* An infusion of the bark is considered a remedy for diabetes.

**Smilax zeylanica** Linn. (Liliaceae)

*V.n.:* Janglui-aushbah (H), Kumarika (B), Nagaubuai (N)

*Distribution:* Throughout India.

*Uses:* Leaf and stem juice are given to patent suffering from diabetes.

**Swetenia mahagoni** Jacq. (Meliaceae)

*V.n.:* Mahagoni (H)

*Distribution:* Planted throughout India.

*Uses:* The seed powder one teaspoonful with water is given once daily in the morning to diabetic patient.

**Syzygium cumini** (Linn.) Skeels (Myrtaceae)

*V.n.:* Jamun (H), Jam (B), Jambu (S)

*Distribution:* Throughout India.

*Uses:* The seeds are regarded as most effective in diabetes. It reduces sugar in urine. A liquid extract prepared from the fresh seeds is considered as a medicine for diabetes. Powdered seeds are also used for curing this disease.

**Trigonella foenum-graecum** Linn. (Fabaceae)

*V.n.:* Methi (H, B, S)

*Distribution:* Wild in Punjab & Kashmir; cultivated in many parts of India.

*Uses:* Seeds are soaked in water and allowed to sprout till the seedlings are about 5-7.5 cm long; they are eaten raw and found useful in diabetes. The decoction of flowers is also given to diabetic patients.

## ACKNOWLEDGEMENT

We are thankful to Director, Botanical Survey of India, Kolkata for giving facilities and encouragement. Thanks are also due to tribal people of different areas.

## REFERENCES

CHOPRA, R.N., S.L. NAYAR & I.C. CHOPRA 1956. *Glossary of Indian Medicinal Plants.*

JAIN, S.K. 1991. Dictionary of Indian Folk medicine and Ethnobotany.

NATH, A. & G.G. MAITY 2003. Ethnobotany of Barak Valley (Southern Assam) with special references to folk medicine. *J. Econ. Taxon. Bot.* 27 (4) : 964-971.

PAL, D.C. & S.K. JAIN 1998. *Tribal Medicine.*

SINGH, R.K. & SANJAY SINGH 2003. Is *Parthenium hysterophorous* Linn. an obnoxious weed?. *J. Econ. Taxon. Bot.* 27 (4) : 904-907.

SUR, P.R., R. SEN, A.C. HALDER & S. BANDYOPADHYAY 1987. Observation on the Ethnobotany of Malda West-Dinajpur District, West Bengal-I. *J. Econ. Taxon. Bot.* 10 (2) : 395-401.

———, ———, ——— & ——— 1992. Ethnobotanical study of Purulia District, West Bengal, India. *J. Econ. Taxon. Bot.* Addl. Ser. 10 : 259-264.

# ETHNOMEDICINAL STUDIES OF PATAN DISTRICT, NORTH GUJARAT

P.K. PATEL

*Dept. of Biology, P.S. Science College, Kadi*

## ABSTRACT

A brief account is presented about the ethnomedicinal plants of Patan district, North Gujarat. Thirty one species which are commonly used by people are enumerated.

## INTRODUCTION

Gujarat State geographically is divided in to five regions, i.e. North Gujarat, Central Gujarat, South Gujarat, Saurashtra & Kutchh. North Gujarat has 4 districts, i.e. Patan, Banaskantha, Sabarkantha & Mehsana. Patan district lies between 23°28' lat. & 72°76' long. People of Patan district are mainly farmers, with the tribals called Koli, Deviputra, Maldhari, Adivasi etc.

Ethonobotanical studies have received due consideration in Gujarat as evident from the literature. The notable works are those of Thaker (1926), Joshi *et al.* (1980), Joshi (1988), Shah *et al.* (1981), Shah & Gopal (1982, 1985, 1986), Bhatt & Sabnis (1987), Bhattacharya (1996), Mitaliya (1998), Punjani (1998), Bhatt & Mitaliya (1999), Jadeja (1999), Ant (2000), Bhatt *et al.* (1999, 2000, 2001), Silori & Rana (2000), Patel & Reddy (2000), Mitaliya *et al.* (2001, 2003), Kakrani & Saluja (2001), Punjani & Kumar (2003) etc. However, no ethnobotanical data from Patan district was available to the author. As such, keeping in view the importance of decreasing and depleting traditional knowledge of folklore, the present study is undertaken for providing data about sustainable utilization and conservation of resources.

## MATERIAL AND METHODS

Regular field trips to patan district were made to collect plant specimens of ethnobotanical value and for interviews with tribal and rural people who regularly use plants of this area. The identification of plants was done with the help of Floras of Shah (1978) and Bole & Pathak (1988). During field work, information about the local name of the plant, part used, mode of preparation of drug and its application were especially recorded and verified with cross interviews with other persons at different localities. The collected data on this subject is enumerated with plant names arranged in alphabetical order.

## OBSERVATIONS

**Abrus precatorius** L. (Fabaceae, *Chanothi*).

Roots and leaves decoction is given for cough and cold. Leaves are used in rheumatism and scabies.

**Acacia nilotica** (L.) Willd. ex Del. (Mimosaceae, *Deshi baval*).

Leaves are given to pregnant women for fair skin or white progeny. Twigs are used as tooth brushes.

**Aegle marmelos** (L.) Corr. (Rutaceae, *Bili*).

Unripe or half ripe fruits are digestive, astringent and stomachic, used for diarrhoea and dysentery. Leaf-juice is given in diabetes. Ripe fruit pulp is used in summer for preparation of cold drinks.

**Ailanthus excelsa** Roxb. (Simaroubaceae, *Arduso*).

Bark is anthelmintic, febrifuge, used for bronchitis, dysentery and asthma. Leaves and shoots are used as fodder.

**Aloe vera** (L.) Burm. f. (Liliaceae, *Kunwarpathu*).

Pulp of leaf is employed externally on boils, burns, piles and swellings. Fresh juice of leaves is also used in eye troubles & liver complaints.

**Argemone mexicana** L. (Papaveraceae, *Darudi*).

Yellow juice of plant is applied on eczema, boils, syphilis, ulcer and scorpion-sting.

**Argyreia nervosa** (Burm. f.) Boj. (Convolvulaceae, *Samudrasoka*).

Leaves are eaten as a vegetable; applied externally in cases of eczema, itch and skin troubles. Plant is also useful in gonorrhoea and chronic ulcers.

**Asparagus racemosus** Willd. (Liliaceae, *Satavari*).

Fresh root juice is mixed with honey and given for dyspepsia. Tuberous root's powder is employed in rheumatism, leucorrhoea, acidity and sexual weakness.

**Azadirachta indica** A. Juss (Meliaceae, *Limdo*).

Bark is used in skin troubles. Leaf-juice is given internally in fever, piles, jaundice. Leaf paste is applied on ringworm, wounds, eczema etc. Fresh tender twigs are used to clean teeth particularly in pyorrhoea.

**Balanites aegyptiaca** (L.) Del. (Balanitaceace, *Ingoriyo*).

Unripe fruit is used for whooping cough and skin troubles. Roasted seeds are given in colic and cough.

**Boerhavia diffusa** L. (Nyctaginaceae, *Satodi*).

Decoction of root powder is employed in swellings, obesity, calculi, piles, rheumatism and oedema.

**Catharanthus roseus** (L.) G. Don (Apocynaceae, *Barmasi*).

Infusion of leaves is used in menorrhagia and juice is applied for relief of pain due to wasp stings.

**Clerodendrum multiflorum** (Burm. f.) O. Ktze. (Verbenaceae, *Arani*).

Juice of leaves is given in worms, dropsy, piles and obesity.

**Clitoria ternatea** L. (Fabaceae, *Gami*).

Roasted seed powder is given in dropsy, abdominal pain and liver complaints.

**Cuscuta reflexa** Roxb. (Convolvulaceae, *Amarvel*).

Purgative; used in liver complaints and externally used for itch.

**Jatropha gossypifolia** L. (Euphorbiaceae, *Vilayati nepalo*).

Decoction of leaf is applied on swellings, scabies, boils and eczema. Latex is used for ulcers.

**Justicia adhatoda** L. (Acanthaceae, *Ardusi*).

Leaf-juice is used in diarrhoea, dysentery, cough, asthma and bronchitis.

**Leptadenia reticulata** (Retz.) Wight & Arn. (Asclepiadaceae, *Dodi*).

Leaves and roots are used in skin affections. Leaves and unripe fruits are employed in asthma and for improving eyesight.

**Nyctanthes arbor-tristis** L. (Oleaceae, *Parijatak*).

Paste of leaves is applied on eczema, scabies and ringworm. Leaf-juice is given in diabetes.

**Pedalium murex** L. (Pedaliaceae, *Motu gokhru*).

Decoction of fresh plant in cold water is employed in spermatorrhoea, calculi, dysuria and burning micturition. Fresh and dried fruits are given for removal of kidney stones.

**Pergularia daemia** (Forsk.) Chiov. (Asclepiadaceae, *Chamar dudheli*).

Decoction of leaves is used in asthma, piles and boils. Juice of leaves is applied to rheumatic swellings.

**Phyllanthus reticulatus** Poir. (Euphorbiaceae, *Kamboi*).

Juice of leaves is used in diarrhoea. Fresh and young twigs are used as tooth brushes.

**Pongamia pinnata** (L.) Pierre (Fabaceae, *Karanj*).

Leaves paste is applied on bleeding piles and eczema. Young twig is used as tooth brush.

**Salvadora persica** L. (Salvadoraceae, *Piludi*).

Decoction of leaves is used in asthma, cough, piles and stomatitis. Young and fresh twig is used as tooth brush.

**Senna auriculata** (L.) Roxb. (Caesalpiniaceae, *Aval*).

Paste of leaves is applied on swellings and boils. Flowers are used in diabetes. Seeds used in eye troubles and diabetes.

**Senna occidentalis** (L.) Link (Caesalpiniaceae, *Kasundro*).

Seeds are used as a substitute for coffee. Roasted and powdered seeds are given in cough and whooping cough.

**Solanum nigrum** L. (Solanaceae, *Piludi*).

Decoction of leaves is useful in dysentery, piles, stomachache and eye complaints.

**Tephrosia purpurea** (L.) Pers. (Fabaceae, *Sarpankho*).

Decoction of roots is given in diarrhoea, dyspepsia, rheumatism, asthma and urinary disorders.

**Tinospora cordifolia** (Willd.) Miers ex Hk. f. & Th. (Menispermaceae, *Galo*).

Juice of fresh stem is used in jaundice, diabetes, fever and liver complaints.

**Tridax procumbens** L. (Asteraceae, *Pardeshi bhangro*).

Leaves used in dysentery and diarrhoea. Juice of leaves is applied to stop bleeding from cuts, wounds and ulcers.

**Vitex negundo** L. (Verbenaceae, *Nagod*).

Leaves are used as tonic and smoked for relief in headache. Roots are also used as tonic, diuretic and in rheumatism and dyspepsia.

**Withania somnifera** (L.) Dunal (Solanaceae, *Ashwagandha*).

Roots are used in hiccup, cough, dropsy, rheumatism.

## ACKNOWLEDGEMENTS

The author is thankful to tribal and rural people of this area for co-operation in providing information about the ethnomedicinal plants. Thanks are also due to Dr. D.C. Bhatt, Head, Department of Marine Sciences, Bhavnagar for giving valuable information and suggestions.

## REFERENCES


ANT, H.M. 2000. *Ethnobotanical study of Angiosperms of Aravalli hills.* Ph.D. Thesis, Bhavnagar University, Bhavnagar.

BHATT, D.C. & K.D. MITALIYA 1999. Ethnomedicinal plants of Victoris Park (Reserved forest) of Bhavnagar, Gujarat, India. *Ethnobotany* 11 : 81-84.

————, ———— & S.K. MEHTA 2000. Plant twigs as tooth brushes. *Adv. Plant Sci.* 13 (1) : 19-22.

————, M.A. NURANI, K.D. MITALIYA & U.S. BAXI 2001. Some ethnomedicinal plants of Barda hills of Saurashtra, Gujarat, India. *Adv. Plant Sci.* 14 (2): 427-431.

————, S.K. MEHTA & K.D. MITALIYA 1999. Ethnobotanical plants of Shetrunjara hill of Palitana, Gujarat. *Ethnobotany* 11:22-25.

BHATT, R.P. & S.D. SABNIS 1987. Contribution to the ehtnobotany of Khedbrahma region of North Gujarat. *J. Econ. Taxon. Bot.* 9(1): 139-145.

BHATTACHARYA, G. 1996. Medico-ethnobotanical value of Saurashtra weeds. *J. Econ. Taxon. Bot.* Addl. Ser. 12 : 166-168.

BOLE. P.V. & J.M. PATHAK 1988. *Flora of Saurashtra,* Vols. 2 & 3. Botanical Survey of India, Kolkata.

JADEJA, B.A. Plants used by tribe Rabari in Barda hills of Gujarat. *Ethnobotany* 11: 42-46.

JOSHI, M.C. 1988. Pharmaceutically important medicinal plants of Gujarat forests. *Bull. Medico-Ethno. Bot. Res.* 10 (4) : 372-373.

————, M.B. PATEL & P.J. MEHTA 1980. Some folk medicines of Dangs, Gujarat State. *Bull. Medico-Ethno. Bot. Res.* 1 : 8-24.

————, ———— & ———— 1998. In Goyal, H.R. (Ed.) *A report on medicinal plants of Kachchh* (Gujarat). CCRA & S, New Delhi.

MITALIYA, K.D. 1998. *Ethnomedicinal study of Angiosperms of Bhavnagar.* Ph.D. Thesis, Bhavnagar University, Bhavnagar.

————, D.C. BHATT, D.M. PATEL & P.N. JOSHI 2001. Medicinal value of some selected stem-bark used by tribals and rural folk in Gujarat. *Adv. Plant Sci.* 14 (1) : 255-259.

————, ————, J.N. PATEL & S.K. PATEL 2003. Observations on plants used for joint disorders by rural and tribal people of Gujarat. *J. Econ. Taxon. Bot.* 27 (1) : 212-216.

PATEL, K.C. & A.S. REDDY 2000. Ethnomedicinal uses of some common plants of Danta forest, North Gujarat State. *Seminar on Recent trends in plant Science,* Talod (Abstract no. 5).

PUNJANI, B.L. 1998. *An ethnobotanical study of tribal areas of district Sabarkantha (North Gujarat).* Ph.D. Thesis, North Gujarat University, Patan.

SHAH, G.L. 1978. *Flora of Gujarat State,* 2 Vols. Sardar Patel University, Vallabh Vidhya nagar (Gujarat).

———— & G.V. GOPAL 1982. An ethnobotanical profile of the Dangies. *J. Econ. Taxon. Bot.* 3 : 355-364.

————, A.R. MENON & G.V. GOPAL 1981. An account of ethnobotany of Saurashtra in Gujarat State. *J. Econ. Taxon. Bot.* 2:173-181.

SILORI, C.S. & A.R. RANA 2000. Indigenous knowledge of medicinal plants and their uses in Narayan Sarovar Sanctuary, Kutchch. *Ethnobotany* 12 : 1-7.

THAKER, J.I. 1910. *Flora of Bharda hills, Vanaspati Shastra.* Pravin Pustak Bhandar, Rajkot.

———— 1926. *Plants of Kutch & their utility.* Kutch Darbar, Kutch.

# ETHNOMEDICINAL STUDIES ON CUCURBITS OF AGRA DISTRICT

K.P. SINGH & REKHA AGARWAL

*Department of Botany, R.B.S. College, Agra (U.P.)*

**ABSTRACT**

Cucurbits are well known as a source of human nutrition and are of major economic importance. The present communication deals with cucurbits of ethnomedicinal importance used by local people of Agra district. The study revealed the uses of 15 cucurbits as new to the existing knowledge in this field. People use various parts of the plants for curing fever, cough, cold, headache, eczema, skin burns, heart problems, nervous disorders and urinary troubles. Some cucurbits are also used as tonic for old persons.

## INTRODUCTION

The cucurbits are widely distributed in tropical and warm temperate regions of the World, being represented by about 118 extant genera and 825 species. Most species are components of native ecosystems and are often little known beyond basic descriptive taxonomic accounts. A few species are cultigens of major economic importance. They are sources of human nutrition and useful products. They are ubiquitous components of household gardens. The great variety of foods and sundry kinds of oil obtained from this family are well known. In addition to their value as food plants, these vegetable cucurbits may be important elements in traditional medical practice (Chakravarty, 1966, 1976, 1982). However, faith and usage of such cures is gradually on decline due to modernization and erosion still continues.

Ethnomedicines play a very important role in phytotherapy and increase the resource base and probability of drug discovery. This is because they have proved their efficacy by generating thousands of herbal formulations, diverse health foods and customs suited for different ecosystems and agro-climates.

Agra district is situated between 26°44' and 77°55' N latitude and 78°26' and 78°32' E longitude on the south-west corner of Uttar Pradesh. Yamuna passes through the district for a distance of nearly 71 km.

On the one side of the river Yamuna, in north-east is Etmadpur subdivision of the district, while the other five subdivisions Agra, Kiroli, Fatehabad and Bah are in south-west. The total area is 4836.5 sq km.

The soil of the district is alluvial except for residual soils which occur in a narrow strip in south and south-west. The climate is of semi-arid nature. The daily maximum and minimum temperature, and mean temperature averaged 38.41°C, 4.12°C and 23.5°C respectively. The average relative humidity is 78% and the annual rainfall is 607.1 mm.

The present communication mentions various interesting and reportedly effective uses of some species of Cucurbitaceae in Agra district.

## MATERIALS AND METHODS

Explorations in various tehsils of Agra district were made to find out cucurbits in the household remedies that have been used by men and women to cure various ailments. Morethan 50 villages were visited. Voucher specimens and ethnomedicinal information were collected following the methods of Jain (1965) and Jain & Rao (1977). Knowledgeable persons or medicine men were interviewed for recording local name, part used, method of preparation of drug and dosage. Plant names have been arranged alphabetically, followed by local name and medicinal uses. The specimens have been deposited in the herbarium of Dept. of Botany, R.B.S. College, Agra.

## OBSERVATIONS

**Benincasa hispida** (Thunb.) Cogn. 'Kumera'

The fruits are used for medicinal and vegetable purposes. The flesh of fruits is used in making sweet called 'Petha' which provides instant energy in summers. The decoction of the fruits is used to check bleeding, urinary and respiratory troubles. The extracts of the leaves is used for wound healing.

**Coccinia grandis** (L.) Voigt 'Kundru'

The decoction of roots is used in burning sensation during urination and uterine troubles. The flesh of ripe fruits is used in skin diseases. The decoction of fresh fruits is used in fever and asthma. The chutney of fruits and leaves is given in jaundice and diabetes.

**Citrullus lanatus** (Thunb.) Mats. & Nakai var. **lanatus** 'Tarbooz'

The flesh of mature fruits is used for cooling repast in summer. The fruit juice along with salt is used in curing fever in children. The seeds are grind with sugar candy to make powder and used for cooling purposes and urinary disorders. The oil of seeds is also extracted by the women and used as an illuminant in night.

**Citrullus lanatus** (Thunb.) Mats. & Nakai var. **fistulosus** (Stocks) Chakr. 'Tinda'

The unripe fruits are used as vegetable. The fruits are crushed and the paste is used for cooling and refreshing purposes. The pulp of fruits is used in curing urinary troubles. The decoction of seeds is applied externally over skin to cure burns. The seeds are grind and used as a tonic for old person and pregnant ladies and in the treatment of heart disorders.

**Cucumis melo** var. **melo** 'Snake gourd, Phute, Sand' (=*C. melo* L. var. *culta* Kurz, var. *momordica* Duthie & Fuller, var. *utilissimus* Duthie & Fuller; concept of Chakravarty, 1982 has been followed as regard the varieties)

The fruit slices are used in quenching thirst and for nourishment. The fruit pieces are also used in salad and considered good for gastric disorders. The seeds are eaten raw and used as a tonic. The seeds along with sugar are also used in curing of frequent urination (for var. *utilissimusa* Duthie & Fuller).

The flesh of fruits is crushed and used in gastric disorders. The ripe and unripe fruits are used in the treatment of jaundice. The fruit pulp along with rice is used for the preparation of 'Phute khir' by muslim population which is given to the children for treatment of urinary disorders. The pulp of unripe fruits is used as vegetable. The village people use mature fruits in their breakfast in curing of kidney problems and for blood purification (for var. *momordica* Duthie & Fuller).

The ripe fruits are used for the treatment of urinary disorders. The paste of fruits is given twice regularly for 7 days as a remedy for the treatment of skin diseases. The decoction of ripe fruits is used for the cure of asthma (for var. *culta* Kurz).

**Cucumis sativus** L. 'Khira'

The fruit pulp is used as a tonic and in urinary troubles. The juice of fruit is extracted and used in the treatment of respiratory disorders and jaundice. The seeds are crushed into a paste and used in burning sensation and fever. The decoction of seeds is nourishing and given in urinary troubles.

**Cucurbita pepo** L. 'Kaddu'

The seed oil is extracted and used for edible purposes. The pulp of fresh seeds are used in fever. The leaves are boiled in water and the decoction is applied externally over skin for curing of burns. The seeds are grind to make paste which is used as a tonic for children.

**Diplocyclos palmatus** (L.) Jeffrey 'Shivalingi'

The decoction of plant parts is applied over skin to cure burns. The fruit paste is used for nutritive purposes. The decoction of the whole plant is used for the treatment of accumulation of gas in the stomach and other gastric disorders.

**Lagenaria siceraria** (Molina) Standl. 'Bottle Gourd'

The immature fruits are boiled in water and used to cause vomiting. The fruit juice mixed with lime juice is used for pimples. The oil of seeds is extracted and applied externally in headache. The decoction of leaves is used for curing of baldness by local man.

**Luffa acutangula** (L.) Roxb. 'Ara tori'

The extract of leaves is used in curing of piles. The fruit pulp is used in the treatment of urinary disorders. The decoction of the fruits is used as tonic and also given for nutritive purposes. The seeds are bitter in taste and used to cause vomiting and to cure skin infections.

**Luffa cylindrica** (L.) M. J. Roem. 'Ghiatorui'

The boiled pulp of immature fruits is used in treatment of urinary disorders, while the pulp of ripe fruits is given in curing gas accumulation in the stomach, heart problems and cancer. The seeds are bitter in taste and used to cause vomiting. The decoction of seeds mixed with diluted curd is given in curing of skin infections. The juice of fresh leaves is dropped into the eyes of children in curing ailments.

**Momordica charantia** Linn. 'Karela'

The pulp of mature fruits is used in gastric disorders. The decoction of fruits is used to cure tension and for cooling purposes. The fruits, leaves and roots are grind and made into a paste which is used in the treatment of diabetes. The juice extract of leaves is used to cause vomiting and rubbed in burning of soles of the feet.

**Momordica dioica** Roxb. ex Willd. 'Kakora'

The boiled immature fruits are given in gastric disorders, cough and heart diseases. The roots are made into a paste and applied over the body for curing fever. The juice extract of leaves is used in curing urinary disorders.

**Mukia maderaspatana** (L.) Roem. 'Bilari'

The paste or juice extract of the whole plant is used to cause vomiting. The decoction of fresh leaves is applied externally over skin against burning sensation. The roots are washed and grind with teeth to relieve toothache.

**Trichosanthes anguina** L. 'Chachinda'

It is commonly called Snake gourd and used as vegetable. The decoction of fruits is given in gas accumulation in stomach. It is also used for curing of burning sensation and skin diseases. The seeds are crushed and mixed with diluted curd to make a paste which is used for the treatment of intestinal disorders.

## RESULT AND DISCUSSION

The 15 cucurbits used in folk medicines have been documented as ethnomedicinal resources of rural areas of Agra district. They are reported to treat skin diseases, heart diseases, gastric troubles, cough and other severe cases of infection. These plants serve as first aid treatment and also life saving drugs in remote areas.

Idu & Omorui (2003) reported that *Coccinia grandis* (L.) Voigt used with *Tamarindus indica* L. fruits to prepare potage for the treatment of hymenia. *Trichosanthes cucumerina* L. is powdered and its decoction is taken to cure hyperacidity and migrain (Kshirsagar *et al.,* 2003). Panthi & Chaudhary (2003) reported that fruit juice of *Momordica charantia* L. is effective in controlling diabetes, and also used as drink mixed with sugar candy to control acidity and to increase appetite. Its fruits are powdered and also used in fever. The leaf and root of *Luffa cylindrica* (L.) Roem. are crushed together and applied to cure leprotic wound (Sen & Behera, 2003).

Rajendran *et al.* (2001) reported the unripe fruits of *Cucumis callosus* (Rottl.) Cong. is used for the treatment of skin diseases like vitiligo and scabies; fruit paste is used in skin itching. Root and leaf of *Zehneria mysorensis* (Wt. & Arn.) Arn. are used in headache; leaf powder snuffed for acute cold and cough. According to Jain & Sharma (2000), the plant of *Momordica balsamina* L. is anthelmintic, laxative, stomachic, tonic and used in haemorrhoids. Kar *et al.* (1999) reported that seeds of *Trichosanthes cucumerina* L. control blood glucose, lowering effect upto the normal fasting level. The seeds of *Diplocyclos palmatus* (L.) Jeffrey are used to prevent miscarriage and ensure conception in Gujarat (Bhatt & Mitaliya, 1999). Katewa & Sharma (1998) reported that the fruits of *Mukia maderaspatana* (L.) Roem. are tried over swollen finger in the treatment of paronichia. Girach *et al.* (1997) reported that the seed oil of *Luffa acutangula* (L.) Roxb. is applied on skin infection.

It is hoped that present study may give clues to various pharmaceutical industries and research organisations engaged in biological screening programmes on plant products for extracting active compounds in order to develop new drug molecules.

## ACKNOWLEDGEMENT

The authors are thankful to University Grants Commission for providing financial assistance for the present work.

## REFERENCES

BHATT, D.C. & K.D. MITALIYA 1999. Ethnomedicinal plants of Victoria park (Reserved forest) of Bhavnagar, Gujarat, India. *Ethnobotany* 11 (1 & 2) : 81-84.

CHAKRAVARTY, H.L. 1966. *Monograph on the Cucurbitaceae of Iraq. Tech. Bull. No. 133* : 86-102. Ministry of Agriculture, Baghdad.

——————— 1976. *Plant Wealth of Iraq*. I. Ministry of Agriculture and Agrarian Reform, Baghdad.

——————— 1982. Cucurbitaceae. In Thothathari, K. (ed.) *Fascicle of flora of India*. Fasc. 11, Botanical Survey of India, Howrah.

GIRACH, R.D., AMINUDDIN, M. BRAHMAN & M.K. MISRA 1997. Observations on Ethnomedicinal plants of Bhadrak district, Orissa, India. *Ethnobotany* 9 (1 & 2) : 44-46.

IDU, M.D. & O.M. OMORUI 2003. Some ethnomedicinal plants of Higgi tribe from Adamawa State, Nigeria. *Ethnobotany* 15 (1 & 2) : 48-50.

JAIN, S.K. 1965. Medicinal Plantlore of tribals of Bastar. *Econ. Bot.* 19 : 236-250.

——————— & P.P. SHARMA 2000. A review of the parallel or unique indigenous uses of twenty five plants in India and some African countries. *Ethnobotany* 12 : 51-55.

——————— & R.R. RAO 1977. *A Hand Book of Field and Herbarium Methods*. Today and Tomorrows Publishers, New Delhi.

KAR, A., B.K. CHOUDHARY & N.G. BANDOPADHYAY 1999. Evaluation of a few Indian folk medicinal plants less known for their hypoglycaemic activity. *Ethnobotany* 11 (1 & 2) : 18-21.

KATEWA, S.S. & R. SHARMA 1998. Ethnomedicinal observations from certain watershed areas of Rajasthan. *Ethnobotany* 10 (1 & 2) : 46-49.

KSHIRSAGAR, S.R., M.H. PARABIA & M.N. REDDY 2003. Ethnobotany of coastal areas in South Gujarat. *Ethnobotany* 15 (1 & 2) : 60-63.

PANTHI, M.P. & R.P. CHAUDHARY 2003. Ethnomedicinal plant resources of Arghakhanchi district, West Nepal. *Ethnobotany* 15 (1 & 2) : 71-86.

RAJENDRAN, S.M., K. CHANDRA SEKAR & V. SUNDARESAN 2001. Ethnomedicinal lore of Seitur hills-Southern Western Ghats, Tamil Nadu. *Ethnobotany* 13 (1 & 2) :101-109.

SEN, S.K. & L.M. BEHERA 2003. Ethnomedicinal plants used against skin diseases in Bargarh district in Orissa. *Ethnobotany* 15 (1 & 2) : 90-96.

# ETHNOMEDICINAL STUDY OF SOME IMPORTANT PLANTS USED IN THE TREATMENT OF HAIR AND BOILS IN DISTRICT PULWAMA OF KASHMIR

IQBAL CHAK, R.K. AGARWAL & MAJEED KAK*

*Department of Botany, Bundelkhand University, Jhansi (U. P.) India*
**Retd. Sr. Lecturer, Department of Botany, Islamia College of Science & Commerce, Srinagar*

## ABSTRACT

The present paper reports the ethno-medicinal uses of 10 plant species belonging to 7 families, being practised by the tribes and rural people of the district for the treatment of hair ailments and 13 plant species belonging to 11 families for the treatment of boils. The information on correct botanical identity, family, local name, part used and method of preparation of drugs are given for each plant species. The study includes hitherto unknown medicinal claims which are already known as medicinal.

## INTRODUCTION

Carved out of Anantnag district, the Pulwama district came into being in 1979. The district, situated 32 km from Srinagar in South Kashmir, is surrounded in the north by Srinagar, in the west by Poonch and Budgam and in the east by Anantnag. The total geographical area of the district is 1398 sq km, of which 1011 sq km are covered by forests. The topography is almost level with some undulation and elevations towards southeast and north which extensively support almond, walnut and apple orchards. Mean elevation is 1500-1750 m above msl. A chain of mountains of Pir Panjal range of the inner Himalaya lies on east and southeast.

The district is divided into four tehsils, six community development blocks, six notified area committees and 554 villages. Majority of the population, i.e. 88%, speaks Kashmiri language while the rest speak Phari and Gujri. Agriculture is the main occupation of the people with a small number have cattle for livelihood. The main crops in the district include Paddy, Maize, Mustard and Pulses. Fodder is also cultivated in the winter. The World famous saffron fields adorn the district and the cherished apple, Ambree, are also grown here. The tribes of the area are Bakerwals and Gujars. These people and the people of adjacent rural areas depend upon the local plants for the treatment of various diseases as folk-medicines. The district enjoys a predominantly dry temperate climate. The areas at higher altitudes are cooler in summer and inaccessible during winter due to heavy snowfall. During winter, there is severe cold and precipitation in the form of snow. The mean daily minimum temperatures of the coldest months December and January are 0.7°C and -3.5°C. The spring is pleasant with frequent rains and gradual increase in both minimum and maximum temperatures (soaring upto 36°C). The mean maximum temperature of the hottest month July is 29.5°C. The autumn is mild with little rainfall and experiences a gradual decrease in both maximum and minimum temperatures till the mean reaches 3.7°C and 1.5°C respectively in November. A perusal of literature (Wali & Tiku, 1964; Kachroo & Navi, 1976; Kapur & Sarin, 1982; Dar *et al.,* 1984; Sharma & Singh, 1989; Navchoo & Kachroo, 1995 etc) revealed that no systematic study and survey from ethnobotanical point of view has been carried out in the proposed area of investigation.

## MATERIALS AND METHODS

Ethnobotanical survey was conducted in mostly two tehsils of the district, namely Pulwama and Shopian. During March 2006 to December 2006, following the procedures as described by Jain (1967) and Croom (1983). Medicinal plants used for the purpose of the hair and boil treatments have been collected from paddy fields, grasslands, slopes, roadsides, mountains of Pir Panjal range of the inner Himalaya etc. However, during the field trips, a number of elderly people of rural and tribal communities including many practising as herbalists, senior Hakims were contacted and through discussions and interviews, information about the plants used for the treatment of hair and boils was recorded. As strategy to confirm the informations, it was rechecked with other herbalists. Repeated queries were also made to get data verified and confirmed.

The information recorded was further screened by consulting important works (Anonymous, 2003; Kirtikar & Basu, 1935; Prajapati *et al.,* 2004) to distinguish between the already known and hitherto unknown uses of plants.

## RESULTS AND DISCUSSION

The enumeration plants has been done in tabular form in alphabetical order, followed by the family name, local name, the part used and mode of administration of each species. The plants marked with an arterisk (*) indicate the uses reported for the first time.

Exhaustive and repeated study shows that some of the important and well known medicinal plants of the investigated area, such as *Colchicum luteum, Dioscorea deltoidea, Podophyllum hexandrum, Aesculus indica, Saussurea simpsoniana, Rheum emodi, Angelica glauca, Cydonia oblonga* and *Jurinea dolomiaea,* mostly growing at high altitudinal ranges of district, have been catagorised as threatened. The depletion of bioresources is due to the rapid deforestation, population growth, unrestricted grazing year after year by goats, sheep, horses in these areas. Some local healers of the district collect the above mentioned medicinal plants very unscientifically, as such further germination does not take place in the same natural environment.

Table 1. Medicinal plants used in the treatment of hairs and boils.

| Species & Family | Local name | Parts used and mode of administration |
|---|---|---|
| **Treatment of hair** | | |
| **Aesculus indica* (Wall. ex Camb.) Hook. f. (Hippocastanaceae) | Handoon | The seed-powder is made into paste with mustard oil, applied on head twice a day before and after hair wash against dandruff. |
| **Allium cepa* L. (Liliaceae) | Gande | The extract of the fleshy leaves is applied on hairs 2-3 times a day for 10-20 days for growth of healthy hairs. |
| **Allium sativum* L. (Liliaceae) | Rhoon | About 3 cloves of garlic are rubbed three times a day for 5 days on head continuously. It is very useful to regenerate hair on the areas of baldness appearing in patches on face and head (*Alopecia areate*). |
| **Brassica rapa* (L.) subsp. *campestris* (L.) Clapham. (Brassicaceae) | Tilgogal | The seed-oil is applied on hair daily after hair wash as a hair tonic; it enhances hair growth, prevents dandruff and hair falling. |
| **Colchicum luteum* Baker (Liliaceae) | Whirkin posh | The powder of bulbs (2-5 mg) is mixed with cow-butter (10 mg) and applied on hair thoroughly at night for 10-20 days; it prevents dandruff and kills lices. |
| **Cuscuta reflexa* Roxb. (Convulvulaceae) | Kukli pot- | The fresh juice (10 ml) of the whole plant is applied once a day for 4-10 days on head for nourishing hair root and it fights strongly against dandruff. |
| **Geranium pratense* L. (Geraniaceae) | Karipati | The flowers are crushed with cow butter, applied on head daily for one month for the growth of healthy hair. |
| **Juglans regia* L. (Juglandaceae) | Doon | The pericarp of the fruit (10 mg) are boiled thoroughly in 3 litres of milk, till the milk remains approximately 1 litre. This extract is used as a hair dye which turns grey hair into natural shiny black. |
| **Raphanus sativus* L. (Brassicaceae) | Muj | The fresh juice of the whole plant (5 ml) is used daily after hair wash for about 5-10 days; it prevents hair falling and dandruff. |
| *Urtica dioica* L. (Urticaceae) | Soi | An infusion of leaves and roots is used for washing head once a day for 3 days as hair stimulent, cleaning hair wax and dandruff. |
| **Treatment of boils** | | |
| **Allium cepa* L. (Liliaceae) | Gand | The fresh juice of onion (fleshy leaves) is applied externally on boils; helps in ripen them, break them and evacate the pus. |
| **Arisaema jacquemontii* Bl. (Araceae) | Haput Gogej | The tubers (2-3) are made into a paste, mixed with human saliva and applied to painful boils with beneficient results. |
| **Bergenia ligulata* (Wall.) Engl. (Saxifragaceae) | Palpashand | The fine powder of roots is made into paste, mixed with the resin of *Cedrus deodara* and applied on fresh boils; it enables them to ripen and burst. |

| | | |
|---|---|---|
| **Calendula officinalis* L. (Asteraceae) | Hamesh bahar | The fresh leaves are mixed with milk, made into paste and applied on boils for pain relief. |
| **Datura stramonium* L. (Solanaceae) | Datur | The fresh leaf is gently warmed till it becomes soft. It is then coated with a layer of mustard oil. It should be replaced every five hours. After a few applications the skin boil will rupture, draining out all the purulent matter. The oiled leaf should be spread over the inflammed part. |
| **Iris decora* Wall. (Iridaceae) | Sonsal | The root powder is mixed with turmeric powder, made into paste and applied on big boils for speeding up the ripening process. |
| **Jurinea dolomiaea* Boiss. (Asteraceae) | Dhupa | The whole plant is thoroughly fried in mustard oil and the extract is applied to boils; it enables them to ripe and burst. |
| **Podophyllum hexandrum* Royle (Podophyllaceae) | Wanwangun | The fresh leaf is warmed till it becomes soft; it is applied to boils for quick recovery. |
| **Potentilla nepalensis* Hook. f. (Rosaceae) | Panzpater | The leaves (4-5) are thoroughly boiled in milk and applied to boils to ripen them. |
| *Rheum emodi* Wall. ex Meissn. (Polygonaceae) | Pambchalen | Warm poultice prepared from the dried roots is applied on unhealing boils. |
| *Salvia moorcroftiana* Wall. ex Benth. (Lamiaceae) | | The leaves are made into a paste which is applied on hard boils to soften them. |
| **Saussurea simpsoniana* (Field. & Gard.) Lipschitz (Asteraceae) | Zoog padsha | The whole plant is made into a paste, mixed with pine resin and applied to unhealing boils. |
| **Vitis venifera* L. (Vitaceae) | Dush | A fresh leaf is gently warmed till it becomes soft. It is wrapped with a woollen piece of cloth and applied to the boils as a poultice. The leaf is changed after twelve hours; it enables the boils to ripe and burst. |

The pattern of use poses a definite threat to the genetic stock and to the diversity of medicinal plants. Threat assessment exercises as per latest IUCN guidelines for Southern and Northern India have already listed around 200 species of medicinal plants that are rare, endangered and threatned (Shankar & Ved, 2003). Keeping this list in view, the Director General, Foreign Trade (DGFT), New Delhi, vide the Foreign Trade Development Act 1992 has prohibited exports of some medicinal plants including *Podophyllum hexandrum, Dioscorea deltoidea* etc.

## CONCLUSION

It can be concluded from this study that the District is mainly agrarian and most of the plants grow luxiariantly as weeds in wastelands, fallowlands, cultivated fields etc. A good number are locally used in medicine, as vegetables and in folklore. The medicinal use of plants by the tribals and local people in day-to-day life for the treatment of boils and hair ailments, have posed a threat to the survival of some of these plants. In near future these valuable species may become extinct or will be endangered. Therefore, an urgent action in this direction is required as these forests are under tremendous human pressure. The Government should organize training programmes on the conservation and cultivation of MAPs to the interested herb grovers/farmers/members of village forest committees/Forest Department officials etc.

## ACKNOWLEDGEMENT

The authors are grateful to the Field Director, Shree Tej Kumar, Regional Research Laboratory, Bonera (Pulwama) for providing valuable suggestions and encouragement.

## REFERENCES

ANONYMOUS 2003. The Wealth of India (Raw Materials). CSIR, New Delhi.

CROOM, E.M. 1983. Documenting and evaluating herbal remedies. *Econ. Bot.* 37 : 13-27.

DAR, G.H., VIRJEE, P. KACHROO & G.H. BUTH 1984. Ethnobotany of Kashmir-I. Sind Valley. *J. Econ. Taxon. Bot.* 3 : 667-675.

JAIN, S.K. 1967. Ethnobotany : its scope and study. *Indian Musc. Bull.* 2 : 39-43.

KIRTIKAR, K.R. & B.D. BASU 2001. *Indian Medicinal Plants with Illustrations,* 2nd edn. Oriental Enterprises, Dehra Dun.

KACHROO, P. & I.M. NAVI 1987. Ethnobotany of Kashmir. In Singh, G. & P. Kachroo (Eds.) *Forest Flora of Srinagar,* pp. 239-263. Bishen Singh Mahendra Pal Singh, Dehra Dun,

KAPUR, S.K & Y.K. SARIN 1982. Economical useful plant of Tarikuta hills (J & K State). *Indian Jour. Forestry* 5 (2) : 100-110.

NAVCHOO, I.A. & P. KACHROO 1995. *Flora of Pulwama (Kashmir).* Bishen Singh Mahendra Pal Singh, Dehra Dun, India.

PRAJAPATI, N.D., S.S. PUROHIT, A.K. SHARMA & T. KUMAR 2004. *A hand book of Medicinal plant.* Agrobios (India), Jodhpur.

SHANKAR, DARSHAN & D.K. VED 2003. A balanced perspective for management of Indian Medicinal Plants. *Indian Forester* 129(2): 275-287.

SHARMA, P.K. & V. SINGH 1989. Ethnobotanical studies in North West Himalaya-V. Ethnoveterinary medicinal plants used in Jammu & Kashmir, India. *J. Ethnopharm.* 27 : 63-70.

WALI, M.K. & S.K. TIKU 1964. Conmtribution to the Flora of Kashmir (Lolab Valley). *Bull. Bot. Surv. India* 6 : 141-149.

# ETHNOMEDICINAL USES OF SOME WEEDS IN SOME AGRO-ECOSYSTEMS OF DOON VALLEY

LOKENDER SINGH*, NEELAM SHARMA, S.P. JOSHI, R.K. MANHAS* & VENITA JOSHI

*Ecology Research Laboratory, Botany Department, D.A.V. (P.G.) College, Dehradun- 248 001*
**Forest Ecology and Environment Division, Forest Research Institute,*
*New Forest Dehradun-248006; Email : lscbotdav@yahoo.co.in*

## ABSTRACT

The present paper describes association of 51 species of weeds belonging to 45 genera under 26 families with crops in Doon Valley. The ethnobotanical uses, especially ethnomedicinal, prevalent in the area have been documented along with vernacular names, family, phenological data, besides parts of plants used and ailments.

## INTRODUCTION

Plants have been used as healers and health rejuvenators since time immemorial. Even now, they play an important role in the healthcare of the man. Although, every plant has a peculiar quality but weeds are known for their erratic behaviour in crop fields, gardens, orchards etc. Due to this, man has always discriminated such unwanted plants and eradicated and destroyed them. Weeds always exist on land in large number and more vigour because of their wide adaptability, even under extreme biotic stress. They are the pioneers, which increases the diversity of agro-ecosystems.

In ancient literature it is clearly mentioned that every plant on this earth is useful for human beings, animals and also for other plants (Oudhia, 1999). It means that these unwanted plants, i.e. weeds, which affect the cropland ecosystem, are also useful for human beings. Who knows when or where which plant may give a treatment for incurable disease. Therefore, the knowledge of farmers and local people regarding the various uses of plants present in their surroundings should be used for the betterment of the human beings. In this paper, the authors have emphasized on the ethnomedicinal uses of some weedy herbs. The observations revealed the agro-ecosystems may be a valuable source of raw materials for drug industry. As such, human's destructive nature towards these species may lead to the extinction or disappearance of some of them.

## MATERIALS AND METHODS

The study was conducted in some of the villages of Doon valley namely Mazra, Kalagaon, Mathurowala, Shyampur, New Forest (FRI), Dakpathar and Vikasnagar, lying between 29°30' and 30°32' N latitude and 77°39' and 78°18' E longitude in Uttaranchal, India. These areas are flat and located at a mean height of about 450 to 650 m above sea level.

The weed species were collected during the year 1999 - 2000 from different crops, viz. Wheat, Paddy, Pea, Potato, Radish and Coriander. Specimens of the species were critically examined and identified with the help of available literature (Babu, 1977; Gaur, 1999; Jain, 1994; Bhakat & Pandit, 2003). The knowledge of local people was documented to record ethnobotanical and ethnomedicinal uses. Herbarium specimen of the collected species were deposited in Ecology Research Laboratory, Botany Department, D.A.V. (P.G.) College, Dehradun. The species are arranged alphabetically in the present work in tabular form.

## RESULTS AND DISCUSSION

A total of 51 species, belonging to 26 families and 45 genera, which have medicinal properties or used for other purposes as food, vegetable and fodder etc have been documented. Crop wise association of these weeds shows that paddy crop sustains maximum 35 species of weeds and wheat crop 16 species. Minimum number of weed species was recorded in coriander and pea crops, i.e. 03 and 07 respectively (Fig. 1). According to life forms, the herbs are the most common; maximum number of herbs (28) was reported from paddy crop and minimum from coriander (03) and pea crops (06) respectively. Maximum grasses and sedges, 4 and 3 species respectively, were reported from paddy crop, followed by radish and other crops.

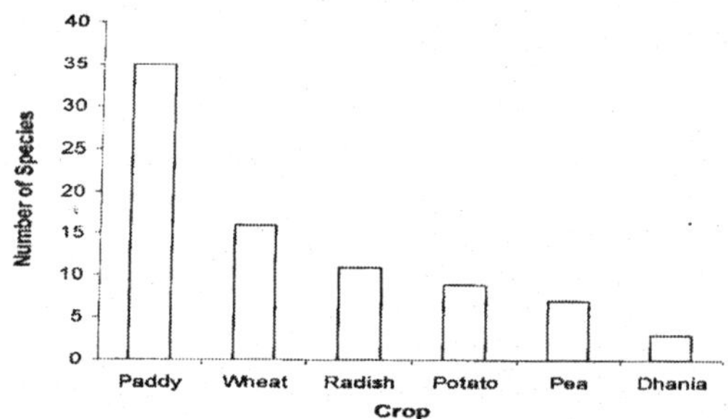


Fig. 1. Crop-weed association in Doon Valley.

Table 1. Ethnomedicinal weeds associated with crops in Doon Valley.

| Botanical name | Family | Crop | Flowering & fruiting time | Plant parts used | Ethno-medicinal uses |
|---|---|---|---|---|---|
| *Achyranthes aspera* Linn. Vern. *Latjiri* | Amaranthaceae | Paddy | Mar.-Dec. | Root, leaf & plant | Root-infusion taken in malaria; leaf supposed to facilitate delivery; decoction of plants used in dropsy & bronchitis; root-powder used in making the local |

| | | | | | |
|---|---|---|---|---|---|
| | | | | | beverage. |
| *Aerva sanguinolenta* (Linn.) Bl.<br>Vern. *Safed phulia* | Amaranthaceae | Paddy | Sep.-May | Whole plant | Plant used as diuretic and demulcent. |
| *Ageratum conyzoides* Linn.<br>Vern. *Gundria* | Asteraceae | Paddy, Potato, Wheat | Jan.-Dec. | Plants, leaves & root | Plant-paste with the roots of *Thalictrum foliosum* is applied on sores, cuts & skin elements; leaves are antipyretic, also used to cure wounds & ulcer. |
| *Alternanthera sessilis* (Linn.) R. Br. ex DC.<br>Vern. *Gudresaag* | Amaranthaceae | Paddy, Pea | Feb.-Oct. | Leaves | Leaves used as vegetable during scarcity. |
| *Amaranthus spinosus* Linn.*<br>Vern. *Kantelu, Cholai* | Amaranthaceae | Paddy, Radish & Pea | July-Dec. | Leaves | Infusion of leaves with salt is given in dismenorrhoea, leaves also used as vegetable. |
| *Anagallis arvensis* Linn.*<br>Vern. *Jonkmari* | Primulaceae | Wheat, Dhania | Feb.-Sep. | Whole plant | Plant-paste or extract used in leprosy, dropsy and cerebral affections, also used to detach leeches from the body. |
| *Argemone mexicana* Linn.*<br>Vern. *Pilikateli* | Papaveraceae | Paddy, Wheat | Apr.-Oct. | Seeds & root | Seeds are emetic and narcotic; roots chewed in leucorrhoea, also used in manufacturing of soap; non-edible oil extracted from the seeds is used as purgative & illuminant. |
| *Bacopa monnieri* (Miller) Penn.<br>Vern. *Brahmi* | Scrophulariaceae | Paddy | Throughout year | Whole plant | Decoction of the plant is considered as diuretic, nervine tonic and useful in insanity, epilepsy, hoarseness. |
| *Bidens biternata* (Lour.) Merr. & Sherff.<br>Vern. *Kumur* | Asteraceae | Paddy | Mar.-Aug. | Whole plant | Plant extract with honey is used in cough and bronchitis; the vegetable of the herb is considered useful in leucoderma. |
| *Boerhavia diffusa* Linn.<br>Vern. *Pundera* | Nyctaginaceae | Paddy | Aug.-Sep. | Root, leaf & plant | Root chewed as energy tonic; leaf extract used in eye complaints; infusion of plant used in asthma and bronchitis; leaves and tender shoots sometimes used as vegetable. |
| *Capsella bursa-pastoris* (Linn.) Medik.*<br>Vern. *Tuntkya* | Brassicaceae | Wheat, Potato | Feb.-Oct. | Root, leaf & plant | Root and leaf extract used in gonorrhoea and urinary troubles; plant paste applied on cuts and wounds. |

| | | | | | |
|---|---|---|---|---|---|
| *Chenopodium album* Linn.* Vern. *Bathua* | Chenopodiaceae | Paddy, Radish | Jan.-Dec. | Leaves | Tea of the leaves is used to relieve the stomach pain; leaves used as pot vegetable; bread prepared from the grinded grains. |
| *Cynodon dactylon* (Linn.) Pers. Vern. *Dub grass* | Poaceae | Paddy, Radish, Pea | Jan.-Dec. | Root | Root taken in fever and internal injury; the plants are used in several religious ceremonies. |
| *Cyperus halpan* Linn.* | Cyperaceae | Paddy | Aug.-Nov. | Plant | Used as fodder; salt is prepared from the ash of the plant. |
| *Cyperus iria* Linn.* Vern. *Dila grass* | Cyperaceae | Paddy | Aug.-Nov. | Whole plant | Plant is considered stimulant, tonic, astringent and useful in stomachache. |
| *Cyperus rotundus* Linn.* Vern. *Motha* | Cyperaceae | Radish, Potato, Paddy | July-Dec. | Plant | Plant extract used as diaphoretic and astringent; dried underground part used in perfumes. |
| *Dicliptera bupleuroides* Nees Vern. *Kulartore* | Acanthaceae | Wheat | Jan.-Dec. | Leaf | Leaf-paste is applied on wounds and to check bleeding; leaf-juice used in cough and gastro enteritis; flowers used as a source of bee forage in apiary. |
| *Echinochloa colona* (Linn.) Link* Vern. *Jangali jhongora* | Poaceae | Paddy | July-Oct. | Grains | Grains edible; leaves used as fodder. |
| *Echinochloa frumentacea* Link* | Poaceae | Paddy | July-Oct. | Grains | Food of famine and fodder. |
| *Eclipta prostrata* (Linn.) Linn. Vern. *Bhangri* | Asteraceae | Paddy | Jan.-Dec. | Whole plant | Decoction of plant is given in throat pain; plant-juice is used in asthma, fever and constipation; plant is a source of black stain used to colour the hairs. |
| *Euphorbia hirta* Linn. Vern. *Duchi* | Euphorbiaceae | Paddy | Jan.-Dec. | Plant & latex | Plant used in bronchial infection and asthma; latex is applied on warts. |
| *Fumaria indica* (Haussk.) Pugsley* Vern. *Pit papra* | Fumariaceae | Wheat | Jan.-May | Plant & leaves | Plant diuretic, diaphoretic and aperients; paste of leaves promote healing of wounds; decoction of plant is given in suppressed urination. |
| *Galinsoga parviflora* Cav.* Vern. *Marchiya* | Asteraceae | Paddy, Potato, Radish, Dhania | Apr.-Oct. | Plant | Plant is rubbed on the skin as an antidote to bettle stings; young plants are cooked as vegetable. |

| | | | | | |
|---|---|---|---|---|---|
| *Gnaphalium polycaulon* Pers. | Asteraceae | Wheat, Potato, Dhania | Apr.-Oct. | Leaves | Leaf-paste is used in pulmonary troubles. |
| *Hedyotis corymbosa* (Linn.) Lam.* Vern. *Daman papar* | Rubiaceae | Paddy | July-Nov. | Plant | Plant-extract used in fever and liver diseases. |
| *Hibiscus cannabinus* Linn. | Malvaceae | Paddy | Sep.-Dec. | Leaves, flowers | Leaves and flowers are medicinally used; stem yields a strong fiber; tender twigs used as fodder. |
| *Lathyrus odoratus* Linn.* Vern. *Motar* | Fabaceae | Wheat | Dec.-Mar. | Pods | Pods are consumed as food. |
| *Lindernia crustacea* (Linn.) F. v. Muell.* | Scrophulariaceae | Paddy, Radish | July-Nov. | Whole plant | Plant is sometime used to cure earache. |
| *Malvastrum coromandelianum* (Linn.) Gray Vern. *Suchi* | Malvaceae | Pea | Throughout year | Leaf | Decoction of leaves is given in dysentery and leaf-paste applied on wounds; stem yields a fiber. |
| *Mazus pumilus* (Burm. f.) Steenis Vern. *Mastyar* | Scrophulariaceae | Paddy, Potato | Mar.-Nov. | Leaf | Leaf-poultice is applied on burns; infusion of plant with ginger extract given in colic. |
| *Medicago lupulina* Linn. | Fabaceae | Potato | Nov.-May | Plant | Used as fodder, as well as fertilizer. |
| *Medicago polymorpha* L. | Fabaceae | Wheat, Potato, Radish | Oct.-Nov. | Leaves | Leaves are used as vegetable, also as a green manure. |
| *Mentha arvensis* Linn. Vern. *Paudina* | Lamiaceae | Paddy | July-Sep. | Leaf | Leaf-extract used in vomiting and indigestion; leaves used as flavouring agent. |
| *Monochoria vaginalis* (Burm. f.) K.B. Presl* Vern. *Nanka* | Pontederiaceae | Paddy | Jun.-Nov. | Rhizome | Rhizomes are used in asthma and liver ailments. |
| *Murdannia nudiflora* (Linn.) Brenan* Vern. *Kansura* | Commelinaceae | Paddy | Aug.-Nov. | Plant | Plant-paste is applied on burns. |
| *Nepeta hindostana* (Heyne ex Roth) Haines Vern. *Bililotan* | Lamiaceae | Wheat, Pea | Jan.-Dec. | Plant | Plant used as a cardiac tonic, also given in fever and gonorrhoea. |
| *Oxalis corniculata* Linn. Vern. *Bhilmori* | Oxalidaceae | Paddy | Throughout year | Leaf | Leaf-juice is dropped in eyes for conjunctivitis. |
| *Persicaria barbata* (Linn.) Hara | Polygonaceae | Paddy | July-Dec. | Leaves | Sap of pounded leaves is applied on wounds. |
| *Phalaris minor* Retz.* Vern. *Phullas* | Poaceae | Wheat, Paddy | Mar.-June | Root | Root-paste is applied on wounds. |

| | | | | | |
|---|---|---|---|---|---|
| *Phyllanthus urinaria* Linn. Vern. *Lal bur* | Euphorbiaceae | Paddy | Throughout year | Plant | The bitter plant is used as a tonic, astringent, febrifuge; leaves used to intoxicate fishes. |
| *Polygonum plebeium* R. Br. Vern. *Dondya* | Polygonaceae | Paddy | Jan.-Dec. | Root | Root-extract is applied on head to avoid baldness; plant is used for manure. |
| *Portulaca oleracea* Linn. Vern. *Lunya* | Portulacaceae | Radish, Pea | Apr.-Sep. | Plant | Plant-extract is used in liver troubles, it is also used as blood purifier; leaves and young branches are used as a pot vegetable. |
| *Ranunculus arvensis* Linn. Vern. *Chambal* | Ranunculaceae | Wheat | Feb.-May | Plant | Herb given in intermittent fever and asthma, also the paste is applied on skin ailments. |
| *Ranunculus sceleratus* Linn. Vern. *Jal dhania* | Ranunculaceae | Wheat | Feb.-Aug. | Plant | Plant is used as vermifuge, anodyne and in skin disorders; different parts of the plant cause blisters, used by beggars to attract the sympathy. |
| *Rumex dentatus* Linn. Vern. *Jangali palak* | Polygonaceae | Wheat, Radish | Feb.-May | Leaves | Leaves are scarcely used as vegetable. |
| *Sida cordifolia* Linn. Vern. *Balu* | Malvaceae | Paddy | Throughout year | Seed & root | Seed powdered is given in dyspepsia; roots are astringent, diuretic and tonic; stem yields a coarse fiber. |
| *Sisymbrium officinale* Linn.* | Brassicaceae | Pea | Mar.-May | Plant | The herb is used as stimulant, irritant, diuretic, resolvent and anti-scorbutic. |
| *Solanum nigrum* Linn. Vern. *Makoi* | Solanaceae | Paddy, Wheat | Throughout year | Plant & fruits | The extract of the plant is used in piles and dysentery; fruits used in diarrhoea, fever and eye ailments; fruits edible; occasionally roots added to the local beverage. |
| *Sporobolus indicus* (L.) R. Br. var, *flaccidus* (R. & S.) Veldk. Vern. *Sitya* | Poaceae | Paddy | Mar.-Sep. | Leaves | Crushed leaves are applied on bums and pimples; supposed to be a good soil binder. |
| *Stellaria media* (Linn.) Vill.* Vern. *Badyalu* | Caryophyllaceae | Paddy, Radish, Potato | Jan.-June | Plant | Plant-paste is applied on burns, boils and wounds; plant is used as green vegetable, as well as fodder. |
| *Veronica agrestis* Linn. | Scrophulariaceae | Radish | Mar.-Oct. | Leaves | Leaves are consumed in salad. |

Sarin (2003) mentioned 54 weed species from agricultural land, gardens, orchards and roadsides, which have great demand in drug industry. Some of these are very selective as regard habitat and tend to disappear with a slight change in the edaphic, hydrological and other site conditions. In our study 20 species highlighted with asterisk (*) are strictly confined to agro-ecosystem, including fallow lands (Table 1).

Weeds are prone to decline in growth and spread due to a change in the site condition or through competition from aggressive intruders (Sarin, 2003). Bentley *et al.* (2002) observed that where herbicides and chemical fertilizers are used in the fields, the number of weeds is less (9-14) and mostly grasses; whereas, in the fields where manure is used and no herbicide, comparatively more number of weeds (18-26) grow. This shows that continuous use of the chemicals may disappear some of the valuable plant species in future. Singh *et al.* (1999) also made an ecological assessment of weeds of Paddy crop in different regimes of fertilizers and reported a clear variation in structure of weed flora in the manure and fertilizer used fields.

## REFERENCES

BABU, C.R. 1977. *Herbaceous Flora of Dehra Dun.* CSIR, New Delhi.

BALICK, M. J. & PAUL ALAN COX 1996. *Plants people and culture : The science of ethnobotany.* Scientific American Library, New York.

BENTLEY, JEFFERY W., MORAG WEBB, SILVIO NINA & SALOMON PEREZ 2002. *Using ethnobotany to find the demand for weed research in the Bolivian Andes.* Agricultural Anthropologist, Casilla, Cochabamba, Bolivia.

BHAKAT, R.K. & P.K. PANDIT 2003. Role of a sacred grove in conservation of medicinal plants. *Indian Forester* 129 (2) : 224-232.

GAUR, R.D. 1999. Flora of District Garhwal North West Himalaya (with ethnobotanical notes). Transmedia Publisher, Srinigar (Garhwal), India.

JAIN, S.K. 1994. *Medicinal plants of India.* National Book Trust of India, New Delhi.

OUTHIA, 1999. Medicinal weed in rice field of Chhatisgarh, India. *International Rice Research Notes* 1491 : 40.

SARIN, Y.K. 2003. Medicinal plant raw materials for Indian drug and pharmaceutical Industry-I. An Appraisal of Resources. *Indian Forester* 129 (1) : 3-24.

SINGH, LOKENDER, S.P. JOSHI, ARUN GUPTA, R.K. MANHAS & NARESH K. VERMA 1999. Demography, diversity and distribution pattern of weeds in response to different regimes of fertilizers in rice crop of Doon valley. *World Weed* 6 : 133-142.

# ETHNO-MEDICO-BOTANICAL STUDIES IN PENCH TIGER RESERVE, MADHYA PRADESH

S.K. Masih

*State Forest Research Institute, Polipathar, Jabalpur (M. P.)*

**ABSTRACT**

Pench Tiger Reserve is situated in the Satpura hills of Seoni and Chhindwara districts of Madhya Pradesh. About 93 species having ethno-medico-botanical uses in and around the Reserve have been identified during present study. The correct botanical names, local names, name of family and medicinal uses have been presented in a systematic manner in this communication, besides general information about the area, forest types, villages and villagers, their occupation etc. It is hoped that present study will be helpful in sustainable utilization of resources, better management of the Reserve and conservation of biodiversity and traditional folk knowledge.

## INTRODUCTION

Forest resources have direct and indirect relationship with the life of local population, tribals, forest dwellers and many other backward inhabitant groups. The sociological systems, customs, cultures and life patterns of these groups are also closely related with forests. They utilized forest products for food, fodder, medicine, fuel, gum, agriculture implements, aromatic oils, basketry works, charcoal, decoration, defence equipments, dye, fencing, fishing, furniture, house building, hunting equipments, musical instruments etc for their sustenance and daily needs.

Forests are not only the source of major and minor forest products but also provide and fulfill the basic needs and demands directly and indirectly in life pattern of these forest dwellers. They use an enormous range of wild plants and have developed a unique understanding of the forest resources and pass on these traditions, taboos, totems, folklore, traditional medicinal remedies and other knowledge by word of mouth from one generation to other generation. Today, ethnobotany has become an important and crucial area of research and development in resource management, conservation of biodiversity and socio-economic development of the region. The scope of ethnobotany has now greatly enlarged, both in terms of its theoretical contributions to an understanding of plant human relationship as well as for practical applications and utilization of biological knowledge of tribals and their medicine systems.

The task of present investigation is to investigate the existing traditional knowledge of local population, forest dwellers etc inhibiting in the adjoining areas of buffer zone of the Pench Tiger Reserve, Seoni (M.P.).

## STUDY SITE

Pench Tiger Reserve (National Park) is situated in the Satpura hills of Seoni and Chhindwara districts. Its name is given after the river PENCH that flows through the middle of this vast forested tract. It was in 1977 that this forest area was declared as a Sanctuary. An area of 11873 sq km was kept as protected area and 449.39 as Sanctuary. An area of about 293 sq km including the core area of Sanctuary was given the status of National Park in 1983. The Park is situated between 79°9' and 79°22' longitude and 21°38' and 21°5' latitude at an elevation of 650 m above m.s.l. in Seoni and Chhindwara districts. Total area of the Pench Tiger Reserve is 757.897 sq km, out of which National Park covers only 293 sq km and reaming area 465.04 sq km is Buffer Zone. The details of the areas under Reserved, Protected and Revenue categories are given below.

| S.n. | Particular of the Area under Category | Area under National Park | Area under Buffer Zone | Total |
|---|---|---|---|---|
| 1 | Reserved Forest Area | 283.814 | 333.040 | 616.854 |
| 2 | Protected Forest Area | 6.262 | 102.400 | 108.662 |
| 3 | Revenue Area | 2.781 | 29.600 | 32.381 |
| | Total | 292.857 | 465.040 | 757.897 |

## Forest Resources

Forest types of Core zone of PTR and adjoining Buffer zone are as under :

### *1. Southern Tropical Dry Deciduous Teak Forest (Type - 5A/C1b)*

Forests of this type are mostly available along the Pench river and on the hill slopes. Teak is the dominant species (25% to more than 50%) in this forest type. Species namely Dhawda, Lendia, Saja, Salai, Moyan, Mahul, Mahua, Dhoban, Mokha, Siras, Tendu, Bija, Tinsa, Semal, Haldu, Koha, Jamun, Aonla, Dhaman, Kumbhi, Ghont, Baranga, Amaltas, Ashta, Bans, Dhawai, Karonda, Palas, Palasbel and Lantana are the common associates in this forest type.

### *2. Southern Tropical Dry Deciduous Mixed Forest (Type - 5A/C3)*

This forest type is prominent in the Core and Buffer zone areas. The presence of Teak in this forest type varies from place to place (less than 20%). Species namely Dhaora, Salai, Saja, Harra, Bahera, Bhirra, Moyan, Tendu, Mahul, Achar, Mahua, Rohan, Siras, Dhoban, Tinsa, Rinjha,

Kakar, Bel, Haldu, Mundi, Makha, Koha, Kassai, Kullu, Khair, Aonla, Paras, Bans, Ber, Jamun, Marorphal, Koilar, Kakai, Dhawai, Karonda, Jangli tuar, Satavar, Safedmusli, Kalimusli, Dioscorea sps., Chiraya, Chhind, Chakoda, Makoi, Ramdatun, Lantana etc associate the teak and enrich the resource value of the forest with teak.

## Profile of Fringe villages

Pench Tiger Reserve has been divided in three Ranges and 9 Circles for better supervision, functioning and proper management.

| Name of Range | Name of Circles |
|---|---|
| Karmajhiri Range | a) Karmajhiri, b) Alikatta and c) Khamrith |
| Kurai Range | a) Murer, b) Khamrith (Kurai) and c) Kurai |
| Gumtara Range | a) Gumtara, b) Pulpuldoh and c) Dudhgaon |

They are further divided in 43 beats and compartments for protection, conservation and development. A total of 99 villages are situated in the adjoining Buffer zone of the study site. The distribution of these villages under two districts is given below.

| District | Number of Village |
|---|---|
| Chhindwara | 43 |
| | 5 |
| Seoni | 23 |
| | 10 |
| | 18 |
| Total (2) | (99) |

The villages are inhabited by different castes, namely Gonds, Pradhan, Aheer, Mehra, Mukasi, Chamar, Kalar, Lodhi, Raghuwansi, Harijan, Dhemar, Sahu, Kirars, Musalman, Nai, Lohar, Mahar, Daheriya, Teli, Gadri, Gupta, Patels, Shrivastava and Choudhry. The most dominating caste is Gonds and their sub-groups which comprise more than 60% population.

The agriculture crops commonly grown by the local population are Makka, Dhan, Chana, Matar, Urad, Chaura, Soyabean, Mungphali, Masoor, Ganna, Jowar, Bajra, Kodo, Kutki, Jagni, Tuar, Bajra, Rahar etc.

Various vegetable crops are also grown by the local population for their day to day needs and for sale in local weekly markets, like Bhata, Alu, Chulai bhaji, Maithi bhaji, Bathua bhaji, Palak bhaji, Lal bhaji, Karela, Kakoda, Kundru, Gilki, Sem, Barbati etc.

The common diseases found in the villages are Gastric problem, Back pain, Fever, Cold, Cough, Skin diseases called - Khaj and Khujli, Dehydration, Vomiting, Diarrhoea, Dhat, Loose-motion, Wounds, Stomach pain, Weakness etc. The cases of serious diseases like Cancer, T.B., Blood pressure, Aids are negligible.

The local population in most of the villages take treatment from private physicians and Governmental Health Care Centers covering 4 to 5 villages. They have full faith and confidence on allopathic medicines for quick relief from the ailments.

The person having traditional knowledge of medicine, locally called Medicinemen, Ojha, Guniya and termed as Resource Person for this study, are very less in numbers in villages.

## METHODOLOGY

The present investigation is conducted through direct approach that involves field surveys in villages adjoining to Pench Tiger National Park. Total 99 villages scattered on the periphery in Seoni and Chhindwara districts were surveyed during 2001-2002. All 5 ranges namely Rukhad, Kurai, West Khawasa, Bicchua and Chaurai were visited periodically for the collection of primary data pertaining to traditional knowledge of ethnobotanical importance. The important villages surveyed under different forest ranges are as under.

*Rukhad range :* Vijaipani, Paraspani, Tikadi Rayat, Tikadi Mal., Barelpar, Sarrhirri, Teveri, Saleh, Bhedki, Ghatkohka, Panjra, Sindaria, Dhutera, Alesur, Patrai, Mohgaon Titri, Aerma and Karmajhiri.

*Kurai range :* Agri, Mohgaon Sdak, Harduli, Pindipar, Raiarao, Kodajhiri, Pipariya, Setewani, Potia and Mohgaon Yadav.

*West Khawasa range :* Durgapur, Jhirewada, Ambadi, Santosa, Awardhani, Teliya, Kohka, Turia, Kuppitola, Khamba, Bijaipani, Amajhiri, Mudiarith, Nayagaon, Pachdhar, Kothra, Murrer, Gandatola, Katangi rayat, Niwari and karhya.

*Bichhua range :* Antra, Bandhan Rayat, Bandhan Mal, Bisanpur, Bordi, Boriya, Chargaon, Chirrewani, Deori, Dholpur, Dongargaon Kala, Dongargaon Khurd, Dudhgaon, Dunda seoni, Ghatkamti, Gumtara, Kanha sagar, Kanhargaon, Karhaiya, Kheranj, Khamarpani, Khamariya Mal, Khursipar, Kokiwada, Kundai, Marjatpur, Mohgon Khurd, Naharjhir, Patra Kala, Patra Khurd, Pathri, Pulpuldoh, Saliwada Mal, Sawri, Silota Kala, Silota Khurd, Silota Raiyat, Singardeep, Surrewani, Tekapar, Thota Mal, Thota Raiyat and Thuepani.

*Chaurai range :* Banskheda, Kumpani, Davajhir, Rampuri, Jamtara.

Methodology adopted for the documentation of traditional knowledge with reference to ethno-medico-botanical details is followed as prescribed by the Masih (1990, 2000).

The detail information was collected in a designed questionnaire as below :

| S. N. | | Details of information |
|---|---|---|
| 1 | Name of informer | |
| 2 | Status of informer | Name of village, Caste, Tribe, Age, Occupation |
| 3 | Name of plant | Local and vernacular names |
| 4 | Availability of species | Rare/Little/More/Sufficient |
| 5 | Plant parts and product used | Utilization of different parts and products |
| 6 | Mode of use | External or internal |
| 7 | Mode of application | Detail description about mode of application |
| 8 | Utilizable form of medicine | Paste, Powder, Liquid, Extraction, Suspension, Decoction, Mixture, Tablet |
| 9 | Prescription of dose for treatment | Once a day: Morning or night<br>Twice a day: Morning and night<br>Thrice a day: Morning / afternoon / night |
| 10 | Method of dose prescribed | Liquid: 1 Tea spoon / 1 Tea cup<br>Tablet: 1 / 2 / more<br>Powder: 1 to 5 gms per dose<br>Paste: 5 to 50 gms as per need |
| 11 | Duration of dose | 1 Day; 3 days; 1 week; 15 days; 21 days and till cure of ailment |

| | | |
|---|---|---|
| 12 | Mode of storage of medicine | Liquid / Powder / Tablet / Paste / Only Fresh |
| 13 | Prescribed pot / ware for storage | Paper / Cloth / Bottle / Iron box / Grain Pot |
| 14 | Validity of storage period | 1 Week; 15 Days; 1 Month; 6 Months; 1 Year; More than a year |
| 15 | Side effects | Pain / Swelling / Any other |
| 16 | Precautions prescribed | To avoid drinking alcohol / tea<br>To avoid eating prickles / non-vegetarian food<br>To avoid eating rice<br>To avoid eating tobacco<br>To avoid heavy diet<br>To avoid exertion<br>To avoid smoking<br>Required complete bed rest<br>Required light food<br>Required fruit juice |

## ENUMERATION OF MEDICINAL PLANTS

**Abrus precatorius** Linn. (Fabaceae).

*Local names:* Gunja, Ratti, Gumchi, Gunchi.

Powder of seeds is used as abortifacent.

**Abutilon indicum** (L.) Sweet (Malvaceae).

*Local name :* Kanghi.

Leaf-paste is applied on cuts, ulcers, wounds and boils for skin recovery. Seed-powder is used in piles.

**Acacia catechu** (L. f.) Willd. (Mimosaceae).

*Local name :* Khair.

Paste of fresh leaves is given with sugar or gud for the dislocation of bones and bone fracture; 5 to 10 gms of paste twice a day up to one week is given in minor cases where as 10-15 days for major fracture cases.

**Achyranthes aspera** L. (Amaranthaceae).

*Local names :* Chirchita, Latjira.

The leaf-paste is applied on wounds for healing. Seed powder is given in morning and evening up to 15 days in bleeding piles. Branches or twigs are used as tooth-brush and used twice a day morning and evening for tooth pain. Roots are made in to a paste with sugar and given in Dhat (semen discharge) up to 3 days once a day.

**Aegle marmelos** (L.) Corr. (Rutaceae).

*Local name :* Bel.

Ripe fruits fresh or dried pulp is mixed with water and taken morning and evening twice a day up to six days for severe dysentery. Decoction of bark about 25 ml is given thrice (morning / afternoon / evening) in a day up to 10-15 days for intermittent fever. Bark is also made into a paste with sugar and given twice a day up to 5 days for dehydration and loose motion.

**Ageratum conyzoides** Linn. (Asteraceae).

*Local name :* Ghanera.

Fresh juice or decoction of leaves or tender shoots is used in weakness, diarrhoea and dysentery; one teaspoon dose is given twice a day up to three days for this purpose.

**Aiianthus excelsa** Roxb. (Simaroubaceae).

*Local names :* Maharukh, Mahaneem.

Bark is used to make decoction which is taken 10-15 ml twice a day upto 3 days for dysentery.

**Albizia lebbeck** (L.) Benth. (Mimosaceae).

*Local names :* Siris, Sirisha.

Paste of bark is used in boils. Gum is used as adulterant.

**Aloe vera** (L.) Burm. f. (Liliaceae).

*Local name :* Gheekwar.

Paste of the fresh leaves is used on burns; it cools the burning sensation.

**Amaranthus tricolor** L. (Amaranthaceae).

*Local name :* Matibhaji.

Decoction of root piece nearly 50 ml twice a day is given in morning and evening up to 10 days regularly for throat and mouth infection.

**Andographis paniculata** (Burm. f.) Wall. ex Nees (Acanthaceae).

*Local names :* Kalmegh, Kaduchirayton, Bhuineem, Chirayta.

Whole plant or leaves are boiled with water and made in to a decoction which is given 5-10 ml twice a day up to 3 days for fever, malaria, joint pain, liver problem and jaundice.

**Annona squamosa** Linn. (Annonaceae).

*Local name :* Sitaphal.

Leaf-paste is used to kill lice. Roots are chewed and juice is taken in snake-bite.

**Anogeissus latifolia** (Roxb. ex DC.) Wall. ex Guill. & Perr. (Combretaceae).

*Local names :* Dhaoda, Dho, Dhawa.

Gum is used medicinally in swellings, pain and stomach disorders.

**Asparagus racemosus** Willd. (Liliaceae).

*Local names :* Satawar, Satavar, Narboth.

Fresh tubers are boiled with milk and decoction is taken up to 15 days for strength after delivery.

**Baliospermum montanum** (Willd.) Muell.-Arg. (Euphorbiaceae).

*Local names :* Danti, Jamal ghota.

Seed-powder is mixed with sugar and given twice a day for indigestion and other digestive problems. Seed oil or seed powder fried with oil & utilized in swellings, joint-pain and rheumatism.

**Barleria cristata** L. (Acanthaceae).

*Local name :* Katsareya.

Paste or decoction of leaves, root-bark or roots is given in cough; prescribed dose is 1 teaspoon twice a day up to 3 days.

**Barleria prionitis** Linn. (Acanthaceae).

*Local name :* Jhinti.

Juice and paste of the leaves are used in cold, cough, boils and toothache; medicine is given one teaspoon twice a day up to 3 days for the treatment of above diseases.

**Bauhinia purpurea** Linn. (Ceasalpiniaceae)

*Local name :* Keolar.

Bark decoction or paste is used in diarrhoea.

**Bauhinia racemosa** Lam. (Caesalpiniaceae)

*Local names :* Kachnar, Kachnal, Apta.

Paste or juice of the leaves and bark is given in dysentery and diarrhea, 1 teaspoon twice a day up to 3 days.

**Bombax ceiba** Linn. (Bombacaceae).

*Local names :* Semul, Semal, Shembal.

Bark is made in to a paste and mixed with small quantity of water and given one teaspoon twice a day up to 3-5 days for smooth regulation or discharge of the menses in women.

**Boswellia serrata** Roxb. ex Colebr. (Burseraceae)

*Local name :* Salai.

Paste of bark is used against diarrhoea and skin diseases. Gum is used in stomachache and as anti-dysenteric.

**Butea monosperma** (Lamk.) Taub. (Fabaceae).

*Local names :* Tesu, Palas.

Decoction of flowers or seeds about 25-50 ml is given two times morning and evening up to 3 days for worms. Bark decoction is also given in severe piles as tonic up to one month once a day. Seed-paste is used in skin diseases.

**Butea superba** Roxb. ex Willd. (Fabaceae).

*Local names :* Palas lata, Betra palas.

Decoction of shoot or bark is given in piles, 2 times per day up to 15 days.

**Caesalpinia crista** L. *emend.* Dandy & Excell (Caesalpiniaceae)

*Local name :* Gataran.

The powdered seeds are taken about 5-10 gms twice a day up to 3 days for stomach disorders and pain.

**Calotropis gigantea** (Linn.) R. Br. (Asclepiadaceae).

*Local names :* Ak, Madar, Akauva.

Leaves coated with oil are heated and applied on the swellings and dislocated parts having pain, twice a day till relief.

**Calycopteris floribunda** (Roxb.) Poir. (Combretaceae).

*Local names :* Kakaray, Ukshi.

Decoction of leaves is used in swellings, pain, cuts inside the digestive system and stomach; dose prescribed is 1 teaspoon twice in a day up to 7 days.

**Capparis zeylanica** Linn. (Capparaceae).

*Local names :* Ardand, Govindi.

Decoction of root-bark is given in stomach pain and throat pain during cough; 1 teaspoon is given twice daily up to 3 days for this purpose.

**Cassia fistula** Linn. (Caesalpiniaceae).

*Local names :* Amaltas, Jagrajadi.

Root-paste with Gud is made in to tablets (up to 1 gm each) and given in paralysis twice a day up to 30 days.

**Catharanthus roseus** (L.) G. Don (Apocynaceae).

*Local name :* Sadabahar.

Juice of the fresh leaves with milk is given to cure diabetes; a teaspoon juice is given every day up to 7 days for the treatment.

**Celastrus paniculatus** Willd. (Celastraceae).

*Local names :* Malkanghni, Malkangni.

Bark is made in to a paste and mixed with oil and used to destroy lice. Decoction of leaves and shoots is utilized in colic, dysentery and diarrhoea; 1 teaspoon dose is given 3 times in a day up to 3 days for this purpose.

**Chlorophytum arundinaceum** Baker (Liliaceae).

*Local name :* Safed musli.

5-10 gms dried or 2-4 fresh tubers are boiled with milk and given twice a day morning and evening up to one month in impotency and weakness.

**Chloroxylon swietenia** DC. (Rutaceae).

*Local names :* Bhirra, Ghiria, Girya.

Fresh paste of the bark is used in rheumatism.

**Cissus quadrangularis** Linn. (Vitaceae).

*Local names :* Hadjor, Harjudi.

Stem and root paste is applied on dislocation of bones and bone fracture; 10-15 gms paste is applied every day up to 15 days for this purpose.

**Cleistanthus collinus** (Roxb.) Benth. ex Hook. f. (Euphorbiaceae).

*Local name :* Garari.

Powdered leaves, root bark and fruits are used in gastric problems and 1-2 peach of it is taken with water twice daily up to 7 days.

**Colebrookea oppositifolia** Sm. (Lamiaceae).

*Local name :* Binda.

Fresh leaves are made in to a paste and applied on wounds.

**Cordia dichotoma** Forst. f. (Ehretiaceae).

*Local names :* Gondi, Lasora, Lasoda.

Fruits are used medicinally in urinary disorders.

**Costus speciosus** (Koen.) Smith (Zingiberaceae).

*Local name :* Keokand.

1-2 square cm piece of fresh rhizome is chewed and juice is taken twice a day morning and evening up to 7 days for swellings and pain at the body joints. It is also needful in smooth regulation of menses in women.

**Curculigo orchioides** Gaertn. (Hypoxidaceae)

*Local name :* Kalimusli.

Fresh or dry tubers are made in to paste or powder mixed with milk and given to patient having piles and diarrhoea; 1 tuber (1 gm) is allowed to give twice a day up to 7 days for this purpose.

**Curcuma angustifolia** Roxb. (Zingiberaceae).

*Local name :* Tikhur.

Paste or powder of fresh or dry tubers are mixed with milk and utilized for the ulcer and digestive problems. The prescribed dose is 2 teaspoon twice a day up to 7 days.

**Curcuma aromatica** Salisb. (Zingiberaceae).

*Local name :* Jangli haldi.

Paste of the rhizome is applied on joint-pain, dislocations and swellings 2 times a day up to 3 days.

**Cuscuta reflexa** Roxb. (Cuscutaceae).

*Local name :* Amarbel.

Plant paste is applied on white spots (leucoderma) diseases for about 10-15 days twice a day.

**Datura stramonium** L. (Solanaceae).

*Local name :* Datura.

Seed powder about 0.5 to 1 gm is kept in the cavity of teeth to relieve pain and infection, once a day up to 3 days.

**Dioscorea hispida** Dennst. (Dioscoreaceae).

*Local name :* Surankanda.

Fresh paste of tubers is made in water and applied on eczema and other skin diseases twice a day up to 7 days.

**Diospyros montana** Roxb. (Ebenaceae).

*Local names :* Gindu, Goindu, Bistendu.

Paste of leaves and fruits is applied on boils once a day up to 3 days.

**Dodonea viscosa** (L.) Jacq. (Sapindaceae).

*Local name :* Aliar.

Leaf-paste is applied on burns, swellings and wounds twice a day up to 3 days.

**Ficus carica** L. (Moraceae).

*Local names :* Jharphali, Porodumer, Khain.

Fresh or dry fruits are eaten for the cure of cuts, swellings and white spot disease in tongue and inner parts of the mouth; 4-5 fruits are taken twice a day up to 7 days for this purpose.

**Flemingia strobilifera** (L.) Ait. & Ait. f. (Fabaceae).

*Local names :* Kanphute, Bundar.

Decoction or juice of the leaves is used to remove or expel the worms from the small and large intestinal tract; 1 teaspoon dose is given in the night only up to 3 days for this purpose.

**Gardenia gummifera** Linn. f. (Rubiaceae).

*Local name :* Dikamali.

Gum is mixed with Gud and made in to tablets, 1-2 tablets are given 2 times in a day to remove nervous complaints and diarrhoea.

**Gloriosa superba** L. (Liliaceae).

*Local names :* Kalihari, Languri.

Small piece of fresh rhizome is given with fodder to animals having any type of wounds on the body.

**Gmelina arborea** Roxb. (Verbenaceae).

*Local names :* Gumhar, Gambhari.

Decoction is taken in fever, 1-2 teaspoon twice a day.

**Grewia hirsuta** Vahl (Tiliaceae).

*Local names :* Govali, Kakranda.

Fruits and seed powder is utilized in diarrhoea and dysentery, ½ teaspoon powder is given twice a day up to 3 days for the same.

**Gymnema sylvestre** (Retz.) R. Br. (Asclepiadaceae).

*Local names:*Gudmar, Vakundi, Madhunasi.

Fresh or dry powder of leaves and tender shoots is given about 1 gm twice a day up to 7 days for the treatment of diabetes.

**Helicteres isora** L. (Sterculiaceae).

*Local names :* Marorphallii, Marotphal, Aintii.

Fruits as a whole or seed powder about 5-10 gms is given with sugar in treatment of dysentery, stomach disorder and stomach pain thrice a day up to 7 days morning / afternoon / evening.

**Hibiscus rosa-sinensis** L. (Malvaceae).

*Local name :* Jason.

Fresh leaves are made in to a paste with sugar and given in problems like pain and disturbances in urinary duct; one-teaspoon quantity is given twice a day up to 3 days for this purpose.

**Indigofera cassioides** Roxb. ex DC. (Fabaceae).

*Local names :* Baroli, Chirmati, Sakena.

Decoction of leaves is used in cough and cold, about 10-20 ml dose is given twice in a day up to 3 days.

**Ixora pavetta** Andr. (Rubiaceae).

*Local names :* Nevari, Nevali.

Decoction of bark is given in debility, about 10-20 ml twice a day up to 7 days.

**Jatropha curcas** Linn. (Euphorbiaceae).

*Local names :* Ratanjot, Chandrajot, Safed arand.

Fresh collected white latex is applied on wounds, cuts and skin diseases, twice a day up to 3-5 days for quick relief.

**Jatropha gossypifolia** Linn. (Euphorbiaceae).

*Local name :* Chandrajot.

Fresh soft branches or twigs are used as tooth-brush, twice a day in tooth pain and pyorrhoea.

**Leea macrophylla** Roxb. ex Horn. (Vitaceae).

*Local name :* Dinda.

Paste or dry powder of root is applied externally on wounds and cuts. About ½ teaspoon powder with milk is also given to remove worms twice a day up to 3 days.

**Madhuca indica** J. F. Gmel. (Sapotaceae).

*Local names :* Mahua, Mohwa.

The leaves are gently warmed on the fire and put on the swellings and minor injuries. This mode of application is repeated twice a day morning and evening up to 10-15 days.

**Mangifera indica** Linn. (Anacardiaceae).

*Local name :* Am.

Ash of the burnt leaves with coconut oil is applied on the burn parts of body. Fresh bark paste is mixed with limestone (Chuna) and given one teaspoon twice a day in constipation, stomach disorder, vomiting and dehydration.

**Maytenus senegalensis** (L.) Excell. (Celastraceae).

*Local names :* Baikal, Bharatti.

Paste of leaves and tender shoots, fresh or dry, with honey is given in dysentery and diarrhoea ½ teaspoon up to 3 days.

**Merremia gangetica** (L.) Cufod. (Convolvulaceae).

*Local names :* Musakani, Bhuibela.

Fresh or dry leaves powder is given in rheumatism and cough; ½ teaspoon powder is given twice a day up to 3 days with milk or honey for this purpose.

**Mimosa pudica** Linn. (Mimosaceae).

*Local names :* Lajwanti, Lajalu, Chumui.

Decoction of fresh roots is utilized in urinary disorders and stone; 10-20 ml decoction is prescribed twice a day up to 7 days for the same.

**Momordica dioica** Roxb. ex Willd. (Cucurbitaceae)

*Local names :* Jangli karela, Van karela.

Fresh leaves are collected and their juice extracted about one teaspoon which is used twice a day up to 3 days for quick removal of insects and worms present in the digestive system.

**Murraya koenigii** (L.) Spreng. (Rutaceae).

*Local names :* Mithineem, Karipatta.

Decoction of leaves about 10 ml is given twice a day up to 3 days for vomiting, dysentery and in diarrhoea.

**Nyctanthes arbor-tristis** Linn. (Oleaceae).

*Local names :* Harsingar, Parijat.

Paste or dry powder of leaves is used in rheumatic pain and fever. Decoction of bark is given in disorders of alimentary canal and digestive system.

**Ocimum americanum** Linn. (Lamiaceae).

*Local names :* Kali tulsi, Jangli tulsi.

Decoction of leaves is utilized in cough and cold; 10-15 ml decoction is given twice a day up to 3 days for this purpose.

**Ougeinia oojeinensis** (Roxb.) Hochr. (Fabaceae).

*Local names :* Tinsa, Tiwas, Panjan.

Fresh bark paste with sugar and small quantity of limestone (Chuna) is given twice a day up to three days for diarrhoea, dysentery, loose motion and stomach disorder.

**Plumbago zeylanica** L. (Plumbaginaceae).

*Local name :* Chitawar.

The seed powder is applied on cuts, boils and wounds.

**Pongamia pinnata** (L.) Pierre (Fabaceae).

*Local names :* Kanji, Karanj.

Fresh or dried seed-paste is applied on skin eruptions, eczema and other skin diseases.

**Pueraria tuberosa** (Roxb. ex Willd.) DC. (Fabaceae).

*Local names :* Bhenskanda, Sural, Bilaikand, Jaminkand, Patal kumhda.

Small piece of tuber is given with fodder to oxen for strength once in a day up to 7 days. About 1 cm piece of tuber is also chewed for the stimulation of body.

**Pygmaeopremna herbacea** (Roxb.) Moldenke (Verbenaceae).

*Local names :* Bharangi, Bharangmul.

Decoction of fresh leaves and roots is given in rheumatic pain, fever, cough and cold; one teaspoon dose is prescribed twice a day up to 3 days for this purpose.

**Quirivelia frutescens** (L.) M.R. & S.M. Almeida (Apocynaceae).

*Local names :* Kalidudhi, Syamlata.

Powdered root is given in diabetes and used as tonic; one teaspoon is given with milk twice a day up to 7 days for this purpose.

**Ricinus communis** L. (Euphorbiaceae).

*Local names :* Arandi, Andi.

Fresh leaves are slightly warmed on fire and put on the boils to remove pus. Leaf-paste is also applied on burn area to decrease pain and for skin recovery twice a day till relief.

**Scoparia dulcis** L. (Scrophulariaceae).

*Local names :* Chinibhaji, Mithi bhaji.

Fresh paste or dry powder of leaves and tender shoots are given in diabetes one teaspoon twice a day up to 7 days.

**Semecarpus anacardium** L. (Anacardiaceae).

*Local names :* Bhilwa, Bhilma.

Oil of seeds is applied with the help of pointed wooden sticks on joints having swelling and pain.

**Senna tora** (L.) Roxb. (Caesalpiniaceae).

*Local name :* Chakoda.

Fresh new leaves and seed-paste is applied on skin diseases.

**Sida acuta** Burm. f. (Malvaceae).

*Local names :* Atibala, Bala, Kharenti.

Decoction of roots and leaves is given in stomachache, disorder of urinary system, weakness and swelling; 5-10 ml decoction is prescribed once a day up to 7 days for this purpose.

**Smilax zeylanica** L. (Smilacaceae)

*Local name :* Ramdatoon.

Root powder is mixed with sugar and given to women having leucorrhoea one teaspoon once or twice per day up to 3 days.

**Soymida febrifuga** (Roxb.) A. Juss. (Meliaceae).

*Local name :* Rohan.

Fresh collected bark is boiled with water up to half an hour, then solution filtered and about 50 ml is given daily twice (morning and evening) up to 14 days in arthritis, joint swelling, pain and weakness.

**Spatholobus parvinorus** (Roxb. ex DC.) Kuntze (Fabaceae).

*Local names :* Maula, Bando, Palsa.

Decoction of fresh bark is given in digestive problems 2 times every day up to 3 days. Gum is utilized with sugar or Gud, in the form of tablets, 2 tablets twice a day up to 7 days for body pain, strength and arthritis.

**Spermadictyon suaveolens** Roxb. (Rubiaceae).

*Local name :* Mahabel.

Root-paste or powder is given in diarrhea; ½ teaspoon is given with water twice a day up to 3 days.

**Sterculia urens** Roxb. (Sterculiaceae).

*Local names :* Kullu, Kurlu, Gurlu.

Bark paste or powder is mixed with gud and given once a day up to 2-3 days for loose motion to cows, oxen, buffaloes and goats.

**Syzygium cumini** (Linn.) Skeels (Myrtaceae).

*Local name :* Jamun.

Seeds powder up to 5 gm per day twice morning and evening is given in diabetes. Root or bark paste mixed with sugar and small quantity of limestone (Chuna) is given one teaspoon twice a day up to three days for constipation, vomiting and stomach disorders.

**Syzygium heyneanum** (Duthie) Wall. ex Gamble (Myrtaceae)

*Local names :* Jamun, Jaman.

Decoction of bark is used for teeth pain and pyorrhoea. It is applied as gargles thrice a day up to 7 days specially after taking food. Seeds powder up to half teaspoon is taken 3 times per day to control diabetes.

**Tamilnadia uliginosa** (Retz.) Tirveng. & Sastre (Rubiaceae).

*Local names :* Katul, Pindalu, Bhidhara.

Small piece of fresh root is chewed in mouth ulcer (known as Agri or Agra disease) thrice a day up to 7 days.

**Terminalia chebula** Retz. (Combretaceae).

*Local name :* Harra.

Fruits are made in to a powder and 10-15 gm is taken with water after meals twice a day morning and evening up to 15 days for constipation and stomach disorders. Roasted fruits are also given in powdered form in cough.

**Terminalia cuneata** Roth (Combretaceae).

*Local names :* Kalua, Kahu, Koha, Arjun.

Root-paste is mixed with sugar and water and mixture is boiled & filtered, solution is given 50-100 ml twice a day up to 3 days to women having white discharge, known as leucorrhoea.

**Vanda tessellata** (Roxb.) Hook. ex G. Don (Orchidaceae).

*Local name :* Jodtod.

Paste of whole plant is mixed with hen egg and small quantity of water. It is taken once in a day up to three days in any type of bone fractures.

**Vitex negundo** Linn. (Verbenaceae).

*Local names :* Nirgndi, Shivari, Nigudi.

Decoction pf leaves and roots is given in fever, rheumatic pain, dysentery and piles; about 10 ml decoction is prescribed once a day up to 3 days.

**Wendlandia heynei** (R. & S.) Sant. & Merch. (Rubiaceae).

*Local names :* Padgira, Birsa.

Decoction of bark is given in urinary complaints and swellings; 5-10 ml decoction is prescribed twice a day up to 3 days for the same.

**Woodfordia fruticosa** (L.) Kurz (Lythraceae).

*Local names :* Dhawai, Dhoran, Chilbelii.

The stem bark is boiled and paste is applied on burnt parts. Leaf-juice is given orally for drying of burnt areas quickly. Leaf-paste is also applied in skin diseases particularly on eczema for quick recovery. Decoction of leaves and flowers are given after the delivery for strength.

## ACKNOWLEDGEMENTS

The author thankfully acknowledge the Director, State Forest Research Institute, Polipathar, Jabalpur for guidance, moral and technical support provided during the entire course of investigation. Thanks are also to Shri R.G. Soni, Field Director, Shri L.K. Choudhry, Dy. Director, SDO's and Range Officers with their Field Staff of Pench Tiger Reserve, Seoni (M.P.) for their kind co-operation, facilitation and providing best hospitality during the field visits in the study areas.

## REFERENCES

MASIH, S.K. 1990. *Ethnobotanical studies in Tribal areas of Bastar and Jabalpur, M.P.* Ph.D. Thesis, R.D.V.V., JBP (M.P.).

———— 2000. Quantitative ethnobotanical observations in submergence areas of Sardar Sarovar Dam. *My Forest* 36 (3):221-236.

# IMPORTANT MEDICINAL PLANTS AND THEIR USES IN ME-GAD WATERSHED AREA OF LAHUAL & SPITI

YOGENDRA SHARMA, ARVIND KUMAR, ARCHNA SINGH & J.C. RANA

*National Bureau of Plant Genetic Resources, Regional Station, Phagli, Shimla — 171 004 (H.P.)*

## ABSTRACT

Largely plant based, traditional medical systems continue to provide primary health care to more than three quarters of the world's population. Medicinal plants are locally used in treating and preventing specific ailments and diseases; people use these medicines deliberately on account of faith, the local myth, tradition and their socio-cultural milieu. The present paper highlights some important medicinal plants and their uses either alone or in combination with other plants/part(s) to treat various diseases in Me-gad watershed area of Lahual and Spiti district of Himachal Pradesh.

## INTRODUCTION

Me-gad watershed area lies between 32° 36' to 32° 42' north latitude and 76° 36' to 76° 42' east longitude in Lahual and Spiti districts of Himachal Pradesh. The elevation varies from 2700 to 5700 meter. A watershed can be defined as the drainage basin or catchment area of a particular stream or river or area from where the water to a particular drainage system, like a river or stream comes from. A watershed provides a natural environment unit for planning a developmental initiative. People and their environment are interdependent. Any change in the surrounding environment directly affects the people living therein. A degraded environment results in a degraded quality of life of the people. Thus, efforts to reduce poverty and improve the standard of living of the people must aim at improving the environment they live in.

All the plants on the earth are important for humanity, as human beings are influenced in various ways by plants and their products. Existence of mankind is impossible without plant kingdom as they provide balanced ecosystem, major source of nutrition, preventive aspects of medicine, primitive aspects of health and curative aspect of disease (Kumar & Dhan, 1998). Largely plant based, traditional medical systems continue to provide primary health care to more than three quarters of the world's population. Plant derived medicines are used in all cultures and, as such, plants have always played a key role in health care. Srivastava *et al.* (1995) defined "Medicinal plants are those that are commonly used in treating and preventing specific ailments and diseases, and that are generally considered to play a beneficial role in healthcare". Plants used in traditional medicine, therefore, had an important role in the maintenance of health all over the world (Bhattarai, 1998). Rural areas don't have access to modern medicines in many developing countries and traditional medicines being locally produced, are readily available. However, several people use these medicines deliberately on account of faith, the local myth, tradition and their socio-cultural milieu. These ethnobotanical plants do not only provide a range of nutritional constituents to supplement the staple diet but also they had the properties of insuring health by providing some medicinal active principles, curing silently any impeding health problem resulting from irregular activities in human conduct.

## IMPORTANT MEDICINAL PLANTS AND THEIR USES

**Abies spectabilis** (D. Don) Mirle (Syn. *A. webbiana* Lindl.) (Pinaceae)

*Local / Common name(s)*: Tosh, Talispatra

*Uses:* The leaves in the form of decoction or infusion are used in chronic bronchitis and other pulmonary affections. Infusion of the leaves is given to infants suffering from fever and chest infection. It is considered poisonous in large doses. It is also reported as antispasmodic, appetizer, astringent, expectorant, carminative, digestive and diuretic.

**Aconitum heterophyllum** Wall. ex Royle (Ranunculaceae)

*Local / Common name(s):* Atvisha, Atish, Patish.

*Uses:* Roots are non-poisonous and are used in diarrhoea, dysentery and gastric pain. A small dose of root powder is given for stomach-ache and if mixed with tobacco, is given to cure cold.

**Arnebia benthamii** (Wall. ex G. Don) Johnston (Boraginaceae)

*Local / Common name(s)*: Ratanjot, Laijari.

*Uses:* The plant is considered expectorant and used for cardiac disorders. Aqueous extracts (syrup, sharbat and jam) prepared from the flowering shoots are considered useful in diseases of the tongue, throat and also useful in fever. Used as a colouring matter in hair oils, cookeries and for dyeing of silk. Root is frequently used as an antiseptic and antibiotic in healing of the wounds and also for subsiding swellings of the organs. Its paste made in water is applied to fire burns and is a quick healer.

**Arnebia euchroma** (Royle) Johnston (Boraginaceae)

*Local / Common name(s)*: Ratanjot, Dimo.

*Uses:* The roots have anti-cancer activity. It has been reported to be an antipyretic, antimicrobial and anti-inflammatory. Roots are used in toothache, ear-ache, eye diseases, as healer of cuts, wounds and fire burns. The purple roots are used as hair tonic by the local people and often dug out for sale.

**Artemisia maritima** L. (Asteraceae)

*Local / Common name(s):* Seinski, Nurcha, Santonin plant.

*Uses:* The drug is used for the loss of appetite, week digestion and stomach complaints. As a carminative, it is one of the strongly aromatic spice herbs used particularly with meat dishes, especially fatty roast goose, where taste and digestibility is improved. Also used in asthma and diseases of brain.

**Betula utilis** D. Don (Betulaceae)

*Local / Common name(s)*: Bhujyarh, Bhojpatra

*Uses:* The bark is useful in bronchitis, diseases of blood and the ear, leprosy etc. It is also antiseptic, carminative, given in anaemia, cough, obesity, urinogenital diseases and various infections. The papery bark was once used as writing paper. The bark of the tree is also used as bandage in bone fracture.

**Bunium persicum** (Boiss.) Fedtsch. (Apiaceae)

*Local / Common name(s):* Kala-jirah, Krishnajeerak.

*Uses:* Its fruits / seeds are used as spice in cooking for flavouring of dishes. The fruits are useful in curing diarrhoea, dyspepsia, fever, flatulence, stomach-ache, haemorrhoids and obstinate hiccup.

**Cannabis sativa** L. (Cannabaceae)

*Local / Common name(s):* Bhang.

*Uses:* Mature leaves and flowering top of the female plants are used medically. Indian hemp mixed with tobacco is used to treat asthma, haemorrhages, tetanus, depressia, bladder inflammation, gonorrhoea and nervous disorder. Seeds are fed to the birds and milk-cattle as a lactagogue. Paste of fresh leaves is used to resolve tumors. Leaf juice removes lice and dandruff. Leaf powder is a useful dressing for wounds or sores. Seeds are roasted and eaten as culinary by the hill people. Bhang leaves mixed with other ingredients make sweet drink which is refreshing in summer season.

**Carum carvi** L. (Family : Apiaceae)

*Local / Common name(s)*: Caraway, Jira.

*Uses:* Fruits are useful for essential oil which is used in the manufacture of 'Kummel', a strong carminative, stimulant and aromatic. It is used in the form of poultices for swellings in the breast and the testicles. It is a mild stomachic and occasionally used in flatulent colic.

**Chenopodium album** L. (Chenopodiaceae)

*Local / Common name(s)*: Bathua.

*Uses:* The young leaves are used as vegetable and given to patients suffering from leucoderma. The juice of the leaves is also applied on the depigmented skin spots for about a month. It is said to be effective in early stages of the disease. The decoction of seeds is given in large doses to induce abortion in women. In small doses, it is supposed to be effective in cough and coryza. The flour prepared out of its seeds is utilized as a catalyst for preparing "soor", a local alcoholic beverage. It is mildly laxative. A tea of the leaves and plant is used to relieve stomach pains.

**Cynodon dactylon** (Linn.) Pers. (Poaceae)

*Local / Common name(s):* Doorva, Doob, Drub, Doob-grass.

*Uses:* Decoction of the root is used as diuretic, in dropsy and in secondary syphilis. Infusion of the root is useful in stopping the bleeding from piles. Crushed roots mixed with curd are used in chronic fleet. Fresh juice of the plant is used to heal cuts and wounds, also useful in dropsy, hysteria, epilepsy, insanity, chronic diarrhoea, dysentery and ophthalmia. It is also regarded anti-abortive, conceptive, demulcent, haemostatic and suppurative. Useful in marks, blemishes, ulcers and wounds. Locally, few drops of the leaf juice are put into nostrils to stop nose bleeding, called *nathar* or *nakseer.* Roots are taken with mishri as memory sharpner by the students and scholars.

**Dactylorhiza hatagirea** (D. Don) Soo (Orchidaceae)

*Local / Common name(s):*Panja, Salampanja.

*Uses:* The roots are used as food, a nervine tonic and aphrodisiac. Mucilage jelly is nutritious and useful in diarrhoea, dysentery and chronic fevers. A decoction of salep with sugar and flavoured with spices is a useful drink for the sick. It is used as medicine in seminal debility, chronic diarrhoea, paralytic affections and general weakness in women after delivery. The root powder is used to relieve fever and against urinary troubles.

**Ephedra gerardiana** Wall. ex Stapf (Gnetaceae)

*Local / Common name(s):* Somlata, Soma, Chhedang.

*Uses:* The drug made from the dried twigs is used in asthma, hay fever and rashes of allergic origin. Liquid extract of the plant is used for controlling asthmatic paroxysms. Juice of berries is used in affections of respiratory passages. Nasal spray prepared from the drug is used in asthmatic attacks and inflammation of mucous membranes. Together with leaves of *Rhododendron hypenanthum* Balf. f., it is used in hydrosele, weakness and palpitation of heart.

The plant contains ephedrine and pseudophedrine. To relieve congestion and swelling, ephedrine salts are prescribed as nasal spray. This is also employed simultaneously to prevent hypotension during anesthesia. Pseudoephedrine is also taken orally which is very effective as nasal decongestant. Moreover, various preparations from *Ephedra* are used to treat malarial and other fevers, colds, coughs and headache etc.

**Fagopyrum esculentum** Moench (Polygonaceae)

*Local / Common name(s):* Ogla, Kathu, Buckwheat.

*Uses:* Due to the presence of the glucoside rutin, it is used in modern medicines as antihaemorrhagic. Decoction of the root is used in rheumatic pains, lung diseases and typhoid, while the juice is useful in urinary disorders. Rutin is valued as a cure for capillary fragility of the heart.

**Heracleum lanatum** Michx. (Family : Apiaceae)

*Local / Common name(s)*: Padara, Patrala.

*Uses:* The xanthotoxin isolated from the plant is highly efficacious in the treatment of leucoderma and psoriasis. It is also used in the preparation of Sun-tan lotions.

**Hippophae rhamnoides** Serv. (Elaeagnaceae)

*Local / Common name(s):* Sarla, Tirku, Seabuckthom.

*Uses:* Fruits are used for preparation of jelly, syrup and used against pulmonary complaints. The plant is also used in tumors, as emollient in prevention of eye region skin wrinkles, in sunburn preventing preparations and other cosmetic preparations.

**Humulus lupulus** Linn. (Cannabaceae)

Local/Common name(s): Hops.

*Uses:* This plant has tonic properties and also regarded as diuretic, diaphoretic and anthelmintic. It stimulates the appetite and helps digestion. It is said to increase vital energy and vigour of the organ. It is used as a sedative and depurative in homeopathic medicine. It is also used for treating fever, intestinal worms and rheumatism. The fruiting catkin is used for brewing.

**Hyoscyamus niger** L. (Solanaceae)

*Local / Common name(s):* Khurasaniajvayan, Henbane.

*Uses:* The drug is useful for relieving certain painful spasmodic conditions of muscles, to counteract the gripping action of purgatives and to relieve spasms in the urinary tracts. It dilates the pupil of the eye and has sedative effects on brain. Also used in nervous affections and irritable conditions, asthma, hysteria and whooping cough. It soothes the cough and induces sleep. Young leaves and fruit tops are the source of drug used in asthma. The seeds are pasted and applied locally on pains. In case of animals, it is used as a urinary sedative.

**Inula racemosa** Hook. f. (Asteraceae)

*Local / Common name(s):* Poshkar, Poskarmul, Pushkarmool, Manu.

*Uses:* The roots possess strong anthelmintic properties. The seeds are bitter and aphrodisiac. Root extract acts as anti-inflammatory, antipyretic, antiseptic and antiasthmatic.

**Jurinea dolomiaea** Boiss. (Asteraceae)

Local/Common name(s): Dhoop, Jari-dhoop.

*Uses:* The roots are given as stimulant in fever after child birth. A decoction of the root is given in colic. The bruised roots are applied to skin eruptions. Aromatic oil from the roots is useful in gout and rheumatism. The aromatic roots are used as incense and form a chief ingredient of dhoop industry.

**Mentha longifolia** (L.) Huds. (Lamiaceae)

*Local / Common name(s):* Jungli pudina, Horsemint.

*Uses:* The leaves are used for rheumatic pains and painful swollen joints. A decoction of the plant is used in fever and heat apoplexy. The plant is taken in the form of chutney. The oil is used as a snuff to remove headache. A poultice placed over the lumber and renal region acts as a good diuretic and emollient. It is useful in the stomach disorders, dyspepsia, vomiting and pain in the liver or womb. The plant is stated to be useful in scorpion-sting and snake-bites.

**Ocimum tenuiflorum** Linn. (Lamiaceae)

*Local / Common name(s):* Tulsi, Basil.

*Uses:* The oil extracted from leaves and tender parts of shoots possess antibacterial and insecticidal properties. Seeds are useful in complaints of the urinary systems. The juice or infusion of the leaves is used in catarrh and bronchitis, applied to skin in ringworm and other cutaneous diseases and dropped into the ear to relieve earache. An infusion of the leaves is used as stomachic in gastric disorders of children. A decoction of the roots is given as a diaphoretic in malarial fevers, while the decoction of leaves is given in common colds.

**Picrorthiza kurooa** Royle ex. Benth. (Scrophulariaceae)

*Local / Common name(s):* Kutki, Karru, Tikta.

*Uses:* Root and stem are considered to be anthelmintic, blood purifier, blood pressure reducer, cardiac, expectorant and is useful in dropsy, asthma, arthritis, urticaria, cold and cough, bile trouble, jaundice, leprosy, constipation and stomach troubles. Leaves are crushed and 1-2 drops of the juice are put in the nose to stop bleeding. Rhizomes of the plant are administered against fever and gastralgia. The powdered root-stock is used in abdominal pain, liver complaints, anaemia and jaundice.

**Pinus wallichiana** A.B. Jackson (Pinaceae)

*Local / Common name(s)*: Kail, Blue pine.

*Uses:* Its oil has antibacterial property. The resin of young saplings is used for healing the cuts and wounds. The bark of pole size plants is used as a bandage/plaster on the dislocated/ fractured bones of humans, cattle, sheep and goats. The resin and cuticular bark tied on the organ keeps the bones in place and facilitates calcification and unification of fractural pieces.

**Plantago major** Linn. (Family : Plantaginaceae)

*Local / Common name(s):* Yakram, Jharam Naram, Isabgol.

*Uses:* Seeds are used in gastric complaints, burning sensation in stomach and dysentery. Leaves are applied to bruises.

**Podophyllum hexandrum** Royle (Podophyllaceae)

*Local / Common name(s):* Bankakri, Giriparpat, Rikhpat, Papra.

*Uses:* The resin or podophyllin is given with Belladona and *Hyoscyamus* to avoid gripping. Podophyllin has been tried on various diseases like warty lesions of the skin and neoplasms of the body parts. It is an effective vermifuge, first stimulating, then paralyzing and finally killing ascaris. The limiting factor has been the severe gastro-intestinal discomfort followed by its use in high doses. The cytotoxic effect causes mitotic arrest, nuclear fragmentations and other cell damage. Podophyllotoxin and peltatins have high potency in reducing damage in sarcoma, carcinoma and melanoma (Cancerous compounds). The tumor damaging potency of the podophyllin compounds is said to be closely associated with their stereo-chemistry.

**Saussurea costus** (Falc.) Lipsch. (Asteraceae)

*Local / Common name(s):* Kuth, Kooth, Costus.

*Uses:* It is useful in bronchial asthma, cough, colic, bronchitis, dental trouble, diarrhoea, dysentery, fever, flatulence, headache, hiccough, hysteria, in some cases of heart troubles, menstruation trouble, rheumatism and promotes urination. It is regarded useful as hair-wash, against marks and blemishes and also in skin diseases.

**Taraxacum officinale** Weber (Asteraceae)

*Local / Common name(s):* Dulal, Kanphul, Dudhli.

*Uses:* The drug promotes the secretion of gall and is a useful remedy for chronic disorders of kidney and liver. The juice of the fresh plant is effective against liver diseases, chronic hepatitis, visceral congestion, intermittent fever and hypochondria. Roots are used to increase urine flow, as a laxative and tonic, to treat liver and spleen ailments and to stimulate appetite. Tea made by boiling flowers is used to treat heart trouble. Cooked young leaves are eaten to purify blood. The whole plant is crushed into a mesh and given internally in snake bites. The paste is applied externally on the wound. Leaves are effectively used for the remedy of swollen parts, boils and sprains.

**Taxus baccata** L. ssp. **wallichiana** (Zucc.) Pilger (Taxaceae)

*Local / Common name(s):* Thuno, Thuner, Birmi.

*Uses:* The young shoots are used for the treatment of headache, giddiness, feeble and falling pulse, coldness of the extremities, diarrhoea and severe biliousness. The leaves are employed for the treatment of hysteria, epilepsy and nervousness. 'Taxol' obtained from the needles is reported to cure uterine and breast cancers.

**Thymus linearis** Benth. (Lamiaceae)

*Local / Common name(s)*: Banajwain.

*Uses:* The plant contains volatile oil from which an antiseptic is extracted. It is effective for the treatment of nervous disorders. Infusion of plant is used in skin diseases. Powder of the flowers is mixed in gur and given as a vermicide. The decoction of the plant is used as an efficacious home remedy in cold, cough, fever and stomach problem. The leaves are good for kidney and eye and regarded as blood purifier. The leaves and floral shoots are employed for the treatment of whooping cough, epilepsy, suppression of urine and menstrual catarrh. An infusion of the leaves is used in the treatment of itch and skin eruptions. The leaves are used in biliousness and their juice considered emetic. The seeds are used in the hills for the remedy of stomach complaints, cough and colds.

**Verbascum thapsus** L. (Scrophulariaceae)

*Local / Common name(s):* Bhootlankra, Jangli tambaku.

*Uses:* The crushed leaves are made into pills and given in constipation and allied stomach pains. Decoction of leaves along with onion is useful for stomach problems. The infusion of plant is given in snake-bite; a paste of leaves is also applied locally on the bitten part. The crushed leaves are warmed in mustard oil and used as massage for curing inflammations of the body. Leaves are also useful for local applications in haemorrhoids and sun burn. Tea prepared from leaves is used to treat cold and dysentery.

## ACKNOWLEDGEMENTS

The authors express hearty gratitude to Dr. Nisha Mehandiretta, Scientist-D,

Department of Science and Technology, New Delhi for providing financial assistance. Sincere thanks are due to Dr. B.S. Dhillon, Director, N.B.P.G.R, New Delhi and Dr. V.D. Verma, Officer-in-Charge, N.B.P.G.R., Regional Station, Shimla for providing necessary facilities.

## REFERENCES

BHATTACHARJEE, S.K. 1998. *Handbook of Medicinal Plants.* Pointer Publishers, Jaipur.

BHATTARAI, N.K 1998. Traditional medicines & global prospectives. In Gautam, P.L., R. Raina, U. Srivastava, S.P. Raychaudhuri & B.B. Singh (Eds.) *Prospects of Medicinal Plants,* pp. 96-104. ISPGR, New Delhi.

CHANDEL, K.P.S., G. SHUKLA & N. SHARMA 1996. Biodiversity in Medicinal and Aromatic Plants in India, Conservation and Utilisation. NBPGR, New Delhi.

CHAUHAN, N.S. 1999. *Medicinal and Aromatic Plants of H.P.* Indus Publishing Co., New Delhi.

CHOPRA, R.N., S.L. NAYAR & I.C. CHOPRA 1996. *Glossary of Indian Medicinal Plants.* CSIR, New Delhi.

DUTT, U.C. 1995. *The Materia Medica of the Hindus.* Mittal Publications, New Delhi.

KARNICK, C.R. 1994. *Pharmacopoeial Standards of Herbal Plants,* Vol. I-II. Sri Satguru Publications, Delhi.

KIRTIKAR, K.R. & B.D. BASU 1975. *Indian Medicinal Plants,* Vol. I-IV. Bishen Singh Mahendra Pal Singh, Dehra Dun.

KUMAR, R. & R. DHAN 1998. Traditional medicines and global prospectives. In Gautam, P.L., R. Raina, U. Srivastava, S.P. Raychaudhuri & B.B. Singh (Eds.) *Prospects of Medicinal Plants,* pp. 92-95. ISPGR, New Delhi.

SHARMA, S.K. & K.C. CHUNEKAR 1998. *Medicinal Plants Used in Ayurveda.* RAV Publications, New Delhi.

SRIVASTAVA, J., J. LAMBERT & VIETMEYER 1995. *Medicinal plants : A growing role in development.* Agriculture and Natural Resources Department, Agriculture and Forestry System, The World Bank Washington, D.C.

TEWARI, D.N. 1994. *Important Plants of India.* International Book Distributors, Dehra Dun.

# IMPORTANT MEDICINAL PLANTS IN MOOLBARI WATERSHED AREA OF HIMACHAL PRADESH

ARCHNA SINGH, YOGENDRA SHARMA, ARVIND KUMAR & J.C. RANA

*National Bureau of Plant Genetic Resources, Regional Station, Phagli, Shimla (H.P.) 171 004*

## ABSTRACT

The Moolbari watershed is located in district Shimla on Shimla-Bilaspur highway about 12 km away from Shimla. It lies between 77°05' to 77°54' E longitude and 31°09' to 31°11 N latitude and at an altitudinal range of 1400-2000 m in Himachal Pradesh. Himalaya with its varied climate, soil and different land forms support a rich medicinal plant heritage, which has been used from time immemorial as herbal remedy for amelioration of people sufferings. In the present communication about 66 species have been recorded which are widely used by the locals against various ailments.

## INTRODUCTION

Plants have always played a major role in the treatment of human traumas and diseases. The existence of man and animals on this planet has been directly related with the vital role played by the plant kingdom in sustaining life in symbiotic manner. Plants have been used in prevention and cure of various diseases for maintenance of health from time immemorial in addition to food, clothing and shelter. Plants with medicinal properties enjoyed the highest reputation in the indigenous system of medicine all over the World, and still constitute one of the major sources of drugs in modern as well as traditional systems of medicine. India possesses the richest source of medicinal plants in the World. Over 9000 economic plants in the World are recorded, of which India has 5000 species of economic importance with a little over 1500 species of medicinal importance (Agarwal & Ghosh, 1985).

There are large number of plants, which are being used medicinally in villages, need documentation. User people do not divulge the information, and it is difficult to determine the authenticity of such ethno botanical information. However, such information opens up new vistas in research and may provide valuable new drug sources. The major concern is that about 95% of the medicinal plants are collected from the wild populations and over 70% of the plant collections involve destructive harvesting mainly because of the use of the plant parts like roots, bark, wood and whole plant (Tiwari, 1994b). For developing conservation strategies, it is essential to study their extent of occurrence and assess their degree of threat. The information endorsed in this paper was collected during the floristic studies in the area. The study is based on field work, review of available literature, and partly on our personal observations and interviews with local people. Important literature consulted is cited under references for economy of spaces. Data is furnished in tabular form and plant names alphabetically. Besides medicinal uses, local/ English names and habit and habitat are also provided.

## ENUMERATION

| Botanical name/ Family | English/ Local name | Habit/Habitat | Uses |
|---|---|---|---|
| *Achyranthes aspera* L. Amaranthaceae | Latjira, Prickly chaff flower | A herb | Useful in dropsy, piles, boils, eruptions of the skin etc; the seeds and leaves are considered emetic and are used in hydrophobia and snake-bite. |
| *Aesculus indica* (Wall. ex Camb.) Hook. f. Hippocastanaceae | Horse-chestnut | Common tree in himalayan forests | Fruits used to cure colic pain of horses; oil extracted from seeds is used to cure rheumatism; bark is used to cure intermittent fever; leaves are used to cure whooping cough. |
| *Anagallis arvensis* L. Primulaceae | Scarlet pimpernel, Nilphuli | Small annual herb | Used as diaphoretic, diuretic, resolvent for gall stones, liver cirrhosis, lung ailments, also used in cerebral affections, leprosy, hydrophobia, dropsy and in nervous disorders. |
| *Asparagus adscendens* Roxb. Liliaceae | Satavar | Scandant prickly shrub, wild in forests | Root bark powder is aphrodisiac, used as tonic; root powder is given in diarrhoea, dysentery and general debility. |
| *Barleria cristata* L. Acanthaceae | Tadrelu | Herb, wild in forests, also grown as ornamental in gardens | Seeds are used as antidote for snake-bite; roots are used in anaemia; roots and leaves are used to reduce swellings and an infusion is given in cough. |
| *Bauhinia variegata* L. Caesalpiniaceae | Mountain ebony, Kachnar | Tree, in forests and in gardens. | Bark is used in skin diseases, ulcers, scrofula, buds in piles. |
| *Berberis aristata* DC. Berberidaceae | Rasaut, Chilra | Spiny shrub with yellow wood, wild on hills and in forests. | Juice of root bark is used in skin diseases, diarrhoea and jaundice; decoction of root bark is used in intermittent fever; root extract, known as rasaut, is highly esteemed as a febrifuge and as local application in eye diseases. |
| *Berberis asiatica* Roxb. ex DC. Berberidaceae | Kilmora | Spinous shrub, wild on hills and in forests. | --do-- |

| | | | |
|---|---|---|---|
| *Bergenia ligulata* (Wall) Engl. Saxifragaceae | Pashanbheda | A rhizomatic perennial herb, on rocky and shaddy places | Rhizomes and roots taste bitter, act as astringent, diuretic, tonic, also used in fever and applied to boils and opthalmia; it is an important drug for dissolving kidney and bladder stones; leaves are kept on head under the cap for relieving fever. |
| *Boerhavia repens* L. var. *diffusa* (L.) Hook. f. Nyctaginaceae | Punarnava, Horse-purslane | Diffusely branched herb, weed of gardens and waste places. | Root decoction is good for jaundice, bronchitis and leaf juice is used in opthalmia and eye wounds. |
| *Buddleja asiatica* Lour. Buddlejaceae | Butterfly bush, Neema | Shrub with thin grey bark, wild in forests, often cultivated. | Flowers are used in skin diseases and as an abortifacient. |
| *Cannabis sativa* L. Cannabaceae | Indian hemp, Bhang | Tall annual herb, wild in waste places and by road sides, cultivated also. | Flowers are used as stomachic, used in pain and as an aphrodisiac. |
| *Capsella bursa-pastoris* (L.) Medic. Brassicaceae | Shepherd's purse | Hairy herb, grows as a weed of cultivation. | Plant is diuretic, used in diarrhoea. |
| *Carissa opaca* Stapf ex Haines Apocynaceae | Jungli karaunda | Evergreen shrub | Root is ground and applied in various infected sores of animals; also enters into the composition of purgatives. |
| *Cedrus deodara* (Royle ex D. Don) G. Don Pinaceae | Devdaru, Devdar | A large evergreen coniferous tree | The wood is carminative, diaphoretic and diuretic, given in fever, gastric problems, heart palpitations, paralysis, pulmonary troubles, dropsy and in urinary diseases like gravel. |
| *Celtis australis* L. Ulmaceae | Southern nettle tree, Bingal | Tree, in forests, also planted. | Fruits are used in amenorrhoea and colic. |
| *Chenopodium album* L. Chenopodiaceae | Bathua | Herb, found as a weed of cultivated fields and waste places. | Plant is good blood purifier and oil kills stomach worms; leaf and root are used in anaemia. |
| *Cissampelos pareira* L. Menisperamaceae | Velvet leaf, Akanadi | Climbing shrub, wild in forests. | Roots are used in the stomach pain, diarrhoea, dropsy, cough, urinary troubles and as antidote for snake-bite. |
| *Commelina benghalensis* L. Commelinaceae | Kanchara | A straggling herb | Used in skin inflammations. |
| *Convolvulus arvensis* L. Convolvulaceae | Small bindweed, Hiranpadi | Herb with a creeping root stock, common on sandy or rocky ground | Roots are used as purgative. |

| | | | |
|---|---|---|---|
| *Curcuma longa* L.<br>Zingiberaceae | Turmeric,<br>Haldi | Perennial herb | Used as carminative and aromatic stimulant; powdered root stock is used externally for recent bruises and wounds. |
| *Cynodon dactylon* (L.) Pers.<br>Poaceae | Couch grass,<br>Doob grass | Perennial grass, extremely variable in habit, in lawns, gardens or in forests. | Roots are used in piles and syphilis; plant is astringent, diuretic, used in cuts, wounds and opthalmia. |
| *Cyperus rotundus* L.<br>Cyperaceae | Nut grass,<br>Motha | Herb, found as a weed of wastelands and moist places. | Root is astringent and used in jaundice, bowel complains and snake-bite. |
| *Datura metel* L.<br>Solanaceae | Dhattura | An annual herb | The plant is beneficial in bronchial asthma, cough, fever, diarrhoea and menorrhagia; seeds are poisonous; the paste of roots mashed in cow's urine or the juice of the leaves is applied externally to alleviate oedema and pain in sciatica, mumps, lumbago, neuralgia etc; leaf paste mixed with lukewarm mustard oil is tied over the joints to get relief from rheumatism. |
| *Datura stramonium* L.<br>Solanaceae | Dhatura | An annual herb | It is narcotic, antispasmodic and is used chiefly to relieve the spasm of the bronchi in asthma; the leaves are applied to boils and sores; the juice of flowers is used for ear ache; the juice expressed from the fruits is applied to the scalp for curing dandruff and falling of hair. |
| *Dioscorea deltoidea* Wall. ex Griseb.<br>Dioscoreaceae | Kildari | Tuberous herb, cultivated and wild in forests. | Rhizome is used in the preparation of various steroidal drugs; leaves are used as febrifuge; decoction of leaves is used to kill lice, also used as fish poison. |
| *Euphorbia hirta* L.<br>Euphorbiaceae | Dudhi | Herb, common in waste places, roadsides and gardens | Plant anthelmintic, also used for boils, cough, diarrhoea, dysentery, lactation and skin diseases. |
| *Euphorbia royleana* Boiss.<br>Euphorbiaceae | Sihund, Surai | Glabrous cactus like shrub, wild on exposed rocky slopes. | Latex is cathartic and used for destroying or expelling worms. |
| *Fumaria indica* (Haussk.) Pugsley<br>Fumariaceae | Pitpapra | Pale green, diffused herb, a common weed of cultivated fields | Plant is used in fever and influenza. |
| *Geranium nepalense* Sweet<br>Geraniaceae | Laljari | A perennial diffuse herb | Plant is considered as an astringent and also used in the disease of kidneys; due to red colouring matter, it is used to colour various medicinal preparations and oils. |
| *Grewia optiva* Drumm. ex Burret<br>Tiliaceae | Phalsa | Small tree | Root bark is used in rheumatism; fruits have astringent, stomachic and cooling properties. |

| | | | |
|---|---|---|---|
| *Hypericum perforatum* L. Hypericaceae | Saint johnswort | Perennial herb | Used in diseases of the digestive tract, nerve pains, gout, stomachache and kidney troubles, against intestinal worms and for healing wounds. |
| *Ipomoea nil* (L.) Roth Convolvulaceae | Kaladana | Herb, wild in hedges and waste places, also occurs as a weed of cultivated fields. | Seeds used as purgative and substitute of jalap. |
| *Juglans regia* L. Juglandaceae | Akhrot | A large deciduous tree. | The bark is used as an anthelmintic and detergent; the leaves are astringent and tonic; the fruit is also believed to have an alterative effect in rheumatism. |
| *Leucas aspera* (Willd.) Link Lamiaceae | White deadnettle, Gumma | Annual herb, common in fields, along roadsides and in forests. | Plant is used as insecticidal; twig is used as toothbrush; flower is used in cold. |
| *Leucas cephalotes* (Roth) Spreng. Lamiaceae | Gumma | Pubescent herb, common weed in cultivated fields. | Plant is used as stimulant, diaphoretic; fresh-juice is used in scabies. |
| *Martynia annua* L. Martyniaceae | Devil's claw | Herb, along road sides in waste places. | Plant used in scorpion sting; leaf antiseptic and in tuberculosis. |
| *Mirabilis jalapa* L. Nyctaginaceae | 4 o'clock plant, Rangbano | Herb with tuberous roots | Used as an aphrodisiac and purgative and for abscesses, boils, bruises, inflammation and itching. |
| *Oxalis corniculata* L. Oxalidaceae | Tipatia | Procumbent herb, common in fields, gardens and wastelands. | Fresh plant juices cures indigestion, piles, anaemia and tympanitis. |
| *Phyllanthus amarus* Schum. & Thonn. Euphorbiaceae | Jar-amla | Glabrous herb, weed of fields and cultivated lands. | Plant used as diuretic, astringent, antidropsical, in dyspepsia and colic. |
| *Physalis minima* L. Solanaceae | Sun berry, Jangli rasbhari | Herb, common weed of irrigated fields and bunds. | Fruits used as purgative, diuretic, tonic and in gonorrhoea. |
| *Pinus roxburghii* Sarg. Pinaceae | Chir, Pine | A moderately tall tree | The turpentine oil is valued in medicines; it is feebly antiseptic and as expectorant, useful in chronic bronchitis; it is especially recommended in the treatment of gangrene of the lungs; as enema, the oil is useful in obstinate constipation, tympanitis and seatworm infections; externally, it is used as rubefacient in various thematic affections, such as lumbago, arthritis and neuralgia. |
| *Plantago major* L. Plantaginaceae | Isabgol, Yakram | A small herb | Seeds are used in gastric complaints, burning sensation in stomach and dysentery; leaves are applied to bruises. |
| *Polygonum plebeium* R. Br. Polygonaceae | Mahichi | Diffuse, prostrate herb, common in moist places | Plant is antidiarrhoeal, used in pneumonia. |

| | | | |
|---|---|---|---|
| *Princepia utilis* Royle Rosaceae | Bhekhal | A thorny shrub | Kernel oil is used for cooking purposes and as illuminant in the hills; also reported suitable for hydrogenation and soap making; the oil is applied externally in rheumatism and pains resulting from over fatigue. |
| *Prunus cerasoides* D. Don Rosaceae | Pajja | A middle sized tree | It is considered as an anti-abortifacient, analgesic, carminative, conceptive, expectorant, febrifuge and tonic, useful in stomach trouble, burning sensation of the body, cold and cough. |
| *Rubia cordifolia* L. Rubiaceae | The Indian maddar, Manjeeth | Prickly climber, common in forest of hilly regions. | Root is used as tonic, astringent; stem is antidote for cobra bite and scorpion sting. |
| *Rumex nepalensis* Spreng. Polygonaceae | Jangli palak | Robust herb, common in forest hills | Roots are purgative. |
| *Solanum anguivi* Solanaceae | Poison berry, barhanta | Common spiny herb with sinuate leaves, weed of wastelands and fields. | Root is used as carminative and as an agent that relieves vomiting, antiemetic, expectorant, in colic and toothache. |
| *Solanum nigrum* L. Solanaceae | Black night shade, Makoi | Suffrutescent herb, common weed of wastelands, agricultural fields. | Berry used in diarrhoea, eye diseases, sore throat, hydrophobia; plant is used in liver disorder and piles; leaf is used as diuretic. |
| *Solanum virgini-anum* L. Solanaceae | Kantakari, Kateli | Prickly diffuse herb | Root is much asteemed as an expectorant in cough, asthma, catarrhal fever and chest pain. |
| *Sonchus oleraceus* L. Asteraceae | Milky thistle, Dudhi | Herb with perennial creeping root stock, weed of wastelands and fields. | Gum is used as hydragogue, cathartic; root and leaf are used as tonic, febrifuge. |
| *Stellaria media* (L.) Vill. Caryophyllaceae | Common chick weed, Kokoon | Herb, common weed of wastelands, gardens and fields. | Plant is applied as a plaster for broken bones and swellings. |
| *Taraxacum officinale* Weber Asteraceae | Blow bael, Dulal | Scapigerous herb, having milky juice in all parts, wild in hills and forests. | Root is used as diuretic, as stomachic, as tonic, laxative and in kidney and liver disorders. |
| *Thalictrum foliolosum* DC. Ranunculaceae | Pilzari | Erect, rigid herb, wild in hilly and coastal region. | Roots are used in diarrhoea, opthalmia, in indigestion as tonic, purgative. |
| *Thymus linearis* Benth. Lamiaceae | Banajwain | Prostrate and evergreen perennial herb | The herb possesses antispasmodic, antiseptic, expectorant, carminative, anthelmintic and stimulating properties; the leaves and floral shoots are used for whooping cough, epilepsy, suppression |

| | | | |
|---|---|---|---|
| | | | of urine; an infusion of the leaves is used in the treatment of itch and skin eruptions; it is a popular remedy in hills for stomach complaints, cough and colds. |
| *Toona hexandra* (Wall. ex Roxb.) Roem. Meliaceae | Indian mahogany tree, Toon | Large deciduous tree, in forests and also grown in gardens. | Bark is used in dysentery, as tonic and astringent. |
| *Trifolium pratense* L. Fabaceae | Trepatra | Hairy plant, wild in hilly tracts. | Flower is used in asthma, whooping cough and bronchitis. |
| *Trifolium repens* L. Fabaceae | Shaftal | Creeping herb, naturalized in hills | Flower is used in gout, as astringent; plant is toxic. |
| *Urtica dioica* L. Urticaceae | Common stinging nettle, Bichhubuti | Robust, dioecious herb with stinging hairs on stem, wild in hilly region. | Plant is used as diuretic, astringent, in jaundice, nephritic troubles, haemorrhages from kidney, uterus; flower and seed are used as tonic. |
| *Vernonia cinerea* (L.) Less. Asteraceae | Purple fleabane, Sahadevi | Erect herb, common weed of wastelands and gardens. | Plant is used as tonic in stomachic, astringent, diaphoretic, useful in constipation; root used as antidiarrhoeal. |
| *Viola pilosa* Bl. Violaceae | Banfshah | A glabrous herb | Useful in asthma, bleeding piles, cancer of throat, cough, fever, constipation, headache and skin diseases. |
| *Woodfordia fruticosa* (L.) Kurz Lythraceae | Fire flame bush, Dhai | Large shrub with long arching branches, wild in forest tracts. | Dried flowers used in disorders of mucous membrane, safe stimulant in pregnancy, as astringent tonic. |
| *Zanthoxylum armatum* DC. Rutaceae | Tirmir, Tumbru | An effect prickly shrub | Seeds and bark are used as aromatic tonic in fever, indigestion and cholera; fruit, branches and thorns are used as a remedy for toothache and considered carminative, stomachic and anthelmintic; extract of fruits is useful in expelling round worms; because of their deodrant, disinfectant and antiseptic properties, the fruits are used in dental troubles and their lotion for scabies. |
| *Ziziphus mauritiana* Lamk. Rhamnaceae | Indian jujube, Ber | Small, armed tree, wild in forests, also planted. | Bark powder is used in dysentery and bark paste in boils; bark and leaf are antipyretic; flowers are used in eye troubles. |

## ACKNOWLEDGEMENTS

We express our sincere thanks to Dr. Nisha Mendiratta, Scientist-D, Department of Science and Technology, New Delhi for providing financial assistance. We also thank Dr. B.S. Dhillon, Director, N.B.P. G.R, New Delhi and Dr. V.D. Verma,

Officer-in-Charge, N.B.P.G.R., Regional Station, Shimla for providing necessary facilities.

## REFERENCES

AGARWAL, V.S & R. GHOSH 1985. *Drug Plant of India (Root drugs).* Kalyani Publishers, New Delhi.

BHATTACHARJEE, S.K. 1998. *Handbook of Medicinal Plants.* Pointer Publishers, Jaipur.

CHANDEL, K.P.S., G. SHUKLA & N. SHARMA 1996. Biodiversity in Medicinal and Aromatic Plants in India : Conservation and Utilisation. NBPGR, New Delhi.

CHAUHAN, N.S. 1999. *Medicinal and Aromatic Plants of H.P.* Indus Publishing Co., New Delhi.

CHOPRA, R.N., S.L. NAYAR & I.C. CHOPRA 1996. *Glossary of Indian Medicinal Plants.* CSIR, New Delhi.

JAIN, S.K. 1981. *Glimpses of Indian Ethnobotany.* Oxford & IBH Publishing Co., New Delhi

———— & V. MUDGAL 1999. *A Handbook of Ethnobotany.* Bishen Singh Mahendra Pal Singh Publications, Dehra Dun, India.

KIRTIKAR, K.R. & B.D. BASU 1975. *Indian Medicinal Plants,* Vol. I-IV. Bishen Singh Mahendra Pal Singh, Dehra Dun.

KUMAR, R. & R. DHAN 1998. Traditional medicines and global prospectives. In Gautam, P.L., R. Raina, U. Srivastava, S.P. Raychaudhuri & B.B. Singh (Eds.) *Prospects of Medicinal Plants,* pp. 92-95. ISPGR, New Delhi.

SHARMA, O.P. 1976. Some useful wild plants of Himachal Pradesh. H.P.U., Palampur.

SHARMA, S.K. & K.C. CHUNEKAR 1998. *Medicinal Plants Used in Ayurveda.* RAV Publications, New Delhi.

TEWARI, D.N. 1994a. *Important Plants of India.* International Book Distributors, Dehra Dun.

———— 1994b. Medicinal plants for healthcare. *Yojana* 44 : 8-17.

# *IPOMOEA CARNEA* SUBSP. *FISTULOSA* (CONVOLVULACEAE) - A FRIEND PLANT OF TRIBALS

KAPTAIN K. BAJPAYEE

*Centre for Research in Ethno & Medicobotany, G. F. College (Rohilkhand University), Shahjahanpur-242 001, India*

*(Postal Address : Near Phoolmati Temple, Whiteganj, Hardoi-241001, India)*

## ABSTRACT

In the present investigation, ethnobotany of *Ipomoea carnea* Jacq. subspecies *fistulosa* (Mart. ex Choisy) Austin (Convolvulaceae) has been discussed. The data were recorded in reference to tribal communities of Kanjars, Nuts, Loharbadhaiya, Tharus and Kalabaz etc of Eastern Uttar Pradesh. The random survey and enquiries were made to draw the various uses of plant in welfare of tribals life.

## INTRODUCTION

Ethnobotany is the study of plants of past and present in reference to aboriginals and tribal races of the respective places. After the contributions of (Jain 1981, 1987, 1990) in the field of ethnobotany, a number of books and research papers have been published and during recent years due to growing interest in this field, there has been a flood which is hard to cite here for economy of space.

A perusal of literature showed that ethnobotanical aspect of *Ipomoea carnea* Jacq. subspecies *fistulosa* (Mart. ex Choisy) Austin have not been recorded sofar. Therefore, the present investigation was taken to draw the ethnobotanical studies on its different aspects. The uses recorded in the present paper are new and have not been recorded earlier.

## METHODOLOGY

During project work (1998-2002), the author surveyed the various communities of plants in relation to tribals. The inverviews were made with various tribal persons of the communities like Kanjars, Nuts, Loharbadhaiya, Banjara, Tharus and Kalabaz etc in reference to the ethnobotany of different plants. The enquiries were repeated in other communities to draw the actual inference regarding ethnobotany of the plant. Observations are present in this paper.

## OBSERVATIONS

### Root

The accessory roots of plant are used as brush and broom by tribal women. An itching substance is formed when the root portion becomes submerged in water upto a long period in fibrous ball-shaped structures, called "Khajuhara". These balls after drying are used to cure as irritant in arthritis. It is also used by many tribals as the itching, irritation agent for enemies. The cure of irritation is mustard oil, when applied on affected area.

### Stem

The unbranched stem rods are used for various purposes in tribal life, like in thatching, hedge preparation, fence making, as fuel in cooking etc. Sometimes sold in tribal markets as the support in hut-making, thatching and roof making. The stem is also used as a supportory stick for old persons, called Lathiya. The young stems when dry up, become hallow; such stems are used as Cigar or Bidi to smoke through nostrils in the treatment of migraine.

### Leaves

The leaf poultice is applied overnight on boils. The dorsal surface of leaf is covered by thin film of mustard oil at warm state and the leaf is applied over boils to cure them It is also used in internal hits and sprains. Further, the paste of *Diospyros melanoxylon* Roxb. var. *tupru* (Buch.-Ham.) Singh fruit pulp and Opium in the ratio of 20:1 is spread on the dorsal surface of *Ipomoea* leaf and applied on boils originated from the infection of hair follicles.

### Flower

The delicate petals are chewed to cure anaemia and blood loss. The pollen grains are collected in mouth at early morning and used as mouth wash and throat wash. The use of pollen grains in aphtha is also heared from some tribal persons.

### Fruit and Seed

The fruit latex is used for adhesive purposes. It is also used as adulterant of opium. The seed-paste with equal amount of Linseed paste is applied over boils to cure them.

## ACKNOWLEDGEMENTS

The author is grateful to all tribal communities and persons who took part in survey work and provided information. For valuable suggestions, the thank goes to Dr. S.C. Sharma, Head, Dept. of Botany and Dr. S.N. Agnihotri; the Principal, G.F. College, Shahjahanpur.

## REFERENCES

JAIN, S.K. 1981. *Glimpses of Indian Ethnobotany*. Oxford & IBH Publishing Co., New Delhi.

———— 1987. *A manual of Ethnobotany*. Scientific Publishers, Jodhpur.

———— 1990. *Contribution of Indian Ethnobotany*. Scientific Publishers, Jodhpur.

# MEDICINAL USES OF SOME TERRESTRIAL EXOTIC WEEDS

P.R. SUR, A.M. SAREN & A.C. HALDER

*Central Botanical Laboratory, Botanical Survey of India, P. O. Botanic Garden, Howrah-711 103*

**ABSTRACT**

The paper deals with 25 terrestrial exotic angiospermic weeds which have some medicinal values. The vernacular name, country of origin, distribution in India and important medicinal uses are discussed.

## INTRODUCTION

According to Sir J.D. Hooker (1904) the Indian flora is more varied than that of any other country in the World. It is stated that about 17,000 species (15,000 Angiosperm, 64 Gymnosperm, 1226 Pteridophytes) of vascular plants are available in the Indian region. It is also observed that the Indian flora is the confluence of plants from the neighbourhood countries like Malaya, Tibet, China, Japan, Europe and wide spread countries like America, Africa, Australia etc. In this context, a good number of alien weeds have naturalized in India and have become a permanent part of the Indian flora. Nair (1977) mentioned that the Indian flora has 35% South East Asian and Malayan, 8% temperate, 1% Steppe, 2% African and 5% Mediterrranean elements. It is estimated that establishment of exotic weeds started in India since the time of Portugues. The Portugues introduced many economically important plants from Brazil, Mexico, Africa and other parts of the World. A number of ornamental plants and some medicinal plants have been introduced by the Britishers. Besides this, the seeds of some obnoxious weeds have come to Indian soil through import of goods and tourists. Now many of them have become the part of folk medicines and of the house hold remedies. It is interesting to note that some neotropical weeds like *Parthenium hysterophorus* Linn., *Chromolaena odorata* (Linn.) King & Robinson, *Mikania micrantha* Kunth, *Xanthium strumarium* Linn., *Tridax procumbens* Linn. Etc, which are very aggressive and harmful to native flora, have make their place in folk life. So, it will be very useful to record the therapeutic uses of such weeds.

The authors during their research career, particularly botanical exploration and documentation of ethnobotanical date in some parts of country recorded that exotic weeds are rather less known to the folk for therapeutic values. In the following account such 25 terrestrial angiospermic exotic weeds belonging to 15 families and 25 genera have been discussed. The plant species are arranged alphabetically by their botanical names, followed by families in parenthesis. Vernacular names, country of origin, distribution in India and important medicinal uses are provided. Abbreviations used under Vernacular names (*Vern. name*) are: S=Sanskrit, B=Bengali, H=Hindi, E= English.

## ENUMERATION

**Abelmoschus moschatus** Medic. (Malvaceae)

*Vern. name* : Latakustarika (S), Mushakdana (B), Kasturidana (H).

Plant of paleotropical region; cultivated throughout hotter parts of India; also occurs as a weed.

Seeds are stimulant, antispasmodic, stomachic, carminative and aphrodisiac.

**Achyranthes aspera** Linn. (Amaranthaceae)

*Vern. name* : Apamarga (S), Apang (B), Latjira (H).

Tropical American herb; occurs throughout India as a weed upto 915 m above msl.

Plant is emetic, purgative and diuretic. It is used in piles, skin diseases, boils and also for abortive purpose.

**Adhatoda zeylanica** Medic. (Acanthaceae)

*Vern. name* : Vasaka (S), Basak (B), Adulasa (H).

Tropical Asian plant; occurs as weed all over India upto 1220 m above msl.

Plant is antiseptic and also used in rheumatism.

**Ageratum conyzoides** Linn. (Asteraceae)

*Vern. name* : Dochunty (B).

South American herb; occurs as a weed throughout India upto 1800 m above msl.

Plant is anthelmintic, antilithic, styptic and applied on fresh cuts and wounds.

**Amaranthus spinosus** Linn. (Amaranthaceae)

*Vern. name* : Tanduliya (S), Kanta-notya (B), Katailichaulai (R).

Tropical American herb; occurs as a weed throughout India.

Root is used in menorrhagia, gonorrhoea, eczema, lactagogue and plant as a laxative.

**Argemone mexicana** Linn. (Papaveraceae)

*Vern. name* : Srigalakanta (S), Shial-Kanta (B, H).

Central American (Mexico) herb; occurs throughout India as a weed upto 1500 m above msl.

Plant is laxative, emetic, expectorant, demulcent, alterative and also used in chronic skin diseases and dropsy.

**Boerhavia diffusa** Linn. (Nyctaginaceae)

*Vern. name* : Punar-nave (S), Punarnaba (B), Sant (H).

Pantropical herb; occurs as a weed throughout India upto 2000 m above msl.

Root is diuretic, laxative, expectorant, and commonly used in asthma, jaundice, scanty urine.

**Celosia argentea** Linn. (Amaranthaceae)

*Vern. name* : Vitunna (S), Swet-murga (B), Sufaid murga (R).

Pantropical herb; occurs as a weed throughout India upto 1550 m above msl.

Seeds are used in diarrhoea and to cure mouth sores.

**Chenopodium album** Linn. (Chenopodiaceae)

*Vern. name :* Vastuk (S), Chandan betu, Bethu sag (B), Bethu sag (H).

Mexican herb; occurs as a common weed throughout India.

Plant is laxative and anthelmintic; the sap of the plant is given to increase appetite.

**Chromolaena odorata** (Linn.) King & Robinson Syn. *Eupatorium odoratum* Linn. (Asteraceae)

*Vern. name :* Kalajhar (B).

Neotropical herb; occurs as a weed throughout India.

Bruised leaves are applied on cuts and wounds to stop bleeding.

**Clitoria ternatea** Linn. (Fabaceae)

*Vern. name :* Aparajita (B, S), Aparajit (H).

Paleotropical twiner; occurs throughout India.

Seeds are purgative and diuretic.

**Cynodon dactylon** (L.) Pers. (Poaceae)

*Vern. name :* Dhurva (S), Durba (B), Dhub, Hariali (H).

Tropical American herb; occurs throughout India.

Plant is diuretic and used in syphilis, dropsy, hysteria, epilepsy, insanity; it also checks bleeding.

**Cyperus rotundus** Linn. (Cyperaceae)

*Vern. name :* Musta (S), Motha (B, H).

Pantropical herb; occurs as a weed throughout India.

Tubers are given in amoebic dysentery and stomach disorder.

**Hedyotis corymbosa** (Linn.) Lam. (Rubiaceae)

*Vern. name :* Parpata (S), Khet-papra (B), Daman papar (H).

Pantropical herb; occurs as a weed throughout India.

Plant is used in remitting fever, nervous depression, jaundice, burning sensation of palm and sole.

**Heliotropium indicum** Linn. (Boraginaceae)

*Vern. name :* Hastisunda (S), Hatisur (B, H).

South American herb; occurs as a weed throughout India.

Plant is diuretic and used in boils, ulcer, eye diseases.

**Kalanchoe pinnata** (Lam.) Pers. (Crassulaceae)

*Vern. name :* Asthibhaksha (S), Koppata (B), Zakhmhaiyat (H).

Tropical American herb; occurs as a weed in many parts of India.

Plant is applied to bruises, wounds, boils and to cure dyspepsia.

**Martynia annua** Linn. (Martyniaceae)

*Vern. name :* Baghnokh, Bagnoki (B), Bichu (H).

Mexican herb; occurs as a weed in different parts of India.

Leaf is used in epilepsy, tuberculosis glands on neck region, throat sore, rheumatism and skin diseases.

**Mikania micrantha** Kunth (Asteraceae )

*Vern. name :* Germanlata (B).

Neotropical twiner; occurs as a weed all over India.

Plant is applied on new cuts and wounds.

**Parthenium hysterophorus** Linn. (Asteraceae)

*Vern. name :* Bish-gachh (B, H), Congress grass (E).

Neotropical herb; occurs as a obnoxious weeds throughout India.

Leaf is used as antialergic, antidiabetic and root is used in piles.

**Scoparia dulcis** Linn. (Scrophulariaceae)

*Vern. name :* Mithasagar (B).

South American herb; occurs as a weed all over India.

Plants are used in ague and as emetic.

**Senna occidentalis** (Linn.) Link (Caesalpiniaceae)

*Vern. name :* Kashamarda (S), Kalkashunda (B), Kasondi (H).

South American herb; occurs as a weed throughout India.

Plant is purgative and febrifuge; the seeds are used in skin diseases; plant juice is applied in eczema.

**Sphaeranthus africanus** Linn. (Asteraceae)

*Vern. name :* Mundirika (S), Murmuria (B).

African herb; occurs as a weed throughout India.

Plant is used in urethral discharges as deodorant, anthelmintic, alterative.

**Tridax procumbens** Linn. (Asteraceae)

*Vern. name :* Bisalyakarani (S, H).

Mexican herb; occurs all over India.

Plant is applied on cuts and wounds and used to cure dyspepsia.

**Wedelia chinensis** (Osbeck) Merrill (Asteraceae)

*Vern. name :* Pitabhringi (S), Bangra (B), Bhangra (H).

Austro-Asian herb; occurs as a weed in various parts of India.

Plant is used in cough, cephalalgia, alopecia, skin diseases and uterine haemorrhage.

**Xanthium strumarium** Linn. (Asteraceae)

*Vern. name :* Arishta (S), Banokra (B), Chhota-gokhru (H).

South American herb; occurs as a weed throughout India.

Plant is sedative, pseudorific, used in malaria, stomach diseases and cancer.

## ACKNOWLEDGEMENT

We are grateful to the Director, Botanical Survey of India, Kolkata for his encouragement and giving facilities.

## REFERENCES

CHOPRA, R.N. *et al.* 1956. *Glossary of Indian Medicinal Plants.* New Delhi.

HOOKER, J.D. 1904. A sketch of the flora of British India. Oxford.

JAIN, S.K. 1991. Dictionary of Indian Folk Medicine and Ethnobotany. New Delhi.

NAYAR, M.P. 1977. Changing pattern of Indian Flora. *Bull. Bot. Surv. India* 19 : 145-154.

VAVILOV, N.I. 1926. Studies on the origin of cultivated plants. *Bull. Appl. Bot. Pl. Breed.* 26 (2) : 1-248.

# MESTA DIVERSITY : A POTENTIAL SOURCE TO IMPROVE RURAL ECONOMY IN NORTH COASTAL ANDHRA PRADESH

N. SIVARAJ, S.R. PANDRAVADA, M.V. NAIDU*, G. SRINIVASA REDDY, N. SUNIL, V. KAMALA, BABU ABRAHAM & K.S. VARAPRASAD

*National Bureau of Plant Genetic Resources, Regional Station, Rajendranagar, Hyderabad-500 030, Andhra Pradesh.*
**Agricultural Research Station, Acharya NG Ranga Agricultural University, Seethampeta-532 443, Srikakulam district, Andhra Pradesh.*

## ABSTRACT

*Mesta* is the bast fibre of commercial importance obtained from *Hibiscus cannabinus* L. and *H. sabdariffa* L., belonging to the family Malvaceae. These two species having wide adaptability to poor soils, are drought tolerant and come up luxuriantly even under rainfed conditions. Andhra Pradesh stands first in both the area and production with 0.8 lakh hectares and 6.8 lakh bales respectively. It is traditionally grown in the Srikakulam and Vizianagaram districts of north coastal Andhra Pradesh for fibre, and rich endemic genetic diversity is reported. Mesta being a renewable natural fibre, which is safe, non-toxic, biodegradable and environment friendly, has wide applications, acceptance and export potential. Culinary varieties are used for the preparation of curry, pickle and chutneys. Several alternate uses such as for paper-making, potting mix and food colourant etc are also reported.

Keeping in view the importance of this crop, concerted efforts have been made by NBPGR Regional Station to collect and conserve the Mesta germplasm. A total of 560 Mesta germplasm accessions of fibre and culinary types belonging to both the species were collected from the State. Good diversity was observed for leaf lamina, vein and petiole colour, leaf length and width, stem pubescence, stem (colour, diameter), branching habit, flower colour, capsule pubescence, seed (colour, shape,size and weight) and yield (900-1800 Kg/ha) etc. *Erra gogu, Gongura, Kasi janumu, Katti naara, Koora gogu, Manchi gogu, , Puntikura, Tella gogu* etc are some of the landraces collected.One collection of high yielding *H.sabdariffa* L., designated as UJWALA, is in the pre release stage by Acharya NG Ranga Agricultural University (ANGRAU). This paper further describes the status of diversity in

Mesta, strategies for the future research and socio-economic aspects of Mesta cultivation in relation to livelihood of people.

## INTRODUCTION

Mesta is the bast fibre of commercial importance obtained from *Hibiscus cannabinus* L. (Kenaf) and *H. sabdariffa* L. (Roselle), belonging to the family Malvaceae. These two species have wide adaptability to poor soils, are drought tolerant and come up luxuriantly even under rainfed conditions. Roselle occupies 90% of cultivated area owing to its hardiness, drought tolerance and ability to grow in soils low in fertility. Rich endemic genetic diversity is reported in these two species for fibre in the north coastal region and all other areas in the State for culinary types. However, due to marketing problems, improvement in the irrigation facilities and other socio-economic reasons, this crop of importance for the poor and marginal farmers in the rural economy till now is being replaced by more remunerative crops. Hence, keeping in view the importance of this crop and the accumulated germplasm variability in the Mesta improvement programmes, concerted efforts have been made by NBPGR Regional Station to collect and conserve the Mesta germplasm.

## AREA AND PRODUCTION

Reported from the African and Hindustani Centers of Diversity, Kenaf or cultivars (cvs) theirof are reported to tolerate laterites, low pH and some water logging or temporary inundations (Duke, 1978, 1983). Many cvs have been developed. Some are *simplex, viridis, tuber, purpureus,* and *vulgaris.* All trial crosses between the varieties have been successful. Diploid ($2n = 36$) and tetraploid ($2n = 72$) races have been found. Both self- and cross-pollination is found among the varieties. Probably native to Africa, East Indies, Asia or Australia, the species are now naturalized in Africa and Asia, generally cultivated in India, Indochina, Taiwan, Indonesia and North and South America.

Mesta is grown in India in more than 12 States, covering about 2 lakh hectares with a production of 11.2 lakh bales. Andhra Pradesh stands first in both the area and production with 80,000 hectares and 53% of country's production respectively. In Andhra Pradesh, it is concentrated in Srikakulam and Vizianagaram districts due to factors like adaptability of crop to wide range of soil and climatic conditions, its ability to grow in areas unsuitable for jute and availability of retting facilities.

## UTILITY OF MESTA

Mesta being a renewable natural fibre which is safe, non-toxic and biodegradable and environment friendly has wide applications, acceptance and export potential. Culinary varieties are used for the preparation of curry, pickle and chutneys. Several alternate uses such as for paper-making, potting mix and food colourant etc are also being reported. Fiber strands, 1.5–3 m long, are used for rope, cordage, canvas, sacking, carpet backing, and fishing nets. Seeds of Mesta contain about 20% oil, which is used for salad, cooking and as lubricant oils. Oil is also used in the manufacture of soap, linoleum, paints and varnishes, and for illumination. In 1968, Florida used 600 ha growing Kenaf for bean poles (Whitely, 1981). Recently, the pulp has been used in paper-making. Leaves are used as a pot-herb. A concentrated food for cattle in the form of seed-cake comes from the residue after oil extraction. Africans use soot from the stems as a black pigment. They also use a piece of the stem as a base for drilling fire.

Reported to be anodyne, aperitif, aphrodisiac, fattening, purgative and stomachic, Kenaf is a folk remedy for bilious conditions, bruises, fever and puerperium (Duke & Wain, 1981). Powdered leaves are applied to guinea worms in Africa. Africans use peelings from the stems for anaemia, fatigue, lassitude etc. In Zambia, the leaf infusion is used for coughs. Ayurvedics use the leaves for dysentery and biliousness, blood and throat disorders. Seeds are applied externally to aches and bruises. Medicinally, juice of the flowers with sugar and black pepper is used in biliousness with acidity. Seeds are considered aphrodisiac and fattening.

## GERMPLASM COLLECTION AND CONSERVATION

National Bureau of Plant Genetic Resources (NBPGR) Regional Station, Hyderabad with a mandate to collect and conserve germplasm of agri-horticultural crops has conducted surveys for the collection of Mesta germplasm in this region. A total of 560 Mesta germplasm accessions of fibre and culinary types belonging to both the species were collected from the State by the Bureau. Depending on the availability of the germplasm to be collected in the cultivated field or in wild state, samples were drawn either from a population or from an individual plant. Method of sampling in most cases has been bulk and in some cases selective or random sampling method was followed. Market samples were also drawn to conserve the available genetic diversity in the surveyed areas. Some of the important local landraces collected include *Erra gogu, Erra gongura, Gomulu, Gongura, Kasi janumu, Katti naara, Koora gogu, Manchi gogu, Mattu gongura, Pulla gogu, Puntikura, Ranasthalam gogu,Tella gogu, Tella janumu* etc. Several varieties of Mesta have been released in Andhra Pradesh directly as pure line selections (AMV-1 and AMV-2 etc) or by using as parents from the local landraces and hence, this collected germplasm material will be of utmost use in the Mesta breeding programmes. Out of the above material collected, one accession of *H. sabdariffa* L. (SS-1569) a leafy vegetable type which is high yielding and late flowering is designated as UJWALA and is in the mini kit trials of Acharya NG Ranga Agricultural University and is in the pre release stage.

The collected germplasm from this region is being conserved in the national gene bank at NBPGR, New Delhi and in Medium term module at NBPGR Regional Station, Hyderabad. A set of collections is being maintained in the Agricultural Research Station, Acharya NG Ranga Agricultural University, Amdalavalasa for utilization in the crop improvement programmes.

## VARIABILITY IN MESTA GERMPLASM

Good diversity was observed in these two species for leaf lamina, vein and petiole colour (green/red), leaf length and width, stem pubescence (smooth/ hairy/ prickly), stem colour (green/red), stem diameter, branching habit (weak/intermediate/ strong), flower colour, capsule pubescence (smooth/hairy/ bristled), seed colour (brown/reddish-brown/ grey), seed shape (reniform/ angular), seed size, seed weight and yield (900-1800 kg/ha) etc.

## ANALYSIS

- Mesta is being grown in the surveyed areas due to factors like adaptability of crop to wide range of soil and climatic conditions, its drought tolerance, its ability to grow in areas unsuitable for jute and availability of retting facilities.
- Cultivation requires low initial investment which is suiting to socio-economic conditions of the local farmers.
- With the current utilization of Mesta for its bast fibre alone is not very paying to the growers. The crop has to be modified into a multifarious crop for commercial diversification purposes in addition to normal manufacturing of gunny bags and twines.
- Kenaf is considered one of the most promising alternatives to virgin soft and hard woods for paper production. As it is the source for most promising non-wood fibre for pulp and paper production, establishing one or two paper mills (one in each district, i.e. Srikakulam and Vizianagaram) at a small scale would benefit the rural poor. Mesta paper production would save critical forest habitat and is considerably less polluting and environmentally destructive than wood-based paper production.
- Since Mesta is also used in the preparation of pickles and other delicacies, it is essential to identify the culinary types.
- The fleshy calyx of Roselle is in demand as a natural colouring in food industry. It has greater export potential. Hence, attempt has to be initiated in bringing in new varieties.
- Creating artificial ponds for retting facilities would enhance the quality of fibre, thereby increasing the chances of farmers getting more returns.

## REFERENCES

DUKE, J.A. 1978. The quest for tolerant germplasm, ASA Special Symposium 32, Crop tolerance to suboptimal land conditions. *Am. Soc. Agron*. Madison, WI. pp. 1–61.

———— 1983. *Handbook of Energy Crops*. Unpublished.

———— & K.K. WAIN 1981. *Medicinal plants of the world*. Computer index with more than 85,000 entries, 3 Vols.

WHITELY, E.L. 1981. Kenaf, *Hibiscus cannabinus*, In McClure, T.A. & E.S. Lipinsky (Eds.) *CRC handbook of biosolar resources, Vol. II-Resource materials,* pp. 259-268. CRC Press, Inc., Boca Raton, FL.

# NATURAL AND CULTIVATED DYE YIELDING PLANTS AND THEIR UTILIZATION BY VILLAGERS IN SOUTHERN DISTRICTS OF TAMILNADU

K. RAJENDRAN*, S. BABURAJ & M. JOTHI BASU

*Department of Botany, Thiagarajar College, Madurai - 625 009, Tamilnadu, India*
**E-mail: kuppurajendran@rediffmail.com*

**ABSTRACT**

Survey and observations were carried out in remote villages of southern districts of Tamil Nadu to identify the wild and cultivated dye yielding plants and their utilization by the local people. This paper deals with about 59 dye yielding plant species of 54 genera belonging to 29 families and provides information on their botanical names, vernacular names, families, plant parts producing dye, colour of the dye etc.

## INTRODUCTION

India is a veritable emporium of medicinal and aromatic plants. It has been estimated that out of 17,000 higher plants occurring in India, 9000 are commercially useful, of which 7500 are medicinal, 3900 edible, 700 culturally important, 525 used for fiber, 400 fodder, 300 for pesticide and insecticide, 300 for gum, resin and dye and 100 for incense and perfume. Usage of dye yielding plants had been a long practice by man from ancient times, which is evident from the literature. This practice of using plants for preparation of dyes is still prevailing throughout the World. However, the use of vegetable dyes continued in few pockets of India as a small-scale industry (Farooqi, 2003).

India is in an advantageous position since the country holds a rich reservoir of natural products. Chemical evaluation and purification of dyes and cultivation of potential dye yielding plants in around 7.5 million hectares of marginal and wastelands is available in India (Kumar *et al.*, 2004).

The main aim of the present study is the survey of the dye yielding plants for sustainable utilization of plant resources and to provide data for biochemical and pharmacological studies. It is also aimed to encourage the farmers to go for cultivation of suitable wild dye yielding plant species in this agro- climatic region. As such, survey and observations were carried out in remote villages to identify the wild and cultivated dye yielding plants and their utilization by the people in southern districts of Tamil Nadu.

## MATERIALS AND METHODS

The study area selected in Madurai, Dindigul and Theni districts of Tamil Nadu is located at 10°05' N latitude and 78°16' E longitude at an elevation of 132 m above msl and the temperature ranges from 28-38°C. Average annual precipitation ranges between 750 and 850 mm. The climate is sub-humid to semi-arid. Soil is sandy-loam to black cotton soil with a pH range of 6.5-7.5. Organic content of the soil is 0.50-0.62%. The nitrogen, phosphorus, potassium, calcium and magnesium contents are 0.3-0.4%, 0.07-0.09%, 0.42-0.47%, 0.5-0.7% and 0.18-0.25%, respectively.

The information regarding the existing dye yielding plants and their local name, uses, plant parts used and method of use were recorded with the help of aged farmers. The collected specimens have been deposited in the herbarium of Department of Botany, Thiagarajar College, Madurai, Tamil Nadu.

## ENUMERATION

Table 1. Dye yielding plants with their botanical and vernacular name and plant parts used.

| Family/Botanical name | Vernacular name (Tamil) | Colour | Plant part Used |
|---|---|---|---|
| **Acanthaceae** | | | |
| *Justicia adhatoda* L. | Adathoda | Yellow | Leaves |
| **Anacardiaceae** | | | |
| *Lannea coromandelica* (Houtt.) Merrill | Udhiyan | Reddish brown | Bark |
| *Semecarpus anacardium* L. f. | Senkottai | Black | Fruit |
| **Apocynaceae** | | | |
| *Allamanda cathartica* L. | Manjal patti | Yellow | Flower |
| *Tabernaemontana divaricata* (L.) R. Br. ex R. & S. | Nanthiavettai | Red | Flower |
| *Wrightia tinctoria* (Roxb.) R. Br. | Veppalai | Blue | Leaves |
| **Asteraceae** | | | |
| *Bidens tripartita* L. | Manjapoovu | Black | Entire plant |
| *Eclipta prostrata* (L.) L. | Karisalan-kanni | Black | Leaf |
| *Helianthus annuus* L. | Sooriakanthi | Yellow | Flowers |
| *Tagetes erecta* L. | Sento-malli, Vedipunaripu | Yellow | Flowers |

| | | | |
|---|---|---|---|
| *Wedelia chinensis* (Osbeck) Merr. | Manjal-karisalan-kanni | Black | Root |
| **Basellaceae** | | | |
| *Basella rubra* L. | Pasalaikeerai | Red | Fruit |
| **Berberidaceae** | | | |
| *Berberis aristata* DC. | Chithirai-poovu | Yellow | Root |
| **Caesalpiniaceae** | | | |
| *Bauhinia variegata* L. | Mantharai | Brown | Bark |
| *Caesalpinia pulcherrima* (L.) Swartz | Mailkonnai | Red | Flower |
| *Cassia fistula* L. | Sarakkonnai | Brown | Bark |
| *Delonix regia* (Bojer) Rafin. | Gumohar | Orange | Flower |
| *Senna alata* (L.) Roxb. | Seemai-agathi | Yellow | Flower |
| *Senna auriculata* (L.) Roxb. | Aavarai | Yellow | Flower |
| *Senna occidentalis* (L.) Link | Payavarai | Yellow | Pod and seed |
| *Senna tora* (L.) Roxb. | Thagarai | Yellow, Brown, Red or Blue | Pod and seed |
| *Tamarindus indica* L. | Puli | Yellow | Leaf |
| **Casuarinaceae** | | | |
| *Casuarina equisetifolia* L. | Savukku | Brown | Bark |
| **Chenopodiaceae** | | | |
| *Beta vulgaris* L. | Beet-root | Red | Root and leaf |
| **Combretaceae** | | | |
| *Terminalia chebula* (Gaertn.) Retz. | Batham | Yellow, Brown | Fruit |
| **Euphorbiaceae** | | | |
| *Chrozophora rotlleri* (Geis.) A. Juss. & Spreng. | - | Blue | Fruit |
| *Jatropha curcas* L. | Kattamanakku | Blue | Bark |
| *Jatropha gossypifolia* L. | Adalai | Blue | Bark |
| *Mallotus philippensis* (Lam.) Muell.-Arg. | Thirisalukkai | Red | Fruit |
| *Phyllanthus amarus* Schum. & Thonn. | Keelanelli | Black | Stem and leaf |
| *Phyllanthus emblica* L. | Nelli | Dark brown | Fruit |
| **Mimosaceae** | | | |
| *Acacia farnesiana* (L.) Willd. | Vedavel | Brown | Flower and bark |
| *Acacia nilotica* (L.) Willd. ex Del. | Karuvel | Brown | Bark |
| *Adenanthera pavonina* L. | Anaikundumani | Red | Wood |
| *Albizia lebbeck* (L.) Benth. | Vakai | Brown | Bark |
| **Fabaceae** | | | |
| *Clitoria ternatea* L. | Sanku-pushpam | Blue-green | Flower |

| | | | |
|---|---|---|---|
| *Erythrina variegata* L. var. *orientalis* (L.) Merr. | Kalyanamuruigai | Brown and Yellow | Wood and bark |
| *Indigofera tinctoria* L. | Avuri | Blue | Leaf |
| *Pterocarpus marsupium* Roxb. | Vengai | Brown | Bark |
| **Lauraceae** | | | |
| *Cassytha filiformis* L. | Varilla-kothu | Brown | Stem |
| **Lythraceae** | | | |
| *Lawsonia inermis* L. | Maruthani | Orange, Red | Leaf |
| **Malvaceae** | | | |
| *Hibiscus rosa-sinensis* L. | Semparuthi | Black | Flower |
| *Thespesia populnea* (L.) Soland. ex Corr. | Poovarasu | Yellow | Flower and fruit |
| **Melastomataceae** | | | |
| *Memecylon edule* Roxb. | Kayankulai | Yellow | Leaf |
| **Moraceae** | | | |
| *Ficus religiosa* L. | Arasu | Red | Bark and root |
| **Moringaceae** | | | |
| *Moringa oleifera* Lamk. | Murungai | Indigo | Wood |
| **Myrtaceae** | | | |
| *Syzygium cumini* (L.) Skeels | Naval | Brown | Bark |
| **Nyctaginaceae** | | | |
| *Mirabilis jalapa* L. | Anthi-mantharai | Crimson Yellow | Flower, bark and stem |
| **Phytolaccaceae** | | | |
| *Rivinia humilis* L. | Rathakkai | Red | Fruit |
| **Rhamnaceae** | | | |
| *Ziziphus mauritiana* Lam. | Elanthai | Reddish-brown, Red-Pink | Leaf |
| **Rubiaceae** | | | |
| *Catunaregam spinosa* (Thunb.) Tirveng. | Madu-karai | Yellow | Fruit |
| *Morinda citrifolia* L. | Nuna | Yellow | Root |
| *Rubia cordifolia* L. | Ottukodi, Shevelli, Manjitti | Purple | Root |
| **Rutaceae** | | | |
| *Aegle marmelos* (L.) Corr. | Vilvam | Yellow | Fruit |
| *Chloroxylon swietenia* DC. | Porasu | Yellow | Bark |
| **Sapotaceae** | | | |
| *Mimusops elengi* L. | Maramalli | Brown | Bark |

| | | | |
|---|---|---|---|
| **Dipterocarpaceae** | | | |
| *Shorea robusta* Gaertn. f. | Sal | Red | Bark |
| **Verbenaceae** | | | |
| *Tectona grandis* L. f. | Thaekku | Yellow | Wood and leaf |
| **Zingiberaceae** | | | |
| *Curcuma longa* L. | Manjal | Yellow | Rhizome |

## RESULTS AND DISCUSSION

By understanding the important dye yielding plants being utilized by people traditionally for various products available in the market and the raw material required for it, we can evolve a system to grow them in farm lands, wastelands, lakes and river-banks which will reduce the pressure on the natural forests and thereby conserve the biodiversity and improve the socio- economics of the rural poor.

The data were obtained through direct field visits and interviews with local men and farmers. In the present study, 59 species of 54 genera belonging to 29 families of important dye yielding plant species were identified and their uses are described in Table 1. Among all the species, *Curcuma longa, Erythrina variegata, Jatropha curcas, Moringa oleifera, Indigofera tinctoria* and *Morinda citrifolia* are some of the most useful multipurpose and dye yielding plants growing in this agro-climatic zone. In view of greater importance of dye yielding plants, many of the progressive farmers are switching for cultivation of these plants as they found it to be more profitable than traditional crops. Apart from the above species, there is much scope to cultivate many other dye yielding plants in this agro-climatic region. To bring more dye yielding plant species in cultivation and for their increased productivity, further research is needed as regards systematic and scientific cultivation methods, which include organic farming, irrigation, harvesting and preservation and marketing. To enhance the economic condition of rural poor and progressive farmers, it is essential to impart necessary training to them in mass cultivation practices.

## REFERENCES

FAROOQI, M.I.H. 2003. Indian vegetable dyes: possiblilities and prospects. *MFP News* 13(2): 4-9.

KUMAR, S., PARIDHI NAGIYAN & SANTOSH KUMAR 2004. Studies on dye yielding plants of Uttaranchal. *Ann. For.* 12 (1) : 87-98.

# NOTE ON AQUATIC MACROPHYTES OF GUJARAT AND THEIR ECONOMIC VALUES

K.D. MITALIYA, D.C. BHATT & B.B. MAITREYA*

*Department of Marine Sciences, Bhavnagar University, Bhavnagar-364002, Gujarat*
**Sir P. P. Institute of Science, Bhavnagar University, Bhavnagar*

## ABSTRACT

In the present study, about 104 species belonging to 80 genera of different groups likely algae, pteridophytes and angiosperms are recorded from aquatic and marshy habitats of Gujarat, along with their economic uses.

## INTRODUCTION

As per the modified IUCN definition, all the submerged or water saturated lands natural or man-made, inland or coastal, permanent or temporary, static or dynamic, vegetated or non-vegetated which necessarily have a land-water interface are defined as wetlands. These wetlands include check dams, canals, ditches, marshes, rivers, streams, ponds, lakes, drains, pits, puddles and marine waters of the shoreline.

Wetlands are highly productive ecosystems which serve as habitats for a variety of plants and animals. Many wetlands serve as the winter habitat for migratory birds. Wetlands play essential functions for the human society, including flood control, natural sewage, famine, stabilization of shore lines against the wave erosion and support rich biodiversity.

Gujarat, the westernmost State of India, is divided geographically into five major regions, i.e. Kutch, Saurashtra, Central Gujarat, North Gujarat and South Gujarat which sustain arid, semi-arid and evergreen dry wet locations.

Several floristic works have been carried out on the flora of Gujarat after Shah (1978) & Bole & Pathak (1988). However, little attention has been paid on the aquatic and marshland plants and their economic potential. A few to mention include Subramanyam (1962), Cook (1996), Mazid (1998), Chavan & Sabnis (1961) etc. During recent years, a number of research papers have appeared in different journals on the ethnobotany and ethnomedicinal plants of Gujarat, but therein again aquatic macrophytes have been neglected. Therefore, the present study has been taken up to fill up the gaps in floristics and ethnobotany of Gujarat.

## MATERIAL AND METHODS

In order to study aquatic macrophytes, frequent trips were made from 1992 to 2004 to various wetlands of Gujarat, i.e. lakes, ponds, puddles, ditches, canals, swamps, marshy tracts, riverine tracts, streams, check dams etc. During these thirteen years of survey, the wetland plants occurring in different water-saturated areas of Gujarat have been collected, photographed and identified. Besides these, their nature of growth, habit, habitat and uses were noted from local rural and tribal people of different areas of Gujarat. In order to acquire detailed knowledge on the utilization of plant resources, senior citizens, old and experienced persons, village heads and farmers were also contacted, besides making personal observations on the spot. The specimens, after proper identification, have been deposited in the herbarium unit of Marine Sciences at Bhavnagar.

The plants are enumerated alphabetically with latest botanical names, in parenthesis its family and local names (in italics), short description of species and uses.

## ENUMERATION

**Aeschynomene indica** L. (Fabaceae, *Ikardi*)

An erect and branched annual marsh-loving herb. Commonly found in ditches etc.

*Uses :* Whole plant used as green manure.

**Ageratum conyzoides** L. (Asteraceae, *Ajagandha*)

A hairy, annual herb; leaves opposite. Common in wet places.

*Uses :* Leaf-juice is given in bronchitis and abdominal pains.

**Alternanthera pungens** Kunth (Amaranthaceae, *Pani-bhaji*)

A prostrate herb; leaves obovate, petiolate. Commonly found in moist areas.

*Uses :* Leaves are used as vegetable.

**Alternanthera sessilis** (L.) R. Br. ex DC. (Amaranthaceae, *Pani-bhaji*)

A prostrate herb; leaves sessile, obovate. Common in wet and marshy areas.

*Uses :* Young shoots and leaves are used as vegetable.

**Alysicarpus monilifer** (L.) DC. (Fabaceae, *Samervo*)

A perennial, prostrate herb. Common in wet and waste places.

*Uses :* Decoction of whole plant is used in coughs, bronchitis etc.

**Ammannia baccifera** L. (Lythraceae, *Jal-agio*)

An erect, annual herb; stem red. Common in wet and marshy places.

*Uses :* Paste of whole plant is applied in skin diseases.

**Ammannia multiflora** Roxb. (Lythraceae, *Zino-agio*)

An erect, annual herb. Commonly found in low-lying areas.

*Uses :* Commonly used as cattle fodder.

**Aponogeton natans** (L.) Engl. & Krause (Aponogetonaceae)

A submerged aquatic herb. Common in ditches.

*Uses :* The tuberous root-stocks are edible.

**Azolla pinnata** R. Br. (Azollaceae)

A minute free-floating aquatic fern. Commonly found in ponds.

*Uses :* Commonly used as manure.

**Bacopa monnieri** (L.) Wettst. (Scrophulariaceae, *Jal-brahmi*)

A prostrate, succulent herb. Commonly found in wet areas.

*Uses :* Juice of leaves is given in nervous disorders.

**Bergia ammannioides** Roxb. ex Roth (Elatinaceae)

A prostrate, annual herb; leaves opposite. Rarely found in wet places.

*Uses :* Paste of whole plant is applied in skin diseases.

**Blumea lacera** (Burm. f.) DC. (Asteraceae, *Kapurio*)

An annual, erect, hairy herb. Common in ditches, canals etc.

*Uses :* Juice of leaves is given in fever, cholera, asthama etc.

**Boerhavia repens** L. var. **diffusa** (L.) Hook. f. (Nyctaginaceae, *Satodi*)

A diffuse, perennial herb with stout root-stock. Common in wet places.

*Uses :* Decoction of roots is given in asthma, swellings, calculi etc.

**Bulbostylis barbata** Clarke (Cyperaceae)

A tufted annual sedge. Commonly found in rivers.

*Uses :* Used as cattle feed.

**Canscora diffusa** (Vahl) R. Br. ex R. & S. (Gentianaceae, *Zinku-kariyatu*)

A small, annual herb; leaves opposite. Common in wet places.

*Uses :* Decoction of whole plant is given in fever, insomnia etc.

**Centella asiatica** (L.) Urban (Apiaceae, *Brahmi*)

A herb with creeping stem, rooting at the nodes. Common in S. Gujarat.

*Uses :* The whole plant is used as a brain tonic.

**Centipeda minima** (L.) A. Br. & Aschers. (Asteraceae, *Nak-chhikni*)

A prostrate, branched, annual herb. Rarely found in South Gujarat.

*Uses :* Paste of leaves is applied in skin diseases.

**Ceratophyllum demersum** L. (Ceratophyllaceae)

A perennial, submerged herb; leaves verticillate, linear. Rarely found in ponds.

*Uses :* Juice of plant is employed in fever.

**Ceratopteris thalictroides** (L.) Brongn. (Parkeriaceae)

A succulent, semi-aquatic fern. Rarely found in ponds, ditches etc.

*Uses :* Leaves are used as vegetable and fodder.

**Chara zeylanica** Willd. (Characeae)

A submerged, dirty green alga. Commonly found in ditches.

*Uses :* Paste of plant is applied in skin diseases.

**Cleome chelidonii** L. f. (Cleomaceae, *Jal-talavni*)

An annual, branched herb. Commonly found near water-bodies.

*Uses :* Juice of leaves is dropped into ears in case of earache.

**Coldenia procumbens** L. (Boraginaceae)

A prostrate, annual herb; leaves deeply emborsed. Common in drying ponds.

*Uses :* Paste of fresh leaves is applied in skin diseases.

**Colocasia esculenta** (L.) Schott (Araceae, *Alavi*)

An erect, cormy herb. Rarely found as an escape near marshy places.

*Uses :* Tuberous corms and fresh leaves are used as vegetable.

**Commelina benghalensis** L. (Commelinaceae, *Shishmuli*)

A succulent, creeping herb. Commonly found in wet places.

*Uses :* Leaves are used as vegetable and also as fodder for cattle.

**Corchorus aestuans** L. (Tiliaceae, *Chhunchh*)

An annual, branched herb; fruit 3-winged. Common in moist areas.

*Uses :* The leaves and fruits are used as vegetables.

**Corchorus fascicularis** Lam. (Tiliaceae, *Chhunchh*)

An annual, glabrous herb. Common in wet fields.

*Uses :* Leaves are used as vegetable.

**Chrozophora prostrata** Dalz. & Gibs. (Euphorbiaceae, *Betho-okharad*)

An annual, prostrate, hairy herb. Commonly found near drying soils.

*Uses :* Paste of whole plant is applied in skin diseases.

**Cynodon dactylon** (L.) Pers. (Poaceae, *Dharodiu*)

A perennial grass, rooting at nodes. Common in wet places.

*Uses :* Juice of fresh plant is given in anaemia, indigestion etc.

**Cyperus alulatus** Kern (Cyperaceae)

A tufted annual sedge. Occasionally found in marshy areas.

*Uses :* The plant is used as fodder.

**Cyperus bulbosus** Vahl (Cyperaceae, *Thek*)

A small, perennial sedge with bulb-like tubers. Common in marshy soils.

*Uses :* Roasted tubers are edible.

**Cyperus difformis** L. (Cyperaceae)

An annual, erect, tufted sedge. Occasionally found in ditches, ponds etc.

*Uses :* The plant is used as fodder.

**Cyperus esculentus** L. (Cyperaceae, *Bhadri-moth*)

A perennial, erect sedge. Occasionally found near water bodies.

*Uses :* Rhizomes are used in abdominal disorders.

**Cyperus rotundus** L. (Cyperaceae, *Moth*)

A rhizomatous perennial aromatic sedge. Occasionally found near water bodies.

*Uses :* Rhizomes are given in flatulence, indigestion, colic etc.

**Dentella repens** (L.) J. R. & G. Forst. (Rubiaceae, *Dantiphaliu*)

A much-branched prostrate herb, rooting at the nodes. In moist areas.

*Uses :* Juice of fresh plant is given in calculi, strangury etc.

**Echinochloa colona** (L.) Link (Poaceae, *Samo*)

An erect, annual grass. Commonly found in marshy places.

*Uses :* Grains are used as food during fast.

**Echinochloa crus-galli** (L.) P. Beauv. (Poaceae, *Adbau-samo*)

An erect, stout annual grass. Commonly found in wet places.

*Uses :* Grains are used as famine food.

**Eclipta prostrata** (L.) L. (Asteraceae, *Bhangro*)

An erect or prostrate herb. Common in wet and marshy places.

*Uses :* Leaves are used in liver disorders and as hair tonic.

**Eichhornia crassipes** (Mart.) Solms. (Pontederiaceae, *Jal-kumbhi*)

A floating herb with spongy stolons. Commonly found in water bodies.

*Uses :* Commonly used as manure and fodder.

**Eleocharis atropurpurea** (Retz.) J. & K. Presl (Cyperaceae)

A tufted, filiform, annual sedge. Occasionally found in wet places.

*Uses :* Plant is grazed by the cattle.

**Equisetum debile** Roxb. (Equisetaceae)

A pteridophytic herb. Rarely found in marshy places.

*Uses :* The plant is used as cattle fodder.

**Evolvulus alsinoides** (L.) L. (Convolvulaceae, *Bhuri-shankhavali*)

A prostrate, pubescent or hairy herb. Found in moist soils.

*Uses :* Powder of whole plant is given in nervous disorders.

**Fimbristylis bisumbellata** (Forssk.) Bub. (Cyperaceae)

A tufted, annual sedge. Occasionally found in moist places.

*Uses :* Underground parts are used as cooling medicine.

**Fimbristylis ferruginea** (L.) Vahl (Cyperaceae)

A tufted perennial sedge. Occasionally found in marshy places.

*Uses :* The plant is used as fodder.

**Glinus lotoides** L. (Molluginaceae, *Mitho-okharad*)

A prostrate, tomentose, annual herb. Commonly found near ponds.

*Uses :* Leaves are used as vegetable.

**Grangea maderaspatana** (L.) Poir. (Asteraceae, *Zinki-mundi*)

A prostrate, hairy, annual herb. Common near drying ponds.

*Uses :* The whole plant is given in digestive disorders.

**Heliotropium supinum** L. (Boraginaceae, *Bethi-hathisundhi*)

A prostrate, much-branched herb.Commonly found in marshy soils.

*Uses :* Paste of fresh plant is applied on skin diseases.

**Hoppea dichotoma** Heyne ex Willd. (Gentianaceae)

An annual, divaricately branched herb. Commonly found in damp places.

*Uses :* Juice of plant is given in fever, piles etc.

**Hydrilla verticillata** (L. f.) Royle (Hydrocharitaceae)

A submerged aquatic herb. Common in ponds.

*Uses :* Paste of plant is applied in skin diseases.

**Hydrolea zeylanica** (L.) Vahl (Hydrophyllaceae)

An annual, erect to decumbent herb. Occasionally grows in pools of S. Gujarat.

*Uses :* Paste of plant is applied in skin diseases.

**Hygrophila schulli** (Buch.-Ham.) M. R. & S. M. Almeida (Acanthaceae, *Akharo*)

An erect, armed herb. Common in ditches, ponds etc.

*Uses :* Seeds are given in calculi, strangury, rheumatism etc.

**Hygrophila serphyllum** (Nees.) T. Anders. (Acanthaceae, *Sarpat*)

A procumbent, pubescent herb. Common in wet places of S. Gujarat.

*Uses :* Leaves are used as cooling medicine.

**Ipomoea aquatica** Forssk. (Convolvulaceae, *Nadani-bhaji*)

A creeping, marshy herb, rooting at the nodes. Common in shallow waters.

*Uses :* Leaves are used as vegetable. Juice of plant is given in bronchitis.

**Ipomoea carnea** Jacq. ssp. **fistulosa** (Mart ex Choisy) Austin (Convolvulaceae, *Naffatiyo*)

A large, sub-erect, straggling shrub. Commonly found in ditches.

*Uses :* Juice of leaves is used as insecticide.

**Juncellus alopecuroides** (Rottb.) Cl. (Cyperaceae)

A perennial, glabrous sedge. Commonly found in ponds.

*Uses :* Aerial parts are used for making mats.

**Juncellus pygmaeus** (Rottb.) Cl. (Cyperaceae)

A tufted, annual, short sedge. Commonly found in marshy and moist soils.

*Uses :* The plants are grazed by the cattle.

**Juncus bufonius** L. (Juncaceae)

A prostrate, tufted annual herb. Occasionally found in moist places.

*Uses :* The plant is used as a cooling medicine.

**Lemna perpusilla** Torrey (Lemnaceae)

A minute, free-floating aquatic herb. Commonly found in water bodies.

*Uses :* It is used as green manure.

**Limnophila indica** (L.) Druce (Scrophulariaceae, *Tarati*)

A small aquatic herb; leaves dissected. Occasional in marshy places.

*Uses :* The plant is used in skin diseases.

**Limnophyton obtusifolium** (L.) Miq. (Alismataceae)

An emergent aquatic herb; leaves radical, sagittate. Common in pools.

*Uses :* The plant is grazed by the cattle.

**Lindernia ciliata** (Colsm.) Penn. (Scrophulariaceae)

A small herb; leaves serrate. Occasionally found in wet places.

*Uses :* The plant is used in skin diseases.

**Lindernia crustacea** (L.) F. v. Muell. (Scrophulariaceae)

A small, glabrous, annual herb. Occasionally found in moist places.

*Uses :* Paste of plant is applied in skin diseases.

**Ludwigia octovalvis** (Miq.) Raven (Onagraceae, *Pan-lavang*)

An amphibious herb. Occasionally found in low lying areas.

*Uses :* Juice of plant is given in fever, toxiemia etc.

**Ludwigia perennis** L. (Onagraceae)

An erect or decumbent herb. Occasionally found in marshy places.

*Uses :* Juice of plant is given in fever, bodyache etc.

**Marsilea minuta** L. (Marsileaceae)

A slender, small, prostrate herb; leaflets four. Found in marshy places.

*Uses :* The plant is used in blood pressure and insomnia.

**Melochia corchorifolia** L. (Sterculiaceae, *Chunch-khapat*)

An annual, small herb. Occasionally found in marshy places.

*Uses :* Leaves are used as vegetable.

**Merremia gangetica** (L.) Cufod. (Convolvulaceae, *Underkani*)

A prostrate herb, rooting at the nodes. Commonly found in wet areas.

*Uses :* Juice of plant is given in abdominal disorders.

**Monochoria vaginalis** (Burm. f.) Presl (Pontederiaceae)

A perennial rhizomatous herb. Occasionally found in ditches.

*Uses :* The leaves and flowers are used as vegetable.

**Najas marina** L. (Najadceae)

A slender aquatic herb. Common in ponds.

*Uses :* Plant is used as fodder.

**Nelumbo nucifera** Gaertn. (Nelumbonaceae, *Kamal*)

An emergent perennial herb. Occasionally found in ponds.

*Uses :* Rhizomes are used as vegetable. Seeds are edible.

**Neptunia oleracea** Lour. (Mimosaceae, *Jal-lajamani*)

A perennial herb. Occasionally found in water bodies.

*Uses :* The plant is used in diarrhoea, piles etc.

**Nitella furcata** (Roxb.) Ag. (Characeae)

A submerged golden green alga. Occasionally found in ditches.

*Uses :* Paste of plant is applied in skin diseases.

**Nymphaea nouchali** Burm. f. (Nymphaeaceae, *Poynu*)

A perennial aquatic with short root-stocks. Common in water bodies.

*Uses :* The whole plant is used as a cooling medicine.

**Nymphaea pubescens** Willd. (Nymphaeaceae, *Poynu*)

A perennial, aquatic herb, with erect root-stocks. Common in water bodies.

*Uses :* Rhizomes, stem and flowers are used as vegetables.

**Nymphoides hydrophylla** (Lour.) O. Ktze. (Menyanthaceae, *Kumudini*)

An aquatic plant with erect rhizome. Occasional in ponds and ditches.

*Uses :* Paste of leaves is applied in skin diseases.

**Nymphoides indicum** (L.) O. Ktze. (Menyanthaceae, *Kumudini*)

A perennial aquatic herb. Occasionally grows in ditches and ponds.

*Uses :* Paste of leaf is applied on boils and sores.

**Ottelia alismoides** (L.) Pers. (Hydrocharitaceae)

A submerged, rooted herb. Occasional in ponds and ditches.

*Uses :* Leaves are used as vegetable.

**Oryza rufipogon** Griff. (Poaceae, *Jangli-chokha*)

A marshy perennial grass. Occasionally found in marshy places.

*Uses :* Grains are used as food by poor people.

**Oryza sativa** L. (Poaceae, *Chokha*)

An annual cultivated plant. Cultivated as a monsoon crop in farms.

*Uses :* Grains are used as food.

**Oxalis corniculata** L. (Oxalidaceae, *Changeri*)

A prostrate-creeping herb. Common in moist areas.

*Uses :* Leaves are used as vegetable.

**Oxystelma esculentum** (L. f.) R. Br. (Asclepiadaceae, *Jal-dudhi*)

An annual, climbing herb. Occasionally found in moist places near ditches.

*Uses :* Paste of plant is applied in skin diseases.

**Paspalum scrobiculatum** L. (Poaceae, *Kodri*)

An erect, perennial grass, rooting at lower nodes. Found in marshes.

*Uses :* Grains are used as food by poor people.

**Persicaria barbata** (L.) Hara (Polygonaceae)

An erect, annual herb. Occasionally found in ditches and ponds.

*Uses :* Juice of plant is given in abdominal pains.

**Persicaria glabra** (Willd.) Gomez (Polygonaceae)

An erect, glabrous herb. Occasionally found in marshes.

*Uses :* Juice of plant is given in fever.

**Phyla nodiflora** (L.) Greene (Verbenaceae, *Ratvelio*)

A prostrate herb, rooting at the nodes. Commonly found in moist areas.

*Uses :* Plant is given in skin disorders.

**Pistia stratiotes** L. (Araceae)

A small, stoloniferous, floating herb. Occasionally found in ponds and ditches.

*Uses :* Paste of plant is applied in skin diseases.

**Polygonum plebeium** R. Br. (Polygonaceae)

A prostrate herb. Commonly found near drying water bodies.

*Uses :* Juice of plant is given in abdominal disorders.

**Potamogeton crispus** L. (Potamogetonaceae)

A slender submerged herb. Occasional in shallow water bodies.

*Uses :* Leaves are used as cattle fodder.

**Potamogeton nodosus** Poir. (Potamogetonaceae)

A submerged, rooted herb, with floating leaves. Common in ponds.

*Uses :* Leaves are used as cattle fodder.

**Potamogeton pectinatus** L. (Potamogetonaceae)

A profusely branched herb. Occasional in ponds.

*Uses :* Plant is used as cattle fodder.

**Pulicaria foliolosa** DC. (Asteraceae)

An erect, annual herb. Occasional in river-beds.

*Uses :* Paste of leaves is applied on cuts, boils etc.

**Ranunculus sceleratus** L. (Ranunculaceae, *Jal-dhana*)

An erect, annual herb. Occasionally found in river-beds.

*Uses :* Juice of plant is given in bronchitis, coughs etc.

**Rotala indica** (Willd.) Koehne (Lythraceae)

An annual herb, rooting at the nodes. Common near water bodies.

*Uses :* The plant is used as fodder.

**Rumex dentatus** L. (Polygonaceae, *Chuko*)

An erect, annual herb. Occasionally found in moist places.

*Uses :* Plant is used as vegetable.

**Sagittaria trifolia** L. (Alismataceae)

A scapigerous aquatic herb. Occasionally found in ditches and ponds.

*Uses :* Tubers are eaten as vegetable after proper cooking.

**Schoenoplectus articulatus** (L.) Palla (Cyperaceae)

An erect sedge with spongy stem. Common in wet places.

*Uses :* Plant is used as fodder.

**Schoenoplectus roylei** (Nees) Ovezinn & Czukav. (Cyperaceae)

An erect, annual sedge; leaves membranous. Occasional in ponds.

*Uses :* Plant is used as fodder.

**Scirpus tuberosus** Desf. (Cyperaceae, *Kaseru*)

An erect, perennial, rhizomatous sedge. Occasional in ditches.

*Uses :* Tubers are eaten after proper cooking.

**Sonchus oleraceus** L. (Asteraceae, *Dudhali-sonki*)

An erect, annual herb with latex. Common in moist places.

*Uses :* Paste of plant is applied in skin diseases.

**Spirodela polyrhiza** (L.) Schleid. (Lemnaceae)

A minute floating aquatic herb. Common in stagnant water bodies.

*Uses :* Plants are used as fertilizer.

**Tonningia axillaris** (L.) O. Ktze. (Commelinaceae, *Vandlo*)

A diffusely branched, annual herb. Found in marshy soils.

*Uses :* Leaves are used as fodder.

**Trapa natans** L. var. **bispinosa** (Roxb.) Makino (Trapaceae, *Shingoda*)

A floating annual herb; leaves with swollen petiole. Cultivated in water.

*Uses :* Fruits are cooked and then seeds eaten.

**Typha angustifolia** L. (Typhaceae, *Ghabajariu*)

A perennial herb with creeping rhizomes. Common in ditches and ponds.

*Uses :* Inflorescence powder is applied on wounds, cuts etc.

**Utricularia stellaris** L. f. (Lentibulariaceae, *Arkazavar*)

An aquatic, insectivorous herb. Occasionally found in ponds and ditches.

*Uses :* Flowers are used as antiseptic.

**Vallisneria spiralis** L. (Hydrocharitaceae, *Jalsarpolia*)

A submerged herb with creeping rhizomes. Occasional in ponds.

*Uses :* Paste of leaves is applied in skin diseases.

## RESULTS AND DISCUSSION

During the present study, altogether 104 plant taxa were collected. Of the 98 angiospermic species, 48 species belong to dicots and 40 species to monocots. Besides this, 4 species of pteridophytes and 2 species of algae were also recorded which possess some medicinal potential. It is hoped that present data will be helpful in sustainable utilization of aquatic resources in Gujarat.

## ACKNOWLEDGEMENTS

We express our thanks to Himmatbhai Meria, K.D. Sorathia, U.S. Baxi, S.K. Dodia, R.P. Parmar and P.K. Patel for providing help during field trips.

## REFERENCES

BOLE, A.V. & J.M. PATHAK 1988. *Flora of Saurashtra,* Vols. 2 & 3. BSI. Calcutta.

CHAVAN, A.R. & S.D. SABNIS 1961. A study on the hydrophytes of Baroda environments. *J. Indian Botanical Society* 40 : 121-130.

COOK, C.D.K. 1996. *Aquatic and Wetland plants of India*. Oxford University Press, New York.

FASSETT, N.C. 2000. *A Manual of Aquatic plants*. Agrobios (India), Jodhpur.

MAJID, F.Z. 1998. *Aquatic weeds, utility and development*. Agro-Botanica, Bikaner.

SHAH, G.L. 1978. *Flora of Gujarat State*, Vol. I-II. Sardar Patel University, Vallabh Vidyanagar.

SUBRAMANYAM, K. 1962. *Aquatic Angiosperms*. CSIR, New Delhi.

# OBSERVATIONS ON TRADITIONAL HERBAL REMEDIES FOR PARALYSIS IN GUJARAT

B.A. JADEJA, N.K. ODEDRA & D.R. ZALA

*Botany Department, M. D. Science College, Porbandar-360 575, Gujarat*

**ABSTRACT**

Gujarat is rich in ethnomedicinal plants. In the present paper, 63 species used in folk medicine for paralysis have been documented. Arrangement of species has been made according to the preparation of drugs along with the name of family and local name, besides parts used and preparation of drug.

## INTRODUCTION

Gujarat State is located on the west cost of India between 20°-24° and 68°-72°. Most of the tribals live in Dang, Panchmahal, Dahod, Surendranagar, Junagadh, Kutch and Porbandar districts. They belong to Maldhari, Koli, Bhil, Adivasi and some other tribes. They have nurtured their skills in identifying the plants for food, medicine, house building, agricultural operations, domestic uses, trade and barter, magico-religious beliefs, taboos etc. From time immemorial, they have especially accumulated deep, time tested knowledge about the medicinal plants in their habitat which has enabled them to come over from deadly diseases.

In paralysis, person loses sense of touch and energy in the affected body parts and that parts become stiff. Brain centres loss ability to control that part. Paralysis attack occurs more in males than the females, above the age of 50 years. It is a non-hereditary disease. Herbal traditional medicines used to cure paralysis give good results without any side effect.

The important ethnomedicinal works done in Gujarat include Thaker (1910, 1926), Shah (1978), Joshi *et al.* (1980, 1998), Shah & Gopal (1982, 1985), Bhatt & Sabnis (1987), Joshi (1988), Bhattacharya (1996), Mitaliya (1998), Punjani (1998), Bhatt & Mitaliya (1999), Jadeja (1999), Bhatt *et al.* (1999, 2000, 2001), Ant (2000), Silori & Rana (2000), Patel & Reddy (2000), Mitaliya *et al.* (2001, 2003) etc.

However, no comprehensive attempt has been made so far to survey the plants used in paralysis by tribal and rural population of Gujarat State. The present study was conducted to bridge the gap in our knowledge regarding the disease of paralysis and its herbal cure.

## MATERIALS AND METHODS

Medicinal plants were collected during 2003-2004 in flowering state in various parts of Gujarat, especially Junagadh (Gir forest), Porbandar (Barda hills), Valsad, Dang, Kutchh, Surat and Sabarkantha districts.

The first hand data were collect through personal interviews with a large number of tribal informants, elder women and Vaidyraj of herbal medicines of the villages who have knowledge about the therapeutic value of plants in paralysis. Besides uses, the local names of plants were also recorded as given by Vaidyraj. The collected plant materials have been deposited in the herbarium at M.D. Science College, Porbandar, Gujarat. The collected plants were identified with the help of recent floras (Shah, 1978; Bole & Pathak, 1988).

## ENUMERATION

Table 1. Plants used in paralysis in Gujarat.

| Scientific name | Family | Local name | Part (s) used | Mode of administration |
|---|---|---|---|---|
| *Aconitium heterophyllum* Wall. ex Royle | Ranunculaceae | Ativish | Stem | Two part *Pluchea arguta* seeds, one part of *Fagonia schweinfurthii* seeds and parts of other plants as mentioned in this table are equally taken and made in to a powder, decoction is prepared in half liter water with 25 gram powder and boiled for sometime, the decoction is divided into two parts and one part in the morning and other in the evening after adding one table spoon caster oil for one month is given to cure paralysis. |
| *Adhatoda zeylanica* Medic. | Acanthaceae | Aradushi | Leaves | |
| *Anethum graveolens* L. | Apiaceae | Suva | Seed | |
| *Asparagus racemosus* Willd. | Liliaceae | Satavari | Seed | |
| *Argyreia nervosa* (Burm. f.) Boj. | Convolvulaceae | Vardharo | Stem | |
| *Barleria prionitis* L. | Acanthaceae | Kantasherio | Fruit | |
| *Cassia fistula* L. | Caesalpiniaceae | Garmalo | Fruit | |
| *Coriandrum sativum* L. | Apiaceae | Dhana | Seed | |
| *Cyperus rotundus* L. | Cyperaceae | Nagarmoth | Leaves | |
| *Fagonia schweinfurthii* (Hadidi) Hadidi ex Ghafoor | Zygophyllaceae | Dhamaso | Seed | |
| *Hordeum vulgare* L. | Poaceae | Vaj | Seed | |
| *Piper longum* L. | Piperaceae | Lindi pipar | Leaves | |
| *Pluchea arguta* Boiss. | Asteraceae | Rasna | Seed | |
| *Sida cordifolia* L. | Malvaceae | Baldana | Fruit | |
| *Solanum incanum* L. | Solanaceae | Bethi Ringani | Fruit | |
| *Solanum indicum* L. | Solanaceae | Ubhi Ringani | Fruit | |
| *Terminalia chebula* Retz. | Combretaceae | Harde | Seed | |

| | | | | |
|---|---|---|---|---|
| *Tinospora cordifolia* (Willd.) Miers. ex Hk. f. & Thoms. | Menispermaceae | Galo | Stem | |
| *Trianthema triquetra* Rottl. ex Willd. | Aizoaceae | Satodi | Stem | |
| *Zingiber officinale* Rosc. | Zingiberaceae | Adu | Rhizome | |
| *Abrus precatorius* L. | Fabaceae | Chanothi | Seed | The parts are taken in equal quantity, crushed and soaked in two liter water, added sesame oil and caster oil and boiled up to total removal of water, the residue is rubbed every day on effected parts of body for 15 days. |
| *Celastrus paniculata* Willd. | Celastraceae | Malkakana | Seed | |
| *Crocus sativus* L. | Iridaceae | Kesar | Seed | |
| *Cyperus rotundus* L. | Cyperaceae | Motha | Seed | |
| *Lycium barbarum* L. | Solanaceae | Kachuro | Leaves | |
| *Papaver somniferum* L. | Papaveraceae | Aphin | Seed | |
| *Rubia cordifolia* L. | Rubiaceae | Majeth | Leaves | |
| *Sesamum indicum* L. | Pedaliaceae | Tal | Seed | |
| *Aegle marmelos* (L.) Corr | Rutaceae | Bili | Fruit, Leaves | Fruit pulp is warmed in oil and rubbed. Two table spoon of leaf-juice is taken orally. |
| *Ageratum conyzoides* L. | Asteraceae | Ajgandho | Stem | Stem mixed with root of *Vigna radiata*, *Mucuna pruriens* and *Ricinus communis* is made into a decoction in which is added 1 gram resin of *Ferula asafoetida* and taken orally. |
| *Allium sativum* L. | Liliaceae | Lasan | Clove (Bulblet) | 50 gram clove, resin of *Ferula asafoetida* and caster oil (about 10 ml) are crushed and taken orally twice a day for 70 to 80 days. |
| *Andrographis paniculata* (Burm. f.) Wall. ex Nees | Acanthaceae | Kariyatu | Seed | Seed powder with powder of clove of *Allium sativum* and rhizome of *Zingiber officinale* are mixed with honey and taken in equal amount twice a day to cure paralysis. |
| *Brassica juncea* (L.) Czern. & Coss. | Brassicaceae | Rai | Seed | 7 to 8 gram powder of seeds is taken orally with seed powder of *Commiphora wightii* and |

| | | | | |
|---|---|---|---|---|
| | | | | seeds of *Myristica fragrans* twice a day for a month. |
| *Caesalpinia crista* L. emend. Dandy ex Excell | Caesalpiniaceae | Kanchki | Root | 3 to 4 gram root-powder is taken orally with honey twice a day for 100 to 120 days. |
| *Calotropis procera* (Ait.) Ait. f. ssp. *hamiltonii* (Wt.) Ali | Asclepiadaceae | Ankado | Leaves | Leaves mixed with leaves of *Moringa oleifera* and leaves of *Nerium indicum* are boiled in water and the vapours are inhaled in paralysis. |
| *Cannabis sativa* L. | Cannabaceae | Bhang | Seed | Seed-powder is mixed with fruit-powder of *Piper nigrum* and 1 gram powder is taken orally with Cow milk after every 12 hour for long time. |
| *Cleome brachycarpa* Vahl ex DC. | Cleomaceae | Kasturi | Seed | 10 gram seeds and excreta of pigeon and 20 gram malbradlead are mixed and crushed to a powder; this powder is stored in a bottle and taken daily three times with honey in paralysis. |
| *Cicer arietinum* L.<br>*Cuminum cyminum* L.<br>*Curcuma longa* L. | Fabaceae<br>Apiaceae<br>Zingiberaceae | Chana<br>Jeeru<br>Haldal | Seed<br>Seed<br>Rhizome | Equal amount of seeds and rhizome are taken to form a powder; 5 gram of this powder is mixed with Ghee and taken with water; after that fruit juice of *Phyllanthus emblica* is taken to cure paralysis. |
| *Cocos nucifera* L. | Aracaceae | Nariyeli | Fruit | The crushed fruits and seeds of *Amygdalus communis* are mixed with Cow milk and consumed to cure paralysis. |
| *Commiphora wightii* (Arn.) Bhandari | Burseraceae | Gugal | Seed | Seed-powder is taken twice a day to give relief in paralysis. |

| | | | | |
|---|---|---|---|---|
| *Datura metel* L. | Solanaceae | Dhaturo | Leaves | Leaves and root of *Nerium indicum* and seeds of *Abrus precatorius* are taken equally to make a powder which is boiled with oil and rubbed in paralysis. |
| *Dolichos lobatus* L. | Fabaceae | Val | Seed | Resin of *Ferula asafoetida* and seeds of *Vigna radiata* are mixed to make a powder which is boiled with oil and taken orally in paralysis. |
| *Phyllanthus emblica* L. | Euphorbiaceae | Aamada | Fruit | |
| *Syzygium caryophyllatum* (L.) Alst. | Myrtaceae | Laving | Seed | The seeds are crushed with seeds of *Myristica fragrans* and seeds of *Nardostachys grandiflora* and in such powder is added fruit juice of *Phyllanthus emblica* to make tablets, which are taken orally in paralysis. |
| *Ferula asafoetida* L. | Apiaceae | Hing | Stem resin | The resin and seeds of *Dolichos lobatus* and seed of *Vigna radiata* are powdered which is boiled with oil and taken orally in paralysis. |
| *Ficus amplissima* J. E. Sm. | Moraceae | Pipar | Root | The powder of roots and clove of *Allium sativum*, rhizome of *Zingiber officinale* and seed of *Andrographis paniculata* is taken twice a day in paralysis. |
| *Glycyrrhiza glabra* L. | Fabaceae | Jethimadh | Seed | The seeds with *Piper longum*, rhizome of *Zingiber officinale*, seeds of *Hordeum vulgare* and rhizome of *Curcuma longa* are crushed to make a powder which is taken orally with 5 gram Ghee in paralysis. |

| | | | | |
|---|---|---|---|---|
| *Macuma pruriens* (L.) DC. | Fabaceae | Kaucha | Seed | The seeds, seeds of *Vigna radiata* and seeds of *Ricinus communis* are crushed and added resin of *Ferula asafoetida,* such a decoction is taken twice a day in paralysis. |
| *Moringa oleifera* Lamk. | Moringaceae | Saragavo | Leaves | Leaves mixed with leaves of *Nerium indicum* and leaves of *Calotropis procera* are boiled in water and the vapors are inhaled in paralysis. |
| *Myristica fragrans* Houtt. | Myristicaceae | Jayfal | Seed | The powder of seeds, seed of *Commiphora wightii* and seeds of *Brassica juncea* are made into small tablets and one tablet is taken twice a day in paralysis. |
| *Nardostachys grandiflora* DC. | Velerinaceae | Jatamasi | Seed | The powder of seeds, and seeds of *Syzygium caryophyllatum* and seeds of *Myristica fragrans* is made and added fruit-juice of *Phyllanthus emblica* to make tablets and one tablet is taken twice a day in paralysis. |
| *Nerium indicum* Mill. | Apocynaceae | Karen | Root | Oil of roots is rubbed everyday in paralysis. |
| *Oryza sativa* L. | Poaceae | Chokha | Seed | Seeds are boiled and added 100 gram powder of clove of *Allium sativum* and rhizome of *Zingiber officinale* in it and taken 5 to 15 gram twice a day. |
| *Vigna radiata* (L.) Wilczek | Fabaceae | Adad | Seed | Seed are soaked in water and boiled in Ghee and taken with jaggery twice a day in paralysis. |

| | | | | |
|---|---|---|---|---|
| *Piper nigrum* L. | Piperaceae | Mari | Seed | The powder of seeds with seeds of *Cannabis sativa* is made and taken 1 gram with Cow milk after every 12 hours in paralysis. |
| *Amygdalus communis* L. | Rosaceae | Badam | Seed | Seed-powder and crushed coconut powder are taken with Cow milk everyday in paralysis. |
| *Ricinus communis* L. | Euphorbiaceae | Aranda | Leaves | Tender leaves are crushed in milk and tied on head in paralysis. |
| *Trachyspermum ammi* (L.) Sprangue | Apiaceae | Ajamo | Seed | In the powder of seeds is added jaggery, boiled and taken orally with milk in paralysis. |
| *Trigonella foenum-graecum* L. | Fabaceae | Methi | Seed | |
| *Trichosanthes dioica* Roxb. | Cucurbitaceae | Parval | Whole plant | Crushed powder of entire plant is taken orally in paralysis. |
| *Vitex negundo* L. | Verbenaceae | Nagod | Leaves | Leaves extract is taken orally in paralysis. |

## RESULT AND DISCUSSION

The present work deals with traditional herbal remedies used to cure paralysis. About 63 species belonging to 61 genera of 38 families are identified to be used by villagers of Gujarat for paralysis. Plants used belong to 34 families of dicots and 4 families of monocots, and Fabaceae is the largest family with 7 species. Leaves of 12 species, fruits of 8 species, stem of 8 species, seeds of 32 species, root of 3 species and whole plant of 1 species provide herbal drugs for paralysis. Some plants are not found in Gujarat but their parts are easily available in market and villagers utilized them for the treatment of paralysis.

## ACKNOWLEDGEMENTS

The authors wish to express their sincere thanks to Dr. D.C. Bhatt, Department of Marine Science, Bhavnagar Univ., Bhavnagar for valuable advice. The authors also thank Prof. S.V. Buch, D.K.V. College, Jamnagar for help in identifying the plants.

## REFRENCES

ANT, H.M. 2000. *Ethnobotanical study of Angiosperms of Aravalli hills.* Ph. D. Thesis, Bhavnagar University, Bhavnagar.

BHATT, D.C. & K.D. MITALIYA 1999. Ethnomedicinal plants of Victoria Park (Reserved forest) of Bhavnagar, Gujarat, India. *Ethnobotany* 11 : 81-84.

———, ——— & S.K. MEHTA 2000. Plant twigs as tooth brushes. *Adv. Plant Sci.* 13 (1) : 19-22.

———, M.A. NURANI, K.D. MITALIYA & U.S. BAXI 2001. Some ethnomedicinal plants of Barda hills of Saurashtra, Gujarat, India. *Adv. Plant Sci.* 14 (2) : 427-431.

———, S.K. MEHTA & K.D. MITALIYA 1999. Ethnobotanical plants of Shetrunjara hill of Palitana, Gujarat. *Ethnobotany* 11 : 22-25.

BHATT, R.P. & S.D. SABNIS 1987. Contribution to the ehtnobotany of Khedbrahma region of North Gujarat. *J. Econ. Taxon Bot.* 9 (1) : 139-145.

BHATTACHARYA, G. 1996. Medico-ethnobotanical value of Saurashtra weeds. *J. Econ. Taxon. Bot.* Addl. Ser. 12 : 166-168.

BOLE. P.V. & J.M. PATHAK 1988. *Flora of Saurashtra,* Vols. 2 & 3. Botanical Survey of India, Calcutta.

JADEJA, B.A. 1999. Plants used by tribe Rabari in Barda hills of Gujarat. *Ethnobotany* 11 : 42-46.

JOSHI, M.C. 1988. Pharmaceutically important medicinal plants of Gujarat forests. *Bull. Medico-Ethno. Bot. Res.* 10 (4) : 372-373.

———, M.B. PATEL & P.J. MEHTA 1980. Some folk medicines of Dangs, Gujarat State. *Bull. Medico-Ethno. Bot. Res.* 1 : 8-24.

———, ——— & ——— 1998. In Goyal, H.R. (Edit.) *A report on medicinal plants of Kachchh* (Gujarat). CCRA & S, New Delhi.

MITALIYA, K.D. 1998. *Ethnomedicinal study of Angiosperms of Bhavnagar.* Ph.D. Thesis, Bhavnagar University, Bhavnagar.

———, D.C. BHATT, D.M. PATEL & P.N. JOSHI 2001. Medicinal value of some selected stem-bark used by tribals and rural folk in Gujarat. *Adv. Plant Sci.* 14 (1) : 255-259.

———, ———, J.N. PATEL & S.K. PATEL 2003. Observations on plants used for joint disorders by rural and tribal people of Gujarat. *J. Econ. Taxon. Bot.* 27 (1) : 212-216.

PATEL, K.C. & A.S. REDDY 2000. Ethnomedicinal uses of some common plants of Danta forest, North Gujarat State. *Seminar on Recent trends in plant Science,* Talod (Abstract no. 5).

PUNJANI, B.L. 1998. An ethnobotanical study of tribal areas of district Sabarkantha (North Gujarat). Ph.D. Thesis, North Gujarat University, Patan.

SHAH, G.L. 1978. *Flora of Gujarat State,* 2 Vols. Sardar Patel University, Vallabh Vidhyanagar (Gujarat).

——— & G.V. GOPAL 1982. An ethnobotanical profile of the Dangies. *J. Econ. Taxon. Bot.* 3 : 355-364.

——— & ——— 1985. Ethnomedicinal notes from the tribal inhabitants of North Gujarat (India). *J. Econ. Taxon. Bot.* 6 (1) : 193. 201.

———, A.R. MENON & G.V. GOPAL 1981. An account of ethnobotany of Saurashtra in Gujarat State. *J. Econ. Taxon. Bot.* 2:173-181.

SILORI, C.S. & A.R. RANA 2000. Indigenous knowledge of medicinal plants and their uses in Narayan Sarovar Sanctuary, Kutchch. *Ethnobotany* 12 : 1-7.

THAKER, J.L. 1910. *Flora of Bharda hills, Vanaspati Shastra.* Pravin Pustak Bhandar, Rajkot.

——— 1926. *Plants of Kutch & their utility.* Kutch Darbar, Kutch.

# PHENOLIC PROFILE IN SOME PETROCROPS

M.S.N. REDDY, M. RADHAKRISHNAIAH* & S. JELANI**

*Dept. of Botany, Sardar Patel College, Secunderabad-500 025, A. P., India*
**Dept. of Botany, Osmania University, Hyderabad-500 007, A. P., India*
***Education Officer, U. G. C., SERO, Hyderabad. A. P., India*

**ABSTRACT**

The present communication deals with the distribution pattern of some simple phenolic acids and phenyl propanoids in some taxa of *Sapotaceae, Apocynaceae, Asclepiadaceae, Convolvulaceae, Euphorbiaceae* and *Urticaceae,* and the taxonomic significance of these chemicals in delimitation of taxa from family to specific level.

## INTRODUCTION

The naturally occurring and taxonomically important phenolic compounds can be broadly categorised in to four groups, namely (i) flavonoids, (ii) phenolic acids, (iii) phenyl propanoids and (iv) phenolic quinines, of which simple phenols are taxonomically less important. Such simple phenolic acids as p-hydroxy benzoic acid, syringic acid, vanillic acid etc are universal in distribution, whereas salycilic acid and protocatchuic acid, characteristically found in *Ericaceae*, are less common (Harborne, 1984). Gallic acid with hydrolysable tannins is mostly confined to the families of Rosales (Margaret & Brain, 1981). However, its dimmer ellagic acid is common in dicotyledones (Harborne, 1984). The phenyl propanoids, as name indicates, are derived from protein amino acid and phenyl alamine. Among them, the most wide spread are the hydroxy sinnamic acids (caffeic, chlorogenic, p-coumaric, ferulic and sinapic acids) that have gained taxonomic significance. Phenolic quinines show rarity in distribution.

The phenolic compounds in general are valued for solving the taxonomic problems at different levels of hierarchy as evident from tribal delimitation of *Commelinaceae* (Singh, 1985), generic level in *Lotus* (Harney & Grant, 1964), *Moringa* (Narayana *et al.,* 1985), infrageneric divisions of *Bauhinia* (Nageshwar, 1986), *Indigofera* (Anuradha *et al.,* 1987). The present communication deals with the distribution pattern of some simple phenolic acids and phenyl propanoids and their taxonomic value in some petrocrops.

## MATERIAL AND METHODS

The taxa under investigation were collected from various localities of Andhra Pradesh, such as Nallamalai hills of Srisailam and its environs of Kuroool district, Mannanoor and Vatvarlapally forest area of Mahabubnagar district, Ananthagiri hills at Vikarabad of Rangareddy district, Narsapur forest of Medak district, Maredumilli forest of West Godavari district, and in and around Hyderabad city (Table 1).

For the detection of phenolic compounds, the leaf and the stem material was used. The shade dried leaves and finely chopped stem bits were heated in 2N Hcl on a water bath for 30 minutes. The acid digest was cooled and filtered. The filter was extracted with a few ml of diethyl ether in separating funnel and left for overnight. The upper supernatent ether layer, which contained the phenolic compounds, was taken for the detection of various phenolic constituents by bi-directional ascending paper chromatography (Harborne, 1984). Many of the constituents could be identified on comparison with the standards, a few that could not be identified are treated as unknown.

Table 1. List of the taxa investigated.

| S.N. | Name of the Taxa | Place of Collection | Voucher Specimen No. |
|---|---|---|---|
| | SAPOTACEAE | | |
| 1. | *Manilkara zapota* (L.) van Royen | Secunderabad | MSR - 1 |
| 2. | *Madhuca longifolia* (Koen.) Mac Bride var. *latifolia* (Roxb.) Chev. | Narsapur | MSR - 2 |
| 3. | *Mimusops elengi* L. | Vikarabad | MSR - 3 |
| | APOCYNACEAE | | |
| 4. | *Alstonia scholaris* (L.) R. Br. | Srisailam | MSR - 4 |
| 5. | *Carissa spinarum* L. | Vikarabad | MSR - 5 |
| 6. | *Wrightia tinctoria* (Roxb.) R. Br. | Narsapur | MSR - 6 |
| | ASCLEPIADACEAE | | |
| 7. | *Cryptostegia grandflora* R. Br. | Narsapur | MSR - 7 |
| 8. | *Gymnema sylvestre* (Retz.) R. Br. ex Schultes | Secunderabad | MSR - 8 |
| 9. | *Tylophora indica* (Burm. f) Merr. | Srisailam | MSR - 9 |
| | CONVOLVULACEAE | | |
| 10. | *Ipomoea carnea* Jacq. | Srisailam | MSR - 10 |
| 11. | *Ipomoea pes-caprae* (L.) R. Br. | Vishakapatnam | MSR - 11 |
| 12. | *Merremia tuberosa* (L.) Rendle | Hyderabad | MSR - 12 |

| | | | |
|---|---|---|---|
| | EUPHORBIACEAE | | |
| 13. | *Euphorbia elegans* Spreng. | Srisailam | MSR - 13 |
| 14. | *Euphorbia neriifolia* L. | Vikarabad | MSR - 14 |
| 15. | *Euphorbia tirucalli* L. | Hyderabad | MSR - 15 |
| 16. | *Givotia rottleriformis* Griff | Srisailam | MSR - 16 |
| 17. | *Jatropha curcas* L. | Narsapur | MSR - 17 |
| 18. | *Jatropha glandulifera* Roxb. | Mahbubnagar | MSR - 18 |
| | URTICACEAE | | |
| 19. | *Elatostemma cuneatum* Wt. | Rajamundry | MSR - 19 |
| 20. | *Pilea microphylla* (L.) Liebm. | Hyderabad | MSR - 20 |

Table 2. Distribution pattern of different phenolic constituents.

| **Chemical compounds** | **Name of the taxa** | | | | | | | | | | | | | | | | | | | |
|---|---|---|---|---|---|---|---|---|---|---|---|---|---|---|---|---|---|---|---|---|
| | 1 | 2 | 3 | 4 | 5 | 6 | 7 | 8 | 9 | 10 | 11 | 12 | 13 | 14 | 15 | 16 | 17 | 18 | 19 | 20 |
| ***Known Compounds*** | | | | | | | | | | | | | | | | | | | | |
| Caffeic acid | - | - | - | + | + | + | - | - | - | - | - | - | - | - | - | + | + | + | - | + |
| Chlorogenic acid | - | - | - | - | - | - | - | + | + | - | + | + | - | - | + | + | + | + | - | - |
| Ferulic acid | + | + | - | - | + | + | - | - | - | - | - | - | + | - | - | - | - | - | - | - |
| Gallic acid | - | - | - | - | - | - | - | + | + | - | - | - | + | + | + | + | - | + | - | - |
| O-Coumaric acid | - | + | - | - | + | - | - | - | + | - | + | + | - | + | + | + | + | + | - | - |
| P-Coumaric acid | - | - | - | + | + | + | - | - | - | + | - | + | + | + | + | + | + | + | - | - |
| P-OH benzoic acid | + | + | + | + | + | + | + | + | + | + | + | + | + | + | + | + | + | + | + | + |
| Protocatachuic acid | + | + | - | - | - | - | + | + | - | + | + | + | + | + | + | + | + | + | - | - |
| Salicylic acid | - | - | + | - | - | - | + | + | + | - | + | + | - | - | - | + | - | - | - | - |
| Vanillic acid | + | + | - | + | + | + | - | - | - | + | + | + | - | - | - | - | - | - | + | + |
| ***Unknown Compounds*** | | | | | | | | | | | | | | | | | | | | |
| 23/60 A | - | - | - | - | - | - | - | - | - | + | - | + | + | + | + | + | + | + | - | - |
| 29/56 B | - | - | - | + | - | + | + | + | + | - | - | + | - | + | + | + | - | + | - | - |
| 39/38 C | + | + | + | - | - | - | + | + | + | + | - | - | - | + | + | + | - | - | + | - |
| 40/39 D | - | - | - | + | + | + | - | - | - | - | - | - | + | + | + | - | - | - | - | - |
| 45/87 E | + | + | - | - | - | - | + | + | + | - | - | - | - | + | - | + | + | + | + | + |
| 54/30 F | + | - | - | - | + | + | - | - | - | - | + | + | + | - | - | - | - | - | + | + |

| | | | | | | | | | | | | | | | | | | | | |
|---|---|---|---|---|---|---|---|---|---|---|---|---|---|---|---|---|---|---|---|---|
| 57/72 G | + | + | + | + | + | - | + | + | + | - | + | - | + | + | + | + | + | + | - | + |
| 60/63 H | + | + | + | - | + | - | - | + | + | - | - | - | + | + | + | + | + | + | + | + |
| 65/74 I | - | - | - | - | - | + | - | - | - | - | - | - | + | + | + | + | + | - | - | - |
| 96/63 J | + | - | - | + | - | - | - | - | - | - | + | + | + | - | - | + | + | + | - | + |

Symbols : 1. *Manilkara zapota,* 2. *Madhuca longifolia* var. *latifolia,* 3. *Mimusops elengi,* 4. *Alstonia scholaris,* 5. *Carissa spinarum,* 6. *Wrightia tinctoria,* 7. *Cryptostegia grandiflora,* 8. *Gymnema sylvestre,* 9. *Tylophora indica,* 10. *Ipomoea carnea,* 11. *Ipomoea pes-caprae,* 12. *Merremia tuberosa,* 13. *Euphorbia elegans,* 14. *Euphorbia neriifolia,* 15. *Euphorbia tirucalli,* 16. *Givotia rottleriformis,* 17. *Jatropha curcas,* 18. *Jatropha glandulifera,* 19. *Elatostemma cuneatum,* 20. *Pilea microphylla.*

Table 3. Conspectus of the distribution pattern of different phenolic constituents.

| **Phenolic compounds** | **Name of the taxa** | | | | | |
|---|---|---|---|---|---|---|
| | 1<br>(3) | 2<br>(3) | 3<br>(3) | 4<br>(3) | 5<br>(6) | 6<br>(2) |
| ***Known Compounds*** | | | | | | |
| Caffeic acid | 0 | 3 | 0 | 0 | 3 | 1 |
| Chlorogenic acid | 0 | 0 | 2 | 2 | 4 | 0 |
| Ferulic acid | 2 | 0 | 0 | 0 | 1 | 2 |
| Gallic acid | 0 | 2 | 0 | 0 | 5 | 0 |
| O-Coumaric acid | 1 | 1 | 1 | 2 | 5 | 0 |
| P-Coumaric acid | 0 | 3 | 0 | 2 | 6 | 0 |
| P-OH benzoic acid | 3 | 3 | 3 | 3 | 6 | 2 |
| Protocatachuic acid | 2 | 0 | 2 | 3 | 6 | 0 |
| Salicylic acid | 1 | 0 | 3 | 2 | 1 | 0 |
| Vanillic acid | 2 | 3 | 0 | 3 | 0 | 2 |
| ***Unknown Compounds*** | | | | | | |
| 23/60 A | 0 | 0 | 0 | 1 | 6 | 0 |
| 29/56 B | 0 | 2 | 3 | 1 | 4 | 0 |
| 39/38 C | 3 | 0 | 3 | 1 | 3 | 1 |
| 40/39 D | 0 | 3 | 0 | 0 | 3 | 0 |
| 45/87 E | 2 | 0 | 3 | 0 | 4 | 2 |
| 54/30 F | 1 | 2 | 0 | 3 | 0 | 2 |
| 57/72 G | 3 | 2 | 3 | 1 | 5 | 1 |
| 60/63 H | 3 | 1 | 2 | 0 | 6 | 2 |
| 65/74 I | 0 | 1 | 0 | 0 | 5 | 0 |
| 96/63 J | 1 | 1 | 0 | 2 | 4 | 1 |

Symbols : 1. Sapotaceae, 2. Apocynaceae, 3. Asclepiadaceae, 4. Convolvulaceae, 5. Euphorbiaceae, 6. Urticaceae. The numbers shown in parenthesis indicate the number of taxa investigated.

## RESULTS AND DISCUSSION

All of the taxa of present study have shown uniform distribution of parahydroxy benzoic acid. The protocatechuic acid and unknown compounds G (57172) & H (60/63) are observed in majority of species. Para coumaric acid, ortho coumaric acid, vanillic acid and unknown compounds like B (29/56), C (39/38), E (45/87) and J (96/63) are noticed in about 50% of taxa under study. The rest of the compounds are rare in their occurrence. The details of distribution is presented in table 2.

Gibbs (1974) studied seven taxa and recorded the presence of caffeic acid in *Alstonia scholaris, Carissa spinarum, Ipomoea pes-caprae* and absence of the same in *Manilkara zapota, Mimusops elengi, Tylophora indica* and *Jatropha glandulifera.* Paracoumeric acid has been observed in *Ipomoea pes-caprae, Jatropha glandulifera* and *Tylophora indica,* where as the same is absent in *Manilkara zapota, Alstonia scholaris* and *Carissa spinarum.* Ferulic acid is traced in *Tylophora indica* and absence of same in the rest of the taxa. Parahydroxy benzoic acid has been identified in *Mimusops elengi, Ipomoea pes-caprae* and *Jatropha glandulifera.*

The present observation concur with those of Gibbs (1974). The present study is with reference to the presence of caffeic acid in *Alstonia* and *Carissa*, and absence of the same in *Manilkara, Mimusops, Tylophora indica,* p-coumeric acid in *Jatropha curcas* and absence of the same in *Manilkara zapota,* parahydroxy benzoic acid in *Manilkara zapota, Mimusops elengi, Ipomoea pes-caprae* and *Jatropha glandulifera.*

An artificial key based on morphological and chemical features is present hereunder :

### Key based on morphological and chemical characters

1a. Monochlamydous perianth is present :
- 2a. Flowers are unisexual. .......... **Euphorbiaceae**
  - 3a. Salicylic acid present. .......... *Givotia rottleformis*
  - 3b. Salicylic acid absent :
    - 4a. Ferulic acid present. .......... *Euphorbia elegans*
    - 4b. Ferulic acid absent :
      - 5a. Gallic acid absent. .......... *Jatropha curcas*
      - 5b. Gallic acid present :
        - 6a. Chlorogenic acid present. .......... *Euphorbia tirucalli*
        - 6b. Chlorogenic acid absent :
          - 7a. Caffeic acid absent. .......... *Euphorbia neriifolia*
          - 7b. Caffeic acid present. .......... *Jatropha glandulifera*
- 2b. Flowers are bisexual. .......... **Urticaceae**
  - 8a. Phenyl propanoids present. .......... *Pilea microphylla*

8b. Phenyl propanoids absent. ........................................................... *Elatostemma cuneatum*

1b. Dichlamydous perianth is present :

9a. Gynoecium with more than two carpels. ........................................................ **Sapotaceae**

10a. Vanillic acid absent. ........................................................................*Mimusops elengi*

10b. Vanillic acid present :

11a. O Coumeric acid present. ......................................*Madhuca longifolia* var. *latifolia*

11b. O Coumaric acid absent. ................................................................ *Manilkara zapota*

9b. Gynoecium bicarpellary :

12a. Ovary with many ovules :

13a. Stigma free, dumbel-shaped. ............................................................... **Apocynaceae**

14a. Ferulic acid absent. ............................................................... *Alstonia scholaris*

14b. Ferulic acid present :

15a. O-Coumeric acid present. ................................................ *Carissa spinarum*

15b. O-Coumeric acid absent. ................................................ *Wrightia tinctoria*

13b. Stigma united with anthers forming gynostegium. ......................... **Asclepiadaceae**

16a. Chlorogenic acid absent. ................................................ *Cryptostegia grandiflora*

16b. Chlorogenic acid present :

17a. Protocatechuic acid present. ........................................... *Gymnema sylvestre*

17b. Protocatechuic acid absent. ........................................... *Tylophora indica*

12b. Ovary with reduced no. of ovules. ............................................... **Convolvulaceae**

18a. P-coumeric acid absent. ............................................................ *Ipomoea pes-caprae*

18b. P-coumeric acid present :

19a. Chlorogenic present. ............................................................ *Merremia tuberosa*

19b. Chlorogenic absent. ................................................................... *Ipomoea carnea*

## ACKNOWLEDGEMENTS

The authors thank the Management and Principal, Sardar Patel College for providing infrastructure facilities and colleagues Dr. B.K. Vijay Kumar, K. Venkateshwarlu, Mrs. Hemalatha Acharya and Mr. V. Narender for their help.

## REFERENCES

ANURADHA, S.M.J., B.K. VIJAYA KUMAR, M. RADHAKRISHNAIAH & L.L. NARAYANA 1987. Chemotaxonomy of some species of *Indigofera* L. *Feddes Rep.* 98 : 499-503.

GIBBS, R.D. 1974. Chemotaxonomy of flowering plants, Vols. I-IV. MC Gill Queens Univ. Press, Montreal and London.

HARBRNE, J.B. 1984. Phytochemical methods. A guide to Modern Techniques of Plant analysis. 2nd edn. Chapman and Hall, London, New York.

HARNEY, M.P. & W.F. GRANT 1964. A chromatographic study of the phenolics of species of *Lotus*, closely related to *L. corniculatus* and their taxonomic significance. *Amer. J. Bot.* 51 : 621-627.

MARGARET, L.V. & V. BRAIN 1981. *Secondary Metabolism.* Macmillan Press Limited, London.

NAGESHWAR, G. 1986. *Chemotaxonomy of some Caesalpiniaceae.* Ph.D. Thesis, Osmania University, Hyderabad.

NARAYANA, L.L., P.S. NARAYANA, V.S. RAJU & M. RADHAKRISHNAIAH 1985. The bearing of phenolic acids on the status and systematic position of Moringaceae. *J. Swamy Bot. Cl.* 2 : 87-92.

SINGH, V. & D.K. JAIN 1987. *Taxonomy of angiosperms.* 2nd ed. Rastogi Publications, Meerut.

# PHYTOCHEMICAL AND ANTIMICROBIAL SCREENING OF THE LEAF EXTRACT OF *OCIMUM GRATISSIMUM* L.

O.I. OKECHUKWU & F. TAHIR

*Biological Sciences Programme, Abubakar Tafawa Balewa University, Bauchi*

## ABSTRACT

The antimicrobial activity of various plant extracts have been recognized for many years. However, few investigations have compared large numbers of extracts using methods that are directly comparable. In this study, the extracts of the plant *Ocimum gratissimum* L. was investigated for activity against *Staphylococcus aureus, Escherichia coli, Proteus mirabilis, Pseudomonas aeruginosa* and *Candida albicans* using the disc diffusion method. The methanolic extract possessed both antibacterial and antifungal activities. The Minimum Inhibitory Concentration (MIC) was between the range of 100mg/ml-750 mg/ml. The phytochemical screening of the extract indicated the presence and absence of some compounds. The extract was further separated by acid/base separation techniques which yielded three subfractions, which were further tested for antimicrobial activities. The results indicated the presence of activity especially with the organic basic and organic neutral portions. Significant differences were observed in the activity of the various extract concentrations, the interactions between the concentrations and the organisms and the varying degrees of response of the organism to the extract. The result support the notion that plant extracts may have a role as pharmaceuticals and preservatives.

## INTRODUCTION

The research into biologically active compounds from the natural sources has always been of great interest for scientists looking for new sources of useful drugs against infectious diseases (Bisignano *et al.,* 2000). *Ocimum gratissimum* L. is widely employed in traditional medicine in Africa for diarrhoea, dysentery, eye diseases, scabies, cold and catarrh etc (Abbiw, 1990). Previous studies (Sofowora, 1984) reported that *Ocimum* sp. when boiled with water to form decoction was devoid of antimicrobial activity.

Other investigations (Oladosu *et al.,* 2002) showed that different extracts of *Ocimum gratissimum* L. exhibited antibacterial and antifungal activity against *Staphylococcus aureus* and *Candida albicans* responsible for common skin infections and candidiasis respectively. While these data are useful, the reports are not comparable due to methodological differences such as choice of plant extract(s), test micro-organism(s) and antimicrobial test method (Janssen *et al.,* 1987).

The purpose of this study was to evaluate the antibacterial and antifungal activities of *Ocimum gratissimum* L. methanolic extract against a diverse range of organisms comprising of a gram-positive organism, three gram-negative organisms and a yeast as well as it's phytochemical analysis.

## MATERIALS AND METHODS

### Organisms and growth conditions

Micro-organisms were obtained from the culture collection of the Microbiology laboratory at the Bauchi specialist Hospital. Organisms were *Proteus mirabilis, Escherichia coli, Pseudomonas aeruginosa, Staphylococcus aureus* and *Candida albicans.* The organisms were maintained on nutrient agar slants (NAS). The inocula was prepared by inoculating approximately 2 ml sterile peptone water with 2-3 colonies of each organism taken from NAS.

### Plant extract

Leaves of *Ocimum gratissimum* L. were collected from the Kari Estate in Bauchi State. The fresh leaves were washed and air dried. The dried leaves were then pounded into powder form. The extract used for this study was obtained using cold maceration method by dissolving the powdered material in the ratio of 1:5 w/v as suggested by Kela *et al.* (1989) and allowed to soak for 72 hours with intermittent changing of the solvent after every 24 hours. The solution gotten after filteration was concentrated using rotary evaporation.

### Disc diffusion method

The antibacterial and antifungal effects were tested by the disc-diffusion method (Bauer *et al.,* 1966) utilizing sterile discs of 6 mm diameter impregnated with 100 mg/ml of the extract. This was done in duplicate plates. The minimum inhibitory concentration (MIC) values were evaluated using the disc diffusion method (Bauer *et al.,* 1966). The disc were impregnated with different concentrations of the extract ranging from 1000 mg/ml-50 mg/ml. The MIC is defined as the lowest concentration of the drug which inhibits the visible growth after 18 hours.

### Phytochemical screening

Phytochemical screening of the plant extract was done to determine presence of certain groups of phytochemicals which include tannins, alkaloids, saponins, reducing sugars, cardiac glycosides, anthraquinones, steroids following standard phytochemical procedures (Pew, 1948; Shoppe, 1964; Harbone, 1973; Trease & Evans, 1983).

### Chemical separation (solvent/solvent extraction)

The crude methanolic extract was subjected to Acid/Base separation technique which is a general purpose protocol for the extraction of natural products soluble in organic solvent.

### Susceptibility testing of fractions from chemical separation

The three sub-fractions (i.e. organic neutral, organic basic and organic acidic) gotten from the chemical separation were concentrated using orotary evaporator and tested for activity against the test organisms using the disc-diffusion method (Bauer *et al.,* 1966).

Table 1. Preliminary screening of plant sample for antimicrobial activity.

| **Test micro-organisms** | **MeOH extract of *Ocimum gratissimum* L. (100 mg/ml)*** |
|---|---|
| *Staphylococcus aureus* | - |
| *Candida albicans* | + |
| *Escherichia coli* | - |
| *Proteus mirabilis* | - |
| *Pseudomonas aeruginosa* | - |
| MeOH (control) | - |

Symbols : *Measured in duplicate; - = Not active; + = Active.

Table 2. Mean diameters of inhibition zones at various concentrations of MeOH extract.

| **Concentration (mg/ml)*** | **Diameters of zone of inhibition (mm) after 24 hours** | | | | |
|---|---|---|---|---|---|
| | ***Staphylococcus aureus* (A)** | ***Escherichia coli* (C)** | ***Candida albicans* (B)** | ***Proteus mirabilis* (D)** | ***Pseudomonas aeruginosa* (A)** |
| 1000a | 23.5 | 10.9 | 17.5 | 12.0 | 9.5 |
| 750b | 17.0 | 7.3 | 12.5 | 7.5 | 7.5 |
| 500c | 11.0 | 6.3 | 11.5 | - | - |
| 350d | 8.0 | 6.0 | 11.0 | - | - |
| 300d | 6.0 | 5.0 | 10.5 | - | - |
| 200d | 5.0 | - | 10.0 | - | - |
| 150e | - | - | 9.5 | - | - |
| 100f | - | - | 9.0 | - | - |
| 50 | - | - | - | - | - |

Symbols : *Measured in duplicate; - = Not active; a-c = means in the same row without a common superscript differ (P>0.05); A-C = means in the same column without a common superscript differ (P>0.05).

Table 3. Minimum inhibitory concentration for test organisms subjected to the methanolic extract of *Ocimum gratissimum* L.

| **Plant extract** | **MIC (mg/ml)* Organisms** | | | | |
|---|---|---|---|---|---|
| | ***Staphylococcus aureus*** | ***Escherichia coli*** | ***Pseudomonas aeruginosa*** | ***Proteus mirabilis*** | ***Candida albicans*** |
| *Ocimum gratissimum* L. | 200 | 300 | 750 | 750 | 100 |

Symbols : *Measured in duplicate.

Table 4. Some phytochemical contents of the MeOH leaf extract of *Ocimum gratissimum* L.

| Groups | Test | Observation | Inference |
|---|---|---|---|
| Alkaloids | (a) Dragendortts | Precipitate observed | Alkaloids present |
| | (b) Picric | Precipitate observed | |
| Tannins | (a) Ferric chloride | Blue green solution was observed | Tannins present |
| | (b) Lime water | No precipitate observed | |
| Saponins | Frothing | No frothing was observed after warming | Saponins absent |
| Reducing sugars | Fehlings solution | A brick-red precipitate was observed at the bottom of test tube on warming in a water bath | Reducing sugars present |
| Cardiac glycosides | Salkowski | A brown ring was observed at the interface between the sulphuric acid layer and acetic acid layer | Cardiac glycosides present |
| Anthraqu-inones | Borntrager | No pink/violet colour was observed in the ammonia layer | Anthraquinones absent |
| Steroids | Salkowski | A reddish brown colour at interface was seen | Steroids present |

Table 5. Activity of fractions upon tested micro-organisms.

| Micro-organisms | Zone of inhibition (mm)* 24 hours Fractions | | | |
|---|---|---|---|---|
| | Organic acidic $C^C$ | Organic basic $H^D$ | Organic basic $B^A$ | Organic neutral $B^B$ |
| *Staphylococcus aureus* | 12.0 | NA | 1.0 | 6.0 |
| *Proteus mirabilis* | NA | NA | 7.0 | 7.0 |
| *Pseudomonas aeruginosa* | 8.0 | NA | 6.0 | - |
| *Escherichia coli* | NA | 9.5 | 10.5 | 6.0 |
| *Candida albicans* | NA | NA | 7.0 | 9.0 |

Symbols : *Measured in duplicate; H = Hexane; C = Chloroform; B = Butanol; NA = No activity; a-c = means in the same row without a common superscript differ (P>0.05); A-D = means in the same column without a common superscript differ (P>0.05).

## Statistical analysis

Simple statistical analysis of Variance (ANOVA) was conducted on the data obtained in order to compare means and established significance of finding. This was carried out using the mini-tab statistical package.

## RESULTS

The results show a wide variation in the antimicrobial properties of the plant extract. Table 1 gives the results for the preliminary screening of the extracts. It was observed that at 100 mg/ml of the extract only *Candida albicans* is susceptible at that concentrations. Table 2 shows the effect of varying concentrations of the extracts on the test organisms. It is observed that of the five test organisms *Staphylococcus aureus* and *Candida albicans* were the most susceptible to the extract in comparison to *Proteus mirabilis* and *Pseudomonas aeruginosa*. Table 3 gives an insight on the various minimum inhibitory concentrations of the extract for each organism and it ranged from 1000 mg/ml-750 mg/ml.

Table 4 shows results of phytochemical analysis conducted on the extract. It was noted that the extract was appreciably positive for some natural products but was not for some. Some phytochemicals present include alkaloids, tannins, reducing sugars etc and some products absent include saponins and anthraquinones.

Different solvents were used for the chemical separation. The fractions obtained were yellowish brown in colour, organic basic (with hexane as extracting solvent) had a light green-brown colouration as against a deep green colour for the organic basic (with butanol as extracting solvent). The organic neutral portion had a light yellow colour. Table 5 contains the results for the antimicrobial activities of the separated portions of the extract. It shows that all the fractions had some activity on the test organisms. The organic basic subfraction (with butanol as extracting solvent) was the most active as indicated by its zone of inhibition and the fact that it showed activity on all the test organisms. The organic basic sub-fraction (in the hexane as extracting solvent) showed some degree of activity as it inhibited *Escherichia coli* but had no activity at all on the other organisms.

## DISCUSSION

The inhibitory activity of the plants crude extract was most active against *Staphylococcus aureus, Candida albicans* and *Escherichia coli* which conform with reports on activity of the plant (Dalziel, 1937; Abbiw, 1990) used in the treatment of many microbial infections. Also observed that *Proteus mirabilis* and *Pseudomonas aeruginosa* were less sensitive to the extract and both organisms are gram negative organisms.

Significant differences were observed in the responses of the organisms to the extract ($P \geq 0.05$), but no significant difference was noted between the responses of *Proteus mirabilis* and *Pseudomonas aeruginosa* to the extract. Significant differences were also noted in activities of the different concentrations of the extract to the organisms ($P \geq 0.05$), suggesting that the concentrations of the active compound(s) were increasing with increasing concentration.

Results for the MIC of the extract showed that generally the MIC for all the test organisms was between 100 mg/ml-750 mg/ml which indicates that the plant extract is active but at higher concentrations. This could be as a result of the mode of extraction where methanol was the only extracting reagent which must have lead to heavy saturation with various compounds (Garreft & Brown, 1964). This suggests the use of more than one solvent for extraction which brings about a finer separation, thus minimizing the saturation of different compounds.

The results from the phytochemical analysis showed that the plant extract tested negative for saponins and anthraquinones. Absence of detection of these compounds in the extract could be due to two reasons; the fact that they were either present in quantities undetectable by the method of analysis or they were completely absent as indicated by the result of the analysis (Tahir, 2001).

Alkaloids, tannins, reducing sugars, cardiac glycosides and steroids were found to be present in the sample (extract). This is in agreement with previous claims on other plants. Claims have been made on tannins and flavonoids as active antifungal agents of plants which explains the use of the plant in treatment of infections caused by *Candida albicans. Ocimum gratissimum* L. is used to treat wounds in traditional medicine. The presence of tannins in the plant justifies its use for wound treatment because tannins encourage the formation of new tissue under the leathery layer formed on broken mucosal surface.

The organic basic portion (with hexane as extracting solvent) had the least activity as against the organic basic portion (with butanol as extracting solvent) which had the highest activity because it inhibited all the organisms. This could be attributed to the fact that the active compounds were in a higher concentration in this fraction, Statistically it showed that their was a significant difference in the activities of the sub-fraction (P>0.05) and there was also a significant difference in the various responses of the organisms to the sub-fractions (P>0.05). Activities of the sub-fractions were observed to be lower than that of the crude extract, it could be that the activity of the active ingredient(s) is enhanced in the presence of other compounds as a mixture (synergism) or alternatively, it could be that the separated fractions yielded lower concentrations of the active ingredients as a result of the separation; thus lowering its activity.

The findings from this study when compared to previously published results may be contradictory due to some reasons. First, the composition of plant extracts is known to vary according to local climatic and environmental conditions (Janssen *et al.,* 1987). Secondly, the method used to assess antimicrobial activity and the choice of test organism(s) vary between publications (Janssen *et al.,* 1987). The method of assay used for this study i.e. disc-diffusion method has it's limitations which include generation of preliminary, qualitative data only, as the hydrophobic nature of most plant extracts prevents the uniform diffusion of these substances through the agar medium (Janssen *et al.,* 1987).

In line with some of the limitations of this work, further research should be done to attempt to isolate and characterize the active ingredients. Further more, other solvents could as well be used for the extraction procedure as well as other techniques to attempt a comparative study of results obtained.

In summary, this study confirms that the plant extract possesses *in vitro* antibacterial and antifungal activity. However, if the plant extract is to be used for medicinal purposes and food preservation, issues of safety & toxicity will need to be addressed.

## ACKNOWLEDGEMENTS

The authors thank the Bauchi Specialist Hospital, Abubakar Tafawa Balewa University (A.T.B.U.) for providing the organisms and materials needed for this study.

## REFERENCES

ABBIW, O.K. 1990. *Useful plants of Ghana,* pp. 118-125. Immediate Technology Publication Ltd., London.

BAUER, S.W., W.M. KIRBY, J.C. SHERRIS & M. THURCHK 1966. Antibiotic susceptibility testing by standardized single disc method. *American J. Pathology* 45 : 491-496.

BISIGNANO, G. *et al.* 2000. Antimicrobial activity of *Mitracarpus scaber* extract and isolated constituents. *Letters Applied Microbiology* 30 : 105-108.

DALZIEL, J.M. 1937. *Useful plants of West Tropical Africa.* The Crown Agents for Over seas, Colonica, London.

GARREFT, E.R. & M.R.W. BROWN 1964. Kinetics and Mechanics of action of antibiotics on micro-organisms. *Journal Pharmaceutical Science* 53 : 179.

HARBONE, J.B. 1973. *Phytochemical Methods,* pp. 81-131. Publisher Chapman and Hall, London & New York.

JANSSEN, A.M., J.J.C. SCHEFFER & A. BAERHEIM SVENDSEN 1987. Antimicrobial activity of essential oils : a 1976-86 literature review. Aspects of the test methods. *Planta Medica* 53 : 395-398.

KELA, S.L. 1998. *Screening of some Nigerian plants for molluscidal activity.* M.Sc. Thesis, Ahmadu Bello University, Zaria.

PEW, J.C. 1948. *American Journal of Chemical Society* 70 : 303 & In Harbone, J.B. 1973. *Phytochemical methods.* Chapman and Hall, London & New York.

SHOPPE, C.W. 1964. *Chemistry of steroids,* pp. 145-148. 2nd edition. Butterwords, London.

SOFOWORA, A. 1984. *Medicinal plants and traditional medicine in Africa.* John Wiley and Sons, Chichester, N. Y.

TAHIR, F. 2001. Screening and characterization of the antimicrobial components of some medicinal plants used in Bauchi. Ph.D. Thesis, ATBU, Bauchi.

TREASE, G.E. & J. EVANS 1983. *Pharmacognosy,* pp. 21-22. 12th ed. Buinlliene Tindall, London.

# PHYTOCHEMICAL STUDIES IN GENERA *TETRASTIGMA* (MIQ.) PLANCH. AND *VITIS* L. (FAMILY VITACEAE) AND THEIR TAXONOMIC SIGNIFICANCE

S.G. PATIL & B.K. HONRAO

*Division of Plant Sciences, Agharkar Research Institute, Agarkar Road, Puna-411 004 (INDIA)*

**ABSTRACT**

Photochemical analysis of nine species of *Tetrastigma* (Miq.) Planch. and 3 species of *Vitis* L. (Vitaceae) revealed presence of raphides and mucilaginous substances and absence of cyanogenic glycosides, quinones, naphto-quinones and aurones compounds in all species. Thirteen tests show presence or absence of chemical compounds with variable reactions and three with doubtful reactions. Presence of raphides in the species suggest the genus *Tetrastigma* (Miq.) Planch. to be primitive. Likewise, other observations also support the taxonomic treatment of the genus *Tetrastigma* (Miq.) Planch. Differential reactions can be used for the identification of these taxa. Presence of such phytochemicals have special significance in plant protection to control various diseases and pests as a natural pesticide; otherwise, in crop improvement to induce genetic resistance through resistant genotypes in cultivated grapes with modern methods of Biotechnology.

## INTRODUCTION

The grape family Vitaceae comprises of thirteen genera and about 800 species (Shetty & Singh, 2000), growing mostly in tropical and subtropical parts of the World. In India, the family is represented by 8 genera and 63 species. The genus *Tetrastigma* (Miq.) Planch. is represented by about 16 species in India (Shetty & Singh, 2000), and *Vitis* L. includes about 3 wild species and several cultivated species and their cultivars (Shetty & Singh, 2000).

Very few reports are available on the phytochemistry of the family Vitaceae (Singh *et al.*, 1986; Patil *et al.*, 1997a, b, 1990, 2000; Patil & Honrao, 2000). Relatively, information on the genus *Tetrastigma* (Miq.) Planch. and *Vitis* L. in family is very meager. Hence, an attempt has been made to study the phytochemical composition of some species of *Tetrastigma* (Miq.) Planch. and *Vitis* L.

## MATERIALS AND METHODS

Nine (9) species of *Tetrastigma* (Miq.) Planch. and 3 species of *Vitis* L. collected from various parts of India under the All India Co-ordinated Research Project on Sub-Tropical Fruits - Grapes and maintained in nursery under identical conditions were studied in present work.

Table 1. Names of the species, distribution and collection sources of different species of *Vitis* L. and *Tetrastigma* (Miq.) Planch.

| Species names | Distribution | Collection area |
|---|---|---|
| *Tetrastigma andamanicum* (King) Suesseng. ex Suesseng. | North Andaman | Andamans |
| *Tetrastigma bracteolatum* (Wall.) Planch. | North East India | Assam |
| *Tetrastigma campylocarpum* (Kurz) Planch. | North East India | Meghalaya |
| *Tetrastigma gamblei* Shetty & Singh | South India | Kerala |
| **Tetrastigma glycosmoides* Planch. | Sri Lanka | Andamans (introduced) |
| *Tetrastigma leucostaphylum* (Dennst.) Alston ex Mabb. | South India & N. E. India | Tamilnadu |
| *Tetrastigma nilagiricum* (Miq.) Shetty | South India | Karnataka |
| *Tetrastigma obovatum* Gagnep. | North East India | Assam |
| *Tetrastigma sulcatum* (M. Lawson) Gamble | South India & North East India | Assam |
| *Vitis kunstleri* King | Malay Peninsula | Andamans (introduced) |
| *Vitis novemefolia* Wall. ex King | Singapore | Kerala (introduced) |
| *Vitis scortechinii* King | Malay Peninsula | Andamans (introduced) |

*Mabberley (Fl. Ceylon 9 : 480. 1995) considered it synonymous to *Tetrastigma nilagiricum* (Miq.) Shetty.

Table 2. Inference of the phytochemical tests in *Vitis* L. and *Tetrastigma* (Miq.) Planch. species.

| Fam. Vitaceae species | Various Phytochemical Tests | | | | | | | | | | | | | | | | | |
|---|---|---|---|---|---|---|---|---|---|---|---|---|---|---|---|---|---|---|
| | 1 | 2 | 3 | 4 | 5 | 6 | 7 | 8 | 9 | 10 | 11 | 12 | 13 | 14 | 15 | 16 | 17 | 18 |
| *Tetrastigma andamanicum* | +++ | + | - | - | +++ | - | + | - | - | - | ++ | - | +++ | ++ | +++ | - | ++ | - |
| *Tetrastigma bractolatum* | - | - | - | - | + | - | + | - | - | - | ++ | - | - | ++ | + | +? | + | - |
| *Tetrastigma campylocarpum* | + | - | + | - | +++ | - | +? | - | + | - | +? | + | ++ | ++ | ++ | - | - | - |
| *Tetrastigma gamblei* | ++ | +++ | - | +? | ++ | - | + | - | - | - | - | ++ | - | ++ | ++ | + | ++ | - |
| *Tetrastigma glycosmoides* | - | - | + | - | ++ | - | -? | - | + | + | ++ | - | ++ | + | + | - | - | - |
| *Tetrastigma leucostaphylum* | ++ | + | - | - | ++ | - | - | - | - | - | ++ | - | - | ++ | ++ | + | ++ | - |

| *Tetrastigma nilagiricum* | ++ | + | - | +++ | ++ | - | - | - | - | - | + | - | + | ++ | ++ | - | ++ | - |
|---|---|---|---|---|---|---|---|---|---|---|---|---|---|---|---|---|---|---|
| *Tetrastigma obovatum* | ++ | +++ | - | - | +++ | - | - | - | - | - | - | - | + | ++ | ++ | + | ++ | - |
| *Tetrastigma sulcatum* | ++ | ++ | - | +++ | +++ | - | ++ | - | - | - | + | + | ++ | + | +++ | - | ++ | - |
| *Vitis kunstleri* | - | + | + | - | ++ | - | + | - | - | - | ++ | - | + | ++ | ++ | + | + | - |
| *Vitis novemefolia* | ++ | ++ | - | +? | + | - | - | - | - | - | ++ | - | - | ++ | + | + | ++ | - |
| *Vitis scortechinii* | ++ | + | - | - | - | - | + | - | - | - | ++ | - | +++ | ++ | ++ | - | +++ | - |

Symbols : 1. Cigarette Test, 2. Hot-Water Test, 3. Oxalic reaction, 4. Leuco-anthocyanin Test, 5. HCL-Methanol Test, 6. HCN Test A, 7. Syringin Test A, 8. Juglone Test A, 9. Juglone Test B, 10. Juglone Test C, 11. Maule Test, 12. Saponin Test A, 13. Tannin Test, 14. Raphide Test, 15. Mucilage Test, 16. Ehrlich Test A, 17. Flavonoid Test, 18. Aurone test. Reactions : + = Very weak, ++ = Weak, +++ = Strong, ++++ = Very strong, ? = Doubtful, - = Negative.

Table 3. Classification of *Tetrastigma* (Miq.) Planch. and *Vitis* L. species based on test/chemical compounds.

| **Tests No.** | **Name of Chemical Compounds** | **Number of species showing reactions** | | | | |
|---|---|---|---|---|---|---|
| | | **Present** | **Absent** | **Variable Quantities** | | |
| | | | | **Stronger** ++++/+++ | **Weaker** ++/+ | **Doubtful** +/- ? |
| 1. Cigarette Test | Polyphenols | 8 | 3 | 1 | 8 | - |
| 2. Hot-Water Test | Polyphenols | 9 | 3 | 2 | 7 | - |
| 3. Oxalic reaction | Acidic compounds | 3 | 9 | - | - | - |
| 4. Leuco-anthocyanin Test | Leuco-anthocyanins | 2 | 8 | - | - | 2 |
| 5. HCL-Methanol Test | Catechol tannins | 11 | 1 | 4 | 7 | - |
| 6. HCN Test A | Cyanogenic glycosides | - | 12 | - | - | - |
| 7. Syringin Test A | Syringin glycosides | 5 | 5 | - | - | 2 |
| 8. Juglone Test A | Quinine, Naptho-quinones | - | 12 | - | - | - |
| 9. Juglone Test B | Flavonoid substance | 2 | 10 | - | - | - |
| 10. Juglone Test C | Aesculin/Coumarin | 1 | 11 | - | - | - |
| 11. Maule Test | Lignins | 9 | 2 | - | - | 1 |
| 12. Saponin Test A | Saponins | 3 | 9 | - | - | - |
| 13. Tannin Test | Tannins | 8 | 4 | - | 6 | - |
| 14. Raphide Test | Raphides (Calcium oxalate) | 12 | - | 2 | 12 | - |
| 15. Mucilage Test | Mucilagenous substances | 12 | - | - | 9 | - |
| 16. Ehrlich Test A | Aucubin like substances | 5 | 6 | 3 | - | 1 |
| 17. Flavonoid Test | Flavonoids | 10 | 2 | 1 | 9 | - |
| 18. Aurone test | Aurenes | - | 12 | - | - | - |
| | Total Tests | 15 | 16 | 6 | 7 | 3 |

Table 4. Phytochemicals showing positive or negative reactions.

| **Phytochemicals** | **No. of species with positive reactions** | **No. of species with negative reactions** |
|---|---|---|
| Polyphenols (Cig. T.) | 9 | 3 |
| Polyphenols (Htw. T.) | 9 | 3 |
| Acidic compounds | 3 | 9 |
| Leuco-anthocyanins | 2 | 8 |
| Catechol tannins | 11 | 1 |
| Syringin glycosides | 5 | 5 |
| Flavonoid substance | 2 | 10 |
| Aesculin/Coumarin | 1 | 11 |
| Lignins | 9 | 2 |
| Saponins | 3 | 9 |
| Tannins | 8 | 4 |
| Aucubin like substances | 5 | 6 |
| Flavonoids | 10 | 2 |

Table 5. The Phytochemicals showing positive reactions with variable quantities.

| **Phytochemicals** | **No. species with positive reactions** | **Stronger reactions (++++/+++)** | **Weaker reactions (++/+)** |
|---|---|---|---|
| Mucilaginous substances | 12 | 2 | 10 |
| Polyphenols (Cig. T) | 9 | 1 | 8 |
| Polyphenols (Htw. T) | 9 | 2 | 7 |
| Catechol tannins | 11 | 4 | 7 |
| Tannins | 8 | 2 | 6 |
| Flavonoids | 10 | 1 | 9 |

Qualitative analysis for detection of phytochemicals was carried out by the methods suggested by Harborne (1973) and Gibbs (1974) from the fresh plant material of stems and leaves.

## RESULTS AND DISCUSSION

In India, about sixteen species of *Tetrastigma* (Miq.) Planch. and three species of *Vitis* L. have been recorded. About nine species of *Tetrastigma* (Miq.) Planch. and 3 species of *Vitis* L. were collected and maintained in nursery. Various phytochemical tests were undertaken and inference for various phytochemicals was recorded for presence or absence or doubtful reactions from stem and leaves. Data has been summarized in Table 2 & 3.

Broadly, the results obtained can be classified into following groups.

### (i) Phytochemicals presents in all the species

Observations were recorded on applying various tests. Raphides and Mucilage tests suggest presence of calcium oxalate and mucilaginous substances in all *Vitis* L. and *Tetrastigma* (Miq.) Planch. species. Raphides were relatively less in *T. glycosmoides* and *T. sulcatum.* According to Gibbs (1954), the raphides have taxonomic significance in evolutionary treats and considered as a primitive character. Thus, the genus *Tetrastigma* (Miq.) Planch. occupies primitive position in family Vitaceae.

Further, *T. andamanicum* and *T. sulcatum* show stronger reactions for mucilage substances and very weaker reactions were observed in *T. bracteolatum, T. glycosmoides* and *Vitis novemefolia.* Smith & Monotgomery (1959) stated that the mucilage may arise from starch and seems to act as an agent for holding water.

### (ii) Phyto-chemicals present or absent in all the species

Cigarette test showed positive reactions for polyphenols in all the species, except *T. bracteolatum, T. glycosmoides* and *V. kunstleri.* Likewise, Hot water test for polyphenols showed negative reactions in *T. bracteolatum, T. campylocarpum* and *T. glycosmoides.* According to Baldi *et al.* (1993), presence of polyphenols and phenoloic compounds have a taxonomic significance in distinguishing different types between or within the species.

Oxalic reaction showed mostly negative results suggesting the absence of acidic compounds in seven *Tetrastigma* and two *Vitis* species, except *T. campylocarpum, T. glycosmoides* and *V. kunstleri.* Leuco-anthocynins were absent in eight species, except *T. nilagiricum* and *T. sulcatum.* While studying systematic distribution of leuco-anthocyanins in leaves, Bate-Smith & Lerner (1954) suggested that a tendency towards the presence of leuco-anthocyanins indicates primitive types, where as its absence shows an advanced character of a phyletic series. The production of leucoanthocyanin is a part of a primitive metabolic pattern associated with, but not essential to a tree or woody habit of growth.

HCL-Methanol test showed presence of catechol tannins in eleven species, except in *Vitis scortechinii.* Gibbs (1974) stated that there is high correlation between positive test for catechol tannins and leucoanthocyanins. Hence, HCL-Methanol test may confirm the presence of leucoanthocyanins. In the present studies, however, leucoanthocynins were absent in eight species. It may be due to variable plant materials. The test for catechol tannins was taken from sapwood and leucoanthocyanins test from fresh leaves respectively.

Syringin glycosides were present in 5 and also absent in five species. Juglone test B represents the presence of flavonoid substances in *T. campylocarpum* and *T. glycosmoides.* While Juglone test C showed the presence of aesculin/coumarin in *T. glycosmoides* only. Lignins were present in nine species, except in *T. gamblei* and *T. obovatum,* saponins in *T. gamblei, T. glycosmoides* and *T. sulcatum* and tannins in eight species except in *T. bracteolatum, T. gamblei, T. leucostaphylum* and *Vitis novemefolia.* Goplen *et al.* (1980) stated that tannins in plants act as an effective repellent to animals and microbial predators or parasites. They protect the plants from birds, insects & pathogenic organisms.

Aucubin like substances have positive reactions with Ehrlich test in five species and flavonoids in ten species, except *T. campylocarpum* and *T. glycosmoides.* According to Moore & Giannasi (1987), the presence of flavonoids in plants has special significance in taxonomic delimitation of taxa establishing & phylogenetic relationship.

**(iii) Phytochemicals present but in variable quantities**

Mucilaginous substances were noted in all species with stronger reactions in *T. andamanicum* and *T. sulcatum.* Cigarette test for *T. andamanicum* and Hot water test for *T. gamblei* and *T. obovatum* indicated stronger positive reactions to polyphenols. Presence of catechol tannins in eleven species with stronger reaction in *T. andamanicum, T. campylocarpum, T. obovatum* and *T. sulcatum* were detected with HCL-Methanol test. According to Gibbs (1974), the differential colour patterns for variable quantities play important role in determining Catechol tannins. Accordingly, dull magenta purple colour indicated strong reaction in four species, whereas rose purple colour for weaker reaction was noted in other seven species. Wood of the *Vitis scortechini* does not show any magenta colour.

Tannins were present in eight species with stronger reactions in *T. andamanicum* and *Vitis scortechinii.* Bachmann & Blaich (1979) reported that the stronger reactions to tannins show more resistance to fungal infections. The tannins react with ferric ammonium citrate and change in purple blue colour. The amount of tannins may vary widely in leaves of different ages or in healthy and diseased specimens. Flavonoids were present in ten species with stronger reactions in *Vitis scortechinii.*

**(iv) Phytochemicals absent in all the species**

In the present study, HCN, Juglone test A and Aurone tests revealed the absence of cyanogenic glycosides, quinones, naptho-quinones and aurones like phytochemicals in stems and leaves of all species of *Tetrastigma* and *Vitis.*

**(v) Phytochemicals present or absent in doubtful reactions**

The *T. gamblei* and *Vitis novemefolia* show doubtful reactions for leuco-anthocynins. It was also recorded for the presence of syringin glycosides in *T. campylocarpum* and absence in *T. glycosmoides.* Similarly, lignins showed positive doubtful reaction in *T. campylocarpum* by Moule test.

*Tetrastigma* and *Vitis* species have been classified based on various tests and given in Table 3. Presence/absence or variable quantities of such phytochemicals have special significance in separation of closely related taxa. The present study supports the taxonomic position of the genus *Tetrastigma* in family Vitaceae.

*Tetrastigma* species also have medicinal importance to cure boils, cough etc. The alcoholic extract of the arial parts of the *T. serrulatum* plants show anti-cancer activity (Anonymous, 1976). Patil *et al.* (1990, 2000) reported that the *Tetrastigma* species provide a good source of resistance against major fungal diseases like downy mildew, powdery mildew and anthrcnose and insects like Flea Beetle.

According to Mahadevan (1979), the phytochemicals like phenols and phenolic compounds, flavonoids, tannins, lignins and leuco-anthocynins have importance to control a number of fungal diseases of cultivated plants. *Tetrastigma* species of the present studies are the wild genotypes of the grape family Vitaceae. Hence, phytochemicals present in these species have obvious significance in protection of commercial grapes from various diseases and pests as natural resources, otherwise in grape improvement programme to improve the genetic resistance with help of somatic hybridization or through advanced technique of Biotechnology.

## ACKNOWLEDGEMENTS

This work has been carried out under the "All India Co-ordinated Research Project on Sub-tropical Fruits-Grapes", sponsored by the Indian Council of Agricultural Research, New Delhi.

## REFERENCES

ANONYMOUS, 1976. *The Wealth of India, Raw Materials. Tetrastigma* Planch. (Vitaceae). 10 : 200. PID, CSIR, New Delhi.

BACHMANN, O. & R. BLAICH 1979. Occurrence of properties of condensed tannins in Vitaceae. *Vitis* 18 :106-116.

BALDI, A., A. ROMANI, N. MULINACCI & F.F. VINCIER 1993. Phenolic compounds in Tudcan varieties of *Vitis vinifera* L. *J. Int. Sci. Vigne. Vin.* 27 : 201-215, 232-233.

BATE-SMITH, C. & L.H. LERNER 1954. Leucoanthocyanins 2. Systematic distribution of leuco-anthocyanin in leaves. *Biochem. J.* 58 : 126-132.

GAMBLE, J.S. 1918. *Flora of* the *Presidency* of *Madras – Vitaceae* 1: 226-240. London.

GIBBS, R.D. 1954. Comparative chemistry and phylogeny of flowering plants. *Trans. R. Soc. Can. (Ser. 3) Sect V.* 48 : 1-47.

———— 1974. *Chemotaxonomy of flowering plants* 1 : 54-80. Mc Gill Queen's University Press, Montreal.

GOPLEN, B.P., R.E. HOWARTH, S.K. SARKAR & K. LESING 1980. A search for condensed tannins in annual and perennial species of *Medicago, Trigonella* and *Onobrychis. Crop Science* 20 : 801-804.

HARBORNE, J.B. 1973. Phytochemical Methods : A Guide to Modern Techniques of Plant Analysis. Chapman and Hall, London, New York.

MAHADEVAN, A. 1979. Biochemical aspects of plant disease resistance. *J. Scient Ind. Res.* 38 : 156-171.

MOORE, M.O. & D.E. GIANNASI 1987. Foliar flavonoids of selected *Vitis* taxa in the south Eastern United States. *Biochemical Systematic & Ecology* 15 : 79-83.

PATIL, S.G., B.K. HONRAO & S.P. KARKAMKAR 1997a. Phytochemical studies in genus *Vitis* L. (Fam. Vitaceae) and their significance. *J. Econ. Taxon. Bot.* 22 : 7-11.

———— & ———— 2000. Phytochemical studies in genus *Cayratia* Juss. (Fam. Vitaceae) and their signficance. *J. Econ. Taxon. Bot.* 24 (3) : 688-694.

————, ————, S.P. KARKAMKAR & M.R. DESHMUKH 2000. Evaluation of wild grape germplasm for infestation of Flea Beetle. *Pestology* 24 (12) : 37-39.

———, ———, V.G. RAO & V.P. PATIL 1990. Screening of grape (*Vitis* species) germplasm for resistance to three major fungal diseases. *Indian J. Agric. Sc.* 60 : 836-838.

———, S.P. KARKAMKAR & B.K. HONRAO 1997. Phytochemical studies in the genus *Leea* L. (Vitaceae) and their significance in taxonomy. *J. Indian Bot. Soc.* 76 : 81-85.

PLANCHON, J.E. 1887. Monographic des Ampelidees Vraies. In De Candolle *Monographiae Phanerogarum* 5 : 305-654.

SHETTY, B.V. & P. SINGH 2000. Vitaceae. In Singh, N.P. *et al.* (Edits.) *Flora India* 5 : 246-324. BSI., Calcutta.

SINGH, B.P., S.K. VARMA & J.V.V. DOORA 1986. Chemotaxonomy of Vitaceae of Bihar-A study of flavonoids. *J. Indian* Bot. *Soc.* 65 : 335-339.

SMITH, F. & R. MONOTGOMERY 1959. *The chemistry of plant gums and mucilages and some related polysacchrides. Plant mucilage.* American Chemical Society, Monograph Series, Reinhold Publishing Corporation, New York, Chapman & Hall Ltd., London.

# PLANT USED IN TRADITIONAL PHYTOTHERAPY FOR INDIGESTION IN GUJARAT, INDIA

B.A. JADEJA, N.K. ODEDRA, J.K. CHAVDA & D.C. BHATT*

*Botany Department, M. D. Science College, Porbandar-360 575, Gujarat, India*
**Department of Marine Science, Bhavnagar University, Bhavnagar-364 002, India*

**ABSTRACT**

This article describe the indigenous knowledge of local communities on the traditional phytotherapy for indigestion in Gujarat. About 90 species which are used as herbal medicine in indigestion are recorded. The species are arranged alphabetically with the name of family, local name, parts used and mode of preparation of drug.

## INTRODUCTION

Gujarat is the fifth largest State in India, situated on the west coast of India between 20° and 25° North latitude and 68° and 75° East longitude, with 1660 km long sea coast and a highly productive off shore continental shelf. Most of the tribals live in Dangs, Panchmahal, Dahod, Surendranagar, Junagadh, Kutch and Porbandar districts. They belong to Maldhari, Koli, Bhil, Adivasi, Maher and some other tribes. These people utilize plant resources to meet their requirements, including herbal medicines for indigestion.

Ethnobotany of Gujarat is known through the earlier works of Thaker (1910, 1926), Shah (1978), Joshi *et al.* (1980, 1998), Shah & Gopal (1982, 1985), Bhatt & Sabnis (1987), Joshi (1988), Bhattacharya (1996), Mitaliya (1998), Punjani (1998), Bhatt & Mitaliya (1999), Jadeja (1999), Bhatt *et al.* (1999, 2000, 2001), Ant (2000), Silori & Rana (2000), Patel & Reddy (2000), Mitaliya *et al.* (2001, 2003) etc. However, the problem of indigestion has been neglected in the State of Gujarat and very limited information is available regarding herbal medicines in the published literature. As such, the present study was taken up to record the data on the above issues and results are presented here.

## MATERIALS AND METHODS

The ethnobotanical extensive field survey was conducted in various parts of Gujarat State, particularly Jamnagar,

Rajkot, Bhavnagar, Junagadh (Gir forest), Dang, Banaskantha, Panchmahal, Kutch, Porbandar, Barda hills, Amreli and Surendranagar during 2002-2004.

Initially, various knowledge rich persons (Farmers, Vidya, Bhagats, Bhuvas) who are popular for their traditional healing practice were identified by enquiring in different villages and nesses. Later, these resource persons were interviewed with the help of plant collections and information on medicinal uses was gathered. The collected plant specimans were identified using floras of Gujarat State (Shah, 1978; Bole & Pathak, 1988). The collected plant materials have been deposited in the Herbarium at M. D. Science College, Porbandar. Such information was subsequently filtered by eliminating the commonly known practices and only the little known practices were recorded and verified. It is believed that the information recorded here will be useful if further scrutiny is made and few practices are targeted for phytochemical and pharmacological studies.

## ENUMERATION

Table 1. Plants used in indigestion in Gujarat.

| Scientific name | Family | Local name | Part (s) used | Mode of administration |
|---|---|---|---|---|
| *Abrus precatorius* L. | Fabaceae | Chanothi | Leaves | Leaf-juice is taken once a day for a week during indigestion. |
| *Abutilon indicum* (L.) Sw. | Malvaceae | Kansaki | Root | The powder of dry roots is taken with boiled water. |
| *Acacia catechu* (L. f.) Willd. | Mimosaceae | Kher | Stem-bark | Boil the stem-bark in water and taken twice a day for a week. |
| *Achyranthes aspera* L. | Amaranthaceae | Aghedo | Root | Crushed dry root is taken orally with boiled water. |
| *Aconitum heterophyllum* Wall. ex Royle | Ranunculaceae | Ativish ni kadi | Root | Crushed dry root is taken orally with boiled water. |
| *Acorus calamus* L. | Araceae | Ghoda vaj | Root | Root decoction with seeds of *Piper nigrum* is given twice a day for a week. |
| *Aegle marmelos* (L.) Corr. | Rutaceae | Bili | Fruits | Crushed dry fruit with rhizome of *Zingiber officinale* is given in indigestion. |
| *Alangium salvifolium* (L. f.) Wangerin | Alangiaceae | Ankol | Root-bark | Juice of root-bark is purgative and given in indigestion. |
| *Allium sativum* L. | Liliaceae | Lasan | Clove | Cooked 50 gm cloves are taken twice a day for 5 days. |
| *Alstonia scholaris* (L.) R. Br. | Apocynaceae | Sapta parna | Fruit | The decoction of fruit-pericarp is taken with cow's milk twice a day for 2 days. |
| *Andrographis paniculata* (Burm. f.) Wall. ex Nees | Acanthaceae | Kariyatu | Leaves, Fruit | Decoction of dry leaves and dry fruits is taken orally twice a day for a week. |

| | | | | |
|---|---|---|---|---|
| *Anethum graveolens* L. | Apiaceae | Suva | Seed | Foment seeds is taken orally thrice a day for 5 days. |
| *Argemone mexicana* L. | Papaveraceae | Darudi | Entire-plant | Milky juice of entire plant is purgative. |
| *Asparagus racemosus* Willd. | Liliaceae | Satavari | Root | Crushed dry root is taken orally with boiled water. |
| *Berberis asiatica* Roxb. ex DC. | Berberidaceae | Daru-haldaṛ | Stem | Powder of dry stem is given orally with boiled water. |
| *Boerhavia diffusa* L. | Nyctaginaceae | Satodi | Entire-plant | Entire plant decoction is given orally twice a day for a week. |
| *Buchanania cochinchinensis* (Lour.) Almeida | Anacardiaceae | Charodi | Stem-bark | Decoction of stem-bark is taken orally twice a day for a week. |
| *Caesalpinia crista* L. emend. Dandy & Excell | Caesalpiniaceae | Kankach | Fruit | The mixture of dry fruits and flower-buds of *Syzygium caryophyllatum* is taken with boiled water twice a day. |
| *Calotropis procera* (Ait.) Ait. f. ssp. *hamiltonii* (Wt.) Ali | Asclepiadiaceae | Ankado | Root | Crushed dry root is taken orally with boiled water. |
| *Cassia fistula* L. | Caesalpiniacae | Garmalo | Fruit, Flower | Pericarp of fruit is purgative and flower paste is given in constipation. |
| *Catunaregam spinosa* (Thunb.) Tirveng. | Rubiaceae | Mindhal | Fruit-pericarp | Decoction of fruit-pericarp is taken orally twice a day for a week. |
| *Celastrus paniculata* Willd. | Celastraceae | Mal kangani | Seed | The powder of dry seeds is taken with water once a day for 3 days. |
| *Centella asiatica* (L.) Urban | Apiaceae | Brahmi | Entire-plant | Entire plant decoction is given orally twice a day for a week. |
| *Cinnamomum tamala* Nees | Lauraceae | Tamal patra | Leaves | Mixture of powdered leaves with flower-buds of *Syzygium caryophyllatum* is taken with boiled water. |
| *Citrus aurantium* L. | Rutaceae | Narangi | Fruit | The powder of dried fruit-pericarp is given orally with boiled water. |
| *Clerodendrum phlomidis* L. f. | Verbenaceae | Arani | Leaves, Root | Decoction of leaves and roots is taken orally twice a day for 5 days. |
| *Clitoria ternatea* L. | Fabaceae | Garani | Root | The powder of dry roots is taken with boiled water. |
| *Commiphora wightii* (Arn.) Bhandari | Burseraceae | Gugal | Stem | Powder of dry stem is given orally with boiled water. |
| *Convolvulus prostratus* Forssk. | Convolvulaceae | Shankhavali | Root, Seed | Root and seed decoction is beneficial in indigestion. |
| *Coriandrum sativum* L. | Apiaceae | Dhana | Leaves | Leaf-juice is taken orally twice a day for 5 days. |

| | | | | |
|---|---|---|---|---|
| *Cucumis melo* L. | Cucurbitaceae | Kadvi methi | Seed | Foment seeds are taken orally thrice a day for a week. |
| *Cuminum cyminum* L. | Apiaceae | Jiru | Seed | The powder of seeds is taken orally twice a day for 5 days with curd. |
| *Cyperus rotundus* L. | Cyperaceae | Nagar moth | Root | Crushed dry root is taken orally thrice a day for 3 days with boiled water. |
| *Eclipta prostrata* (L.) L. | Asteraceae | Bhangaro | Entire-plant | Entire plant decoction is given orally twice a day for three days. |
| *Elettaria cardamomum* Maton | Zingiberaceae | Elichy | Fruit | The mixture of dry fruits and flower-buds of *Syzygium caryophyllatum* is given twice a day with boiled water. |
| *Embelia ribes* Burm. f. | Myrsinaceae | Vavding | Fruit | Crushed dry fruit is taken orally with boiled water. |
| *Fagonia schweinfurthii* (Hadidi) Hadidi ex Ghafoor | Zygophyllaceae | Dhamaso | Leaves | Crushed dry leaves are taken orally with boiled water. |
| *Feronia limonia* (L.) Swingle | Rutaceae | Kothu | Stem-bark | Powder of dry stem-bark is given orally with boiled water. |
| *Ficus religiosa* L. | Moraceae | Pipalo | Stem-bark | Boiled stem-bark is taken orally with boiled water. |
| *Fumaria indica* (Haussk.) Pugsley | Fumariaceae | Pit papdo | Leaves | Decoction of leaves is taken orally twice a day for 3 days. |
| *Gossypium herbaceum* L. | Malvaceae | Kapas | Root | Crushed dry root is taken orally with boiled water. |
| *Hemidesmus indicus* (L.) R. Br. | Periplocaceae | Uplasari | Root-bark | Juice of root-bark is given orally twice a day for 4 days. |
| *Holarrhena pubescens* (Buch.-Ham.) Wall. ex G. Don | Apocynaceae | Kado | Bark, Seed | Boiled stem-bark and seed are taken orally with water. |
| *Hygrophylla schulli* (Buch.-Ham.) M. R. & S. M. Almeida | Acanthaceae | Aekhero | Leaves | Decoction of leaves is taken orally twice a day for 5 days. |
| *Justicia adhatoda* L. | Acanthaceae | Aradushi | Leaves | Boiled extract of leaves are given in indigestion. |
| *Launaea sarmentosa* (Willd.) Sch.-Bip. ex Kunze | Asteraceae | Galgibhi | Leaves, Flower | Juice of leaves and flowers is taken orally with boiled water. |
| *Mangifera indica* L. | Anacardiaceae | Ambo | Bark, Leaves | Boiled stem-bark and leaves are taken orally with water. |
| *Melia azedarach* L. | Meliaceae | Limado | Leaves, Flower | Leaf-extract with table salt is given in indigestion. Flower with table salt is given in indigestion. |

| | | | | |
|---|---|---|---|---|
| *Mentha spicata* L. | Lamiaceae | Fodino | Leaves | The mixture of leaves and leaves of *Coriandrum sativum* is given with boiled water twice a day for 5 days. |
| *Momordica charantia* L. | Cucurbitaceae | Karela | Fruit | Leaves or fruits are cooked and given orally twice a day for 3 days. |
| *Moringa concanensis* Nimmo ex Dalz. & Gibs. | Moringaceae | Surgavo | Leaves | Leaves or fruits are cooked and given in indigestion. |
| *Mucuna pruriens* (L.) DC. | Fabaceae | Kaucha | Seed, Root | Crushed root and seeds are given during indigestion. |
| *Myristica fragrans* Houtt. | Myristicaceae | Jaifal | Seed | The powder of dry seeds and flower-buds of *Syzygium caryophyllatum* is given with boiled water. |
| *Nardostachys grandiflora* DC. | Velerianaceae | Jatamasi | Root | Root decoction with seeds of *Piper nigrum* are given twice a day for a week. |
| *Nerium indicum* Mill. | Apocynaceae | Karen | Root | Crushed dry root is taken orally with boiled water. |
| *Ocimum tenuiflorum* L. | Lamiaceae | Tulhi | Leaves | 10-12 leaves are chewed orally twice a day for a week. |
| *Operculina turpethum* (L.) Manso | Convolvulaceae | Nasoter | Root | The powder of dry root is taken with boiled water. |
| *Pergularia daemia* (Forssk.) Chiov. | Asclepiadaceae | Chamar dudheli | Fruit | Crushed fruit with sugar cube is given orally thrice a day for 3 days. |
| *Phyllanthus emblica* L. | Euphorbiaceae | Amala | Fruit | The powder of dried fruit-pericarp is given orally with boiled water. |
| *Piper longum* L. | Piperaceae | Piper | Stem-bark | Decoction of stem-bark is taken orally with boiled water. |
| *Piper nigrum* L. | Piperaceae | Mari | Fruit | The mixture of dry fruits and flower-buds of *Syzygium caryophyllatum* and dry leaves of *Ocimum tenuiflorum* is given with boiled water. |
| *Pluchea lanceolata* (DC.) Cl. | Asteraceae | Rasna | Root | Crushed dry root is taken orally with boiled water. |
| *Plumbago zeylanica* L. | Plumbaginaceae | Chitrak | Root-bark | Crushed dry root-bark is taken orally with boiled water twice a day for 3 days. |
| *Pongamia pinnata* (L.) Pierre | Fabaceae | Karanj | Stem-bark | Crushed dry stem-bark is taken orally with boiled water thrice a day for 2 days. |
| *Prosopis cineraria* (L.) Druce | Mimosaceae | Khijado | Stem-bark | Crushed dry stem-bark is taken orally with boiled water. |
| *Punica granatum* L. | Punicaceae | Dadam | Fruit | Crushed dried fruit-pericarp is taken with boiled water. |
| *Raphanus sativus* L. | Brassicaceae | Mula | Root | Every morning root is eaten. |
| *Ricinus communis* L. | Euphorbiaceae | Divelo | Leaves | Decoction of leaves is given twice a day for 2-3 days. |

| | | | | |
|---|---|---|---|---|
| *Rubia cordifolia* L. | Rubiaceae | Majith | Root | The powder of dry roots is taken early morning with water. |
| *Senna auriculata* (L.) Roxb. | Caesalpiniacae | Aval | Leaves | Leaf-juice is given along with leaves powder of *Ocimum teniuflorum*. Leaves also be chewed. |
| *Senna occidentalis* (L.) Link | Caesalpiniacae | Kasundro | Leaves | Decoction of leaves is given twice a day for 1 or 2 days. |
| *Senna tora* (L.) Roxb. | Caesalpiniaceae | Kuvadio | Seed | The powder of dry seeds is taken early morning with water. |
| *Sesamum indicum* L. | Pedaliaceae | Tal | Seed | Foment seeds are taken orally thrice a day for 5 days. |
| *Solanum indicum* L. | Solanaceae | Ubhi bhonya-ringni | Root, Fruit | The powder of dried roots and dried fruit-pericarp are given orally with boiled water. |
| *Solanum nigrum* L. | Solanaceae | Piludi | Leaves | Juice of leaves is taken orally. |
| *Solanum virginianum* L. | Solanaceae | Bhonya ringni | Leaves | The powder of dry leaves is taken early in the morning with water. |
| *Syzygium cumini* (L.) Skeels | Myrtaceae | Jambu | Fruit, Leaves | The powder of dried fruit-pericarp and dried leaves is given orally with boiled water. |
| *Tamarindus indica* L. | Caesalpiniaceae | Amli | Fruit | Pericarp of fruit is given in indigestion. |
| *Terminalia bellerica* (Gaertn.) Roxb. | Combretaceae | Baheda | Fruit-pericarp | Fruit-pericarp with fruits of *Terminalia chebula* is given twice a day for a week. |
| *Terminalia catappa* L. | Combretaceae | Badam | Seed | Crushed dry seed is taken orally with boiled water. |
| *Terminalia chebula* Retz. | Combretaceae | Harede | Fruit | Fruit-pericarp with fruit of *Terminalia bellerica* is given twice a day for a week. |
| *Terminalia cuneata* Roth | Combretaceae | Arjun | Stem-bark | Decoction of stem-bark is given twice a day for 2 or 3 days. |
| *Tinospora cordifolia* (Willd.) Miers ex Hk. f. & Th. | Menispermaceae | Galo | Entire-plant | Decoction of entire plants is given in indigestion. |
| *Trachyspermum ammi* (L.) Sprangue | Apiaceae | Ajamo | Seed | Decoction of dry seed and resinous gum of *Ferula asafoetida* is given orally twice a day for a week. |
| *Trachyspermum roxburghianum* (DC.) Craib. | Apiaceae | Bodi Ajamo | Seed | Foment seed is taken orally thrice a day for 5 days. |
| *Tribulus terrestris* L. | Zygophyllaceae | Gokharu | Fruit | The powder from dried fruit-pericarp is given orally with boiled water. |
| *Vitex negundo* L. | Verbenaceae | Nagod | Leaves | Decoction of leaves is taken orally twice a day for 5 days. |
| *Vitis vinifera* L. | Vitaceae | Draksh | Fruit | Fruit-juice is taken orally thrice a day for 3 days. |

| | | | | |
|---|---|---|---|---|
| *Withania somnifera* (L.) Dunal | Solanaceae | Ashva gandha | Root | The powder of dry root is taken with boiled water. |
| *Zingiber officinale* Rosc. | Zingiberaceae | Adu | Rhizome | Extract of rhizome is given orally twice a day for 3 days. |

## RESULT AND DISCUSSION

The present work deals with the herbal drugs as a remedy for treatment of indigestion. About 90 species belonging to 83 genera under 48 families are used as traditional phytotherapy by people of Gujarat for treatment of indigestion. Plants belong to 44 families of dicots and 4 families of monocots. Caesalpiniaceae and Apiaceae are the largest families whose 6 species are used. Leaves of 23 species, roots of 23 species, fruits of 21 species, seeds of 13 species, stem of 10 species, entire plants of 5 species, flowers of 3 species, rhizome of 1 species are used as herbal drugs in indigestion. Some plants not found in Gujarat but plant parts easily available in the market are also used by rural folk. Documentation of these remedies is a first hand report to fulfill their needs in rural and tribal areas of Gujarat. Medical practitioners and manufactures can take the help of nature to provide safe and harmless substitute for some harmful chemical products that have side effects.

## ACKNOWLEDGEMENTS

The authors wish to express their sincere thanks to Dr. U.S. Baxi, Department of Botany, M.D. Science College, Porbandar for valuable advice. The authors also thank Prof. S.V. Buch, D.K.V. College, Jamnagar for help during study, particularly in identifying the plants during the field trips.

## REFERENCES

ANT, H.M. 2000. *Ethnobotanical study of Angiosperms of Aravalli hills.* Ph. D. Thesis, Bhavnagar University, Bhavnagar.

BHATT, D.C. & K.D. MITALIYA 1999. Ethnomedicinal plants of Victoria Park (Reserved forest) of Bhavnagar, Gujarat, India. *Ethnobotany* 11 : 81-84.

————, ———— & S.K. MEHTA 2000. Plant twigs as tooth brushes. *Adv. Plant Sci.* 13 (1) : 19-22.

————, M.A. NURANI, K.D. MITALIYA & U.S. BAXI 2001. Some ethnomedicinal plants of Barda hills of Saurashtra, Gujarat, India. *Adv. Plant Sci.* 14 (2) : 427-431.

————, S.K. MEHTA & K.D. MITALIYA 1999. Ethnobotanical plants of Shetrunjara hill of Palitana, Gujarat. *Ethnobotany* 11 : 22-25.

BHATT, R.P. & S.D. SABNIS 1987. Contribution to the ehtnobotany of Khedbrahma region of North Gujarat. *J. Econ. Taxon. Bot.* 9 (1) : 139-145.

BHATTACHARYA, G. 1996. Medico-ethnobotanical value of Saurashtra weeds. *J. Econ. Taxon. Bot.* Addl. Ser. 12 : 166-168.

BOLE. P.V. & J.M. PATHAK 1988. *Flora of Saurashtra,* Vols. 2 & 3. Botanical Survey of India, Calcutta.

JADEJA, B.A. 1999. Plants used by tribe Rabari in Barda hills of Gujarat. *Ethnobotany* 11 : 42-46.

JOSHI, M.C. 1988. Pharmaceutically important medicinal plants of Gujarat forests. *Bull. Medico-Ethno. Bot. Res.* 10 (4) : 372-373.

————, M.B. PATEL & P.J. MEHTA 1980. Some folk medicines of Dangs, Gujarat State. *Bull. Medico-Ethno. Bot. Res.* 1 : 8-24.

————, ———— & ———— 1998. In Goyal, H.R. *A report on medicinal plants of Kachchh (Gujarat).* CCRA & S., New Delhi.

MITALIYA, K.D. 1998. *Ethnomedicinal study of Angiosperms of Bhavnagar.* Ph.D. Thesis, Bhavnagar University, Bhavnagar.

————, D.C. BHATT, D.M. PATEL & P.N. JOSHI 2001. Medicinal value of some selected stem-bark used by tribals and rurals folk in Gujarat. *Adv. Plant Sci.* 14 (1) : 255-259.

————, ————, J.N. PATEL & S.K. PATEL 2003. Observations on plants used for joint disorders by rural and tribal people of Gujarat. *J. Econ. Taxon. Bot.* 27 (1) : 212-216.

PATEL, K.C. & A.S. REDDY 2000. Ethno-medicinal uses of some common plants of Danta forest, North Gujarat State. *Seminar on Recent trends in plant Science,* Talod (Abstract no. 5).

PUNJANI, B.L. 1998. An ethnobotanical study of tribal areas of district Sabarkantha (North Gujarat). Ph.D. Thesis, North Gujarat University, Patan.

SHAH, G.L. 1978. *Flora of Gujarat State,* 2 Vols. Sardar Patel University, Vallabh Vidhya-nagar (Gujarat).

———— & G.V. GOPAL 1982. An ethno-botanical profile of the Dangies. *J. Econ. Taxon. Bot.* 3 : 355-364.

———— & ———— 1985. Ethnomedicinal notes from the tribal inhabitants of North Gujarat (India). *J. Econ. Taxon. Bot.* 6 (1) : 193. 201.

————, A.R. MENON & G.V. GOPAL 1981. An account of ethnobotany of Saurashtra in Gujarat State. *J. Econ. Taxon. Bot.* 2:173-181.

SILORI, C.S. & A.R. RANA 2000. Indigenous knowledge of medicinal plants and their uses in Narayan Sarovar Sanctuary, Kutchch. *Ethnobotany* 12 : 1-7.

THAKER, J.L. 1910. *Flora of Bharda hills, Vanaspati Shastra.* Pravin Pustak Bhandar, Rajkot.

———— 1926. *Plants of Kutch & their utility.* Kutch Darbar, Kutch.

# POTENTIAL ETHNOMEDICINAL PLANTS AT KAPTIPADA FOREST RANGE ORISSA, INDIA AND THEIR USES

K.K. BEHERA, N. MANJARI MISHRA & G.R. ROUT*

*Parija Post-Graduate Department of Botany, Ravenshaw College, Cuttack-753 003, Orissa*
**Regional Plant Resource Centre, Department of Forest and Environment, Bhubaneswar-751 015, Orissa, India*
*E-mail: grout@hotmail.com*

## ABSTRACT

A survey of potential ethnomedicinal plants occurring in Kaptipada forest range in Orissa, India was undertaken. In the present study, 120 species belonging to 51 families were found to have the potential ethnomedicinal value and are being used to cure various ailments by the local tribal people and the pharmaceutical companies. The maximum number of ethnomedicinal plants were represented by family *Asteraceae, Euphorbiaceae, Fabaceae* and *Verbenaceae*. Under enumeration, botanical name, family, use of different parts by specific tribes (S-Santal, Ba-Bathudi, Ko-Kolha, K-Kondha, Bh-Bhumija, G-Gond, Lo-Lodha, Su-Saunti etc) and information about the diseases have been given.

## INTRODUCTION

The collection of ethnomedicinal herbs dates back to antiquity. The use of herbal drugs in the indigenous system of medicine has been in existence for several centuries and India is a biggest exporter of medicinal plants and related products (Mitra & Jain, 1991). During the last few decades scientists all over the World are paying much more attention to the studies of a newer branch of science, ethnobiology, specially to tribal medicine or ethnomedicine. Floristic survey of medicinal plants were recorded by different researchers (Abraham, 1981; Choudhuri *et al.,* 1982; Pal, 1984; Das & Mishra, 1987; Brahmam & Saxena, 1989; Brahamam *et al.,* 2000 etc), but no record has been found for potential ethnomedicinal plants in Kaptipada forest range, Orissa, India. Kaptipada forest range is under Kaptipada Sub-division and a part of Similipal Bioreserve.

The total population of the Sub-division is 2,77,961, out of which scheduled castes are 17,330 and scheduled tribes

1,85,782 (1991 census). It is a mixed deciduous forest and lies between 21°19' and 21°44' North latitude and between 86°23' and 86°42' East longitude. The area covered by forests is 875 square miles. About 580 square miles are under reserved forests and 252 square miles under protected forests. The climate of the Kaptipada forest range is characterized by an oppressive hot summer, high humidity and well distributed rain fall as mentioned in Table 1.

## METHODOLOGY

The plant specimens were collected randomly from different sites of the forest pockets and taken into laboratory for identification and preservation in the herbarium section at the P.G. Department of Botany, Ravenshaw College, Cuttack.

The sequence of species is arranged alphabetically. Flora of Orissa (Saxena & Brahmam, 1994-1996) has been used for identification of taxa. Nomenclature has been brought up-to-date as far as possible.

While conducting survey in Kaptipada forest range, some interesting plant species were encountered which are potential ethnomedicinal plants. The ethnobotanical information presented in this paper are based on actual field observations, data from herbarium collections and published literature on the flora of Orissa (Satpathy & Bramam, 1996, 1999).

Table 1. Average rainfall (mm), Temperature (°C) and relative humidity (%) during study period (1995 -2002).

| **Month** | **Temperature (°C)** | | **Rainfall (mm)** | **Relative Humidity (%)** | |
|---|---|---|---|---|---|
| | **Max.** | **Min.** | | **Mor.** | **Eve.** |
| January | 25 | 11 | 023 | 64 | 52 |
| February | 28 | 14 | 018 | 66 | 44 |
| March | 34 | 18 | 013 | 51 | 30 |
| April | 36 | 21 | 066 | 54 | 37 |
| May | 37 | 23 | 123 | 58 | 47 |
| June | 32 | 23 | 218 | 74 | 70 |
| July | 31 | 23 | 272 | 80 | 81 |
| August | 29 | 23 | 252 | 85 | 83 |
| September | 30 | 22 | 196 | 83 | 83 |
| October | 30 | 20 | 117 | 78 | 78 |
| November | 30 | 15 | 007 | 70 | 65 |
| December | 25 | 11 | 008 | 69 | 55 |

Table 2. Potential ethno-medicinal plants in Kaptipada forest range and their uses.

| **Plant species** | **Family** | **Parts used** | **Medicinal uses** |
|---|---|---|---|
| *Abrus precatorius* L. | Fabaceae | Leaf, Seed, Root | Arthitis, Menorrhagia, Mumps, Swelling. (S, Lo, Ba, Su) |

| | | | |
|---|---|---|---|
| *Abutilon indicum* (L.) Sw. | Malvaceae | Root, Bark, Seed | Leprosy, Piles, Haematuria. (S, G, Ko, Ba) |
| *Acalypha indica* L. | Euphorbiaceae | Root, Stem, Flower bud | Asthma, Bronchitis, Haemoptysis. (Ba, K, Mu, Bh, Su) |
| *Achyranthes asper* L. | Amaranthaceae | Leaf, Plant | Dropsy, Piles, Skin eruptions. (S, Ba, Bh, Su, Lo) |
| *Ageratum conyzoides* L. | Asteraceae | Root | Bed wetting by children. (Lo, S, Su, Ba, K) |
| *Alstonia scholaris* (L.) R. Br. | Apocynaceae | Bark | Chest pain. S, Lo, Ba, Su) |
| *Alternanthera sessilis* (L.) R. Br. ex DC. | Amaranthaceae | Leaf, Plant | Fevers, Ophthalmia, Gonorrhoea, Pruritis. (S, G, Ko, Ba) |
| *Amaranthus viridis* L. | Amaranthaceae | Plant, Stem | Bronchitis, Piles, Leucorrhoea, Breast abscess, Menorrhagia. (Ba, S, K, Ld) |
| *Andrographis paniculata* (Burm. f.) Wall. ex Nees | Acanthaceae | Plant, Leaf | Fever, Tonic. (S, Lo, Ba, Su) |
| *Annona reticulata* L. | Annonaceae | Bark, Fruit | Diarrhoea, Indigestion. (S, Lo, Ba, Su) |
| *Annona squamosa* L. | Annonaceae | Fruit | Diarrhoea, Tumor, Antihelminthic. (S, Ba, Bh, Su, Lo) |
| *Arenaria serpyllifolia* L. | Caryophyllaceae | Leaves, Plant, Fruit | Diseases of bladder, Acute-chronicystis. (S, B, Bh, Su, Lo) |
| *Argemone mexicana* L. | Papaveraceae | Fruit, Soft root, Plant | Skin disease, Asthma, Leprosy. (S, Lo, Ba, Su) |
| *Asparagus racemosus* Willd. | Liliaceae | Root, Plant | Aphrodisiac, Diarrhoea, Tonic, Galactogenic. (S, Lo, Ba, Su) |
| *Azadirachta indica* A. Juss | Meliaceae | Leaf | Boil, Diabetes, Scabies, Gonorrhoea. (S, Lo, Ba, Su) |
| *Bauhinia variegata* L. | Caesalpiniaceae | Seed, Root | Diarrhoea, Blood dysentery. (S, Ba, Bh, Su, Lo) |
| *Blumea lacera* (Burm. f.) DC. | Asteraceae | Root, Leaf | Astringent, Febrifuge, Anthelmintic. (S, Lo, Ba, Su) |
| *Boerhavia diffusa* L. | Nyctaginaceae | Leaf, Plant | Colic pain, Menorrhagia. (S, G, Ko, Ba) |
| *Bombax cieba* L. | Bombacaceae | Root, Flower, Bark | Boil, Burning sensation, Menstrual disorder. (S, Ba, Bh, Su, Lo) |
| *Borassus flabellifer* L. | Arecaceae | Inflorescence, Leaf, Fruit | Flatulence, Hiccups, Scabies. (Ba, K, Mu, Bh, Su) |
| *Buchanania cochinchinensis* (Lour.) Almeida | Anacardiaceae | Fruit, Leaf | Mumps, Burnt wounds. (S, G, Ko, Ba) |
| *Calotropis gigantea* (L.) R. Br. | Asclepiadaceae | Leaf, Latex | Earache, Scabies. (S, Lo, Ba, Su) |
| *Capparis zeylanica* L. | Capparaceae | Leaf, Bark, Stem, Root | Leucoderma, Rheumatism, Long standing wounds. (Ba, K, Mu, Bh, Su) |
| *Caryota urens* L. | Arecaceae | Stem | Lactation. (S, Ba, Bh, Su, Lo) |
| *Catharanthus roseus* (L.) G. Don | Apocynaceae | Leaf, Flower | Diabetes, Colic pain. |

| | | | |
|---|---|---|---|
| *Celosia argentea* L. | Amaranthaceae | Leaf, Root-bark | Diarrhoea, Vision, Blood disease, Eye disease. (Ba, K, Mu, Bh, Su) |
| *Centella asiatica* (L.) Urban | Apiaceae | Leaf, Plants | Memory, Leprosy. (S, G, Ko, Ba) |
| *Cleome gynandra* L. | Cleomaceae | Bark, Stem | Round worm, Rheumatism. (Lo, S, Su, Ba, K) |
| *Cleome viscosa* L. | Cleomaceae | Bark, Flower | Fever, Diarrhoea, Round worms, Purulent discharges from the ear. (S, G, Ko, Ba) |
| *Clerodendrum indicum* (L.) O. Ktze. | Verbenaceae | Root, Plant, Pod, Leaf | Asthma, Breast abscess, Burning sensation. (S, Lo, Ba, Su) |
| *Clitoria ternatea* L. | Fabaceae | Seed, Leaf, Root | Ascites, Fevers, Chronic bronchitis, Irritation of bladder and urethera. (Ba, K, Mu, Bh, Su) |
| *Commelina benghalensis* L. | Commelinaceae | Leaf, Plant | Laxative, Demulcent, Leprosy. (S, Ba, Bh, Su, Lo) |
| *Conyza canadensis* (L.) Cronq. | Asteraceae | Root, Bark, Leaf | Diarrhoea, Dysentery, Dropsy, Eczema, Uterine haemorrhage. (S, G, Ko, Ba) |
| *Corchorus olitorius* L. | Tiliaceae | Leaf, Seed, Plant | Demulcent, Tonic, Diuretic, Cystitis. (S, Lo, Ba, Su) |
| *Cuscuta reflexa* Roxb. | Cuscutaceae | Plant, Seed | Tonic, Anthelmintic, Blood purifier. (Ba, K, Mu, Bh, Su) |
| *Cyperus rotundus* L. | Cyperaceae | Tuber | Diarrhoea, Dysentery, Dyspepsia, Cholera, Gonorrhoea. (S, Lo, Ba, Su) |
| *Datura metel* L. | Solanaceae | Leaf, Seed, Root | Dandruff, Earache. (S, Ba, Bh, Su, Lo) |
| *Desmodium gangeticum* (L.) DC. | Fabaceae | Leaf, Flower-bud | Diarrhoea. (S, Ba, Bh, Su, Lo) |
| *Desmodium heterocarpon* (L.) DC. | Fabaceae | Flower-bud | Blood dysentery. (S, G, Ko, Ba) |
| *Dillenia aurea* Sm. | Dilleniaceae | Fruit | Loss of appetite. (Ba, K, Mu, Bh, Su) |
| *Dillenia pentagyna* Roxb. | Dilleniaceae | Carpel | Pimples. (Lo, S, Su, Ba, K) |
| *Dioscorea bulbifera* L. | Dioscoreaceae | Tuber | Bronchitis, Dyspepsia. (S, G, Ko, Ba) |
| *Diospyros melanoxylon* Roxb. | Ebenaceae | Flower, Bark | Leucorrhoea, Amoebic dysentery. (S, Ba, Bh, Su, Lo) |
| *Echinops echinatus* Roxb. | Asteraceae | Inflorescence, Seed, Plant | Dyspepsia, Scrophula, Syphilis, Hysteria, Impotence. (S, Lo, Ba, Su) |
| *Eclipta prostrata* (L.) L. | Asteraceae | Root, Seed | Jaundice, Spleen enlargement, Tonic, Liver disorder. (S, Lo, Ba, Su) |
| *Elephantopus scaber* L. | Asteraceae | Root, Plant | Diarrhoea, Flatulence. (S, Ba, Bh, Su, Lo) |
| *Euphorbia dracunculoides* Lam. | Euphorbiaceae | Seed, Root | Removes warts. (S, Lo, Ba, Su) |
| *Euphorbia helioscopia* L. | Euphorbiaceae | Bark, Seed, Leaf, Root | Anthelmintic, Cholera, Rheumatism. (Lo, S, Su, Ba, K) |
| *Euphorbia neriifolia* L. | Euphorbiaceae | Plant | Breast abscess, Foot crack, Dry skin. (Lo, S, Su, Ba, K) |

| | | | |
|---|---|---|---|
| *Evolvulus nummularius* (L.) L. | Convolvulaceae | Root, Leaf, Plant | General fever. (Ba, K, Mu, Bh, Su) |
| *Ficus benghalensis* L. | Moraceae | Bark, Fruit, Leaf | Fertility, Leucorrhoea. (Ba, K, Mu, Bh, Su) |
| *Ficus hispida* L. f. | Moraceae | Root, Fruit, Latex | Diabetes, Amoebic dysentery, Gonorrhoea, Mumps, Impotence. (Ba, K, Mu, Bh, Su) |
| *Ficus religiosa* L. | Moraceae | Bark, Fruit, Leaf | Debility, Fertility, Leucorrhoea, Vomiting. (S, Ba, Bh, Su, Lo) |
| *Fumaria indica* (Haussk.) Pugsley | Fumariaceae | Leaf, Bark, Root, Stem | Dyspepsia, Jaundice, Leprosy, Scrofulous, Skin infection. (S, G, Ko, Ba) |
| *Garcinia xanthochymas* Hook. f. ex T. Anders. | Clusiaceae | Leaf | Mouth ulcer. (S, Lo, Ba, Su) |
| *Gloriosa superba* L. | Liliaceae | Root | Worm infection. (S, G, Ko, Ba) |
| *Gmelina arborea* Roxb. | Verbenaceae | Fruit | Hair growth. (Lo, S, Su, Ba, K) |
| *Gnaphalium leuto-album* L. | Asteraceae | Leaf, Flower | Fevers, Gout. (S, Ba, Bh, Su, Lo) |
| *Hedyotis corymbosa* (L.) Lam. | Rubiaceae | Leaf, Bark, Root | Gastric disorders, Jaundice, Liver problem. (Ba, K, Mu, Bh, Su) |
| *Helicteres isora* L. | Sterculiaceae | Pod, Leaf | Round worm infection, Earache. (S, G, Ko, Ba) |
| *Heliotropium indicum* L. | Boraginaceae | Leaf | Epilepsy. (Ba, K, Mu, Bh, Su) |
| *Heliotropium strigosum* Willd. | Boraginaceae | Root, Leaf, Bark | Snake-bite, Gum boils, Wounds, Ulcers. (S, Lo, Ba, Su) |
| *Indigofera linnaei* Ali | Fabaceae | Root, Leaf, Plant | Chronic venereal disease. (S, G, Ko, Ba) |
| *Justicia adhatoda* L. | Acanthaceae | Leaf, Flower, Bark | Nasal bleeding, Eczema, Whooping cough. (S, Ba, Bh, Su, Lo) |
| *Lantana camera* L. | Verbenaceae | Leaf | Asthma. (Lo, S, Su, Ba, K) |
| *Leucas aspera* (Willd.) Link | Lamiaceae | Leaf, Inflorescence, Plant | Skin eruptions, Cough, Cold, Rheumatism. (Ba, K, Mu, Bh, Su) |
| *Mangifera indica* L. | Anacardiacae | Embryo, Bark | Diarrhoea, Peptic-ulcer. (S, G, Ko, Ba) |
| *Martynia annua* L. | Martyniaceae | Fruit, Bark | Gonorrohoea. (S, Ba, Bh, Su, Lo) |
| *Mimosa himalayana* Gamble | Mimosaceae | Leaf | Toothache. (Lo, S, Su, Ba, K) |
| *Mimosa pudica* L. | Mimosaceae | Leaf, Root | Haemorrhoids, Toothache. (Ba, K, Mu, Bh, Su) |
| *Mollugo cerviana* (L.) Ser. | Molluginacae | Bark, Seed | Fever, Gout, Rheumatism. (Lo, S, Su, Ba, K) |
| *Nelumbo nucifera* Gaertn. | Nelumbonaceae | Flower, Leaf, Petal, Tuber | Fever, Fresh wounds, Boil, Dysmenorrhoea. (S, Lo, Ba, Su) |
| *Neolamarckia cadamba* (Roxb.) Bosser | Rubiaceae | Bark | Burning sensation. (Ba, K, Mu, Bh, Su) |
| *Nyctanthes arbor-tristis* L. | Oleaceae | Leaf, Bark | Malaria, Fresh wounds. (S, Ba, Bh, Su, Lo) |
| *Ocimum gratissimum* L. | Lamiaceae | Leaf, Seed | Cold & Cough, Conjuctivitis, Eczema. (Lo, S, Su, Ba, K) |

| | | | |
|---|---|---|---|
| *Ocimum tenuiflorum* L. | Lamiaceae | Leaf | Cough, Fever. (S, Ba, Bh, Su, Lo) |
| *Opuntia stricta* (Haw.) Haw. | Cactaceae | Flower, Bud, Stem | Foot crack. (S, G, Ko, Ba) |
| *Oroxylum indicum* (L.) Vent. | Bignoniaceae | Bark | Arthritis. (Ba, K, Mu, Bh, Su) |
| *Oxalis corniculata* L. | Oxalidaceae | Stem, Bark, Root | Dyspepsia, Piles, Anaemia, Tymapnitis. |
| *Pedalium murex* L. | Pedaliaceae | Leaf, Seed | Dysuria, Gonorrhoea. (Ba, K, Mu, Bh, Su) |
| *Phyla nodiflora* (L.) Greene | Verbenaceae | Leaf, Bark | Stoppage of boils, Fever. (S, Lo, Ba, Su) |
| *Phyllanthus emblica* L. | Euphorbiaceae | Root, Leaf | Bronchitis, Burning sensation, Insect bite, Amoebic dysentery, Leucoderma. (S, G, Ko, Ba) |
| *Phyllanthus fraternus* Webster | Euphorbiaceae | Root, Plant, Shoot | Dysentery, Disease of genito-urinal tract, Diarrhoea, Jaundice. (S, G, Ko, Ba) |
| *Plumbago indica* L. | Plumbaginaceae | Leaf, Stem, Fruit | Leucorrhoea, Abortion, Bronchitis. (S, Ba, Bh, Su, Lo) |
| *Plumbago zeylanica* L. | Plumbaginaceae | Fruit, Stem | Bronchitis, Leucoderma, Abortion. (Ba, K, Mu, Bh, Su) |
| *Psidium guajava* L. | Myrtaceae | Leaf, Young shoot | Diarrhoea, Amoebic dysentery. |
| *Pterospermum acerifolium* (L.) Willd. | Sterculiaceae | Flower | Abdominal pain. (S, Lo, Ba, Su) |
| *Rauvolfia serpentina* (L.) Benth. ex Kurz | Apocynaceae | Root | Blood pressure. (S, G, Ko, Ba) |
| *Salvia santolinaefolia* Boiss. | Lamiaceae | Leaf, Seed | Gonorrhoea. (Lo, S, Su, Ba, K) |
| *Santalum album* L. | Santalaceae | Wood | Boil, Pimples. (Lo, S, Su, Ba, K) |
| *Schleichera oleosa* (Lour.) Oken. | Sapindaceae | Seed, Bark | Eczema. (Lo, S, Su, Ba, K) |
| *Scoparia dulics* L. | Scrophulariaceae | Leaf, Plant | Fever, Toothache. (S, G, Ko, Ba) |
| *Senna tora* (L.) Roxb. | Caesalpiniaceae | Fruit, Stem, Leaf | Skin disease, Arthritis, Ringworms. (S, Ba, Bh, Su, Lo) |
| *Sesamum indicum* L. | Padaliaceae | Leaf, Seed | Infection of kidneys & bladder, Piles, Respiratory problem. (S, G, Ko, Ba) |
| *Shorea robusta* Gaertn. f. | Dipterocarpaceae | Seed, Bark, Leaf | Scabies, Diarrhoea, Earache. |
| *Sida acuta* Burm. f. | Malvaceae | Stem | Dental care. (S, Lo, Ba, Su) |
| *Sida cordifolia* L. | Malvaceae | Root, Leaf | Dysentery, Urinary troubles, Hematuria. (Lo, S, Su, Ba, K) |
| *Solanum nigrum* L. | Solanaceae | Root, Leaf, Fruit | Piles, Diarrhoea, Hydrophobia, Cirrhosis of liver, Dysentery. (Ba, K, Mu, Bh, Su) |
| *Solanum violaceum* Orteg. | Solanaceae | Root | Stoppage of delivery bleeding. (S, Ba, Bh, Su, Lo) |
| *Soymida febrifuga* (Roxb.) A. Juss | Rutaceae | Bark | All types of fever. (S, G, Ko, Ba) |

| | | | |
|---|---|---|---|
| *Spondias pinnata* (L. f.) Kurz | Anacardiaceae | Plant, Bark | Menstrual disorder, Arthitis. (Lo, S, Su, Ba, K) |
| *Streblus asper* Lour. | Moraceae | Leaf, Latex, Bark, Root | Eczema, Skin diseases, General wounds, Labour pain. (S, Ba, Bh, Su, Lo) |
| *Strychnos nux-vomica* L. | Strychnaceae | Bark, Stem | Cholera, Dysentery, Intestinal problem. |
| *Syzygium cumini* (L.) Skeels | Myrtaceae | Bark, Fruit | Diarrhoea, Dysentery. (S, Lo, Ba, Su) |
| *Tephrosia purpurea* (L.) Pers. | Fabaceae | Seed, Leaf, Bark | Dog biting, Diarrhoea, Dyspepsia, Asthma, Rheumatism, Urinary disorders, Piles, Liver problem. (S, G, Ko, Ba) |
| *Terminalia bellirica* (Gaertn.) Roxb. | Combretaceae | Fruit | Debility, Graying of hairs, Stomach disorder. (S, Ba, Bh, Su, Lo) |
| *Terminalia chebula* Retz. | Combretaceae | Fruit | Stomach disorder, Dyspepsia, Toothache. (S, G, Ko, Ba) |
| *Terminalia cuneata* Roth | Combretaceae | Bark, Fruit | Cranial-infection, Heart-palpitation, Long standing wounds. (S, G, Ko, Ba) |
| *Tinospora cordifolia* (Willd.) Meirs ex Hook. f. & Thoms. | Menispermaceae | Leaf, Bark | Obesity, Long standing wounds. (Lo, S, Su, Ba, K) |
| *Tridax procumbens* L. | Asteraceae | Leaf, Plant | Fresh wounds. (S, Ba, Bh, Su, Lo) |
| *Tylophora indica* (Burm. f.) Merrill | Asclepiadaceae | Stem, Leaf, Bark | Asthma, Bronchitis, Whooping cough, Diarrhoea, Rheumatism. (S, G, Ko, Ba) |
| *Verbascum thapsus* L. | Scrophulariaceae | Leaf | Piles, Heart stimulant, Cramps. (S, Lo, Ba, Su) |
| *Vitex negundo* L. | Verbenaceae | Shoot, Leaf | Induce abortion, Mouth ulcer. (S, Ba, Bh, Su, Lo) |
| *Withania somnifera* (L.) Dunal | Solanaceae | Root, Leaf, Plant | Cough, Dropsy, Ulcers, Scabies, Fevers, Debility. (Lo, S, Su, Ba, K) |
| *Woodfordia fruticosa* (L.) Kurz | Lythraceae | Bark | Bear bite. (S, G, Ko, Ba) |
| *Xanthium indicum* Koen. | Asteraceae | Leaf, Root, Seed | Malaria, Urino-genital diseases, Cancer. (S, Lo, Ba, Su) |
| *Zingiber officinale* Rosc. | Zingiberaceae | Rhizome | Epilepsy, Sore tongue, Waist pain, Cold & Cough. (S, G, Ko, Ba) |
| *Ziziphus mauritiana* Lam. | Rhmnaceae | Bark, Leaf | Diarrhoea, Hair loss. (S, G, Ko, Ba) |
| *Ziziphus oenoplia* (L.) Mill. | Rhmnaceae | Root, Bark | Burnt wound. (S, Ba, Bh, Su, Lo) |
| *Ziziphus rugosa* Lam. | Rhmnaceae | Bark | Flatulence. (Lo, S, Su, Ba, K) |

## RESULTS AND DISCUSSION

The Kaptipada forest range, which is considered one of the richest botanical treasures in Orissa, is under intense biotic pressure owing to forest fire, over-exploitation of medicinal plants for drugs and conversion of forest to agriculture lands. These factors of degradation result not only in the depletion of certain valuable interesting taxa but also lead to substantial reduction in the density and extent of forest cover. This rapid destruction of forest causes severe damage to natural habitats, thus threatening the very survival of several indigenous species. The present investigation is carried out by the intensive field explorations and interaction with tribal people for nearly two years to identify the potential ethnomedicinal plants distributed in the Kaptipada forest range. The data when cross-checked with the published literature, was found that most of the uses are either less known or not recorded earlier. During the survey, 120 species belonging to 102 genera and 51 families have been collected, critically studied and mentioned in Table 2. The present study is unique in nature so far as the information is concerned. All the medicinal herbs reported in Table 2 are being employed as herbal drugs in various pharmaceutical preparations (Anonymous, 1976; Kapoor & Mitra, 1979; Kritikar & Basu, 1984; Saxena, 1986; Mitra & Jain, 1991).

Few species such as *Plumbago indica, Gmelina arborea, Abrus precatorius* and *Argemone mexicana* are found to be discontinuous in distribution and presently occur only in the selected habitats which are rather becoming endangered/rare either on account of habitat loss or any other factor. Plants like *Gloriosa superba, Pterospermum acerifolium* and *Tinospora cordifolia* have been badly depleted due to over exploitation for medicinal uses. So, this study will help in conservation of the potential medicinal plants which appear to be the most threatened.

**ACKNOWLEGEMENT**

The authors wish to acknowledge to the Head, Post-Graduate Department of Botany, Ravenshaw College, Cuttack to provide the laboratory facility.

**REFERENCES**

ABRAHAM, Z. 1981. Ethnobotany of the Todas, the Kolas and the Irulas of Nilgiris. In Jain, S. K. (ed.) *Glimpses of Indian Ethnobotany,* pp. 308-320. New Delhi.

ANONYMOUS 1976. *Medicinal Plants of India.* C.S.I.R., New Delhi.

BARUAH, P. & G.C. SHARMA 1987. Studies on the medicinal uses of plant by the northeast tribes : III. *J. Econ. Taxon. Bot.* 11 (1) : 71-76.

BRAHMAM, M. & H.O. SAXENA 1989. Ethnobotany of Gandhamardan Hills. Some noteworthy Folk medicinal uses. *Int. Conf. Rec. Adv. Med. Arom. Spices Crops.* New Delhi (Abst. 1919).

————, K.B. SATPATHY & P. DASH 2000. *Ethnobotanical studies among Kolha Tribes of Keonjhar District, Orissa.* Souvenir National Seminar on Med. Pl. in our environment, V. N. College, Jajpur Road.

CHAUDHURI RAI, R.N., M.A. MOLLA, D.C. PAL & D.C.B. ROY 1982. Plants used in traditional medicines by some tribals of Jalpaiguri district, West Bengal. *Bull. Bot. Surv. India* 24 : 87-90.

DAS, P.K. & M.K. MISHRA 1987. Some medicinal plants used by the tribals of Deomali and adjacent areas of Koraput district, Orissa, India. *Indian J. For.* 10 (4) : 301-303.

KAPPOR, S.L. & R. MITRA 1979. *Herbal drugs in Indian Pharmaceutical Industry.* Economic Botany Information Service, National Botanical Research Institute, Lucknow, India.

KRITIKAR, K.R. & B.D. BASU 1984. (Ed.) *Indian Medicinal Plants.* Bishen Singh Mahendra Pal Singh, Dehra Dun.

MITRA, R. & S.K. JAIN 1991. Medicinal plants research in India - A review. *Ethnobotany* 3 : 65-77.

PAL, D.C. 1984. Observations on ethnobotany of tribals of Subansiri, Arunachal Pradesh. *Bull. Bot. Surv. India* 26 : 26-37.

SATPATHY, K.B. & M. BRAHMAM 1996. Some medicinal plants used by the tribals of Sundargarh district, Orissa. In Jain, S.K. (Ed.) *Ethnobotany in human welfare,* pp. 153-158. Deep Publication. New Delhi.

——— & ——— 1999. Some interesting phytotherapeutic claims of tribals of Jajpur district, Orissa, India. *J. Econ. Taxon. Bot.* 23 (2) : 467-472.

SAXENA, H.O. 1986. Observations on ethnobotany of Madhya Pradesh. *Bull. Bot. Surv. India* 28 (1-4) : 149-156.

——— & M. BRAHMAM 1994. *Flora of Orissa,* Vol. I-IV. Orissa Forest Development Corporation, Bhubaneswar.

———, ——— & P.K. DUTTA 1981. Ethnobotanical studies in Orissa. In Jain, S.K. (Ed.) *Glimpses of Indian Ethnobotany,* pp. 232-244. New Delhi.

# POTENTIAL FODDER SPECIES OF LADAKH HIMALAYA

ANJU VERMA, O.P. CHAURASIA, BASANT BALLABH & B. RAUT

*Field Research Laboratory, Leh-Ladakh (J & K)*

**ABSTRACT**

Plants play an important role in sustaining and development of animal husbandry. The heavy population of animals (about ten times more in comparison to human beings) has to survive and sustain on wild fodder plants. In Ladakh, alfalfa is only crop cultivated on limited scales, but the barren mountains have potential to grow a number of wild fodder species mainly during summers every year.

In view of the above, an effort has been made to study the potential fodder yielding plants of Ladakh Himalaya which are either being grazed or harvested for domestic uses at the time of maturity. The present paper deals with 22 most important fodder plants which are being commonly used by the local people. Besides this, nutritional values have also been presented for fodder plants..

## INTRODUCTION

Ladakh, situated in the Himalayas on the far northern frontiers of India, is one of the coldest regions in the World. For about six months every year it remains cut-off from the rest of the country. All the agricultural land becomes frozen during this time and nothing can be grown. It has an area of 68,321 sq km and falls approximately between latitude 32°30' N and longitude 74°30' to 80°30' E (Singh *et al.*, 1996).

The area is now divided into two districts of Leh and Kargil. The area suffers an extremely harsh climate, with winter temperatures as low as -30°C at Leh and -50°C at Dras and the summer temperatures go up to 35°C (Jul. & Aug.). The annual precipitation varies from 9 cm (Leh) to 23.9 cm (Kargil), with extremely low humidity.

In this region the cultivation of fodders is restricted to the lower hills, Beyond this zone, the cultivation of fodder is not practical because of adverse climatic and the land constraints, Almost all the available cultivable land, which is around 10% of the total geographical area, is used for the cultivation of cereals, pulses and oil seeds. Rangelands and forests, comprising 50 to 70 per cent of the total geographical area, are the major sources of forage for sheep, goats, rabbits, cattle and other animals.

Nomadic grazers (Khadwals, Gaddis, Gujjars, Khas) use sub-alpine and alpine pastures for rearing their livestock. Due to ever-increasing demand of animal products, the livestock population has increased manifold, thereby increasing the grazing pressure on these pastures. This has resulted in the deterioration of the grass cover as well as valuable forest species. As a result of this, a significant area of this natural resource has been covered by noxious plant .species like *Anemone, Adonis, Aquilegia, Cincifuga, Clematis, Lepidium, Artemisia, Ranunculus, Stipa, Sorbinia* and *Sambucus*. All these factors have led to the decrease in carrying capacity of these pastures.

## COLD DESERT PLANTS AND FODDER

Flora of alpine and high alpine zones prevails in this region. The lower alpine zones are covered with lovely meadows, pasturelands, moraines, screes and marshes. The higher reaches, which are covered with snow for more than six months in a year and which has a bright sunshine (with more quantity of UV rays), is more amicable to vegetation during the brief summer (Polunin *et al.,* 1989). The constraints at high altitudes, which need to be recognized are:

| | |
|---|---|
| *Physical* | Short growing season, low plant growth rates, therefore limited potential. |
| *Soils* | Often shallow, acidic low P availability and degraded. |
| *Ecological* | Fragile ecosystems, ingress of weeds, loss of forest. |
| *System* | Traditional systems, often migratory, pastorals and farmers conservative and averse to risk, lack of cash flow to fund improvement and communal grazing of range. |

Here the principle vegetative growth starts with the beginning of summer in April when the melting of the snow provides moisture to the herbaceous vegetation to go through its complete course in a speeded up time frame. The flora is in full bloom by August and starts to degenerate by end September (Kachroo *et al.,* 1977). These plants have a brief and intense life.

The high altitude systems, temperate and alpine systems at altitudes of >2500 to < 5000 m, include the entirely pastoral migratory goat/sheep systems. There is a range of temperate grasses with species which are adaptable to the sub-humid and semi-arid zones of the Himalaya and which potentially could be used as fodder crops on arable land or improve alpine and sub alpine range some of which are degrading through over-use.

In the lower Himalayan region also the natural grasslands are under heavy grazing pressure and have been badly degraded. Insignificant land area is under fodder cultivation and there is least scope of increasing this. However, integration of forage crops in the existing crop sequences, growing food cum fodder crops, intensive forage cultivation, raising improved grasses on terrace-risers, developing suitable cropping systems for rain fed and irrigated areas, improving silvi and horti pastoral systems in addition to natural pasture and wasteland improvement are some of the approaches to solve the forage problem. Introduction of improved grass/legume species in the native vegetation not only helps in increasing total biomass production but also improve the quantity of the produce. However, the introduction of improved species in the existing vegetation, particularly with seed, is not an easy task, unless initial competition from the native vegetation is checked by some means.

"Ole" is the most commonly cultivated crop in this region. It is a cross pollinated crop, and as such a mixture of Alfalfa is grown. It produces highly nutritive and excellent quality hay. Its deep root system enables it to tolerate the harsh climatic conditions for many years, making it most economical to grow. There are three species of Alfalfa (*Medicago falcata, M. lupulina* and *M. sativa*) found in Ladakh. The crop grown in this region is more resistant to cold and drought.

## MATERIALS AND METHODS

The all seasons study was conducted through various field surveys regularly during May 1993 to Sept. 2003, covering far-flung areas in all the valleys, mountains and river belts of Ladakh Himalaya. On the spot, information was gathered from local people, *Amchis,* village elders, old aged persons and shepherd's/grazers. Species were collected in triplicate and identified through literature (Usher, 1984), monographs and with the help of herbarium lodged at Field Research Laboratory, Leh and finally confirmed in the herbaria like of BSA, DD and herbarium of J & K. Fodder samples were also collected from all possible places in triplicate and then nutritional analysis has been done at IGFRI, Jhansi.

## OBSERVATIONS & ENUMERATION

Fodder species have been described in alphabetical order alongwith botanical name, family, vernacular names, habitat and uses in the enumeration (Chaurasia & Singh, 1996-2001). In addition, the nutritional value of some important fodder species has been appended in Table 1.

**Aconogonum tortuosum** (D. Don) Hara (Polygonaceae)

*Local name :* Nyalo, Serpalulu, Itimeng (L)

*Habitat :* On dry stony slopes and along roadsides and water streams in Zanskar & Changthang valleys, 13,500 ft.

*Uses :* The important wild fodder species which contains 8.63 per cent crude protein. It is harvested during August-September and the dry mass is stored on the roofs after making loose bundles for prolonged winter uses.

**Avena fatua** Linn. (Poaceae)

*Local name :* Wild oat (E)

*Habitat :* Commonly found in wheat fields; Alchi village, 10,500 ft.

*Uses :* Wild oat contains 6.79 per cent crude protein; it is mixed with wheat straw and fed to the domestic animals.

**Capparis spinosa** Linn. (Capparaceae)

*Local name :* Caper bush (E), Ber (H), Kabra (L)

*Habitat :* Found along the roadside; Nimmu,10,500 ft.

*Uses :* The shoots and leaves are grazed by the animals. It contains 17.87 per cent crude protein.

**Chenopodium botrys** Linn. (Chenopodiaceae)

*Local name* : Sagani (L)

*Habitat :* Weed in vegetable garden; Murtse village, 11,500 ft.

*Uses :* The plant contains high crude protein content upto 26.40 per cent; it is best for young and milking animals.

**Cicer microphyllum** Benth. (Fabaceae)

*Local name :* Wild gram (E), Seri (L)

*Habitat :* Found growing in alpine pastures; Zingral, South Pulu, Namika La, Namshang La, 14,000-15,000 ft.

*Uses :* The whole plant is grazed by animals. This is commonly collected for fodder which contains 15.30 per cent crude protein content.

**Ephedra gerardiana** Wallich ex Stapf (Ephedraceae)

*Local name :* Asmania (H), Chhapat (L)

*Habitat :* Commonly found on stony slopes, gravel terraces and in dry areas; Hemis village, 11,600 ft.

*Uses :* The whole plant is grazed by wild animals at the time of scarcity of fodder during the winters. It contains 9.27 per cent crude protein content.

**Fagopyrum esculentum** Moench (Polygonaceae)

*Local name :* Buck wheat (E), Dayat (L)

*Habitat :* Grows as a weed in wheat fields; Alchi village, 10,500 ft.

*Uses :* The whole plant is used as fodder. The high protein content 25.19 per cent is considered best for young and milking animals.

**Fagopyrum tataricum** (Linn.) Gaertn. (Polygonaceae)

*Local name* : Dayat (L)

*Habitat :* Cultivated fields; Tangtse village, 14,500 ft.

*Uses :* The whole plant is used as fodder. It contains 20.12 per cent crude protein.

**Hippophae rhamnoides** Serv. (non Linn.) var. **turkestanica** Rousi. (Elaeagnaceae)

*Local name :* Seabuckthorn (E), Shermung (L)

*Habitat :* Grows along Indus river, road side and waste lands; Leh, 11,500 ft.

*Uses :* The shoots and leaves are grazed by animals. The high crude protein content upto 23.93 per cent is considered the best quality fodder value.

**Iris lactea** Pallas (Iridaceae)

*Local name :* Tesmamentok (L)

*Habitat :* Common around Leh city, 11,400 ft.

*Uses :* Iris is considered good quality fodder for lactating animals. The plant contains 10.62 per cent crude protein content.

**Lepidium latifolium** Linn. (Brassicaceae)

*Local name :* Dittander (E), Shangsho (L)

*Habitat :* Wasteland and stony slopes; Darbuck village, 14,400 ft.

*Uses :* The plant contains 24.62 per cent crude protein which is considered best for young and milking animals. The plant is usually fed to the animals as green fodder.

**Medicago falcata** Linn. (Fabaceae)

*Local name :* Alfalfa (E), Ole (L)

*Habitat :* Cultivated fodder crop in the valley.

*Uses :* Alfalfa is one of the most ancient fodder crops of Ladakh. It is only fodder available throughout Ladakh as green and dried. The crude protein content found in the plant is about 15.85 per cent.

**Medicago lupulina** L. (Fabaceae)

*Local name :* Lucerne (E), Ole (L)

*Habitat :* Cultivated fodder crop in the valley.

*Uses :* As *M. falcata.* The plant contains 17.50 per cent crude protein.

**Medicago sativa** Linn. (Fabaceae)

*Local name :* Alfalfa or Lucerne (E), Ole (L)

*Habitat :* Commonly cultivated fodder crop in the valley.

*Uses :* As *M falcata.* The plant contains 18.22 per cent crude protein.

**Melilotus officinalis** (Linn.) Pallas (Fabaceae)

*Local name :* Yellow sweet clover (E), Ole (L)

*Habitat :* Common on irrigated land, growing mixed along with *Medicago* spp. throughout the valley.

*Uses :* This species is used as a substitute of Alfalfa which is also an ancient fodder crop of the region. It is cultivated mixed with Alfalfa as fodder crop. The high protein content 24.01 per cent is considered best for young and milking animals.

**Prangos pabularia** Lindey (Apiaceae)

*Local name :* Prangos (E), Plans (L)

*Habitat :* Common around the stony slopes; Dras (Suru valley), 10,200 ft.

*Uses :* The plant contains 12.66 per cent crude protein and is used as most important fodder in Dras regions.

**Rheum webbianum** Royle (Polygonaceae)

*Local name :* Rhubarb (E), Lacchu (L)

*Habitat :* On way to Penzila (Zanskar valley), 14,000 ft.

*Uses :* The whole plant is used as fodder which contains 12.80 per cent crude protein.

**Rumex patientia** Linn. ssp. **tibeticus** Rech. f. (Polygonaceae)

*Local name* : Shoma (L)

*Habitat :* Common on wet places, near water channel; Mulbeck (Suru valley), 10,000 ft.

*Uses :* The whole plant is used as fodder. The high crude protein content upto 23.57 per cent is considered best for milking animals.

**Salix babylonica** Linn. (Salicaceae)

*Local name :* Weeping Willow (E)

*Habitat :* Planted near habitations and along the rivers in lower reaches of Ladakh; Leh, 11,500 ft.

*Uses :* The plant provides an excellent reserve of feed having 18.12 per cent crude protein for livestock in summer months.

**Trifolium pratense** Linn. (Fabaceae)

*Local name :* Red clover (E)

*Habitat :* Found along the moist places; Pandras, 10,500 ft.

*Uses :* The plant contains 17.56 per cent crude protein which is grazed by animals along with other grasses.

**Trifolium repens** Linn. (Fabaceae)

*Local name :* White clover (E)

*Habitat :* On moist places; Padam, 11,500 ft.; Dras, 10,200 ft.

*Uses :* The plant is grazed by animals which contains 12.60 per cent crude protein.

**Ulmus wallichiana** Planch. (Ulmaceae)

*Local name :* Elm (E), Umbuk (L)

*Habitat :* Rare, on moist places; Nubra valley, around 10,500 ft.

*Uses :* The leaves are used as fodder. It contains 19.02 per cent crude protein.

Table 1. Nutritional profile of wild fodder plants.

| S.N. | Plant name | Part used | Moist. | DM% | CP% | NDF% | ADF% | L% | HC% | C% |
|---|---|---|---|---|---|---|---|---|---|---|
| 1. | *Aconogonum tortuosum* | Whole plant | - | 88.87 | 8.63 | 51.04 | 38.22 | 12.63 | 12.82 | 25.59 |
| 2. | *Avena fatua* | Whole plant | 09.09 | 90.91 | 6.79 | 38.35 | 28.96 | 04.60 | 09.39 | 24.34 |
| 3. | *Capparis spinosa* | Leaves | 09.20 | 90.80 | 17.87 | 06.83 | - | - | - | - |

| | | | | | | | | | |
|---|---|---|---|---|---|---|---|---|---|
| 4. *Chenopodium botrys* | Leaves | 07.10 | 92.90 | 26.40 | 07.10 | - | - | - | - |
| 5. *Cicer microphyllum* | Whole plant | 08.60 | 91.40 | 15.30 | 36.38 | 32.34 | 05.90 | 04.04 | 26.44 |
| 6. *Ephedra gerardiana* | Leaves & Fruits | 08.19 | 91.81 | 09.27 | 45.23 | 42.48 | 21.28 | 02.75 | 21.20 |
| 7. *Fagopyrum esculentum* | Whole plant | 05.70 | 94.30 | 05.19 | 07.70 | 17.28 | 7.15 | 13.26 | 30.13 |
| 8. *Fagopyrum tartaricum* | Whole plant | 03.22 | 96.78 | 20.12 | 06.80 | - | - | - | - |
| 9. *Hippophae rhamnoidse* | Leaves | 08.99 | 91.09 | 23.93 | 22.72 | 20.21 | 98.47 | 02.51 | 11.74 |
| 10. *Iris lactea* | Leaves | 09.63 | 90.37 | 10.62 | 47.28 | 42.35 | 04.01 | - | - |
| 11. *Lepidium latifolium* | Whole plant | 08.26 | 91.74 | 24.62 | 11.20 | - | - | - | - |
| 12. *Medicago falcata* | Whole plant | - | - | 15.85 | 20.86 | - | - | - | 25.80 |
| 13. *Medicago lupulina* | Whole plant | - | - | 17.50 | 20.58 | - | - | - | 25.70 |
| 14. *Medicago sativa* | Whole plant | - | - | 18.22 | 20.65 | - | - | - | 25.90 |
| 15. *Melilotus officinalis* | Whole plant | 10.05 | 89.95 | 24.01 | 20.94 | 18.53 | 04.71 | 02.41 | 13.82 |
| 16. *Prangos pabularia* | Whole plant | 10.57 | 89.43 | 12.66 | 35.78 | 32.69 | 05.12 | 03.09 | 27.57 |
| 17. *Rheum webbianum* | Leaves | 08.70 | 91.30 | 12.80 | 14.95 | - | - | - | - |
| 18. *Rumex patientia* ssp. *tibeticus* | Whole plant | - | 88.38 | 23.57 | 30.89 | 21.00 | 7.29 | 9.89 | 13.71 |
| 19. *Salix babylonica* | Leaves | 08.97 | 91.03 | 18.12 | 36.35 | 24.68 | 09.28 | 11.67 | 15.40 |
| 20. *Trifolium pratense* | Whole plant | 10.32 | 89.68 | 17.56 | 39.14 | 32.56 | 06.23 | 06.58 | 26.35 |
| 21. *Trifolium repens* | Whole plant | 10.53 | 89.47 | 12.60 | 41.80 | 32.75 | 06.87 | 09.05 | 25.88 |
| 22. *Ulmus wallichiana* | Leaves | - | 89.70 | 19.02 | 28.98 | 22.74 | 6.05 | 6.24 | 16.69 |

Abbreviations : DM=Dry matter; NDF=Neutral detergent fibre; ADF=Acid detergent fibre; CP=Crude protein; L=Legnin; HC=Hemi cellulose; C=Cellulose.

## CONCLUSION

The information gathered reveals that Alfalfa is the only cultivated fodder crop. As such, there is an urgent need to introduce other species, both wild and cultivated, to enhance fodder availability in Ladakh. With the increase in the domestic animal population in the area and the necessity of higher animal product yields, there has been a progressive increase in both the quantity and nutritive quality of fodder. This can only be met by introduction of better fodder species in Ladakh, to be increasingly cultivated alongwith traditional fodder. A specific intervention programme aimed at spreading awareness, as also developing fodder species amicable in the area, needs to be implemented on priority.

## ACKNOWLEDGEMENTS

We wish to convey our sincere thank for the kind cooperation and technical help provided by Field Research Laboratory (DRDO), Leh and The Indian Grassland and Fodder Research Institute, Jhansi, without which this work would not have been possible.

## REFERENCES

CHAURASIA, O.P. & B. SINGH 1996-2001. *Cold Desert Plants,* Vol. I-V. Field Research Laboratory, Leh, Ladakh.

KACHROO, P., B.L. SAPRO & U. DHAR 1977. *Flora of Ladakh.* Scientific Publishers, Jodhpur.

POLUNIN, O. & A. STAINTON 1989. *Flowers of the Himalaya.* Oxford Univ. Press, New Delhi.

SINGH, B. & D.P. CHAURASIA 1996. An ethnobotanical study of Indus Valley (Ladakh). *J. Econ. Taxon. Bot.* Addl. Ser. 12 : 92-101.

USHER, G. 1984. *A Dictionary of Plants.* CBS Publishers & Distributors, Delhi.

# RITUAL PLANTS OF NEERU VALLEY, BHADARWAH, JAMMU AND KASHMIR, INDIA

HARISH CHANDER DUTT & SHASHI KANT

*Department of Botany, University of Jammu, Jammu*

**ABSTRACT**

The present paper deals with few plant species which are used in rituals by the natives of Neeru Valley. The plant species have been arranged alphabetically with their specific utility in rituals. The local names of the species are also given in parenthesis.

## INTRODUCTION

Since pre-historic times, man has been using plants either for food, medicine, shelter or rituals, religious rites and folk ceremonies and this knowledge has been passed on from generation to generation which is evident from the various folklore, rock art, archaeological remains, and in very recent times, through ancient scripts. Due to many problems like food scarcity, side effects of allopathic medicine etc, man again turned to Mother Nature and manure the ethnobotany in the whole World. Ethnobotany in wider context denotes the entire realm of useful relationship between plants and man. In India, organized ethno-botanical research started about five decades ago. Within the last thirty five years, organized field work among the tribals of different regions has been undertaken and continued by more than 20 Indian institutions. Several institutions are also engaged in studies of allied interdiscip-linary aspects like ethno-pharmacology and ethno-paleobotany. Botanists at BSI initi-ated ethnobotanical studies in fourteen States of India covering more than 30 different tribes (Jain & Mudgal, 1999).

Bhadarwah, with an estimated area of 64,024 hectares, is situated between latitude 33°8'24"-32°52'26" North and longitude 75°48'38"-75°32'41" East in North-west Himalayas and is a place with an amalgam of cultures from Himachal Pardash, Kashmir and Duggar belt. Two permanent tribes of the area are Gaddies and Shippies while Gujjars and Bakerwals represent the dominating migratory tribes. These tribes migrate to the upper reaches of Udhampur in summer. This may be attributed to which their respective culture gets transferred from one region to the other. Besides these tribes, other inhabitants of the area also use plant species in their day-to-day life. A perusal of literature revealed that ethnobotanical studies in Jammu and Kashmir have not received due attention. Therefore, several workers are actively engaged in the ethnobotanical survey of Jammu and Kashmir. The frequent reference to plants in folk tales, their use in ritual practices of festivals, birth and death in Bhadarwahi and Duggar culture prompted us to conduct the present study. Literature on such ethnobotanical studies in India is scanty (Agrawal, 1981; Choudhuri & Pal, 1981; Grigson, 1949; Gupta, 1965; Jain, 1963; Vartak & Gadgil, 1981), but it is now in consideration of so many workers, *i.e.* Changikija & Kumar (1996), Chauhan (1999), Dixit (1997), Dhiman (2003), of modern society. The present communication lucidly furnishes the information collected from the native people and tribes of Neeru Valley, i.e. Bhadarwah. The botanical names of 21 species of the ritual plants with their use in ritual ceremonies are enlisted below.

## OBSERVATIONS

**Acorus calamus** L. (Naglachh); Araceae

The roots of the plant are cooked as vegetable in Dinner on the eve of the New Year (Vikrami). The leaves of the plant are also used in worships at that night.

**Angelica glauca** Edgew. (Chori); Fam.: Apiaceae

The roots of the plant are burnt to repel entry of snakes in human dwellings.

**Artemisia maritima** L. (Sayski); Asteraceae

Just like *Ocimum tenuiflorum* L., the shrub is planted in front of the main door of the house and worshipped daily in the morning.

**Brassica rapa** L. ssp. **campestris** (L.) Clapham. (Sarsoon); Brassicaceae

The seeds of the plant are wrapped in the piece of cloth and tied in the neck of children to avoid evil spirits.

**Cannabis sativa** L. (Bhang); Cannabaceae

The leaves of the plant are crushed and mixed to the rice pudding (Kheer) on the occasion of Shiv Ratri. Crushed leaves of the plant with milk are offered as Charen-Amrit in temples of Shiva on the day of Shiv Ratri.

**Cedrus deodara** (Royle ex G.Don) G.Don (Deodar); Pinaceae

The wood is used for burning exclusively in Havana. The fire-sticks of the wood of *Cedrus deodara* are formed in the morning of Utrayain i.e. Macker-Sakranti (next day to Lorhi) and burnt during the night of Fitter-Amavasya.

**Cynodon dactylon** (L.) Pers. (Droob); Poaceae

This graminaceous herb is used in wedding rituals to put mustard oil on the forehead of bride and the bridegroom. The grass is also pasted on the pots containing paddy grains for germination (seedlings).

**Datura innoxia** Mill. (Datura); Solanaceae

The seeds of the plant are collected and used in the preparation of the Charen-Amrita on the festival of Shiv Ratri. The leaves are smoked by the Sadhus as Pershada (Offering) of Shiva.

**Desmostachya bipinnata** (L.) Stapf (Darb); Poaceae

The grass is harvested in the morning of Kusha-Amavasya, it is dried as such and stored for the use in coming year. The sticks of this dried plant are placed in the grain stores and utensils to avoid the bad effect of the solar-eclipse and lunar eclipse.

**Ficus palmata** Forssk. (Fago); Moraceae

One log of the plant is used in pyre at the time of cremation of dead bodies.

**Hordeum vulgare** L. (Jou); Poaceae

The seeds of the plant are used in the worship of Agni (fire).

**Juglans regia** L. (Akhrot); Juglandaceae

The kernel of the plant are used for the worship of Lord Shiva on the festival of Shiv-Ratri. Whole of the ftuit is offered as Aahuti during worship of Fire of Lorhi.

**Jurinea dolomiaea** Boiss. (Dhooph); Asteraceae

The roots of the plant are used as incense in all the temples and in worship/prayers at home.

**Ocimum tenuiflorum** L. (Tulsi); Lamiaceae

The plant is worshiped everyday. Once in the year in the month of Kartik the marriage ceremony of the plant is celebrated. It is local myth that at the time of its marriage the Tulsi Deity bless whole of the family.

**Pinus roxburghii** Sargent (Chir); Pinaceae

The immature needles of the plant are collected and tied to form a broom; this small broom is used in sacred places of deities, temples and kitchen.

**Prinsepia utilis** Royle (Jintoi); Rosaceae

The flowers and leaves of the plant are used in worshipping Lord Shiva on the festival of Kunchoth (Gori Tritya). On this day, married women worship for long life of their husbands (the day is other than the Karwachoth). The leaves are generally used by the locals for the worship of Satya Naryan Bhagwan.

**Rubus ellipticus** Sm. (Kryer); Rosaceae

Twigs of the plant are tide to the doors at the night of Shiv-Ratri. It is believed that the evil souls connot enter the house during that night by doing so.

**Sesamum orientale** L. (Til); Pedaliaceae

The seeds of the species are considered as the holy seeds and used at the time of cremation of dead bodies.

**Tagetes patula** L. (Gotta); Asteraceae

The flowers are grind and their paste is used as Tilak in temples. The flowers are used for the worship of all the Gods and Deities.

**Vigna mungo** (L.) Happer (Mah); Fabaceae

The seeds of the plant are used in worships in the morning of Mecker-Sakranti.

**Ziziphus mauritiana** Lam. (Ber); Rhamnaceae

The sticks of the plant are kept in the hands of male children when they are supposed to adopt Janaoo/Yagopavit (Sacred thread) during the ceremony of Yagopavitum.

## REFERENCES


AGRAWAL, S.R. 1981. Trees, flowers and fruits in Indian folk songs, folk proverbs and folk tales. In Jain, S.K. (Ed.) *Glimpses of Indian Ethnobotany,* pp. 3-11. Oxford & IBH Publishing Co., New Delhi.

CHANGIKIJA, S. & Y. KUMAR 1996. Ethnobotanical folk practices and beliefs of the Ao-Nagas in Nagaland, India. *Ethnobotany* 8: 26-30.

CHOUDHURI, R.N. & D.C. PAL 1981. Plants in folk religion and mythology. In Jain, S.K. (Ed.) *Glimpses of Indian Ethnobotany,* pp. 59-68. Oxford & IBH Publishing Co., New Delhi.

CHAUHAN, N.S. 1999. *Medicinal and aromatic plants of Himachal Pradesh.* Indus Publishing Company, New Delhi.

DHIMAN, A.K. 2003. *Sacred Plants and their medicinal uses.* Daya Publishing House, Delhi

DIXIT, G. 1997. Fire Sacrificial Plants. *Geobios News Reports* 16 (8) : 47-48.

GARIGSON, W.V. 1949. *The Maria Gonds of Bastar.* London (1st ed. 1938; revised 1949 with a supplement by Auther and V. Elwin).

GUPTA, S.S. 1965. *Tree symbol worship in India.* Calcutta.

JAIN, S.K. 1963. Magico-religious beliefs about plants among the tribals of Bastar, Madhya Pradesh. *Q. J. M. S.* 54 (3) : 73-94.

———— & V. MUDGAL 1999. *A hand book of Ethnobotany.* Bishen Singh Mahendra Pal Singh, Dehra Dun, India.

VARTAK, V. D. & M. GADGIL 1981. Studies on sacred grooves along the Western Ghats from Maharashtra and Goa : Role of Beliefs and Folklore's. In Jain, S.K. (Ed.) *Glimpses of Indian Ethnobotany,* pp. 277-278. Oxford & IBH Publishing Co., New Delhi..

# SOME ETHNOMEDICINALLY IMPORTANT NEW PLANT REPORTS FROM MELGHAT

P.Y. BHOGAONKAR & V.D. DEVARKAR*

*Department of Botany, Govt. Vidarbha Institute of Science and Humanities, Amravati-444 604*
**EIRA Division, National Environmental Engineering Research Institute, Nagpur-440 020*

## ABSTRACT

During ethnomedicinal survey of Melghat six plant species, hitherto unreported from the region were studied and presented here.

## INTRODUCTION

The flora of Melghat has been studied by Bhogaonkar & Devarkar (1999), Dhore & Joshi (1988), Dhore (1988), Patel (1968) etc. During ethnomedicinal survey of Melghat, the authors came across six species of flowering plants which find wide medicinal uses among the tribe Korkus and these uses are restricted to them only and have not been recorded earlier. The taxa are arranged under their respective families in the present work along with Korku names in italics in parenthesis. Short diagnostic features of the taxa along with locality of occurrence and plant parts used have been discussed, besides ailments.

## ENUMERATION

PAPAVERACEAE

**Argemone ochroleuca** Sweet (*Safed Darudi*)

Erect, prickly herbs with yellow, milky juice. Leaves sinuate - pinnatifid, prickly on nerves. Flowers sessile; petals pale yellow or white; stigma 5-lobed, lobes spreading, dark red. Capsules ovate-lanceolate, clothed with spines.

*Locality:* Gadga river bed, *PYB* 210.

*Uses :* Root-paste is used for scorpion bite. Latex from stem is put into eyes (very small quantity) to cure cataract. It is believed to clear off within a week. Seeds are given in asthma. Smoke of plant produces hypnotic effect.

ASTERACEAE

**Tithonia diversifolia** (Hemsley) A. Gray (*Ran Zendu*)

Robust undershrubs. Leaves ovate-rhomboid, lower ones palmately three lobed, hairy on both surfaces. Heads yellow, solitary on long peduncles. Marginal flowers ligulate, central tubular. Anthers black with yellow tips. Fruits trigonous, obovoid achenes. Pappus broad, scaly.

*Locality:* Near Dharni, around villages, *VDD* 328.

*Uses :* Leaf-juice or whole plant decoction is given in acidity. Also used to reduce body heat.

ACANTHACEAE

**Haplanthodes verticillata** (Roxb.) Majumdar (*Ghamasa*)

Robust, glandular-pubescent herbs. Leaves ovate. Cladodes axillary, quadrangular, stout, *ca* 1.2 cm long. Flowers blue-violet, sessile, in axillary spikes. Corolla 2-lipped. Fruits oblong, pointed capsules.

*Locality:* Frequent in forest undergrowth, *VDD* 56.

*Use :* Shade dried plant powder is given in chronic fever.

COMMELINACEAE

**Commelina suffruticosa** Blume (*Ghodi ghas*)

Perennial herbs, 28-50 cm high. Leaves cauline, sessile or subsessile, lanceolate or elliptic. Spahtes pedunculate, broadly ovate-turbinate, margins hirsute. Flowers white. Capsules obcordate, laterally flattened. Seeds filling the locules.

*Locality:* Dhargad, Koktu, Narnala, Gullarghat, *VDD* 9.

*Uses :* Root-paste is given orally as well as applied over snake-bite. Root-powder is given in piles, fistula and in menorrhegia.

LILIACEAE

**Chlorophytum glaucum** Dalz. (*Musali*)

Annual herbs; roots cylindric. Leaves oblanceolate, acute, glaucous. Flowers white, few, arranged in cymose pairs in long racemes. Fruits globose, emarginate, sharply 3-winged capsule.

*Locality:* Dhargadi Gullarghat range, *PYB* 230.

*Use :* Tubers are nutritious. Also used in sexual impotency.

**Iphigenia magnifica** Ansari & Rolla Rao (*Lasni*)

Erect herbs, 7-20 cm tall; tubers subglobose. Leaves few, linear, grass-like, acute or acuminate. Flowers few, in small, terminal racemes. Perianth reddish or brown-purple; filaments flattened, glabrous. Capsules oblong, ellipsoid; seeds many, brown.

*Locality:* Chikhaldara, Semadoh core area, *VDD* 202.

*Uses :* Oil extracted from bulbs is used as pain reliever, especially in migraine. Oil is also snuffed to avoid sleep after snake-bite. Oil applied externally on snake-bite.

## REFERENCES


BHOGAONKAR, P.Y. & V.D. DEVARKAR 1999. Additions to the flora of Melghat (Some rare and uncommon plants) *Tech. Bull.* No. VII. The Directorate Project Tiger, Melghat, Amravati.

DHORE, M.A. 1988. *Flora of Amravati district with special reference to trees.* Ph. D. Thesis, Nagpur University, Nagpur.

———— & P.A. JOSHI 1988. Flora of Melghat Tiger Reserve. *Tech. Bull.* No. I. Project Tiger Melghat, Amravati.

PATEL, R.I. 1968. *Forest Flora of Melghat.* Dehra Dun.

# SOME IMPORTANT WEST HIMALAYAN HIGH ALTITUDE MEDICINAL PLANTS

SANJAY K. SINGH* & V. P. UNIYAL

*Wildlife Institute of India, Dehra Dun, Uttranchal 248 001*
**C/o Anand Singh Dhanik, Lohariasal Talia, P.O. Katgharia, Haldwani, Nainital, Uttranchal-263 139*

## ABSTRACT

The present article presents a brief account of some medicinal plants of Western Himalaya. Uncontrolled exploitation, other biotic factors and physical factors have endangered many of these very important species. These plants immediately need protection since they are heavily exploited for commercial use and are under severe threat for their existence. The medicinal uses of some important species are provided in this communication.

## INTRODUCTION

The lofty clouded, freezing peaks, vast river system, varied topography, climate, soil, different vegetation types, important religious shrines and beautiful tourist resorts in this great Himalayan mountain range have remained the source of inspiration not only to the local community but also to the outsiders. There is great deal of variation in the altitude, topography and exposition in Himalayas. The average annual precipitation varies from 1000-2000 mm/annum (Gaston *et al.,* 1983). The species richness in the area is attributed mainly to the altitude aspect, slope and soil types (Singh, 1999). During last four-five decades the modern system of medicine has established its lead over indigenous systems with the result several allopathic drugs have been discovered to treat several disorders, but often it is observed that many of these drugs have deleterious side effects. Therefore, keeping in view the various harmful side effects and other metabolic reactions caused by these synthetic drugs, the western countries are now gradually inclining towards the use of herbal drugs, which are considered to be effective and much safe.

## MATERIAL AND METHODS

Medicinal plant surveys were conducted in the Great Himalayan National Park (GHNP) located in the Kullu district of

Himachal Pradesh. Information on the local uses of wild plants was collected using structured questionnaires and informal interviews with the locals including villagers, shepherds, herb and mushroom collectors during 1995-1998. Elderly persons who had more knowledge about the forests, medicinal plants and their distribution in the Park were interviewed and discussions were held about the availability of useful plants and their uses in the past and present. The villages selected for more intensive medicinal plant surveys included Kharongcha (2200), Ropa (1800 m), Shakti (2000 w), Maror (2100 m), Lappah (2000 m), Bathad (1800-2400 m), Majharna (2000 m) and Kunder – Majharn (2400 - 2600 m). The plants were identified in the field with the help of available floras (Aswal & Mehrotra, 1994; Chowdhery & Wadhwa, 1984 and Rau, 1975). Local field assistants, herb collectors and shepherds also helped in verifying the identity of useful plants and their local names during field surveys. Unidentifiable plants were brought to headquarters (Dehra Dun) and later identified with the help of voucher specimens available at Wildlife Institute of India, Forest Research Institute and Botanical Survey of India located at Dehra Dun, Uttranchal.

## PECULIARITIES OF HIGH ALTITUDE PLANTS

The high altitude flora is generally referred to the species that grow above "timberline" (Rau, 1975). Tree-line or timberline in the Western Himalaya generally starts above 3,600 m a.m.s.l., which is dominated by herbaceous communities impounded with few patches of *Quercus semecarpifolia, Abies spectablis, Betula utilis, Rhododendron campanulatum* etc, besides species of *Berberis, Lonicera, Juniperus, Rosa* and *Salix*. The efficacy of the medicinal plants growing here is perhaps as old as these mountains themselves. The therapeutic value of some of these plants is mentioned in the oldest Hindu scriptures like Rigveda that is said to be the source of ancient medicine. The system of the Ayurveda also describes the Himalaya as the best habitat for the medicinal herbs and about 3,000 vegetational drugs have been incorporated in Ayurvedic Materia Medica, a large number of which occur in Himalayas. The significant publications appeared during the last four decade dealing with high altitude flora are of Rau (1975), Nair (1977), Kachroo *et al.* (1977), Dhar & Kachroo (1983), Naithani (1984), Polunin & Stainton (1964).

At present about 214 species of plants, mainly the high altitude herbs, inhabiting this region have been listed in the "Red Data Books" by Botanical Survey of India. Today not only the Ayurvedic but also Homeopathic, Unani and Allopathic medicines are the extract of these medicinal herbs. There is an increasing demand of these medicinal herbs in the Western countries; as such, the medicinal companies have accelerated their production resulting in over exploitation and depletion and even extinction of some of the species. According to Singh (1999), about 832 plant species, 31 mammal species and more than 300 bird species have been identified from this area. Out of 832 plant species, about 250 species are being used as ethno-medicine and about 60 species are being commercially exploited, of which 23 species are listed in the "Red Data Books" of IUCN (Singh, 1999).

Drugs obtained from the high altitude plants cure almost all known diseases. The dreadful disease like cancer is now being cured by "Taxol" extracted from *Taxus baccata* ssp. *wallichiana*. One day it may be possible to find out a plant to cure AIDS also.

## ENUMERATION

Table 1. Some commonly used plants of Western Himalaya.

| Botanical name/ Family | Local name | Zo | Uses/Parts used |
|---|---|---|---|
| ***Aconitum heterophyllum*** Wall. ex Royle (Ranunculaceae) | Patis | A | Aphrodisiac, tonic, diarrhoea, fever and headache (Roots). |
| ***Aconitum violaceum*** Jacq. ex Stapf (Ranunculaceae) | Mita patish | A | Stomachache in children, gastrointestinal complaints and renal pain (Roots). |
| ***Acorus calamus*** L. (Araceae) | Bach | T | Used in mental disorders, dysentery and in bronchitis (Rhizome & roots). |
| ***Actaea spicata*** L. (Ranunculaceae) | Pholee | T | Used against snake-bite, nerve diseases and rheumatic fever (Whole plant). |
| ***Ainsliaea aptera*** DC. (Asteraceae) | Sath jalari | T, UT | Used as diuretic and stomach-ache (Roots). |
| ***Ajuga bracteosa*** Wall. ex Benth. (Lamiaceae) | Neelkant | T | Cuts and boils treatment (Leaves). |
| ***Allium humile*** Kunth (Liliaceae) | Farn | A | Indigestion, spices and flavouring food (Leaves). |
| ***Anemone rivularis*** Buch.-Ham. ex DC. (Ranunculaceae) | Ratan | T, SA | For producing vomitting, oil extract is used in rheumatism, headache, fever and ear complaints (Leaves & stem). |
| ***Angelica glauca*** Edgew. (Apiaceae) | Chora | T, UT | For treatment of stomachache, gastric complaints and flavours food items (Roots). |
| ***Arisaema jacquemontii*** Bl. (Araceae) | Basair | T | Skin disease, ringworm removal (Tuber). |
| ***Arnebia benthamii*** (Wall. ex G. Don) I. M. Jonston (Boraginaceae) | Ratan jot | SA | Hair tonic, fever, headache, cuts and wounds (Roots). |
| ***Artemisia maritima*** L. (Asteraceae) | Seski | UT | Anti-helminthic and cuts (Stem & leaves). Whole plant as laxative, floral tops are vermifuge. |
| ***Berberis asiatica*** Roxb. ex DC. (Berberidaceae) | Mosholi | T | Jaundice, snake-bite and eye complaint (Roots). |
| ***Berberis lycium*** Royle (Berberidaceae) | Kasmal, Kirmora | T | Piles and jaundice (Roots). |

| | | | |
|---|---|---|---|
| ***Bergenia ciliata*** (Haw.) Sternb. (Saxifragaceae) | Sabia | T | Kidney stone, asthma, cuts, wounds, fever tuberculosis also to clean eyes of livestock (Roots). |
| ***Bergenia stracheyi*** (Hk. f. & Th) Engl. (Saxifragaceae) | Sabla | UT | Gall stone, cuts and wounds (Roots). |
| ***Caltha palustris*** L. (Ranunculaceae) | Peeloo | SA, A | The leaves are powdered and used to keep maggots out of cattle wounds. The roots are used as cure for gonorrhoea. Local people sometimes use the green leaves to clean their hands. Roots are considered to be poisonous and used in warts and anaemia (Roots & leaves). |
| ***Codonopsis ovata*** Benth. (Campanulaceae) | Sirdandi | SA, A | Used in the form of poultices for the treatment of ulcers, wounds and stimulant (Roots & leaves). |
| ***Corydalis govaniana*** Wall. (Fumariaceae) | Bhuot jato | T | Tonic, stomachache (Roots). The plant is also used for the treatment of leprosy and rheumatism. |
| ***Dactylorhiza hatagirea*** (D. Don) Soo (Orchidaceae) | Hathpanja | A | Wart and boils (Tuber). |
| ***Daphne papyracea*** Wall. ex Steud. (Thymeliaceae) | Gandiri | T | Poisonous to animals, used for cuts (Leaves). |
| ***Delphinium denudatum*** Wall. ex Hk. f. & Th. (Ranunculaceae) | Nirbishi | A | Tonic, stimulant, tooth ache (Roots). |
| ***Dioscorea deltoidea*** Wall. ex Kunth (Dioscoreaceae) | Singli-Mugali | LT, T | The rhizomes are one of the commercial source of diosgen, which is used as starting material for synthesis of steroidal drugs employed in sex hormone and cortico steroids. Tubers also used for soap making for washing wool, silk and hair; it is also used for fish poisoning and in dysentery (Tuber). |
| ***Geranium pratens*** L. (Geraniaceae) | Laljahri | T | Headache, toothache and cuts and wounds. A black dye is made by boiling the root with a sulfurous raw mineral called 'Naktsur'. The roots are popularly applied as a poultice to bruises and local Vaidyas use the plant for stomach trouble (Roots). |
| ***Girardinia diversifolia*** (Link) Friis (Urticaceae) | Beechi-booti | LT | Stimulant and contraceptive (Roots). |
| ***Hedychium spicatum*** Buch.-Ham. ex Smith (Zingiberaeeae) | Ban haldi | T | Liver complaints (Roots). Rhizomes are used in Ayurvedic system of medicine as febrifuge, carminative, also used in preparation of attar. |
| ***Hippophae rhamnoides*** Serv. ssp. ***salicifolia*** (D. Don) Serv. (Elaeagnaceae) | Soori | T | Cuts and wounds, rich in Vitamin C. The oil, extracted from seeds, pulp, tender branches and leaves is used for making different kinds of life saving medicines and drugs for combating numerous diseases and for healing wounds, treating burns, frostbite, eczema, ulcer, radiation burns of the skin, malnutrition, oral mucositis, rectal mucositis and vaginal mucositis (Fruits). |
| ***Hypericum perforatum*** L. (Hyperiaceae) | Basanti | LT | Hypertension (Leaves). |

| | | | |
|---|---|---|---|
| ***Iris kumaonensis*** D. Don ex Royle (Iridaceae) | Cherachi | SA, A | Anti-poisonous (Roots & leaves) |
| ***Juglans regia*** L. (Juglandaceae) | Akhrot | LT, T | The root bark is used for the cleaning of teeth and it prevents decaying (Roots). |
| ***Jurinea dolomiaea*** Boiss. (Asteraceae) | Dhoop | A | For incense, antiseptic, fever. Decoction of the roots is given in the colic (Roots). |
| ***Meconopsis aculeata*** Royle (Papaveraceae) | Poppy | SA, A | Narcotic, backache and renal pain (Roots). |
| ***Megacarpaea polyandra*** Benth. (Brassicaceae) | Ban-Mooli | SA, A | Intestinal injury and fever (Roots & leaves). |
| ***Nardostachys grandi-flora*** DC. (Valerianaceae) | Nahni | A | Ulcer, stimlant, stomachache, dysentery (Roots). |
| ***Oxalis corniculata*** L. (Oxaliacaeae) | Khati-buti | T | Cooling and controls irritation (Whole plant). |
| ***Oxyria digyna*** (L.) Hill (Polygonaceae) | Jagali palak | A | Digestion of food (Leaves). |
| ***Phlomis bracteosa*** Royle ex Benth. (Lamiaceae) | Chukhari | A | Used for burns and boils and cuts (Roots). The nectar of the flowers is put into the eyes and is popularly considered as eye tonic. |
| ***Phytolacca acinosa*** Roxb. (Phytolaccaceae) | Jharka | T | Eaten after being cooked and are stimulant and used as contraceptive (Leaves). |
| ***Picrorhiza kurrooa*** Royle ex Benth. (Scropholariaceae) | Karoo | SA, A | Roots and rhizomes are used as adulterant of gentians, cholagogue, laxative. Also used for stomachache. Recently diuretic and hepato protective property of its root and rhizomes has also been established (Whole plant). |
| ***Pistacia chinensis*** Bunge ssp. ***integerrima*** (Stewart) Rech. f. (Anacardiaceae) | Kokar singi | LT | Fruit ash is used to treat cough, asthma, powdered and fried in ghee given in dysentery, antidotes to snake bite and scorpion stings (Fruits). |
| ***Pleurospermum densiflorum*** (Lindl.) Clarke (Apiaceae) | Lossar | A | Massage with ghee on various body joints to relieve pain. Alternative to tea (Whole plant). |
| ***Podophyllum hexandrum*** Royle (Podophyllaceae) | Galakada | T-A | Used in cough, fever, skin diseases, wounds and breathing problems (Roots & fruits). |
| ***Polygonatum verticill-atum*** (L.) All. (Liliaceae) | Salam Misri | T, UT | Aphrodisiac (Tuber). The rhizomes are considered to be one of the eight ingredients of 'Astavagra' of Ayurveda. |
| ***Polygonum amplexicau-lis*** D. Don (Polygonacaeae) | Dora | T-A | Cuts and wounds (Roots). |
| ***Rheum australe*** D. Don (Polygonaceae) | Dolu | T, SA | Purgative, tooth powder, dysentery, fever (Roots & rhizome). |
| ***Rhododendron arboreum*** Smith (Ericaceae) | Brass | T | Used for treatment for blood dysentery in children. Soft drink prepared to beat the heat (Flowers). |
| ***Rhododendron campanu-latum*** D. Don (Ericaceae) | Kashmiri patta | T, SA | Used in skin disease, cough and fever (Leaves & twigs). |

| | | | |
|---|---|---|---|
| ***Rhododendron hypen-anthum*** Balf. f. (Ericaceae) | Talsi | SA | For preparation of tea during winter to protect from the cold (Leaves). |
| ***Rubia manjith*** Roxb. ex Fleming (Rubiaceae) | Kasoos | T | Blood clotting agent (Whole plant). |
| ***Saussurea graminifolia*** Wall. ex DC. (Asteraceae) | Googhee | A | Cough and cold (Whole plant). |
| ***Saussurea obvallata*** (DC.) Edgew. (Asteraceae) | Brahma kamal | A | Hydroseal (Flowers). |
| ***Saussurea simpsoniana*** (Field & Gard.) Lipsch. (Asteraceae) | Googhee | A | Male reproductive disorders, fever (Flowers & roots). |
| ***Selinum vaginatum*** (Edgew.) Clarke (Apiaceae) | Bhoe, Bura | SA, A | Hypertensive, analgesic (Whole plant). |
| ***Skimmia laureola*** Sieb. & Zucc. ex Walp. (Rutaceae) | Kashtori pata | T | Cure in small pox (Leaves). |
| ***Taraxacum officinale*** Weber (Asteraceae) | Dudhalee | LT-A | Cure kidney and liver disorders. It is also used as a blood purifier (Roots). |
| ***Taxus baccata*** L. ssp. ***wallichiana*** (Zucc.) Pilger (Taxaceae) | Rakhal | T, UT | Cancer (Leaves). Bark is used for preparation of tea of medicinal value. |
| ***Thalictrum foliolosum*** DC. (Ranunculaceae) | Mimiri | T | Diuretic and purgative (Roots) |
| ***Thymus linearis*** Benth. (Lamiaceae) | Ban ajwain | SA, A | Shoots for flavouring, leaves as beverages, seeds as vermifuge, leaf and floral tops as aromatic, stomach, kidney and liver problems. The plant extract is used to cure troubles caused after childbirth (Whole plant). |
| ***Urtica dioica*** L. (Urticaceae) | Kagoos | LT-A | Injury, contraceptive and the plant is used for the treatment of gout by the local Vaidyas (Leaves & roots). |
| ***Valeriana jatamansi*** Jones (Valerianaceae) | Muskbala, Nahani | T | Roots and rhizome for incense and for treatment of hysteria and hypochondria. The root is an excellent substitute for valerian, and is used as an ingredient in many medicines, while the oil obtained from the root is used in many medicinal preparations and is well known as a hair tonic (Roots). |
| ***Verbascum thapsus*** L. (Scrophulariaceae) | Ban tambacoo | T | Used to cure diarrhoea in cattle and in man (Leaves). |
| ***Viola canescens*** Wall. ex Roxb. (Violaceae) | Banafsha | T | Rhizomes internally used as an infusion or as a component of herbal tea for relieving cough and bronchitis. Diuretic, emetic, externally used in gargles. Flowers emollient, |

| | | | |
|---|---|---|---|
| | | | demulcent, used in lung troubles. Petals made into syrup and used as a remedy for infantile disorders, considered good for throat. The fresh flowering herb is used in homeopathy in treatment of diseases of skin, eye and ear. It is also used as blood purifier (Whole plant). |
| ***Viola pilosa*** Bl. (Violaceae) | Banafsha | T | Flowers and leaf for treating wounds and fever. Leaves as pain killer and perfume. Roots emetic and antipyretic (Whole plant). |

Abbreviations used : Zo=Climatic zones : (LT=Lower Temperate 1500-2000 m), (T=Temperate 2000-2800 m), (UT=Upper Temperate 2800-3300 m), (SA=Sub-Alpine 3300-3600 m), (A=Alpine >3600 m).

## ACKNOWLEDGEMENTS

The authors are thankful to the anonymous referees for giving valuable comments in the draft copy of this article. Authors are also thankful to the Director, Wildlife Institute of India, Dehra Dun and Director, Great Himalayan National Park, Kullu, Himachal Pradesh.

## REFERENCES

ASWAL, B.S. & B.N. MEHROTRA 1985. Ethnobotanical studies on the flora of Lahaul Valley (North -West Himalaya).In Sharma & Gupta (Eds.) *Recent Advances in Plant Sciences,* pp. 16-130. BSMPS, Dehra Dun.

CHOWDHERY, H.J. & B.M. WADHWA 1984. *Flora of Himachal Pradesh.* Vols. 1-3. Botanical Survey of India, Calcutta.

DHAR, U. & P. KACHROO 1983. *Alpine Flora of Kashmir.* Jodhpur.

GASTON, A.J., P.J. GARSON & M.L. HUNTER 1983. Status and Conservation of Forest Wildlife in Himachal Pradesh, Western Himalaya. *Biol. Cons.* 27 : 291-314.

KACHROO, P., B.L. SAPRU & U. DHAR 1993. *Flora of Ladakh.* BSMS, Dehra Dun.

NAIR, N.C. 1977. Flora *of the Bashahr Himalaya.* Today and Tomorrow Publication, Delhi.

NAITHANI, B.D. 1984. *Flora of Chamoli,* Vols. 1-2. New Delhi.

POLUNIN, O. & A. STAINTON 1984. *Flowers of Himalaya.* Oxford University Press, New Delhi.

RAU, M.A. 1975. *High Altitude Flowering Plants of Western Himalaya.* Botanical Survey of India, Calcutta.

SINGH, S.K. 1999. A study on the plant community composition and species diversity of Great Himalayan National Park, Western Himalaya. Ph. D. Thesis, Kumaun University, Nainital.

# SOME MEDICINAL PLANTS OF GORUMARA NATIONAL PARK, JALPAIGURI DISTRICT, WEST BENGAL

LOPAMUDRA MOHANTY

*Central National Herbarium, Botanical Survey of India, Howrah-711 103*

## ABSTRACT

During botanical survey of Gorumara National Park, situated in Jalpaiguri district of West Bengal, about 70 species belonging to 38 families were collected which have medicinal value. The medicinal importance of these plants are described in this paper.

## INTRODUCTION

In India so far 92 National Parks have been established in different geographical regions. Among them Gorumara National Park is one which is situated in Jalpaiguri district of West Bengal. First it was declared as a Wildlife Sanctuary in 1949, comprising an area of 2129 acres. In view of its contiguity with Upper Tondu Reserves and Chapramari Sanctuary which extends up to Kalimpong forests, it was declared as Gorumara National Park in 1992 covering an area of 80 sq km. The altitude of this area varies between 130-150 metres. Gorumara National Park has a diverse floral assemblages, mainly comprising of terrai-grasslands interspersed with riverine forests, dry mixed forests, wet mixed forests and seral sal forests.

The present paper is an endeavour to bring out the importance and value of the medicinal plants available in this National Park. The specimens collected were deposited in the Central National Herbarium (CAL), Howrah.

Botanical names of the plants have been arranged in alphabetical order with family, local names and uses. The local names are abbreviated as B, E, H and O for Bengali, English, Hindi and Oriya respectively.

## ENUMERATION OF SPECIES

**Abelmoschus moschatus** Medik. (Malvaceae)

*Local names:* Musk mallow (E), Mushakdana (B, H)

*Fls. & Frts.:* June - December.

*Uses* : Seeds are biter, cooling, aphrodisiac and carminative; they cure intestinal problems, stomatitis, dyspepsia, leucoderma and itch, diseases of heart and used in snake-bite. They allay thirst and check vomiting. Roots and leaves are used for gonorrhoea.

**Abutilon indicum** (L.) Sweet (Malvaceae)

*Local names:* Potari (B), Jhampi (H), Nakachana (O)

*Fls. & Frts.* : Throughout the year.

*Uses* : Root cures uterine haemorrhagic discharges. Root infusion is a cooling agent in fever and is used in leprosy, stanguary & haematuria. Bark is diuretic, febrifuge, anthelmintic, alexiteric and it allays vomiting, lessens perspiration. Seeds are aphrodisiac and used in chest troubles, bronchitis and piles. Leaves are cooked and eaten in bleeding piles. A decoction of the leaves is used in gonorrhoea, in catarrhal bilious diarrhoea, fever, inflammation of the bladder and in toothache.

**Ageratum conyzoides** Linn. (Asteraceae)

*Local names:* Dochunty (B), Boksungal (O)

*Fls. & Frts.* : Throughout the year.

*Uses* : Plant is externally applied in ague and its juice is used for prolapsus. Root juice is antilithic. A cold decoction of the root is used in purulent opthalmia as a lotion. Leaves are styptic and applied on wounds to heal quickly. Root and leaves are antidysenteric.

**Alternanthera sessilis** (Linn.) R. Br. ex DC. (Amaranthaceae)

*Fls. & Frts.* : July - February.

*Uses* : Plant is used as a galactogogue, febrifuge, a wash for the eyes and is given in baths for pruritus and phlyctena of young children. Stem and leaves are used in snake bites.

**Aphanamixis polystachya** (Wall.) Parker (Meliaceae)

*Local names:* Pittaraj (B), Harinhara (H)

*Fls. & Frts.* : March - August.

*Uses* : Seeds are anthelmintic, refrigerent and laxative. These are used for curing muscular pain and ulcer. Seed-oil is used as liminent in rheumatism. Bark is an astringent and used in the diseases of liver, spleen, for tumours and abdominal diseases. The pounded bark is used as a poultice in rheumatism.

**Ardisia solanacea** Roxb. (Myrsinaceae)

Local name: Banjam (B)

*Fls. & Frts.* : March - October.

*Uses* : Roots are febrifuge. It is used in diarrhoea and rheumatism.

**Arundo donax** Linn. (Poaceae)

*Local names:* Gaha nal (B), Bara nal (H)

*Fls. & Frts.* : September - December.

*Uses* : Rhizomes are diuretic, sudarific and anti-lactant. They are used in the form of decoction or infusion in dropsy. The decoction of rhizome is emollient, diuretic and stimulates menstrual discharge and diminish the secretion of milk.

**Asparagus racemosus** Willd. (Asparagaceae, some include in Liliaceae)

*Local names:* Satamuli (B, H), Chhotaru (O)

*Fls. & Frts* : September - February.

*Uses* : Roots are aphrodisiac, cooling, appetiser, laxative, expectorant, galactogogue, stomachic, antidiarrhoetic, antidysenteric, antiplasmodic, diuretic. They are also used in the diseases of the kidney, liver, blood, eye, leprosy, throat complaints, tuberculosis and nightblindness.

**Boerhavia diffusa** Linn. (Nyctaginaceae)

*Local names:* Gadhapurni (B, O), Hogweed (E), Sant (H)

*Fls. & Frts.:* Throughout the year.

*Uses* : Plant is bitter, useful in biliousness, blood impurities, leucorrhoea and anaemia. Seeds are expectorant, carminative, purify the blood, hasten delivery, useful in muscular pain, scabies and scorpion sting. Root is expectorant, diuretic, emenetic and is used in jaundice, anasarca and internal inflammation. The root ground with rice water is given internally for snake bite.

**Bridelia montana** Willd. (Euphorbiaceae)

*Local names:* Kantakoi (B), Gondni (H), Nuanali (O)

*Fls. & Frts.* : June - November.

*Uses* : The plant is anthelmintic. The root and the bark are used as an astringent.

**B. retusa** (L.) Spreng. (Euphorbiaceae)

*Local names:* Kantakoi, Geio (B), Khaja (H)

*Fls. & Frts.* : May - September.

*Uses:* Root and bark are astringent. Bark with Gingili oil is used as liminent in rheumatism.

**Callicarpa macrophylla** Vahl (Verbenaceae)

*Local names:* Mathara (B), Daya (H)

*Fls. & Frts* .: June - March.

*Uses* : An aromatic oil extracted from the root is used in disorders of the stomach. The leaves are applied to rheumatic joints by heating.

**Camellia sinensis** (L.) O. Kuntze (Theaceae)

*Local names:* Tea (E), Chaa (B, H, O)

*Fls. & Frts.* : September - February.

*Uses* : Tea is an astringent and gently excitant. It exerts a decided influence over the nervous system. The leaf is hot with a sharp acrid taste, a good appetiser, stomachic, diuretic, aphrodisiac, destroys cough, useful in pain in heart, eye troubles, biles, purifies the blood and brighten the brain.

**Cayratia pedata** (Lam.) Juss. ex Gagnep. (Vitaceae)

*Local name:* Goalilata (B)

*Fls. & Frts.:* April - October.

*Uses* : Leaves are astringent, refrigerent and used for ulcers. Leaf decoction is used to check uterine reflexes.

**Cleome gynandra** L. (Cleomaceae)

*Local names:* Anasarisha (B, O), Safed hulhul (H)

*Fls. & Frts.:* July - August.

*Uses* : The seeds are anthelmintic, rubefacient and applied internally for roundworms and externally as a counter irritant. Root cures fits, tumours, ulcers, earache, spleen enlargement and bilious fevers. Leaves are externally applied to prevent the formation of the pus and used in rheumatism. The plant is used in snake-bite and scorpion-sting.

**Commelina suffruticosa** Bl. (Commelinaceae)

*Fls. & Frts.:* August - September.

*Uses* : The root is applied to sores.

**Costus speciosus** (Koen.) Sm. (Costaceae)

*Local names:* Keu (B, H, O), Spiral flag cane reed (E)

*Fls & Frts.* : August - February.

*Uses* : Root is pungent and bitter, useful in bronchitis, dyspepsia, inflammation, anaemia, rheumatism, hiccough and fever. It is also used as anthelmintic, aphrodisiac and depurative, sometimes used in snake-bite.

**Cynodon dactylon** (L.) Pers. (Poaceae)

*Local names:* Dub (B, O, H), Bahama grass (E)

*Fls. & Frts.* : Throughout the year.

*Uses* : The plant is acrid, sweet, expectorant, useful in cobra bite, burning sensation, diarrhoea, leprosy, skin diseases and fever. The expressed juice is astringent and used on fresh cuts and wounds. Roots are crushed, mixed with curd and used in chronic gleet. Root decoction is diuretic and used in dropsy, in secondary syphilis and catarrhal ophthalmia .A cold infusion stops bleeding from piles. Rhizomes and roots are used as adulterants.

**Desmodium gangeticum** (L.) DC. (Papilionaceae)

*Local names:* Salpani (B, H), Salapami (G)

*Fls. & Frts.:* May - January.

*Uses* : Root is hot and bitter, used as an astringent, aphrodisiac, anthelmintic, alterative, cures typhoid, piles, asthma, bronchitis, prevents death of foetus in the womb. It is used in chronic fever and chronic affections of chest and lungs. It is given with other drugs for the treatment of snake-bite and scorpion-sting.

**D. heterocarpon** (L.) DC. (Papilionaceae)

*Fls. & Frts.* : July - September.

*Uses* : Plant is used in fainting and convulsion. Decoction of plant is considered as tonic and useful in cough.

**D. heterophyllum** (Willd.) DC. (Papilionaceae)

*Fls. & Frts.:* June - September.

*Uses* : Roots are carminative, tonic, diuretic. Leaves are galactogogue. Decoction of the whole plant is given in stomach and abdominal complaints.

**D. pulchellum** (L.) Benth. (Papilionaceae)

*Local names:* Jatsalpan (H), Jatasalaparni (G)

*Fls. & Frts.:* September - February.

*Uses* : A decoction of the bark is used in diarrhoea, haemorrhage and the diseases of the eye. Leaf-juice with ragi flour is made into paste and given to cure menstrual disorders. It also acts as an antiseptic and used to stop bleeding of wounds.

**D. velutinum** (Willd.) DC. (Papilionaceae)

*Fls. & Frts.* : September - February.

*Uses* : The roots are mixed with small hot peppers and used in enema to cure blood in urine.

**Diplazium esculentum** (Retz.) Sw. (Polypodiaceae)

*Uses* : Fronds are antimalarial, used in earache, jaundice and constipation. Rhizome is haemoptysis and used in fever, asthma, cough, stomachache and diarrhoea. It is also an antidysenteric, anthelmintic, insect and pest repellent.

**Eclipta prostrata** (L.) L. (Asteraceae)

*Local names:* Bhringraj (O), Keshori (B), Bnandra (H)

*Fls. & Frts.:* Throughout the year.

*Uses* : The plant is hot and bitter, used as alterative, anthelmintic and alexipharmic. It is good for the complexion, hairs, eyes, teeth and cures inflammation, hernia, eye diseases, bronchitis, asthma, anaemia, diseases of the heart and skin, itching. It is used to prevent abortion, miscarriage and for uterine pains after delivery. Fresh juice of leaves is rubbed on the shaven scalp for promoting the growth of the hairs. It is given one teaspoonful dose in the jaundice and fever.

**Eichhornia crassipes** (Mart.) Solms. (Pontederiaceae)

*Local name* : Kachuri pana (B)

*Fls. & Frts.* : April - November.

*Uses* : The flowers are regarded as the medicine for horses.

**Elephantopus scaber** Linn. (Asteraceae)

*Local names:* Dadisaka (B), Gobhi (H), Prickly leaved, Elephant's foot (E)

*Fls. & Frts* : Throughout the year.

*Uses* : The plant is an astringent, antipyretic, alexipharmic, cures cough, useful in poisoning from bites or from nails of animals and good in. the diseases of blood and the heart, bronchitis and small pox. Leaves are used in pains and piles. The flowers are aphrodisiac, expectorant, cure liver troubles.

**Eleusine indica** (L.) Gaertn. (Poaceae)

*Local name:* Malankuri (H)

*Fls. & Frts.* : June - November.

*Uses* : Root is diaphoretic and antipyretic, much used in liver problems. Plant decoction is given to children for convulsions.

**Emilia sonchifolia** (L.) DC. (Asteraceae)

*Local name:* Sadimodi (B)

*Fls & Frts.:* February - September.

*Uses* : Plant is astringent, antiasthmatic, vulnerary; its decoction is febrifuge and used in bowel complaints. Decoction of the leaves is used as antipyretic. Pure juice of the leaves is dropped into eyes in nightblindness.

**Equisetum ramosisimum** Desf. ssp. **debile** (Roxb. ex Vauch.) Hauke (Equisetaceae)

*Uses* : The plant is a cooling medicine for gonorrhoea.

**Euphorbia hirṭa** Linn. (Euphorbiaceae)

*Local names:* Austalian asthma herb (E), Burakeru (B), Chitakutei (O), Dudhi (H)

*Fls & Frts.* : Throughout the year.

*Uses* : The herb is an astringent, used in chronic diarrhoea and dysentery; also considered as tonic, narcotic and antiasthmatic. Juice of the plant is taken by nursing mothers to increase the flow of milk and applied to destroy warts. Leaves are used in curing sores and their juice is applied to the eyes in eye troubles.

**Hedyotis corymbosa** (L.) Lamk. (Rubiaceae)

*Local names:* Khetpapra (B), Daman papar (H), Parpata (S), Gharpodia (O)

*Fls. & Frts.* : September - December.

*Uses* : The plant is used as an anthelmintic, in jaundice and diseases of the liver. A decoction of the whole plant is used for the remittent fever, gastric irritability, nervous depression. It is applied externally to the surface of the body and internally to cure heat eruptions.

**Hypolepis punctata** (Thunb.) Mett. ex Kuhn (Hypolepidaceae)

*Uses* : Fronds are used as poultice for boils.

**Impatiens balsamina** Linn. (Balsaminaceae)

*Local names:* Haragaura (O), Balsam weed (E), Dupati (B), Gulmendi (H)

*Fls.& Frts* : Throughout the year.

*Uses* : The whole plant is taken internally as an emenetic, cathartic and diuretic. the flower is cooling and tonic; it can be applied to burns and scalds. It is mainly used for joint pains.

**I. tripetala** Roxb. ex DC. (Balsaminaceae)

*Fls. & Frts.:* April - October.

*Uses* : The juice of the root is used in haematuria.

**Ipomoea quamoclit** Linn. (Convolvulaceae)

*Local names:* Kunjalata (O), Cypress vine (E), Tarulata (B), Karnlata (H)

*Fls. & Frts.:* April - September.

*Uses* : Leaves are used as a poultice for carbuncles. Leaf-juice with ghee is given internally for bleeding piles. Plant is having cooling properties.

**Kyllinga tenuifolia** Steud. (Cyperaceae)

*Local name:* Nirbisi (B, H)

*Uses* : Decoction of the root is used to relieve thirst in fevers and diabetes. Oil boiled with root is used to relieve pruritus of the skin.

**Lantana camosa** L. var. **aculeata** (L.) Moldenke (Verbenaceae)

*Local names:* Wild sage (E), Jaikoli (O)

*Fls. & Frts.:* Throughout the year.

*Uses* : Plant is carminative, diaphoretic, vulnerary, antiplasmodic . A decoction is given in tetanus, rheumatism, malaria. Root extract is used in toothache, gargle and in fever. Seed oil is used in rheumatism. Leaf oil is antiseptic and used in skin itches.

**Leea asiatica** (L.) Ridsd. (Leeaceae)

*Local names:* Banchalita (B), Hatikanapatra (O)

*Fls. & Frts.:* June - September.

*Uses* : The tubers are used as a remedy for worms. The bruished leaves are applied to wounds.

**Leea macrophylla** Roxb. ex Horn (Leeaceae)

*Local name:* Dholsamudra (B, H)

*Fls. & Frts:* July - March.

*Uses* : Root is an astringent and alexipharmic. It is used to cure worms and pounded root is given in the obstinate sores to promote cicatrisation. Externally it is applied to allay pain and to prevent bleeding of wounds.

**Leucas aspera** (Willd.) Link (Lamiaceae)

*Local names:* Chotahalkusa (B, H), Gayasa (O)

*Fls. & Frts:* August - February.

*Uses* : The whole plant is used as a tonic, stimulant, diaphoretic, antipyretic, insecticide, used in scabies, cough and cold. Leaves are used in chronic rheumatism and their juice in skin diseases, painful swellings and as an eye-drop. Flower juice is used for cold, cough and catarrh.

**Lindernia ruellioides** (Colsm.) Penn. (Scrophulariaceae)

*Fls. & Frts.* : June - December.

*Uses* : The herb is applied externally for worms in the skin.

**Lygodium flexuosum** (L.) Sw. (Lygodiaceae)

*Fertile* : September - January.

*Uses* : Plants are anti-ovulatory, expectorant and used in jaundice. Roots are anti-fertility, anti-rheumatic and used in ulcers, sprains, scabies, eczema and wounds. Fresh roots are boiled with mustard oil and used as local application to carbuncles. Fronds are anti-bacterial, used in fever and cough.

**Macaranga peltata** (Roxb.) Muell.-Arg. (Euphorbiaceae)

*Fls. & Frts.*: July - December.

*Uses* : The gum is powdered and made into a paste which is externally applied for veneral sores.

**Marsilia minuta** L. (Marsileaceae)

*Local names:* Sunsunia sag (O), Sushni sag (B)

*Uses* : The leaf-juice is given to cure insomnia.

**Mimosa pudica** Linn. (Mimosaceae)

*Local names:* Sensitive plant (E), Lajjabati (B), Lajwanti (H), Lajkuli lata (O)

*Fls & Frts.* : July - December.

*Uses* : The whole plant is diuretic, astringent, antiplasmodic and is used in rheumatism and tumour of the uterus externally. Internally used for vesical calculi. The root is bitter, cooling, alterative, alexipharmic and cures leprosy, dysentery, vaginal and uterine complaints, burning sensation, asthma and small pox. The leaf and stem in combination with other drugs given for treatment of snake-bite and scorpion-sting.

**Osbeckia nepalensis** Hook. (Melastomataceae)

*Fls. & Frts.* : July - November.

*Uses* : The flowers are pounded and applied to sores in children's mouth.

**Paederia foetida** Linn. (Rubiaceae)

*Local names*: Gandhabhadulia (B), Pasaruni (O), Prasarani (H)

*Fls. & Frts.* : May - October.

*Uses* : The whole plant is bitter, aphrodisiac, indigestible and tonic; it cures inflammations, piles, fever, rheumatism and good for diseases of the eye. The root is an emetic. The leaf juice is an astringent and given to children in diarrhoea.

**Persicaria hydropiper** (Linn.) Spach (Polygonaceae)

*Local names:* Packurmul (B), Biting pepper (E)

*Fls. & Frts.* : August - February.

*Uses* : The plant is acrid, stimulant and diuretic. Its juice is carminative, anthelmintic and used for itch. Root is stimulating, bitter and tonic. Bruished leaves are diuretic and used for nephritic complaints and toothache.

**Phyllanthus emblica** Linn. (Euphorbiaceae)

*Local names* : Amlaki (B), Emblic Myrobalan tree (E), Amla (H), Onola (O)

*Fls & Frts.:* July - March.

*Uses* : The seed is sweet, aphrodisiac and antipyretic; it cures asthma, bronchitis and vomiting. The root, bark and ripe fruits are astringent. The leaves are used in the preparation of the antithermic lotions. The flowers are refrigerant and aperient. The fruit is cooling, diuretic, laxative, alterative, carminative and useful in burning sensation, leprosy, constipation, piles, anaemia, ophthalmia, incipient blindness, cough.

**Phyllanthus fraternus** Webster (Euphorbiaceae)

*Local names:* Bhuinamla (B), Badiamla (O), Jaramla (H)

*Fls. & Frts:* Throughout the year.

*Uses* : Fresh roots are given for jaundice. Root, stem and leaves are given internally for snake-bite. Leaves are stomachic. An infusion of the young shoot is given in dysentery. Fruit is bitter and diuretic and used in tubercular ulcers, wounds, scabies, ringworm and useful in bronchitis, leprosy, anaemia, asthma, hiccough.

**Piper longum** Linn. (Piperaceae)

*Local names:* Indian long pepper (E), Pipli (B), Pippali (O), Pipal (H)

*Fls. & Frts.:* September - February.

*Uses* : The root is laxative, anthelmintic, carminative and stomachic; it is useful in bronchitis, tumours, abdominal pains. The unripe fruit is sweet and useful in biliousness. The ripe fruit is sweet, aphrodisiac, alterative, antidiarrhoetic and antidysenteric; also useful in piles, asthma, jaundice, fevers, leprosy and insomnia.

**Piper sylvaticum** Roxb. (Piperaceae)

*Local name:* Paharipipal (B)

*Fls. & Frts:* September - February.

*Uses* : The fruit is used as a carminative.

**Piper trioicum** Roxb. (Piperaceae)

*Fls. & Frts.* : August - December.

*Uses* : The root is an excellent diuretic.

**Polyalthia longifolia** (Sonn.) Thw. (Annonaceae)

*Local names:* Debdaru (B, H, O), Mast tree (E)

*Fls. & Frts:* March - September.

*Uses* : The bark is a febrifuge. Its aqueous extract depresses the heart, lowers the blood pressure, stimulates respiration and the isolated ileum and uterus.

**Pouzolzia zeylanica** (L.) Benn. (Urticaceae)

*Fls. & Frts.:* August - October.

*Uses* : Plant is used in syphilis, gonorrhoea and snake poison.

**Randia cochinchinensis** (Lour.) Merr. (Rubiaceae)

*Fls. & Frts.:* July - November.

*Uses:* Bark is bitter and is given in forest fever. Wood is used in the treatment of paludism.

**Rauvolfia serpentina** Benth. ex Kurz (Apocynaceae)

*Local names:* Patalgaruda (O), Chhotachand (B, H), Rauvolfia root (E)

*Fls. & Ftrs.:* May - September.

*Uses* : The root is bitter, heating, pungent, anthelmintic; it cures ulcers, the poisonous effects of scorpion-sting and snake-bite. Decoction of the root is used in labours to increase uterine contraction. The juice of the leaves is applied to the eye for removal of opacities of the cornea.

**Scoparia dulcis** Linn. (Scrophulariaceae)

*Local name:* Sweet broom weed (E)

*Fls. & Frts.:* June - November.

*Uses* : Root is an astringent, mucilaginous and emollient. Leaves are diuretic. Infusion of bitter leaves is prescribed in febrile condition. Plant decoction is used to cure yellow fever and its juice is given for headache. Plant infusion is used in ague as emetic.

**Senna alata** (L.) Roxb. (Caesalpiniaceae)

*Local names:* Dadmardan (B, H), Ringworm shrub (E), Jadurnari (O)

*Fls. & Frts.:* June - December.

*Uses* : The wood decoction is used as a mild purgative. The leaves are sour, cure itching, cough, fits, asthma, ringworm, skin diseases. Leaves and flowers are used as tonic internally. Fresh leaves are given internally for snake bite. Decoction of the leaves and flowers are used internally in bronchitis and for washing eczematous patches.

**Senna tora** (L.) Roxb. (Caesalpiniaceae)

*Local names:* Chakunda (B, O), Foetid Cassia (E)

*Fls. & Frts.* : July - January.

*Uses* : Root is bitter, tonic, stomachic and used in snake-bite. Leaves are used as antiperiodic, aperient, anthelmintic. Leaf decoction is used as laxative. Leaves and seeds are used for skin diseases, ring worm and itch.

**Shorea robusta** Roxb. ex Gaertn. f. (Dipterocarpaceae)

*Local name:* Sal (B, H, O), Sal tree (E)

*Fls. & Frts.:* March - June.

*Uses* : The bark and the leaves are anthelmintic, alexiteric and cure ulcers, leprosy, itch, enrich the blood, improve the complexion. The oil is good for skin diseases and wounds. Resin is used in dysentery, weak digestion, enlargement of spleen, toothache and purifying the blood. Fruit is sweet, cooling and useful in burning, tubercular ulcers and blood troubles.

**Sida acuta** Burm. f. (Malvaceae)

*Local name:* Bariera (H), Banmethi (B)

*Fls. & Frts.:* August - December.

*Uses* : The root is astringent, diuretic and digestive. It is useful in fever, burning of the body, in nervous and urinary diseases. The leaves made warm and moistened with Gingili oil and employed to hasten suppuration. The plant in combination with other drug is prescribed for the treatment for snake-bite and scorpion-sting.

**Spermacoce hispida** L. (Rubiaceae)

*Local name:* Madanghanti (H)

*Fls. & Frts.* : July - December.

*Uses* : The seeds are stimulant and used as substitute for coffee. Root decoction is used as an alterative. The plant vapour is inhaled to kill tooth worms.

**Synedrella nodiflora** (L.) Gaertn. (Asteraceae)

*Fls. & Frts.* : September - March.

*Uses* : The whole plant is used to cure bodyache. The boiled leaves are used as a laxative.

**Thysanolaena maxima** (Roxb.) Kuntze (Poaceae)

*Local names:* Phulachhanchuni (O), Jhadu ghas (B)

*Fls. & Frts.*: September - May.

*Uses* : A decoction of the root is used as a mouth wash during fever.

**Trichosanthes anguina** Linn. (Cucurbitaceae)

*Local names:* Chichinga (B), Chhachindra (O), Snake gourd (E) , Chachinga (H)

*Fls. & Frts.:* August - December.

*Uses* : The plant is purgative and vermifuge. The roots and seeds are cooling, antidiarrhoeal and used for biliousness. The fruit is tonic, purgative, anthelmintic, emetic and cures thirst, biliousness and increases appetite.

**Urena lobata** Linn. (Malvaceae)

*Local names:* Benochra (B), Bachita (H), Jotyahala (O)

*Fls. & Frts.:* August - December.

*Uses* : The root is diuretic and cures rheumatism. A decoction of the root and stem is used for gastic colic. The leaves are used as emollient. The flowers are pectoral and expectorant. An infusion of the flowers is used as a gargle for apathe and sore throat.

**Vernonia cinerea** (L.) Less. (Asteraceae)

*Local names:* Kalajira (B), Ash coloured Fleabean (E), Dandotpala (H)

*Fls. & Frts.* : September - February.

*Uses* : The plant is cold, sweet, astringent, strong diaphoretic; it cures asthma, bronchitis and is used as a remedy for spasm of the bladder and stangury. The plant juice is given in piles. Seeds are alexipharmic, anthelmintic and effective against worms. The root is given for dropsy. The root decoction is given for diarrhoea and stomachache. Its juice is given for cough and cold. The leaves poultice is used for worms and eczema. The flowers cure fever and conjunctivitis.

**Wrightia arborea** (Dennst.) Mabb. (Apocynaceae)

*Local names:* Dudhi, Dudhali (B), Harida (O), Daira (H)

*Fls. & Frts.:* May - December.

*Uses* : The bark is given in menstrual and renal complaints. The bark and root-bark are used in snake-bite and scorpion-sting. The milky juice of the leaves is used to stop haemorrhage. Leaves with salt are given to cure toothache.

## ACKNOWLEDGEMENTS

I am indebted to Dr. H.J. Chowdhery, Joint Director, Central National Herbarium, B.S.I., Howrah for his helpful suggestions and guidance. I also owe to the Director, B.S.I. for providing me the facilities and the MoEF under whose aegis the project is going on. Thanks are also due to the officers and staffs of the Forest Department of Gorumara National Park.

## REFERENCES

ANONYMOUS 1985. *The Wealth of India, Raw Materials,* Vol. I. New Delhi.

BANERJEE, A.K. & IRA BANERJEE 1986. A survey of Medicinal plants in Shevaroy Hills. *J. Econ. Taxon. Bot.* 8 (2) : 271-290.

CHOPRA, R.N. *et al.* 1950. Glossary of Indian Medicinal Plants. CSIR, New Delhi.

JAIN, S.K. & R.A. DE FILLIPS 1991. *Medicinal Plants of India,* Vol. I & II. U. S. A.

KIRTIKAR, K.R & R.D. BASU 1935. *Indian Medicinal Plants,* Vol. 1-4. Allahabad.

TRIVEDI, G.N. *et al.* 1982. Some Medicinal Plants of Mayurbhanj (Orissa). *Bull. Bot. Surv. India* 24 (1-4) : 117-120.

# STUDIES ON ETHNOBOTANY OF SUNDARBANS WITH SPECIAL REFERENCE TO TRIBAL AND RURAL WOMEN HEALTH CARE

ANKHI SAU, D.C. PAL & N.K. VERMA*

*Botanical Survey of India, Indian Botanic Garden, Howrah - 3*
**Department of Botany, Vidyasagar University, Midanapore, West Bengal*

**ABSTRACT**

This paper deals with 17 angiospermic plant species used as medicinal plants by the women of the area. The botanical names, local names, families and the plant parts used as home remedies and other details e.g. ingredients, methods of preparations, doses, mode of administration and duration etc have been provided.

## INTRODUCTION

Sunderbans is one of the two World Heritage sites of India designated in World Heritage Convention and declared as one of the twelve Biosphere Reserves of the country on 29th March, 1989 by Government of India, to protect the natural ecosystems. A great variety of plants from algae to angiosperms form the mangrove vegetation of Sunderbans - a tropical humid evergreen type of forest, situated in the district of South 24 Parganas and located in the South east corner of West Bengal. It lies in the Ganga delta between 21°31' and 22°37' North latitude and between 88°2' and 89°6' East longitude, encompassing an area of about 9630 sq km. It is the largest and only mangrove reserve on the globe inhabited by the tigers. The Biosphere Reserve covers reserve forest area of 42666 sq km having one tiger reserve area, viz. Sunderbans National Park and three Wildlife Sanctuaries, viz. Sajnekhali, Lothian Island and Haliday Island. This area contains the richest biodiversity among the intertidal forests in the World.

Much of the macrotidal (tidal range >4m) Sunderbans area is uninhabited. It is spread over the Ganga delta where the process of land building is a regular feature. Hoogly, Matla and Bidyadhari rivers carry and deposit quaternary sediments over the area. Tidal estuaries and creeks have fragmented the whole area into a number of Islands of various shape and sizes.

The total land mass area is plain with alluvial and loamy soils and without any hillocks or rocks. The temperature ranges between 13.7°C to 38°C and the annual rainfall varies from 1015.3 mm to 1690.1 mm. The region sustains about 42.2 millions (2001 census) people of different communities, including scheduled casts and tribes. Munda, Oraon, Bhumij and Santals are the main tribes (2%) of the district which have migrated from the adjascent districts. They have different culture and social interactions and live on the periphery of Sunderbans.

Of the total population, 70% people are below poverty level (BPL). They depend directly on forest and forest based resources in their daily life in a variety of ways, including house building, thatching, furniture, boat making, food, drug, beverages, fire wood, alcohol and other materials for cultural needs. The females occupy 51% population, of which the women between 19-45 yrs are 51% and girls between 11-19 yrs about 25%. Their main occupation is fishing, prawn (meen) catching (58%), daily labour/agriculture (39%), sex worker (2%) and 40% of women population are adopting family welfare norms.

Most of the female (*ca* 79%) are suffering from various gynecological problems due to occupational hazards, tremendous work load for household work including cropping, gardening, collection of fuel wood from distant places, domesticating cattle, pigs, chicken and grazing the animal in the field etc. They are also the victims of malnutrition. These people still depend upon ancient methods of treatment and herbs of ethnic value which they find and think most effective and convenient to use.

Earlier workers (Chaudhuri *et al.*, 1976; Mukherjee, 1975; Mukherjee & Mukherjee, 1978; Naskar & Guhabakshi, 1986; Trivedi *et al.*, 1993; Mudgal *et al.*, 1999 etc) have made general reports on medicinal plants of Sunderbans.

Therefore, the study has been undertaken by the authors to identify the plants which have special significance in various gynecological diseases related to women health care, such as gonorrhoea, leucorrhoea, dysmenorrhoea (painful menstruation), irregular menstruation, venereal diseases, stimulant of lactation, facilitation of child birth and stimulant for pregnancy etc.

## MATERIAL AND METHODS

The plants were collected from various localities of man inhabited areas of Sunderbans and herbarium was prepared periodically during 1999-2003. Studies were conducted for home remedies following sample survey method by interviewing Vaidyas, medicine men and older women of various ethnic communities and the other local people. The information on home remedies was compared, verified and compiled. The therapeutic uses of home remedies presented here are not included in the work of Kirtikar & Basu (1935), Chopra *et al.* (1959) etc. Along with botanical name in alphabetical order, family, local name, plant parts used, method of preparation of drug, dose, its ingredients if any, mode of administration, duration of treatment etc are discussed.

## ENUMERATION

**Abroma augusta** (L.) L. f. (Sterculiaceae)

*Local name:* Olat kambal.

Paste of root bark with one pinch of powdered black pepper *(Piper nigrum* L.) is taken twice daily 1-2 days before the start to last date of menstrual period and repeated for 2-3 months to treat dysmenorrhoea (painful menstruation). Further, 1-2 leaves are crushed and soaked in a glass of water, kept for overnight and then it is filtered and taken in empty stomach in morning once daily for 1-2 months to treat leucorrhoea.

**Abrus precatorius** L. (Pabaceae)

*Local name:* Kunch

Poultice of seeds is used to induce abortion. The dried roots or root paste about 5 gm with water is used once daily in gonorrhoea. It is also used for external application on affected parts.

**Acanthus illicifolius** L. (Acanthaceae)

*Local name:* Hargoza, Harkonch Kanta.

The root about 2 cm is worn with white thread by mother during child birth.

**Achyranthes aspera** L. (Amaranthaceae)

*Local name:* Chirchiri, Apang.

Root-paste about 5 gm with water is given twice a day to stop bleeding after abortion. It is also useful in gonorrhoea and applied externally on affected parts.

**Ceriops decandra** (Griff.) Ding Hou (Rhizophoraceae)

*Local name:* Goran.

Juice of fresh leaves 10-15 ml with a pinch of common salt is given to the mother, 2-3 times a day in abdominal pain after child birth.

**Ficus racemosa** L. (Moraceae)

*Local name:* Dumur.

About 15-20 ml of stem-bark decoction is given twice daily for 12-15 days to treat excessive menstruation. 25 ml of ripe fruit juice mixed with a little honey is taken in the morning for 3-5 days in the treatment of menorrhagia.

**Hygrophila phlomoides** Nees var. **roxburghii** Clarke (Acanthaceae)

*Local name:* Pani Kulekhara.

Freshly collected root paste about 5 gm with water is given to women to expedite child birth.

**Hygrophila schulli** (Buch.-Ham.) M. R. & S. M. Almeida (Acanthaceae)

*Local name:* Kulekhara.

Decoction of leaves (½ cup) is given for 10-12 days in a month, once daily, in empty stomach, in morning and repeated for 2-3 months to regulate the irregular menstruation.

**Moringa oleifera** Lam. (Moringaceae)

*Local name:* Sajina.

About 2-3 gm of powdered seed is taken with water once daily for 2-3 days before and during menstrual period to regulate the irregular menstruation and problems related to it. The root bark is used as an abortifacient.

**Pedilanthus tithymaloides** (L.) Poir. (Euphorbiaceae)

*Local name:* Sij.

2-3 cm long root is introduced in vagina and kept for hours to induce abortion.

**Phyla nodiflora** (L.) Greene (Verbenaceae)

*Local name:* Bhuiokra.

Infusion of leaves and tender stem is given orally after delivery to normalize the state of health.

**Pongamia pinnata** (L.) Pierre (Fabaceae)

*Local name:* Karanja.

In the treatment of gonorrhoea, the root-juice (10 ml) mixed with equal volume of coconut milk or water in given internally.

**Saraca asoca** (Roxb.) de Wilde (Caesalpiniaceae)

*Local name:* Ashoke.

Stem-bark (25 gm) is boiled with 1 cup milk and 4 cups of water till the volume is reduced to 1 cup. It is given after adding sugar in empty stomach in morning for 2-3 months in menorrhagia due to uterine fibroids. The stem-bark extract is also used for irregular menstruation with severe pain and leucorrhoea. In this case, 20 ml bark extract is taken twice daily till symptoms disappear.

**Sesuvium portulacastrum** (L.) L. (Aizoaceae)

*Local name:* Nunia.

Leaf-paste (about 5 gm) with water is taken twice daily for 10-12 days in a month and repeated for 2-3 months till recovery in case of gonorrhoea.

**Sida rhombifolia** L (Malvaceae)

*Local name:* Berela.

Powdered root bark (1 gm) is given with milk for few days in leucorrhoea. Seed powder (1 gm) with water or milk is given twice daily for few day in gonorrhoea.

**Tinospora cordifolia** (Willd.) Hook. f. & Thoms. (Menispermaceae)

*Local name:* Gulancha.

Decoction of stem (½ cup) with juice of Zinger *(Zingiber officinale* L.) in 2 : 1 ratio is used once daily in empty stomach in morning for 7-10 days in a month and repeated for 2-3 months till recovery in venereal diseases.

**Xylocarpus granatum** Koen. (Meliaceae)

*Local name:* Dhundul.

Seed oil is used to cure breast tumour as an external application.

## ACKNOWLEDGEMENTS

The authors are grateful to the Director, Botanical Survey of India for providing necessary facilities for the study. Authors are very grateful to the local tribals and the people of other communities of the area who have willingly shared their knowledge on the local medicinal flora.

## REFERENCES

CHAUDHURI RAI, H.N. & G.N. TRIBEDI 1976. On the occurrence of some medicinal plants in 24 Parganas, West Bengal. *Bull. Bot. Surv. India* 18 : 161-165.

CHOPRA, R.N., S.L. NAYAR & I.C. CHOPRA. 1959. *Glossary of Indian Medicinal Plants.* CSIR, New Delhi.

JAIN, S.K. 1991. Dictionary of Indian folk medicines & Ethnobotany. Deep Printers, NewDelhi.

———— & V. MUDGAL 1999. *Handbook of Ethnobotany.* Bishen Singh Mahendra Pal Singh, Dehra Dun.

KIRTIKAR, K.R. & B.D. BASU 1935. *Indian Medicinal Plants,* Vols. 1-4. Allahabad.

MUDGAL, V., D.C. PAL & P.K. HAJRA 1977. Anthropogenic effects on Sundarbans Biosphere Reserve of West Bengal. *J. New Sci. Coll. Rewa (M. P.).*

MUKHERJEE, A.K. 1975. The Sundarbans of India and its biota. *J. Bombay Nat. Hist. Soc.* 72 (1) : 1-20.

MUKHERJEE, B.B. & J. MUKHERJEE 1978. Mangroves of Sundarbans of India. *Phytomorphology* 28 (2) : 217-220.

NASKAR, K. & D.N. GUHABAKSHI 1986. Economic importance of the dominant mangrove family Rhizophoraceae from the Sundarbans delta of 24 Parganas district, West Bengal. *J. Econ. Taxon. Bot.* 8 (2) : 359-372.

PAL, D.C. & S.K. JAIN 1998. *Tribal Medicine.* Naya Prakash, Kolkata.

TRIVEDI, G.N., V. MUDGAL & D.C. PAL 1993. Some less known ethnomedicinal uses of plants in Sundarbans, India. *Bull. Bot. Surv. India* 35 (1-4) : 6-10.

# STUDIES ON FLOWERING AND FRUITING IN VEGETATIVELY PROPAGATED *TAXUS BACCATA* L. PLANTS : A HIGHLY VALUED MEDICINAL PLANT OF THE HIMALAYAS

DEWAN B. BASNET

*Darjeeling Forest Division, Barakakjhora, Darjeeling-734 101*

**ABSTRACT**

*Taxus baccata* L. is a highly valued medicinal plant, used in extracting anti-carcinogenic drug "Taxol". It is an evergreen, dioecious and indigenous plant of the Himalayas. Observations on flowering behaviour showed that vegetatively propagated plants started flowering exactly at the sixth year of planting. The male flowers (mean size 3.94 ± 0.37 x 2.81 ± 0.45 mm) opened completely at 499.71hrs (21 days). Similarly female flowers of size 3.25 ± 1.875 mm took 100.5 days to develop into mature arillate fruits. The (mean) size of seed was 5.72 ± 0.15 x 4.015 ± 0.01, weighing 20.68 ± 0.36 numbers of seeds to a gram. The Indian plant belongs to subsp. *wallichiana* (Zucc.) Pilger.

## INTRODUCTION

Though several works are available, for example on status of *Taxus baccata* L. in Arunachal Pradesh (Chatterjee *et al.,* 1997), successful rooting of cuttings (Dubey, 1997), with application of different growth regulators (Nandi *et al.,* 1996), propagation of *Taxus baccata* L. in relation to type of branch cuttings (Basnet, 2004), *ex-situ* growth performance and ecology (Basnet, 2000), yet the phenological characteristics and chronology of development of male and female flowers and ripening of fruits of the vegetatively propagated plants are lacking. The bright red cup-like fleshy arillus structure enclosing the seeds attracts local birds. They are mostly consumed before they fall on the ground and are rarely seen on the tree branches. This leads to meagre *in-situ* natural regeneration due to the seed scarcity for trial and seed propagation. In the present investigation the phenology of vegetatively propagated seedlings (plants) has been studied. The ratio between male and female flowers has also been recorded to ascertain the successful pollination and fertilization to establish small clonal seed orchard.

## MATERIALS AND METHOD

Stem cuttings up to four year old branches (male and females) were collected from Relling Council Forest of Darjeeling (T) Range, (T) Division in June 1994. The stem cuttings (up to four year old & diam. 03 mm to 15 mm) were planted directly in the polythene tube containing sand, leaf mould and well rotted cow manure in 1:2:1 ratio. The cuttings were embedded half inside the polythene tube maintaining polarity. The irrigation was done as per requirement till setting of roots and afterwards. The observations on growth, flowering habit, flower development, fruiting and finally seeding were carried out for three years. The exact time taken for flower initiation, total number of male and female plants that flowered and the number of male and female flowers were recorded. The mean numbers of fruits per plant were recorded with observation of 22 flowering male and 10 female plants. Ten female flowers were selected at random and measured to ascertain the mean size (mm) in February. Similarly, 10 male flower buds were selected and labelled with initial measurement during the month of March to study the time of opening and liberation of sporangia. The size wise time taken for initiation of opening, peak and completion of male flowers were recorded separately (Period from 05.03.03 - 08.04.03). Similarly observations on development of female flowers were made out on few flowers selected for this purpose. The development of the disc enclosing solitary fruits (>90%) into red, juicy arils was made separately on day-to-day basis (up to 102 days after initial measurement). The fruits considered ripened at this stage. The percentage of red arillus fruits was recorded by actual count and sampling dwarf shoots of natural population of female trees during September. Some fresh arillus fruits collected were measured and weighed separately. The rest of the fresh fruits were hand rubbed to depulp and washed in fresh water and dried in the sun. The seeds were weighed in 10 replicate, irrespective of number of seeds, to find out the actual dry weight of the seeds. For further study and to establish small C. S. Orchard, altogether 22 plants were transplanted in open space of nursery area after first fruiting during 2001.

Table 1. Mean value of vegetatively propagated *Taxus baccata* L. plants at flowering / fruiting time (Values are mean ± SEm).

**(a) Male plants**

| Collar girth (cm) | Plant ht (cm) | Total no. of flowers |
|---|---|---|
| 1.98 ± 0.33 | 88.00 ± 19.9 | 41.5 ± 2.3 |

**(b) Female plants**

| Collar girth (cm) | Plant ht (cm) | Total no. of flowers | | |
|---|---|---|---|---|
| | | Branches | Dwarf shoots | Fruits |
| 1.6 ± 0.348 | 103 ± 14.71 | 37.3 ± 20.02 | 13.8 ± 8.49 | 33.7 |

Table 2. Development of male flowers and complete opening (period of observation 05.03.03 – 08.04.03).

| S.N. | Mean size of Male flowers (mm) | Opening | | |
|---|---|---|---|---|
| | At "0" hr | Start hrs. | Peak hrs. | Completion hrs. |
| 1 | 4.0 x 3.0 | 96.5 | 336 | 456 |
| 2 | 3.5 x 2.0 | 264.5 | - | 354 |
| 3 | 4.5 x 2.5 | 144 | 336 | 720 |
| 4 | 4.0 x 3.5 | 1.5 | 294.5 | 336 |
| 5 | 4.0 x 4.3 | 3.5 | 192.5 | 384 |
| 6 | 5.0 x 2.5 | 24 | 336 | 456 |
| 7 | 4.0 x 2.5 | 384 | 288 | * |
| 8 | 3.2 x 2.3 | 312 | 792 | * |
| 9 | 4.0 x 2.5 | 456 | 696 | * |
| 10 | 3.2 x 3.0 | 168 | 296 | 792 |
| Total | 39.4 x 30.6 | 1854 | 3831 | 3144 |
| Mean | 3.94 x 2.812 | 185.4 | 383.1 | 499.71 |
| S.D ± | 0.52 x 0.637 | 154.2 | 106.2 | 168.69 |
| S.Em ± | 0.37 x 0.455 | - | - | - |

N.B. * Wasted.

Table 3. Chronology of development of female flowers into red arilate fruits.

| Flower S.No. | Number of days Size of flowers (mm) | | | | | | | | |
|---|---|---|---|---|---|---|---|---|---|
| | "0" | "30" | "50" | "80" | "96" | "99" | "100" | "101" | "102" |
| 1 | 2.5x1.5 | 4.5x2.5 | 4.5x3.0 | 6.0x4.5 | * | 10.5 x 8.0** | - | - | - |
| 2 | 3.0x1.5 | 4.0x4.0 | 6.0x4.5 | 7.0x5.0 | - | * | 11.0 x 10.0** | - | - |
| 3 | 3.5x2.5 | 3.5x3.0 | 5.0x3.5 | 7.0x5.0 | * | 10.0 x 8.0** | - | - | - |
| 4 | 3.2x1.87 | 4.0x3.0 | 5.0x3.5 | 6.12x4.37 | - | * | - | 10.16 x 8.5** | - |
| 5 | 4.0x2.0 | 4.0x2.5 | 4.5x3.0 | 4.5x3.0 | - | * | - | - | 9.14 x 8.0** |
| Mean | 3.2x1.87 | 4.0x3.0 | 5.0x3.5 | 6.12x4.37 | - | - | - | - | - |

N.B. *Inititation of red colour in aril. **Completely ripened fruit with red aril.

## RESULT AND DISCUSSION

The "Himalayan Yew" *(Taxus baccata* L.) is an evergreen undergrowth, and grows as a second canopy or with the associate tree species like *Quercus* spp., *Rhododendron grande* Wight, *R. arboreum* Smith, *R. griffithianum* Wt., *Magnolia campbellia* Hk. f. & Th. etc. Isolated examples of large size up to 3.4 m (B.H.G.) and 25 m heights were encountered in Kankibong Block of Darjeeling Forest Division. Mostly this plant is propagated through branch cuttings and occasionally through naturally occurring seedlings.

Rooting success of cuttings depends on the type of cuttings used. However, more than 66% rooting success was recorded within 16 month of planting. The mean P.A.I (height) of vegetatively propagated plants was 0.128 m in normal field condition (Basnet, 2000). But in present case, the mean (P.A.I) height measured at first flowering plants was slightly less, i.e. 0.11 m in female and 0.09 m in male plants. This might be due to close spacing at the time of planting (Table 1a & b).

Further it was observed that out of 322 plants, 28 number of female plants started flowering from the year 2000 (after sixth year). The branching pattern of plants was monopodial with shoots of unlimited growth. The ascending main stem gave forth horizontal expanded lateral branches in rows. The shoots bearing either male or female reproductive structures were all dwarf shoots. The leaves were persistent, almost sessile or sub-sessile, dorsiventral, darker green above, pale green beneath, linear, entire, with pointed apex and single vein. As the plant is dioecious, the size of leaves of dwarf shoots bearing female reproductive structures was 25.0 x 2.67mm, and that of male plants was 25.67 x 2.92 mm. The female dwarf shoots were found drooping downward and the leaves appeared to be narrower than male plants. The scale leaves were brownish and confined to the base of both male and female reproductive shoots. The winter buds (covered with such scale leaves) later on grow into dwarf shoots. The scale leaves were lanceolate to awl-shaped with acute apex and decurved or reflexed inward on drying. The mean lengths were 3.8 mm. The male reproductive structures (flowers) started to appear with winter buds in the lower surface and arose in the leaf axils of the preceding year's dwarf shoots. The mean collar diameter and heights at first year of flowering of male plants were 0.972 cm ± 0.33 cm and 88.45 cm ± 19.69 cm respectively. The mean number of male flowers in each plant was 41.5 ± 21.37 numbers (Table 2). The shapes of male flowers were almost roundish, sub-pedicellate, size (mean) 3.94 ± 0.37 x 2.81 ± 0.45 mm during the month of March. This sized male flowers took 185.4 hrs to start opening and completely opened by 499.71 hrs. (Table 2). The matured cones were pale green or pale yellow in colour. At maturity these became loose and sporangia are dehisced. During the month of January onwards the dehisced flowers liberate numerous pollens at feeble shake or gush of wind.

The female flowers arise in most cases singly from the leaf axils of previous year's dwarf shoots. During second year of flowering the mean collar diameter and height of female plants recorded were 1.6 ± 0.348 cm and 103.7 ± 14.71 cm respectively. The mean number of flower bearing dwarf shoots were 13.8 ± 8.4 numbers, with 33.7 numbers of fruits (Table 1b). Initially, the vegetative buds and reproductive organs (female flowers) were difficult to differentiate. But, later on, the female flowers did not grow as fast as vegetative buds. Instead, they developed gradually attaining a mean size of 3.25 x 1.875 mm during last week of March. It took full 100.5 days to develop into mature fruits with a reddish fleshy arillus structure (Table 3). The percentage of ripened fruits ranged between 2-5% in natural stand during September. The size of arillus fruits was (mean length) 10.16 mm ± 0.40 mm and (mean width) 8.55 mm ± 0.29 mm. The (mean) weight of such single fruit was 0.50 gm ± 0.18 gm. The single green seed was always concealing inside the aril. After complete absorption of aril, a cup-like brownish bract holds the seeds till fall on the tree floor. The cup-like brownish bract attaches up to 4mm length or ½ ways of the seeds. The weight (mean) of single freshly depulped clean seed was 227.5 mg ± 17.7 mg. The mean number of clean and dry seeds per gm were 20.68 ± 0.36. The clean and dry seeds were ovoid subcompressed, testa hard which measured (mean) 5.72 mm ± 0.15 mm (L) x 4.015 mm ± 0.01 mm, and width at compressed side was 4.902 ± 0.041 mm.

## REFERENCES


BASNET, D.B. 2000. Studies on *ex situ* growth performance and ecology of *Taxus baccata* L. at Rinchingtong P. F. Darjeeling, W. B. India. *Ind. Jour. Env. & Eco.* 3 (3) : 753-760.

———— 2004. Propagation of *Taxus baccata* L. in relation to type of branch cuttings. *Env. & Ecology* 22 (Spl. 3) : 548-550.

CHATTERJEE, SUDIPTO & SOUMEN DEY 1997. A preliminary survey of the status of *Taxus baccata* in Tawang district of Arunachal Pradesh. *Indian Forester* 123(8):746-754.

DUBEY, K.P. 1997. Himalayan yew *(Taxus baccata)* conservation : A vegetative approach. *Indian Forester* 123 (12) : 1150-1154.

NANDI, S.K. *et al.* 1996. Chemical induction of adventitious root formation in *Taxus baccata* cuttings. *Plant growth regulation* 19 : 117-122.

# STUDIES ON THE ANTIVENIN POTENTIAL OF THE CRUDE EXTRACTS OF *BALANITES AEGYPTIACA* (L.) DEL. (BALANITACEAE)

B.M. WUFEM, H.M. ADAMU, Y.A. CHAM & S.L. KELA*

*Industrial Chemistry Programme, Abubakar Tafawa Balewa University, PMB 0248 Bauchi, Nigeria*
**Biological Sciences Programme, Abubakar Tafawa Balewa University, PMB 0248 Bauchi, Nigeria*

**ABSTRACT**

The stem bark of *Balanites aegyptiaca* (L.) Del. was sequentially extracted with hexane toluene, chloroform, acetone and methanol. The extracts were administered intramuscularly into albino rats which had been previously injected with 0.25 ml of 0.194 mg/ml of saw-scale viper venom *(Echis carinatus)*. The acetone and methanol extracts were active against the venom when 1.0 ml and 0.75 ml were injected respectively. Phytochemical screening of the extracts showed the presence of saponins, tannins and volatile oils in the plant. The preliminary studies indicate that acetone and methanol extracts of *Balanites aegyptiaca* (L.) Del. have potentials of providing a readily available and cheaper source of antivenins against viper venom.

## INTRODUCTION

Occurrences of snake bites among the low income earners and peasant farmers are on the increase, especially among the rural settlers of most third world countries. Africa has been considered the headquarter of vipers (Reid, 1936) and in Nigeria, cases of snake bites, especially vipers, are prevalent in Gombe, Bauchi, Plateau, Abuja and Enugu States. Sofowora (1982) had earlier reported that of the 2500 species of known snakes, one-third are poisonous as to cause pain or dangerous injury and even death. As a result, many cases of snake bites among the natives had led to serious consequences and even death.

Death of victims has always been the final consequence as a result of accumulation of venoms in the lungs, kidneys, the heart, thymus lymph nodes, liver and the thyroid glands (Houssay, 1991). The bite of

*Echis carinatus* has been reported to result in a serious life threatening haemorrhagic state that may be accompanied by defibrimuont (Jerushalmy *et al.,* 1973). The venom produces defects in coagulation, pulmonary edema and bleeding in the lungs, heart and kidney which may in turn bring about alterations in the cardiac dynamics and renal functions (Gay, 1962). Deaths have been reported to last an average time of three hours after viper bites (Vick *et al.,* 1967). This time is short since most bites occur at a long distance away from hospitals, and even when the victim is right on time to a point of medication, the high cost of the available immunoglobin antivenin drug from horses may not be affordable by the victims. Moreover, it may have some adverse effects that could be lethal (Reid, 1957).

Herbs have always provided a readily available source of medicine for various ailments. *Balanites aegyptiaca* L.) Del. has been reported to be of medicinal value (Abbiw, 1990). It grows well in the Sudan and Sahel zone of the Savanna and it is also called desert date. The root scrapings are used as antidote to snake bite (Sofowora, 1982). The roots are known to be used as antihelmintic, purgative and as a cure for parasitic infections (Abbiw, 1990). The biological activity against microorganisms has been reported by Taniguchi *et al.* (1978). There is virtually no literature on the use of the stem bark extract as antivenom against saw-scale viper (*Echis carinatus*) bites. Moreover, with the advent of the immunoglobin antivenin drug from horses, little attention is paid to alternative drug that is non-lethal and with less adverse effects. This work tries to examine *in vivo* the efficacy of the stem bark of *Balanites aegyptiaca* (L.) Del. on the saw-scale viper venom using albino rats.

## METHODS

### Sampling

The stem bark of *Balanites aegyptiaca* was collected behind Rafin Zurfi in Bauchi, Nigeria in brown paper bags. It was identified by Mallam H.U. Dukku of the Biological Sciences Programme, Abubakar Tafawa Balewa University, Bauchi. The sample was air dried for four weeks before powdered for extraction. The solvents used were BDH Analar grades.

Table 1. Result of Phytochemical screening.

| Extracts | Reactions + = Reactive; - = Not reactive |
|---|---|
| Saponins | + |
| Tannins | + |
| Flavonoids | - |
| Alkaloids | - |
| Glycosides | - |
| Volatile oils | + |

### Test Animals

Both male and female albino rats (8-12 kg), breeded at the National Veterinary Research Institute (NVRI), Vom, Nigeria were used for the study. They were kept for three weeks before the test was carried out on them. Rats were fed with a mixture of groundnut cake and maize husk, and water.

Five live saw-scale vipers (ages not ascertained) between the lengths of 40-60 cm were milked after five weeks of caging. They were fed on live grasshoppers and water.

### Extraction of *Balanites aegyptiaca* (L.) Del.

200 g of the powdered stem bark of the sample was extracted serially using five successive solvents. The sample was extracted for three days each in hexane, toluene, chloroform, acetone and methanol. The extract was then tested on the animals in each case.

### Phytochemical Screening

Phytochemical test was carried out for the presence of alkaloids, flavonoids, saponins, tannins, glycosides and volatile oil using the method of Sofowora (1982) and Matos (1988).

### Preparation of Venom Stock

23.3 mg venom was dissolved in 10 ml of clinical normal saline solution from which 4.29 ml was measured out into 10 ml flask and made to volume to prepare 1.00 mg/ml stock solution, from which 0.25 mg/ml, 0.194 mg/ml, 0.076 mg/ml, and 0.039 mg/ml were prepared.

### Determination of Lethal Dose ($LD_{50}$)

0.25 ml of the prepared venom solution was administered to the three sets of rats each by intramuscular (1M) injection using insulin syringe and needle, and the sets of rats kept in separate wooden cages for observations. $LD_{50}$ was determined using the method as reported by Vidakovic-Bival (1973). Control rats were injected with normal saline only.

Table 2. $LD_{50}$ of the Saw-scaled viper venom.

| Expt. | Venom. Conc. (mg/ml) | Mortality | Observations |
|---|---|---|---|
| 1 | 0.250 | 6 | Body swollen, bleeding at the limb, weakness of the body and nervousness. Death was recorded within 30-45 min. |
| 2. | 0.194 | 6 | Body swollen, bleeding at the limb, weakness of the body. Death follows after 1 hour to 1 hour 45 minutes. |
| 3. | 0.076 | 2 | Body swollen, bleeding and nervousness. Two died after 3 hours. Survivors were active after 24 hours. |
| 4. | 0.039 | 0 | Swollen limbs, weakness of the body and paralysis at the point of injection. All survived and became active after 24 hours. |

### Determination of Effective Dose ($ED_{50}$)

The extract solution (100 mg/ml),which was prepared by dissolving 1.0 mg of the extract in 10 ml of normal saline was injected into the rats IM that had been previously administered the venom while varying the volume of the extract in each group of rats. Control rats were also injected with only the extracts. The rats were kept in separate wooden cages for observation.

## RESULTS

### Phytochemical Screening

The results of phytochemical screening is presented in Table 1. The presence of alkaloids, flavonoids and glycosides was not detected, however, saponins, tannins & volatile oils were detected by these methods.

## Pharmacological screening

### *$LD_{50}$ Testing*

The results for the lethal dose following observations for the viper venom is shown in Table 2. In all the concentrations the same effect is manifested due to the venom administered, except differences in deaths recorded.

### $ED_{50}$ Testing

Table 3 and 4 provide the effective dose of the acetone and methanol extracts needed for half the rat populations to survive. Despite the extract levels, the rats still suffer swollen body, body weakness and to some extent, bleeding and even deaths.

Table 3. $ED_{50}$ of the acetone extract against 0.194 mg/ml Saw-scale viper venom.

| Expt. | Vol. of Extract (ml) | Mortality | Observations |
|---|---|---|---|
| Control | 1 | 0 | Rat survived and was active. |
| 1 | 0.25 | 6 | Swollen body, bleeding at the limb and weakness of the body. All died after 1 hour. |
| 2. | 0.50 | 6 | Swollen body, bleeding at the limb weakness of the body. All died after 1 hour. |
| 3. | 0.75 | 2 | Swollen body, bleeding at the limb and weakness. Two died after 1 hour 20 minutes. Survived rat was active after 3 hours. |
| 4. | 1.00 | 1 | Swollen limbs, weakness of the body, one died after 1 hour 45 minutes. Survived rats were active after 3 hours. |

## DISCUSSION

The results observed from administering venom concentrations above 0.076 mg/ml revealed deaths happening within 30-45 min. However, at lower venom concentrations no deaths were recorded. When 1 ml of the extracts (100 mg/ml) were administered to the rats that had been injected with 0.25 ml of the venom (0.194 mg/ml), only the methanol and acetone extracts were found to be active.

This result shows that the venom of the viper could be so lethal as to cause death within a very short time. Nevertheless, varying the concentration of the venom from 0.25 mg/ml to 0.039 mg/ml (Table 2) reveals a gradual decrease of venom effects. The lethal dose was established at 0.076 mg/ml venom concentration.

Table 4. $ED_{50}$ of the methanol extract against 0.25 ml of 0.194 mg/ml Saw-scale viper venom.

| Expt. | Vol. of Extract (ml) | Mortality | Observations |
|---|---|---|---|
| Control | 1 | 0 | Rat survived and were active. |
| 1 | 0.25 | 6 | Swollen body, bleeding at the limb & weakness of the body. All six died after 1 hour. |
| 2. | 0.50 | 6 | Swollen body, bleeding at the limbs and weakness of the body. All six died after 1 hour 45 minutes. |
| 3. | 0.75 | 2 | Swollen body, bleeding at the limb and weakness. Two died after 1 hour 45 minutes. Survived rat were paralyzed but became active after 3 hours. |
| 4. | 1.00 | 0 | Swollen limbs, weakness of the body. All survived and were active after 2-3 hours. |

On administering the extracts, similar trends (Table 3 and 4) were observed except it took longer time for deaths to occur when methanol extracts were used. The variation in extract volume suggest that the venom effect can be counteracted with increased extract volume as shown by gradual increase in death time with increasing volume of extract. The effective dose, $ED_{50}$ was established to be 1000 mg (1 ml of 100 mg/ml) for the rats. However, the potency seems to favour the methanol extract than the acetone extract (Tables 3 & 4). Our findings now suggest a scientific base for using *Balanites aegyptiaca* (L.) Del. as antidote for snake bites. The acetone and methanol extracts have shown significant activity in counteracting the viper venom effects. The major constituents, saponins, Tannins and volatile oils, may be responsible for the pharmacological activity of these plants. Our focus now will be on the isolation, purification and structural elucidation of the active components of the extracts.

## ACKNOWLEDGEMENTS

The authors wish to thank the National Veterinary Research Institute, Vom for providing the albino rats.

## REFERENCES

ABBIW, D.K. 1990. *Useful plants of Ghana.* Intermediate Technological Publ., London.

GAY, R. 1962. Viper bites. In Russel, F. C. & P. R. Sauchers (ed.) *Animal toxins.* Pergamon Press, New York.

HOUSSAY, B.A. 1991. Hemolytic effect of snake venom on respiration. In Chen-Yaun Lee (ed.) *Snake venom,* Vol. 52. Sigma Chem. Company, USA.

JERUSHALMY, Z., U. SANDBANK & A. DE VRIES 1973. Intra vascular coagulation induced in the rabbit by *Echis carinatus* venom. In De Vries, A. & E. Koch. (eds.) *Toxins of animal and plants origin,* 3 : 1029-1032. Gordon and Breach Science Publishers, New York.

MATOS, F.J.A. 1988. *Introducao a fitoquimica experimental.* Imprensa Universitaria, Universidade Federal do Ceara, Fortaleza.

REID, G. 1936. Viper bites and toxins of animal and animal origin. In A. de Varies & E. Koch. (eds.) *Toxins of animal and plants origin.* Gordon and Breach Science Publ., New York.

REID, H.A. 1957. Antivenene reaction following accidental sea-snake bite. *British Medical Journal* 2 : 26.

SOFOWORA, A.E. 1982. *Medicinal plants and traditional medicine in Africa* Part 2 : 128-146. Pitman Press Ltd., London.

TANIGUCHI, I., KUBO CHAPYA & NAKANISHI 1978. Screening of East Africa plants for antimicrobial activity. *Chem. Pharm.* 26:2910-2913.

VICK, J.A., H.P. CHECITA & J.H. MANTHEI 1967. *Pathophysical studies of ten poisonous snake venoms* 52: 54-55. Sigma Chem. Company, USA.

VIDAKOVIC-BIVAL, V. 1973. Determination of the $ED_{50}$ and Tentative Units in Serum antiviperinum by I.V. Titration in mice. In A. de Vries & E. Koch. (eds.) *Toxins of Animal and Plant Origin* 3 : 863-875. Gordon and Breach, Science Publishers, New York, USA.

# STUDY ON NUTRITIONAL POTENTIAL AND ANTI-NUTRITIONAL CONTENTS IN SEED KERNELS OF *CYCAS CIRCINALIS* L.

S. CHIDAMBARAM PILLAI & V.R. MOHAN

*Post Graduate Department of Botany, V. O. Chidambaram College, Tuticorin-628 008, India*

## ABSTRACT

*Cycas circinalis* L. grows in wild and also as cultivated plant in tribal hamlets. Their seed kernels are used as food source among the Kanikkar tribals of Kanyakumari District, Tamil Nadu. An attempt is made to analyse their proximate composition, mineral composition, vitamins, amino acid profiles and antinutritional factors in the kernel and their mode of use has been discussed.

## INTRODUCTION

There are many under-utilized plants which have tremendous potential for commercial exploitation (Bhag Mal, 1992) and await exploration for food, fodder, energy & industrial purposes (Balogun & Fetuga, 1986; Paroda & Bhag Mal, 1989; Siddhuraju *et al.*, 1993; Vadivel & Janardhanan, 2000). One among them is the genus *Cycas*, whose seeds are historically recorded as a staple food in many parts of the World (Roth, 1901; Thompson, 1949).

A number of *Cycas* species have been included in the Red data list of threatened plants, of which *Cycas circinalis* L. is categorized as 'data deficient' and is considered as 'living fossil' (Internet news, 2003). The endosperm or the Paereenji kernel of this species is made into flour or starch after various treatments to remove poisnous contents. In Fiji, the seeds are boiled before consumption. In Srilanka, the kernels are cut into slices, dried and ground into a meal and the dough is baked into cakes. In Java and Prince's islands, the seed kernels are soaked in frequent changes of water over several days before they are made into cakes and are served as extender with rice (Vasishta, 1996).

In India, the mature seeds of *Cycas circinalis* L. are known to be eaten by Nayadis tribe of North Kerala (Prasad & Abraham, 1984); Kondh tribe of Orissa

(Girach & Aminuddin, 1992) and Palliyar tribes of South Eastern slopes of Western Ghats, Tamilnadu (Arinathan, 2002). During famine period, the Kanikkar tribes of Southern Western Ghats, Tamilnadu, repeatedly soak the kernels in water (7 times) and consume the steam boiled kernel powder along with honey or ghee. Very scanty information is available on the chemical composition of under-utilized wild plants in general and *Cycas circinalis* L. in particular. An attempt has been made to evaluate certain nutritional and anti-nutritional contents of its kernel in the present study.

## MATERIALS AND METHODS

The mature seeds of *Cycas circinalis* L. were collected from the upper hills of Petchiparai, Kanyakumari District, Tamilnadu. Soon after collection, all the kernels were sun dried for 2-3 days and stored in screw-capped containers.

### Proximate composition

Moisture content was determined by the method of Rajaram & Janardhanan (1990). The kernels were powdered in a Willey mill to 60 mesh size and stored in screw-capped bottles at room temperature for further analysis. Nitrogen content was estimated by the microkjeldahl method (Humphries, 1956) and crude protein was calculated (N x 6.25). The contents of crude lipid, crude fibre and ash were estimated by AOAC (1975) methods. Nitrogen Free Extractive (NFE) or crude carbohydrates were obtained by difference (Muller & Tobin, 1980). The energy value of the kernels was estimated by multiplying the percentages of crude protein, crude lipid and NFE by the factors 16.7, 37.7 and 16.7 respectively (Siddhuraju *et al.*, 1996).

The purified total proteins were acid hydrolysed with 6N HCl at 110°C for 24h *in vacuo*. After flash evaporation, the dried residue was dissolved in citrate buffer (pH 2.2). Known aliquots were analysed in a LKB – Bio chrome automated amino acid analyzer Model-4151, Alpha Plus. The contents of different amino acids recovered were compared with the FAO / WHO (1991) reference pattern. The essential amino acid score was calculated as follows.

$$\text{Essential amino acid score} = \frac{\text{grams of essential amino acid in 100 g of test protein}}{\text{grams of essential amino acid in 100g of FAO/WHO (1991) reference pattern}} \times 100$$

### Minerals & Vitamins

Five hundred mg of the ground kernel was digested with a mixture of 10 ml concentrated nitric acid, 4 ml of 60% perchloric acid and 1 ml of concentrated sulphuric acid. After cooling, the digest was diluted with 50 ml of deionized distilled water, filtered with Whatman No.42 filter paper and the filtrate was made up to 100 ml in a glass volumetric flask with deionized distilled water. All the minerals except phosphorus were analysed from triple acid digested sample by Atomic Absorption Spectrophotometer (ECIL-Electronic Corporation of India Ltd) (Issac & Johnson, 1975). Phosphorus content was estimated colorimetrically (Dickman & Bray, 1940) from triple acid digested samples. Ascorbic acid and Niacin contents were extracted and estimated as per the method given by Sadasivam & Manickam (1992).

### Analysis of antinutritional factors

The anti-nutritional factors like total free phenolics (Bray & Throne, 1974), tannins (Burns, 1971) and hydrogen cyanide (Jackson, 1967) were quantified.

## RESULTS & DISCUSSION

The result of proximate analysis of kernels of *Cycas circinalis* L. is shown in Table 1. The crude protein content was close to that of *Canavalia gladiata* (Jacq.) DC., *Neonotonia wightii* (Wt. & Arn.) Lackey var. *coimbatorensis* (Ajita Sen) Karthik. and *Vigna unguiculata* (L.) Walp. subsp. *unguiculata* (Arinathan, 2002). The crude fibre content of *Cycas circinalis* L. was higher than that of other commonly cultivated pulses such as *Vigna mungo* (L.) Hepper (Kuzayli *et al.*, 1966) and *Cajanus cajan* (L.) Millsp. (Nwokolo, 1987).

Table 1. Proximate composition (g $100g^{-1}$)*.

| Component | g 100 $g^{-1}$ flour |
|---|---|
| Moisture | 7.92 ± 0.31 |
| Crude Protein (Kjeldahl N x 6.25) | 14.78 ± 0.44 |
| Crude lipid | 2.23 ± 0.25 |
| Crude fibre | 3.95 ± 0.11 |
| Ash | 2.73 ± 0.55 |
| Nitrogen Free extractive (NFE) (Crude carbohydrates) | 76.31 |
| Calorific Value (KJ 100 $g^{-1}$ DM) | 1605 |

*All values are mean of triplicate determinations expressed on dry-weight basis.
± denotes standard error.

Table 2. Mineral composition and vitamin contents (mg $100g^{-1}$)*.

| Minerals | mg 100 $g^{-1}$ kernel flour |
|---|---|
| Sodium | 27.12 ± 0.11 |
| Potassium | 964.50 ± 0.28 |
| Calcium | 224.00 ± 0.07 |
| Magnesium | 238.10 ± 0.18 |
| Phosphorus | 120.10 ± 0.25 |
| Zinc | 3.28 ± 0.02 |
| Manganese | 1.08 ± 0.01 |
| Iron | 21.18 ± 0.08 |
| Copper | 1.20 ± 0.01 |
| **Vitamins** | |
| Ascorbic Acid | 9.24 ± 0.04 |
| Niacin | 38.25 ± 0.09 |

*All values are mean of triplicate determinations expressed on dry-weight basis.
± denotes standard error.

The mineral and vitamin composition of *Cycas circinalis* L. kernels is shown in Table 2. The kernels were found to be a rich source of minerals such as sodium, potassium, calcium, magnesium, zinc and iron, when compared with a common pulse, *Cicer arietinum* L. (Meiners *et al.*, 1976). The content of iron was higher when compared with Recommended Dietary Allowances (RDA) (NRC / NAS, 1989) which may be due to reduced tannin levels.

The niacin content in the kernels of *C. circinalis* L. registers a higher level when compared with the pulse, *Cicer arientinum* L. (Fernandez & Berry, 1988).

The ascorbic acid content is naturally found to be higher than that of earlier reports on some common pulses like *Vigna mungo* (L.) Hepper, *Vigna unguiculata* (L.) Walp. ssp. *cylindrica* (L.) Van Eseltine,

*Lablab purpureus* (L.) Sweet, *Macrostyloma uniflorum* (Lam.) Verdc. (=*D. biflorus*), *Cajanus cajan* (L.) Millsp. and *Mucuna pruriens* (L.) DC. (Rajyalakshmi & Geervani, 1994).

The amino acid profiles of the purified kernel proteins and the essential amino acid score are presented in Table 3. The contents of valine, threonine, phenylalanine, tyrosine, isoleucine, leucine and lysine of present investigation are found to be higher as compared to the FAO/WHO (1991) requirement pattern. In general, the amino acid profiles are incomplete; it is because some amino acids are destroyed during the preparation of the sample by acid digestion.

When unexploited plants are investigated as a potential food source, the possible presence of antinutritional factors must be considered. These antinutritional factors adversely affect the protein digestibility of food intake (Gupta, 1987). These substances unless destroyed by heat or some other suitable treatments can exert adverse physiological effects when ingested by man and animals (Liener, 1980). In view of this, in the present investigation, an attempt has been made to detect the presence of certain antinutritional factors such as total free phenolics, tannins and hydrogen cyanide (Table 4).

Phenol generally decrease the digestibility of carbohydrates, availability of vitamins and minerals and interact with proteins to make them insoluble (Singh *et al.*, 2000). In the study made here, the content of total free phenolics appears to be lower when compared with *Vigna umbellata* (Thunb.) Ohwi & Ohashi var. RBL 40 and K1 (Rajaram & Janardhanan, 1990).

A low level of tannin is nutritionally significant because tannins are known to inhibit the activities of digestive enzymes (Jambunathan & Singh, 1980; Sathe & Salunkhe, 1984; Akapanyung *et al.*, 2001). Here, the tannin content of the investigated sample is negligible, compared with commonly consumed pulses such as black gram, chickpea, cowpea and green gram (Khan *et al.*, 1979; Rao & Deosthale, 1982).

Hydrogen cyanide toxicity is generally attributed to its ability to inhibit the respiratory chain at the level of cytochrome oxidase (Akapanyung *et al.*, 2001). The content of HCN level in the kernel of *C. circinalis* L. is also far beyond the lethal level i.e. 0.36mg/100/g (Oke, 1975).

Table 3. Amino acid profiles of acid-hydrolysed, purified kernel proteins.

| Amino Acid | g 100 g$^{-1}$ protein | Essential Amino Acid Score | FAO/WHO requirement pattern (g 100 g$^{-1}$ protein) |
|---|---|---|---|
| Aspartic Acid | 9.98 | - | - |
| Glutamic Acid | 12.74 | - | - |
| Alanine | 9.42 | - | - |
| Valine | 7.37 | 147.4 | 5.00 |
| Glycine | 8.89 | - | - |
| Arginine | 8.02 | - | - |
| Serine | 6.19 | - | - |
| Cysteine* | N.D. | - | - |
| Methionine | 1.07 | - | - |
| Threonine | 4.79 | 119.75 | 4.00 |
| Phenylalanine | 3.06 | 122.0 | 6.00 |
| Tyrosine | 4.26 | | |
| Isoleucine | 4.65 | 116.25 | 4.00 |
| Leucine | 9.98 | 142.57 | 7.00 |
| Histidine | 1.29 | - | - |

| | | | |
|---|---|---|---|
| Lysine | 2.00 | 36.36 | 5.50 |
| Tryptophan* | N.D. | - | - |

*Cysteine and Tryptophan are destroyed during acid hydrolysis.
N.D. : Not Detected.

Table 4. Data on antinutritional contents*.

| Component | g 100 $g^{-1}$ kernel flour |
|---|---|
| Total free phenolics | 0.14 ± 0.02 |
| Tannins | 0.04 ± 0.01 |
| Hydrogen Cyanide (mg 100 $g^{-1}$) | 0.22 ± 0.03 |

*All values are mean of triplicate determinations expressed on dry-weight basis.
± denotes standard error.

## CONCLUSION

This unconventional and unfashionable seed food from endangered taxon *Cycas . circinalis* L. consumed by the Kanikkar tribes is rich in crude protein, NFE, minerals and certain essential amino acids like valine, leucine, phenylalanine, tyrosine and threonine. The anti-nutritional factors like tannins, phenolics and hydrogen cyanide are well below their toxic levels. The deleterious effects of their anti-nutritional factors present are eliminated customarily by simple processing like repeated soaking, roasting and decanting their boiled water etc which makes the kernel meals nutritionally safe and their consumption might be popularized among the poor sections of population in many developing countries, including India.

## REFERENCES


AKAPANYUNG, E.G., A.P. UDOH & M.O. ETENG 2001. Antinutrient profile and chemical composition of custard powder produced in Nigeria. *J. Food. Sci. Tech.* 38 : 120-123.

AOAC 1975. *Official Methods of Analysis*. Association of Official Analytic Chemists, Washington, DC.

ARINATHAN, V. 2002. *Studies on tribal food plants of South India.* Ph. D. Thesis, Manonmaniam Sundaranar University, Tirunelveli.

BALOGUN, A.M. & B.L. FETUGA 1986. Chemical composition of some under-exploited leguminous crop seeds in Nigeria. *J. Agric. Food Chem.* 34 : 189-192.

BHAG MAL 1992. Biodiversity utilization and conservation in under-utilized plants: Indian perspective. *Ind. J. Plant Genet. Resour.* 5 : 1-22.

BRAY, H.G. & W.V. THRONE 1954. Analysis of phenolic compounds. *Methods Biochem. Anal.* 1 : 27-52.

BURNS, R.R. 1971. Methods for estimation of tannins in grain of *Sorghum. Agron. J.* 63 : 511-512.

DICKMAN, S.R. & R.H. BRAY 1940. Colorimetric determination of phosphate. *Ind. Eng. Chem. Ed.* 12 : 665-668.

FAO/WHO 1991. *Protein quality Evaluation,* pp. 66. Food and Agricultural Organization of the United Nations, Rome, Italy.

FERNANDEZ, M.L. & J.W. BERRY 1988. Nutritional evaluation of chickpea and germinated chickpea flours. *Plant Foods Hum. Nutr.* 38 : 127-134.

GIRACH, R.D. & P.A. AMINUDDIN 1992. Addition to little known edible plants from Orissa. *J. Econ. Taxon. Bot.* 16 : 589-594.

GUPTA, Y.P. 1987. Antinutritional and toxic factors in food legumes: A Review. *Plant Foods Hum. Nutr.* 37 : 133-136.

HUMPHRIES, E.C. 1956. Mineral components and ash analysis. In Peach, K. & M.V. Tracey (Eds.) *Modern methods of plant analysis* 1 : 468-502. Springer Verlag, Berlin.

INTERNET NEWS 2003. Million-year-old *Cycas* plants may soon be wiped out. *Nat. Prod. Rad.* Sept-Oct. 2 : 277-278.

ISSAC, R. A. & W. C. JOHNSON 1975. Collaborative study of wet and dry ashing techniques for the elemental analysis of plant tissue by atomic absorption spectrophotometer. *J. AOAC.* 58 : 436-440.

JACKSON, M. L. 1967. Cyanide in plant tissue. *Soil Chemical Analysis,* pp. 337. Asia Publishing House, New Delhi, India.

JAMBUNATHAN, R. & U. SINGH 1980. Studies on Desi and Kabuli chickpea (*Cicer arietinum* L.) cultivars. 1. Chemical composition. *Proceedings of the International workshop on Chickpea improvement,* pp 61-76. ICRISAT, Patancheru, Hyderabad (A.P), India. February 28 - March 2, 1979.

KHAN, A. M., I. JACOBSON & O. B. EGGUM 1979. Nutritive value of some improved varieties of legumes. *J. Sci. Food Agri.* 30 : 395-400.

KUZAYLI, M. V., J. W COWAN & Z. I. SABRY 1966. Nutritive value of middle eastern foodstuff II. Composition of pulses, seeds, nuts and cereal products of Lebanon. *J. Sci. Food Agri.* 17 : 82-84.

LIENER, I. E. 1980. Heat-labile anti-nutritional factors. In Summerfield, R. J. & A. H. Bunting (Edits.) *Advances in Legume Science,* pp.157 - 170. Royal Botanic Gardens, Kew, Richmond Surrey, U.K.

———— 1994. Implications of anti-nutritional components in soybean foods. *CRC. Crit. Rev. Food Sci. Nutr.* 34 : 31-67.

MEINERS, C. R., N. L. DERISE, H. C. LAU, M. G. CREWS, S. J. RITCHEY & E. W. MURPHY 1976. The contents of nine mineral elements in raw and cooked mature dry legumes. *J. Agri. Food Chem.* 24 : 1126-1130.

MULLER, H. G. & G. TOBIN 1980. *Nutrition and Food Processing.* Croom Helm Ltd., London.

NRC/NAS 1989. National Research Council Committee on Dietary Allowances. Recommended Dietary Allowances, 10th Edn. National Academy of Science Press, Washington, DC, USA.

NWOKOLO, E. 1987. Lipid composition of tropical seeds used in the Nigerian diet. *J. Sci. Food Agri.* 50 : 253-255.

OKE, O. L. 1975. The role of *Cassava* in the nutrition of Nigerian population. *J. Root Crops.* 1 : 1-15.

PRASAD, P. N. & Z. ABRAHAM 1984. Ethnobotany of the Nayadis of North Kerala. *J. Econ. Taxon. Bot.* 5 : 41-48.

PARODA, R. S. & BHAG MAL 1989. New plant sources for food and industry in India. In Wickens, G. E., N. Haq & P. Day (Edits.) *New crops for Food and Industry,* pp. 135-149. Chapman & Hall, London.

RAJARAM, N. & K. JANARDHANAN 1990. Chemical analysis and nutritional evaluation of certain under-exploited *Vigna* spp. *J. Food Sci. Nutr.* 42 : 213-221.

RAO, P. U. & Y. G. DEOSTHALE 1982. Tannin content of pulses, varietal differences and effects of germination and cooking. *J. Sci. Food Agri.* 33 : 1013-1016.

RAJYALAKSHMI, P. & P. GEERVANI 1994. Nutritive value of the foods cultivated and consumed by the tribals of South India. *Plant Foods Hum. Nutr.* 46 : 53-61.

ROTH, W. E. 1901. Food, its search, capture and preparation. *Ethnography Bull.* No. 3. North Queensland.

SADASIVAM, S. & A. MANICKAM (Eds.). 1992. *Biochemical methods for Agricultural Sciences,* pp. 187-188. Wiley Eastern Ltd., New Delhi,

SATHE, S. K. & D. K. SALUNKHE 1984. Technology of removal of unwanted components of dry bean. *CRC. Crit Rev. Food Sci. Nutr.* 21 : 263-286.

SIDDHURAJU, P., K. VIJAYAKUMARI & K. JANARDHANAN 1993. Genetic resources of tribal pulses. *Pl. Genet. Res. Newslett.* 96 : 47-49. FAO / IBPGR.

———, ——— & ——— 1996. Chemical composition and protein quality of the little known legume, velvet bean (*Mucuna pruriens* (L.) DC.). *J. Agri. Food Chem.* 44 : 2636-2641.

SINGH, U., S. SEHGAL & T. S. TOMER 2000. Influence of dehulling, soaking solution and enzyme treatment on the cooking quality of improved varieties of pulses. *J. Food. Sci. Tech.* 37 : 627-630.

THOMPSON, D. F. 1949. Economic structure and the ceremonial exchange cycle in Arnhem Land. Melbourne Univ. Press, Melbourne.

VADIVEL, V. & K. JANARDHANAN 2000. Chemical composition of the under-utilized legume *Cassia hirsuta* L. *Plant Foods Hum. Nutr.* 55 : 369-381.

VASISHTA, P. C. 1996. *Botany for degree students,* Vol. V-Gymnosperms. Chand & Company Ltd., New Delhi.

# TRADITIONAL ETHNOBOTANICAL KNOWLEDGE OF KIRAN PARGANA, SHIMLA DISTRICT, HIMACHAL PRADESH

JOGINDER SINGH, A. RAJASEKARAN & K.D. SHARMA

*Himalayan Forest Research Institute, Shimla - 171 009*

**ABSTRACT**

Ethnobotanical study of Kiran Pargana, a remote part of Shimla district, Himachal Pradesh, revealed 97 plant species belonging to 50 families used for various purposes. The 97 plant species include three gymnosperms, one fungus and 93 angiosperms. Of the 97 species, herbs and trees contributed 33.0% each followed by shrubs (23.7%), climbing shrubs 5.2%, small trees 4.1% and herbaceous climbers 1.0%. Thirty-nine species are edible, 26 medicinal, 11 used as fuel wood, 7 as veterinary medicine, 4 used for making agricultural implements, 3 as timber and 15 are used for miscellaneous purposes.

## INTRODUCTION

Himachal Pradesh has been considered as a veritable emporium of medicinal and aromatic plants in the Western Himalayan region. Communities in this region are depending on the natural resources for their day-to-day needs and they have traditional rights to collect and sell medicinal herbs, roots, flowers, fruits and aromatic plants from forests. Out of 3,300 plant species recorded from the State, about 1000-1200 species have got medicinal value (Badola & Pal, 2003). Almost all these plants grow wild in valleys, forests, pastures or occurs as weeds near field and village habitations. The traditional knowledge accumulated by the communities through trial and error is getting depleted with the advent of modern education and cultural changes. Hence, documentation of ethnobotanical knowledge is highly essential. Moreover, over-exploitation, coupled with habitat degradation, has already threatened many medicinal plants. Several efforts have been made to document the traditional ethnobotanical knowledge of the people (Gupta, 1964; Uniyal & Chauhan, 1971; Chauhan, 1984; Chauhan & Khosla, 1988; Chauhan & Chauhan, 1988; Kapahi,

1990; Kala & Manjrekar, 1999; Sharma & Rana, 1999; Sood *et al.*, 2001; Meenakshi, 2002 etc). In spite of the above studies, information about the traditional uses of plants is meager, in relation to the floristic diversity. Moreover, some of the areas like Kiran Pargana are still unexplored. Hence, the present study was undertaken with the aim to document the traditional ethnobotanical knowledge of the area.

## STUDY AREA

Kiran Pargana (Tehsil Chopal, District Shimla) lies between 30°38' - 31°3' North latitude and 77°26' -77°54' East longitude. The Kiran is one of the nine Parganas of erstwhile Throach State. It is located in remote parts of Chopal Tehsil of Shimla district and presently form the part of Chopal forest division. It lies on the right bank of Tons, a tributary of river Yamuna. The area is in the form of a triangle, with Jaunsar Pargana of Dehra Dun district of Uttranchal on southern and eastern boundaries and Jubbal on the western side. The altitudinal range varies from 1200 m - 2400 m above mean sea level. The climate of the area is subtropical to temperate. The average annual rainfall of the area is around 1200 mm; the maximum annual average temperature goes up to 33°C in the month of June and average minimum annual temperature falls to 5°C in the month of January. Most parts of the area receive winter snow except for some lower regions. The area is mountainous, having moderate to steep slopes. The main geological formations of the region include micashist, sandstone, quartzite and gneiss. Vegetation of the area is subtropical to temperate type (Anonymous, 1910). Deodar and Chirpine are dominant tree species in the upper and lower regions respectively. The forests have been much damaged by excessive felling in the past. The human populations mainly consist of 'Kanets' and are divided into various classes, such as Rajput, Brahmin, Turi, Koli etc and all are agriculturists. Hindi, Pahari and Kirani (dialect) are main languages spoken in the area. The uses of herbal medicines are still prevalent to some extent. Due to remoteness and lack of motorable roads the area has not been explored for detailed floristic and ethnobotanical studies, which enhances the importance of this detailed study.

## MATERIALS AND METHODS

Extensive survey of remote villages such as Kashak, Telor, Thithrouli and Maneoti of Kiran Pargana of Shimla district was carried out between April, 2004 and January, 2005. Traditional ethnobotanical information was documented by interviewing fourteen people using a semi-structured questionnaire. Men and women of different age groups, especially older, were interviewed. The uses were cross-checked with the informants. Efforts were made to identify these plants in the field itself and those which couldn't be identified in the field were brought to laboratory and identified with the help of regional Floras (Collett, 1921; Chowdhery & Wadhwa, 1984) and herbarium at HFRI Shimla.

## RESULT AND DISCUSSION

During the present study, ethnobotanical informations on 97 plant species belonging to 50 families were documented. The 97 plant species include three gymnosperms, one fungus and 93 angiosperms. Of the 50 families recorded, Rosaceae is the dominant family with 11 species. Of the 97 species, herbs and trees contributed 33% each, followed by shrubs (23.7%), climbing shrubs 5.2%, small trees 4.1% and herbaceous climbers 1.0%. Some of the species are used for more than one purpose. Thirty-nine plants are used for edible purposes, 34 as fodder, 26 for treating different ailments, 11 as fuel wood, 7 considered as sacred and used in different religious ceremonies, 7 are used as veterinary medicine, 4 for making agricultural implements, 3 for timber, ropes are prepared from 3 species and 15 are used for miscellaneous purposes. During the discussion, local people showed interest to share knowledge about plants used for various purposes, but were hesitant to share knowledge about medicinal plants. They believe that if medicinal uses of plants are shared with other people, healing power of these plants will reduce.

Among edible plants, fruits of *Berberis lycium, Ficus palmata, Ficus auriculata, Morus alba, Prunus armeniaca, Prunus persica, Punica granatum, Pyrus pashia, Rhus parviflora, Rubus ellipticus, Zizyphus mauritiana* and *Zizyphus oxyphylla* are eaten by children. Floral buds of *Indigofera pulchella* and *Bauhinia variegata* are also edible. Poor people eat flour prepared from the fruits of *Rhus parviflora* as food.

Among medicinal plants, species such as *Achyranthes aspera, Ajuga parviflora, Angelica glauca, Artemisia* spp., *Berberis lycium, Fragaria vesca, Kalanchoe spathulata, Leucas lanata, Musa paradisiaca, Origanum vulgare, Plantago erosa, Reinwardtia indica, Ricinus communis, Roylea cinerea, Rubus niveus, Thalictrum foliolosum* and *Viola canescens* are important. All these medicinal plants are commonly seen in the area. However, *Angelica glauca* (Chora) is confined to Kunuei locality, where the habitat is temperate humid with sufficient humus and accessibility to the area is limited. Apart from the species growing in the region, people collect medicinal plants growing in the higher areas, especially from Chhajpur forest of Jubbal and Murach forest of Chakrata tehsil of Uttranchal and use them for treating different ailments. The important plants collected from the higher areas include *Aconitum* spp. *(Atish)* and *Picrorhiza kurrooa (Kadu).* Powdered roots of these plants are taken with water to cure stomach ailments. Although these plant species have got high cultural significance in the region and utilized for local consumption, people do not commercially exploit them.

*Quercus leucotricophora* and *Melia azedarach* are most preferred species for making knives and plough respectively. *Cedrus deodara* and *Pinus roxburghii* are main timber species of the region. Also, needles of these conifers along with cow dung are used for making organic manure. Among fodder species, *Grewia optiva* is the most preferred species followed by *Celtis australis* and *Albizia lebbeck.* Ban oak *(Quercus leucotricophora)* and Moru oak *(Quercus floribunda)* are heavily lopped for fodder purposes in higher regions; besides this, Ban oak tree is also felled for making Charcoal. Hence, the population of this species is decreasing rapidly in the study area.

Traditional ethnobotanical knowledge will likely persist in the Kiran Pargana only if sufficient forests near villages are protected from degradation, and if local people retain their value for and traditional knowledge of plant use. Studies on nutritional qualities of edible plants and phytochemical, pharmacological and clinical studies on medicinal plants should be carried out. Such studies are of special significance for India, where the flora is quite rich and a sizeable rural population still depends on these plants for their day-to-day needs. It is, therefore, imperative that the traditional knowledge regarding the uses of plants from ethnobotanically unexplored areas should be documented systematically before it is lost along with the fast dwindling natural resources of such areas due to urbanization and industrialization.

## ENUMERATION

Plant species have been described in alphabetical order with botanical name, family (in parenthesis), vernacular name, part(s) used and uses.

**Achyranthes aspera** Linn. (Amaranthaceae)

Vern. name: *Uang*

*Uses:* Fresh leaves are crushed and paste is applied on cuts.

**Adhatoda zeylanica** Medik. (Acanthaceae)

Vern. name: *Bansei*

*Uses:* Leaves are used as ripening agent for banana.

**Agave americana** Linn. (Agavaceae)

Vern. name: *Kuwar palla*

*Uses:* Plants are used as live fence around agriculture fields. Foam prepared from the crushed leaves is used for washing clothes. The leaf fibres are used for making ropes.

**Ajuga parviflora** Benth. (Lamiaceae)

Vern. name: *Neelkanth*

*Uses:* Decoction of root powder is taken internally to get relief from fever.

**Albizia chinensis** (Osb.) Merr. (Mimosaceae)

Vern. name: *Siris*

*Uses:* Leaves are used as fodder.

**Albizia lebbeck** (L.) Benth. (Mimosaceae)

Vern. name: *Bhandir*

*Uses:* Leaves are used as fodder during winter.

**Allium sativum** Linn. (Liliaceae)

Vern. name: *Lahsun*

*Uses:* Green leaves and bulbs are used as vegetable and as spice.

**Angelica glauca** Edgew. (Apiaceae)

Vern. name: *Chara / Charei*

*Uses:* Root powder is used as carminative and also to treat stomach problems. The aromatic roots are also used as spice. Smoke from the burnt roots is snake repellent.

**Artemisia indica** Willd. (Asteraceae)

Vern. name: *Chhambar*

*Uses:* It is considered as a sacred plant; leaves are used in '*Pooja*'. The fresh leaf-juice is used to cure cuts and wounds.

**Artemisia vestita** Wall. ex DC. (Asteraceae)

Vern. name: *Chhamri*

*Uses:* Decoction prepared from dried leaf powder is useful in curing stomach problems and also to kill worms in the intestine.

**Asparagus adscendens** Roxb. (Liliaceae)

Vern. name: *Shatabari*

*Uses:* Tubers are given to cows to increase lactation.

**Bauhinia variegata** Linn. (Caesalpiniaceae)

Vern. name: *Koral / Kachnar*

*Uses:* Floral-buds are used to make vegetable. Leaves are good source of fodder.

**Berberis lycium** Royle (Berberidaceae)

Vern. name: *Kashmal*

*Uses:* Mature fruits are edible. A very valuable preparation called "*Rasaut*" is prepared from this plant. For preparing Rasaut, the bark of the root and lower part of the stem is boiled in water, strained and evaporated till a semi-solid mass *(Rasaut)* is obtained. *Rasaut* is fairly soluble in water. It is applied externally to the eyelids to cure eye diseases.

**Bombax ceiba** Linn. (Bombacaceae)

Vern. name: *Shaimal*

*Uses:* The leaves are used as fodder.

**Brassica campestris** Linn. (Cruciferae)

Vern. name: *Sarson*

*Uses:* The leaves are used as vegetable.

**Buddleja asiatica** Lour. (Loganiaceae)

Vern. name: *Chinderae*

*Uses:* Leaves are used as fodder for goats and sheep.

**Cannabis sativa** Linn. (Cannabaceae)

Vern. name: *Bhang*

*Uses:* Seeds are edible. Fibre extracted from stem is used for making ropes. Resinous gum (Ganja) is burnt and the smoke inhaled for intoxication.

**Cedrus deodara** (Roxb. ex D. Don.) G. Don (Pinaceae)

Vern. name: *Devdar, Kailon*

*Uses:* Oil is prepared from the wood and used as an insecticide. Timber is extensively used in house construction. Needles mixed with cow dung are used to prepary organic manure.

**Celtis australis** Linn. (Urticaceae)

Vern. name: *Kharik*

*Uses:* Dried branches are collected for firewood. Leaves are the best source of fodder, especially during summer months.

**Chenopodium album** Linn. (Chenopodiaceae)

Vern. name: *Bathua*

*Uses:* The leaves are used as vegetable.

**Cinnamomum tamala** (Buch.-Ham.) Nees & Eberm. (Lauraceae)

Vern. name: *Meetha gulandra*

*Uses:* The aromatic leaves are used as spice in tea and food preparations. Only 2-3 mature trees exist near Mundlah area.

**Cirsium verutum** (D. Don) Spreng. (Asteraceae)

Vern. name: *Buanda*

*Uses:* The raw roots are edible, mostly eaten by children.

**Coriaria nepalensis** Wall. (Coriariaceae)

Vern. name: *Mosori / Gangara*

*Uses:* Leaves are used as fodder for goats and sheep.

**Cotoneaster bacillaris** Wall. ex Lindl. (Rosaceae)

Vern. name: *Riunsh*

*Uses:* The straight branches are used to make walking sticks and handles of different agricultural implements.

**Cuscuta reflexa** Roxb. (Cuscutaceae)

Vern. name: *Akash bel*

*Uses:* Paste prepared from the whole plant is applied on head to kill lice.

**Cyathula capitata** Moq. (Amaranthaceae)

Vern. name: *Sirla*

*Uses:* Powder prepared from the roots is used to cure boils. Leaves are used as fodder for goats and sheep.

**Cynodon dactylon** (L.) Pers. (Poaceae)

Vern. name: *Joob*

*Uses:* Leaves are used for sprinkling holy water during *'Pooja'*.

**Cynoglossum microglochin** Benth. (Boraginaceae)

Vern. name: *Shanathu*

*Uses:* Leaves are used as vegetable.

**Cyperus rotundus** Linn. (Cyperaceae)

Vern. name: *Motha*

*Uses:* Decoction prepared from the root powder is taken internally to cure stomach ailments.

**Dalbergia sissoo** Roxb. (Fabaceae)

Vern. name: *Sisham*

*Uses:* Leaves are used as fodder during scarcity period. Timber from the tree is the best source for making all types of furniture. The logs are used to make containers for storing hay.

**Debregeasia salicifolia** (D. Don) Rendle (Urticaceae)

Vern. name: *Siyar*

*Uses:* The leaves are used as fodder.

**Eleusine coracana** Gaertn. (Poaceae)

Vern. name: *Koda*

*Uses:* Flour prepared from the seeds is used to make *"Roti"*. Seeds are also used to prepare a local beverage known as *"Soor"*.

**Euphorbia royleana** Boiss. (Euphorbiaceae)

Vern. name: *Suruanth*

*Uses:* The pieces of fresh and succulent stems are given to goats as feed.

**Ficus auriculata** Lour. (Moraceae)

Vern. name: *Tyamal*

*Uses:* Ripened fruits are edible. Leaves are used as fodder during winter months.

**Ficus palmata** Forsk. (Moraceae)

Vern. name: *Fedu, Fegra*

*Uses:* Leaves are used as fodder during scarcity. Local people eat the fruits. Stem-bark is used as plaster for broken limbs of goats and sheep.

**Ficus religiosa** Linn. (Moraceae)

Vern. name: *Pipal*

*Uses:* The tree is worshiped. It is believed that stammering problem in children can be cured if rice cooked with milk is taken on leaf plates of this plant. The leaves are also used as fodder.

**Fragaria vesca** Linn. (Rosaceae)

Vern. name: *Bhumbal*

*Uses:* Fruits are eaten by local people. The decoction of root-powder is used to cure stomach problems.

**Geranium wallichianum** D. Don ex Sweet (Geraniaceae)

Vern. name: *Kidi jadi*

*Uses:* Root-paste is used to kill worms in wounds of cattle.

**Gerbera gossypina** (Royle) Beauv. (Asteraceae)

Vern. name: *Kupar*

*Uses:* Leaf-fibres are used as thread in oil lamps. Shepherds create fire using the leaf-fibre without the use of match-boxes by holding the fibres in hand with small piece of stone and strike it with an iron instrument known as 'Akchnaan' and the fibres catches fire.

**Grewia optiva** Drumm. ex Burret (Tiliaceae)

Vern. name: *Buel*

*Uses:* Plant is the best source of fodder especially during winters. Crushed bark of the tree is used for washing woolen clothes. The fibres are extracted through the process of retting and used for making ropes and small bags to store walnut and seeds of apricot. Dried branches are used as firewood.

**Hypericum oblongifolium** Chiosy (Hypericaceae)

Vern. name: *Banbakari*

*Uses:* Leaves are used as fodder for goats and sheep.

**Indigofera cassioides** Rottl. ex DC. (Fabaceae)

Vern. name: *Kathi*

*Uses:* Floral-buds are used to make vegetable. Leaves are used as fodder for goats and sheep.

**Juglans regia** Linn. (Juglandaceae)

Vern. name: *Akhrot*

*Uses:* Leaves are kept with grains in granaries to avoid insect attack. Bark is used for brushing the teeth. Dye prepared from the bark is used to dye woolen clothes. Nuts are edible.

**Kalanchoe spathulata** DC. (Crassulaceae)

Vern. name: *Auva, Nunu*

*Uses:* The succulent leaves are crushed and mixed with butter and applied on boils.

**Leucas lanata** Benth. (Lamiaceae)

Vern. name: *Fulchhidi*

*Uses:* Roots are tied around stomach to cure pain and decoction of roots is used as febrifuge.

**Machilus odoratissima** Nees (Lauraceae)

Vern. name: *Gulandra*

*Uses:* Leaves are used as fodder for goats and sheep.

**Mallotus philippensis** (Lam.) Muell.-Arg. (Euphorbiaceae)

Vern. name: *Kamal*

*Uses:* Leaves are used as fodder for goats and sheep. Branches are used as firewood.

**Melia azedarach** Linn. (Meliaceae)

Vern. name: *Darek*

*Uses:* Wood is used for making agricultural implements, especially for making the plough.

**Meriandra strobilifera** Benth. (Lamiaceae)

Vern. name: *Luladi*

*Uses:* Dried branches are used as firewood.

**Morchella esculenta** (L.) Arg. (Helvellaceae)

Vern. name: *Chiyaun, Guchhi*

*Uses:* Fruit bodies of this mushroom are used as vegetable. They are collected from forest and sold in the market, people earn about Rs. 1800 - 3000 /kg.

**Morus alba** Linn. (Moraceae)

Vern. name: *Kimu, Shahtoot*

*Uses:* Leaves are one of the best source of fodder. Fruits are edible.

**Morus serrata** Roxb. (Moraceae)

Vern. name: *Kanchu*

*Uses:* Leaves are used as fodder.

**Murraya koenigii** (L.) Spreng. (Rutaceae)

Vern. name: *Karri-patta*

*Uses:* The aromatic leaves are used as spice in "*Curries*".

**Musa paradisiaca** Linn. (Musaceae)

Vern. name: *Kela*

*Uses:* Rhizome sap is useful in stomach problem due to heat stroke.

**Myrisine africana** Linn. (Myrsinaceae)

Vern. name: *Pisaran*

*Uses:* Leaves are spread in goats and sheep shed to kill flea.

**Olea ferruginea** Royle (Oleaceae)

Vern. name: *Kawu*

*Uses:* Wood is the best source of fuel. Branches are lopped for fodder for goats and sheep during winter.

**Origanum vulgare** Linn. (Lamiaceae)

Vern. name: *Jokjadi*

*Uses:* Roots are smelled to cure dizziness and also used to cure fever in children.

**Oxalis corniculata** Linn. (Oxalidaceae)

Vern. name: *Almori*

*Uses:* Leaves or whole plant are used for making "*Chutneys*".

**Perilla frutescens** (Linn.) Britt. (Lamiaceae)

Vern. name: *Bhangjir*

*Uses:* Seeds are edible.

**Pinus roxburghii** Sarg. (Pinaceae)

Vern. name: *Saral, Chir*

*Uses:* Needles are used to make brooms. The tree is the best source of timber in lower region. Dried branches are utilized for firewood. Bark is tied around the broken limbs of goats and sheep. Resin is applied to heal cuts of cattle. Seeds are edible.

**Pinus wallichiana** A. B. Jacks (Pinaceae)

Vern. name: *Kailon, Rau*

*Uses:* The tree is one of the best sources of timber in the region. Needles are spread in cattle shed along with cow dung to prepare organic compost.

**Pistacia chinensis** Bunge ssp. **integerrima** (Stewart) Reich. f. (Anacardiaceae)

Vern. name: *Kakrae, Kakkar singi*

*Uses:* The occurrence of the tree is scattered in the region and it is a good source of fuel wood. The branches are lopped for fodder during winters.

**Plantago erosa** Wall. (Plantaginaceae)

Vern. name: *Isabgol*

*Uses:* Seed are crushed into fine powder and decoction prepared from the powder is taken internally to cure stomach ailments.

**Prinsepia utilis** Royle (Rosaceae)

Vern. name: *Bekhal*

*Uses:* Young twigs are used for sprinkling holy water during '*Pooja*'. Local poor people use seed oil as a substitute for ghee.

**Prunus armeniaca** Linn. (Rosaceae)

Vern. name: *Chulli*

*Uses:* Fruits are edible. Oil extracted from the seeds is edible. It is considered as an antiseptic and applied on cuts and wounds. 2-3 drops of oil are poured into ear to get relief from earache.

**Prunus cornuta** (Wall. ex Royle) Steud. (Rosaceae)

Vern. name: *Paja*

*Uses:* Leaves are offered to local deity during '*Shivratri*'. Leaves are also used as fodder during scarcity.

**Prunus persica** (L.) Batsch (Rosaceae)

Vern. name: *Aru*

*Uses:* Fruits are edible and edible oil is extracted from the seeds.

**Punica granatum** Linn. (Punicaceae)

Vern. name: *Daru*

*Uses:* Fruits are edible.

**Pyrus pashia** Buch.-Ham. ex D. Don (Rosaceae)

Vern. name: *Kainth*

*Uses:* Fruits are eaten by local people.

**Quercus floribunda** Lindl. (Fagaceae)

Vern. name: *Moru*

*Uses:* Agricultural implements are made from the wood. The branches are lopped for fodder during winters. It is also a good source of fuel.

**Quercus leucotricophora** A. Camus (Fagaceae)

Vern. name: *Ban*

*Uses:* Small branches are lopped for fodder during winter. Wood is the best source of fuel. Local population fell trees for making charcoal.

Agricultural implements, such as plough and knives, are made from straight branches.

**Reinwardtia indica** Dumort. (Linaceae)

Vern. name: *Pinguni*

*Uses:* Fresh leaf-juice is used to cure cuts and wounds.

**Rabdosia rugosa** (Wall. ex Benth.) Hara (Lamiaceae)

Vern. name: *Chhichhadi*

*Uses:* The branches are used to make the brooms.

**Rhododendron arboreum** Smith (Ericaceae)

Vern. name: *Buras*

*Uses:* Flowers are edible, mostly preferred by children. Juice is prepared from crushed fresh flower petals, sugar and water is added after filtering. The flowers and leaves are fitted in long ropes made of "*Munja*" grass and tied around the houses including temples as decoration during "Baisakhi" festival.

**Rhus parviflora** Roxb. (Anacardiaceae)

Vern. name: *Ninau*

*Uses:* Leaves are good source of fodder for goats and sheep. Fruits are edible. Flour *('Satu')* prepared from fruits is eaten by poor people.

**Ricinus communis** Linn. (Euphorbiaceae)

Vern. name: *Hardei*

*Uses:* Seeds are crushed and paste is applied on back to cure backache.

**Robinia pseudo-acacia** Linn. (Fabaceae)

Vern. name: *Robinia*

*Uses:* It is an introduced tree to the area and used for fodder.

**Rosa moschata** Mill. (Rosaceae)

Vern. name: *Kujain*

*Uses:* Flowers are offered to local deity. It is believed that flowers should be plucked before sunrise.

**Roylea cinerea** (D. Don) Baill. (Lamiaceae)

Vern. name: *Karuain*

*Uses:* It is used as fodder for goats and sheep. Root powder is used to cure stomach problem.

**Rubus ellipticus** Smith (Rosaceae)

Vern. name: *Hinsar*

*Uses:* Fruits are edible.

**Rubus niveus** Thunb. (Rosaceae)

Vern. name: *Achhein*

*Uses:* Fruits are edible. Leaves are crushed and decoction is used as febrifuge.

**Rumex hastatus** D. Don (Polygonaceae)

Vern. name: *Almara*

*Uses:* Leaves are used for preparing *'Chutney'*.

**Rumex nepalensis** Spreng. (Polygonaceae)

Vern. name: *Garash*

*Uses:* Fresh leaf-juice is used to treat skin problem, especially skin eruption on the hands. .

**Solanum khasianum** C. B. Clarke (Solanaceae)

Vern. name: *Jokaru*

*Uses:* Fruit-juice is poured into the nostrils of the livestock to kill Leeches in the nostrils.

**Solanum nigrum** Linn. (Solanaceae)

Vern. name: *Banchulli*

*Uses:* Fruits are edible and considered good for stomach problems.

**Solanum pseudo-capsicum** Linn. (Solanaceae)

Vern. name: *Pipli, Mirch*

*Uses:* Fruits are used as spices. Leaves are crushed and juice is applied on skin eruptions, especially on hands.

**Thalictrum foliolosum** DC. (Ranunculaceae)

Vern. name: *Pili-jarin*

*Uses:* Root powder is considered good against snake-bites.

**Toona hexandra** (Wall. ex Roxb.) Roem. (Meliaceae)

Vern. name: *Toon*

*Uses:* Leaves are good source of fodder. Dried branches are used as firewood. Agricultural implements are made from straight stems.

**Urtica dioica** Linn. (Urticaceae)

Vern. name: *Karla, Bichu-butti*

*Uses:* The young leaves are used to make vegetable. The leaves are mixed with dried inflorescence of *Amaranthus caudatus (Chaulai)* after extracting seeds and cooked and then given to cows as a feed during winter months.

**Verbascum thapsus** Linn. (Scrophulariaceae)

Vern. name: *Jangli Tambacku*

*Uses:* Leaves are used for smoking.

**Viola canescens** Wall. ex Roxb. (Violaceae)

Vern. name: *Banksha*

*Uses:* Decoction of flowers is used to cure cough.

**Vitex negundo** Linn. (Verbenaceae)

Vern. name: *Shualain*

*Uses:* Leaves are used as fodder. Dried branches are used as fuel wood.

**Vitis lanata** Roxb. (Vitaceae)

Vern. name: *Patand luar*

*Uses:* Leaves are used as fodder. Juice from the stem is used to cure eye problems in livestock.

**Woodfordia fruticosa** (Linn.) Kurz (Lythraceae)

Vern. name: *Dhae*

*Uses:* Leaves are used as fodder.

**Zanthoxylum armatum** DC. (Rutaceae)

Vern. name: *Timur*

*Uses:* Leaves are used as spices especially with meat. Local people use small twigs for brushing the teeth, it helps in relieving the toothache.

**Ziziphus mauritiana** Lamk. (Rhamnaceae)

Vern. name: *Borei, Ber*

*Uses:* Fruits are edible. Roots are grounded into powder and mixed with Oat flour, which is used to make local wine known as 'Soar'.

**Ziziphus oxyphylla** Edgew. (Rhamnaceae)

Vern. name: *Densrae*

*Uses:* Fruits are edible.

## REFERENCES

ANONYMOUS. 1910. *Gazetteer of the Simla Hill State : Pb. State Gazetteer,* Vol. III. Indus Publishing Co., New Delhi.

BADOLA, HEMANT K. & MOHINDER PAL 2003. Threatened medicinal plants and their conservation in Himachal Himalayas. *Indian Forester* 129 (1) : 55-68.

CHAUHAN, N. S. 1984. *Medicinal wealth of Pabbar Valley in H. P.* M. Sc. Thesis, Himachal Pradesh Krishi Vishvavidhalaya, Palampur.

———— & P. K. KHOSLA 1988. Commercially important medicinal plants of Himachal Pradesh, Trends in Tree Sciences, pp 81-89. ISTS Publication.

CHAUHAN, V. & N. S. CHAUHAN 1988. Ethnobotany of Trans-Giri area of Sirmour District of Himachal Pradesh. *Bull. Medico. Ethnobot. Res.* 9 (3-4) : 106-122.

CHOWDHERY, H. J. & B. M. WADHWA 1984. *Flora of Himachal Pradesh,* 3 Vols. Botanical Survey of India, Calcutta.

COLLETT, HENRY 1921. Flora Simlensis : A handbook of the flowering plants of Simla. Calcutta & Simla.

GUPTA, R. 1964. Survey record of medicinal and aromatic plants of Chamba Forest Division, H. P. *Indian Forester* 90 : 454-468.

KALA, C. P. & NIMA MANJREKAR 1999. Ethno-medicobotany of Indian Trans-Himalaya : A case study from Spiti. *Journ. Econ. Taxon. Bot.* 23 (1) : 177-181.

KAPAHI, B. K. 1990. Ethnobotanical investigation in Lahaul (H. P). *Journ. Econ. Taxon. Bot.* 14 (1) : 49-55.

MEENAKSHI 2002. *The study on phyto-diversity of Shilly Wildlife Sanctuary, H. P.* M. Sc. Thesis, Dr. Yashwant Singh Parmar University of Horticulture & Forestry, Nauni, Solan.

SHARMA, B. D. & J. C. RANA 1999. Traditional medicinal uses of plants of Himachal hills. *Journ. Econ. Taxon. Bot.* 23 (1) : 173-176.

SOOD, S. K., RAM NATH & D. C. KALIA 2001. *Ethnobotany of cold desert tribes of Lahaul-Spiti (N. W. Himalaya).* Deep Publications, New Delhi.

UNIYAL, M. R. & N. S. CHAUHAN 1971. Traditionally important medicinal plants of Kangra Valley in Dharmasala Forest Circle, Himachal Pradesh. *J. Res. Ind. Med.* 8 (1) : 76-85.

# TRADITIONAL HERBAL VETERINARY MEDICINES FROM TRIBAL AREA OF RAJASTHAN, INDIA

ANITA JAIN, S.S. KATEWA, B.L. CHAUDHARY & PRAVEEN GALAV

*Laboratory of Ethnobotany and Agrostology, Department of Botany, College of Science, M. L. Sukhadia University, Udaipur-313001 (Raj.) India*

**ABSTRACT**

An ethnobotanical survey of tribal areas of Rajasthan was carried out during 2001-2002 for ethnoveterinary herbal medicines. The information collected on herbal medicines is based on the exhaustive interviews with local medicine-men, shepherds and other .knowledgeable persons who prescribe their own herbal preparations for the treatment of domestic animals. During ethnobotanical survey, 46 ethnoveterinary herbal medicines belonging to 34 families have been recorded from Southern Rajasthan. A list of plant species along with their family name, local name, plant part/s used and the mode of administration for effective control in different ailments of domestic animals is given.

## INTRODUCTION

The rich and diversified flora of India provides a most valuable storehouse of medicinal plants. The curative properties of herbs have long been known and are documented in ancient manuscripts, such as the Sanskrit Rig Veda, Garuda Purana and Agni Purana (Holland, 1994; Priyadarsan, 1991). These treatises focus on the potential of plants to cure human ailments and diseases. But, the botanical wealth of India also offers the people who tend livestock a rich reservoir which can be tapped to treat the diseases and ailments. Indian communities have accumulated over the years the indigenous veterinary knowledge which is also worth recording. Ethnoveterinary medicines are of specific value in developing countries where allopathic veterinary medicines are often beyond the reach of livestock holders. During recent years, the scientists have documented various ethnoveterinary practices based on plant drugs (Anzuino, 1999; De, 1967; Dold & Cocks, 2001; Geerlings, 2001; Sebastine, 1984; Sebastine & Bhandari, 1984; Maheswari *et al.,* 1986; Reddy & Sudarsanam, 1987; Joshi, 1995; Singh & Pandey, 1998; Katewa & Chaudhary, 2000

*Acacia leucophloea* (Roxb.) Willd.

*Ampelocissus latifolia* (Vahl) Planch.

*Grewia orientalis* Linn.

*Cissus quadrangularis* Linn.

**Plate - 1**

*Euphorbia fusiformis* Buch.-Ham. ex D. Don

*Dichrostachys cinerea* (L.) Wt. & Arn.

*Cosmostigma racemosum* (Roxb.) Wt.

*Crinum asiaticum* Linn.

**Plate - 2**

*Pandanus odoratissimus* L. f.

*Dendrophthoe falcata* (L. f.) Etting

*Echinops echinatus* Roxb.

*Curcuma amada* Roxb.

**Plate 3**

etc). The present article is intended to contribute to this growing body of knowledge by supplying information on the plant based ethnoveterinary curative techniques found in tribal area of Southern Rajasthan.

## MATERIALS AND METHODS

The traditional knowledge of plant based remedies for the treatment of animals rests with the medicine-men and elderly men and some women who pass their skill and experiences from one generation to the next by word of mouth and guard like secrets. In view of secretiveness of traditional veterinary medicine-men, it was decided to interview also a number of elderly people, village heads, priests and other community laiders who have a great deal of practical know - how about the plants used as veterinary medicines in the native system. The linguistic fluency, personality and social standing are crucial for establishing rapport between the participants involved. It was experienced that the persons mostly above the age of 60 have more accurate information regarding their old traditions. Generally two types of interviews were taken, firstly of individuals and secondly of groups. To determine the authenticity of information collected during fieldwork, repeated verification of data from different informants and in different times was done. Thus, only the specific and reliable information cross-checked with at least 13 informants has been incorporated in the present study. The collected ethnoveterinary herbal plants were identified up to species level at the Herbarium of Forest Research Institute, Dehradun (India). All the collected specimens were deposited in the Herbarium of Laboratory of Ethnobotany and Agrostology, Department of Botany, College of Science, M.L. Sukhadia University, Udaipur (India) for further reference. Numbers of voucher specimens are provided after species name in the table 1, besides botanical name, family name, local name, part/s used and the mode of their administration.

Table 1. Ethnoveterinary herbal medicines of Southern Rajasthan.

| Botanical name, Voucher specimen no. & family | Local name | Plant part/s used | Medicinal value |
|---|---|---|---|
| *Acacia leucophloea* (Roxb.) Willd. 47 (Caesalpiniaceae) | Rijva | Roots | Extract of root is given orally to animals for early cure of fractured bone. |
| *Achyranthes aspera* Linn. 150 (Amaranthaceae) | Chiruieya | Roots & seeds | (i) Half cup of root extract is given orally to animals in hypogalactia.<br>(ii) Extract of seed powder is given to animals in dysentery. |
| *Aeschynomene aspera* Linn. 270 (Fabaceae) | Dhadum | Pods | Mature pods are crushed and given to cows and buffaloes for ousters induction. |
| *Ailanthus excelsa* Roxb. 156 (Simaroubaceae) | Aadusa | Bark | (i) Extract of bark is given to animals orally twice a day for three days to cure asthma, pneumonia & bronchitis.<br>(ii) Warmed extract is given orally to animals in flatulence. |

| | | | |
|---|---|---|---|
| *Ampelocissus latifolia* (Vahl) Planch. 205 (Vitaceae) | Khata limbu | Roots & bark | Extract of both is given orally to animals for early cure of fractured bone. |
| *Anethum graveolens* Linn. 199 (Apiaceae) | Sua | Seeds | Powder of seed is given orally in vulvo-vaginal-uterine-prolapse. |
| *Annona squamosa* Linn. 195 (Annonaceae) | Sitaphal | Leaves & bark | (i) Extract of leaf is given to animals orally in colic, dysentery and diarrhoea. (ii) Extract of bark is given to animals orally to cure bronchitis. |
| *Arisaema tortuosum* (Wall.) Schott 359 (Araceae) | Suran | Tubers | Decoction of tuber is given to animals orally in flatulence. |
| *Balanites aegyptiaca* (L.) Delile 78 (Simaroubaceae) | Hingora | Bark | Paste of bark is given in snake bite as an antidote. |
| *Bombax ceiba* Linn. 202 (Bombacaceae) | Semal | Flowers | The flowers are kept in an earthen pot filled with water for the whole night. Sugar is mixed in this infusion and candy is prepared. This is given orally to animals to cure urinary problems. |
| *Boswellia serrata* Roxb. ex Colebr. 52 (Burseraceae) | Salar | Bark | Infusion of 2 kg powdered bark is divided in three equal doses and each dose is given orally for three days to animals early in the morning to dissolve iron piece engulfed by the animals accidentally. |
| *Butea monosperma* (Lamk.) Taub. 267 (Fabaceae) | Dhola Khakhra | Bark | Warmed bark extract is given orally to animals suffering from gastroenteritis. |
| *Capparis sepiaria* Linn. 307 (Capparaceae) | Kathar | Roots | Extract of root is given to animals orally in diphtheria. |
| *Cissampelos pereira* Linn. 170 (Menispermaceae) | Kalipar | Leaves | Extract of leaves is given orally as antivenom in snake-bite. |
| *Cissus quadrangularis* Linn. 378 (Vitaceae) | Hadjore | Stem | Extract of internode is given orally to the animals for early cure of fractured bone. |
| *Cissus repanda* Vahl 367 (Vitaceae) | Pani Bel | Roots | Root extract is given to animals orally to prevent abortion. |
| *Clerodendrum phlomidis* Linn. f. 115 (Verbenaceae) | Arni | Leaves | Paste is prepared by 250 gm leaves mixed with 250 gm leaves of *Azadirachta indica.* This paste is applied on the skin for killing ectoparasites. |
| *Cosmostigma racemosum* (Roxb.) Wt. 112 (Asclepiadaceae) | Rahidodi | Pods | Powdered pods mixed with curd are given orally twice a day to sheep and goats suffering from gastroenteritis. |
| *Crinum asiaticum* Linn. EA322 (Amaryllidaceae) | Jahari kanda | Tubers | (i) Extract of tuber is given to animals orally in flatulence. (ii) Poultice of crushed tuber is tied on tumours, warts and carbuncles for early cure. |
| *Curcuma amada* Roxb. 323 (Zingiberaceae) | Amba-haldi | Tubers | Extract of tuber is given to animals orally in flatulence. |
| *Curcuma longa* Linn. 221 (Zingiberaceae) | Haldi | Tubers | Extract of tuber is given to animals orally in food poisoning. |

| | | | |
|---|---|---|---|
| *Datura innoxia* Mill. 104 (Solanaceae) | Kala dhatura | Whole plant | Paste of the whole plant is applied against mastitis. |
| *Dendrophthoe falcata* (L. f.) Etting 327 (Loranthaceae) | Salar Bel | Whole plant | Extract is applied on uterus in vulvo-vaginal-uterine-prolapse. |
| *Dichrostachys cinerea* (Linn.) Wight & Arn. 311 (Mimosaceae) | Amado | Seeds | Seeds are boiled and given to animals in hypogalactia. |
| *Diospyros melanoxylon* Roxb. 240 (Ebenaceae) | Timru | Fruits | Crushed fruits are given to the animals orally in dysentery. |
| *Echinops echinatus* Roxb. 196 (Asteraceae) | Utkali | Roots | Extract is applied on uterus in vulvo-vaginal-uterine-prolapse. |
| *Elytraria acaulis* (L. f.) Lindau 314 (Acanthaceae) | KalaGathia | Whole plant | Whole plant extract is given to animals in dysentery. |
| *Euphorbia fusiformis* Buch.-Ham. ex D. Don 330 (Euphorbiaceae) | Jangli Palak | Tubers | 150 gm tuber is crushed and given orally to the animals in sunstroke. |
| *Gloriosa superba* Linn. 222 (Liliaceae) | Kalihari | Roots | Extract is applied on uterus in vulvo-vaginal-uterine-prolapse. |
| *Grewia orientalis* Linn. 241 (Tiliaceae) | Gengchi | Rots | Extract of root is given orally for early cure of fractured bone of animals. |
| *Heteropogon contortus* (Linn.) Beauv. ex R. & S. 93 (Poaceae) | Suva ghas | Seeds | Seeds are crushed and given orally in dysuria. |
| *Hordeum vulgare* Linn. 103 (Poaceae) | Jau | Seeds | The electury made from powdered seeds is given to animals to facilitate easier removal of placenta after delivery. |
| *Jatropha gossypifolia* Linn. 294 (Euphorbiaceae) | Ratanjot | Stem & roots | (i) Extract of stem is given orally for early cure of fractured bone in domestic animals.<br>(ii) Decoction of root is given orally in uterine diseases. |
| *Madhuca longifolia* (Koen.) MacBride 203 (Sapotaceae) | Mahua | Flowers | The electury prepared from flowers are given to animals to facilitate easier removal of placenta after delivery. |
| *Nyctanthes arbor-tristis* Linn. 207 (Oleaceae) | Tamat | Leaves | 50 ml extract of leaves is mixed with water and given orally by drenching tube twice a day for early cure of colic. |
| *Pandanus odoratissimus* L. f. 300 (Pandanaceae) | Kavada | Roots | 250 gm of root extract is given to animals orally to increase potency. |
| *Saccharum officinarum* Linn. 99 (Poaceae) | Hantha | Stem & apical buds | (i) 7-8 kg of sugarcane is given to animals to facilitate easier removal of placenta after delivery.<br>(ii) Apical bud is given to animals in mouth disorder. |
| *Senna tora* (Linn.) Roxb. 181 (Caesalpiniaceae) | Puwadia | Seeds | Boiled seeds are given to animals in hypogalactia. |
| *Sesamum indicum* Linn. 189 (Pedalaceae) | Til | Seeds | 250 gm seed oil is given orally to domestic animals for early cure of foot and mouth diseases. |

| | | | |
|---|---|---|---|
| *Solanum virginianum* Linn. 340 (Solanaceae) | Bhor ringni | Roots | Electury prepared from roots is given orally to animals in acbolis. |
| *Tinospora cordifolia* (Linn.) Meirs ex Hk. f. & Th. 280 (Menispermaceae) | Geloe | Whole plant | Extract of whole plant is given to animals to increase potency. |
| *Trichosanthes anguina* Linn. 234 (Cucurbitaceae) | Charchenda | Roots & fruits | 100 gm root powder is mixed with 100 gm crushed onion, 250 gm curd and 150 gm salt and given orally thrice a day in gastritis. This mixture is also used as antivenom. |
| *Typha angustifolia* Linn. 191 (Typhaceae) | Era grass | Inflorescence & seeds | Poultice is tied over the horn to stop horn bleeding and also pasted over cuts and wounds for early healing. |
| *Urginea indica* (Roxb.) Kunth 402 (Liliaceae) | Kolikanda | Tubers | Decoction is made with cow urine and this paste is applied locally on eczema, carbuncles, wounds and tumour for early cure. |
| *Vitex trifolia* L. f. 331 (Verbenaceae) | Tinpati | Tubers | The paste of tubers is applied on throat tumour. |
| *Vitis vitiginea* Haines 250 (Vitaceae) | Jangli angur | Whole plant | Extract of aerial part is given in bronchitis of animals. |

## RESULTS AND DISCUSSION

In the present study, the ethnoveterinary medicinal uses of 46 plant species (Table 1) belonging to 34 families have been reported from the study area. Some important ethnoveterinary herbal plants which occur in this region are *Acacia leucophloea, Ampelocissus latifolia, Grewia orientalis, Cissus quadrangularis* (Plate 1), *Euphorbia fusiformis, Dichrostachys cinerea, Cosmostigma racemosum, Crinum asiaticum* (Plate 2), *Pandanus odoratissimus, Echinops echinatus, Sesamum indicum, Dendrophthoe falcata* and *Curcuma amada* (Plate 3) which have remarkable superior ethnoveterinary medicinal properties.

On analyzing the indigenous veterinary remedies based on the 46 herbal plants, it was observed that 16 species are used to treat the ailments related to reproduction and birth and 11 species are used to treat gastrointestinal diseases through oral application. Next come the remedies for various skin diseases like tumours, wounds, sores, swelling, lice and ticks and various diseases of horns which may benefit through local application as a poultice or paste. The other ailments which may get benefit by ethnoveterinary remedies are bone fracture, urinary and kidney stone problems, foot and mouth diseases caused by worms, cough and cold and also as antivenom against snake-bite. These highly interesting findings require further research and the efficacy of indigenous veterinary practices need to be subjected to pharmacological validation.

## ACKNOWLEDGEMENTS

One of the authors (Dr. Anita Jain) is thankful to the CSIR, New Delhi (India) for providing financial assistance and to the authorities of Forest Research Institute, Dehradun (India) for providing herbarium facilities.

## REFERENCES

ANZUINO, GUISEPPE 1999. Investigation and validation of ethnoveterinary medicine in the context of the Community Animal Health Programme in Somaliland. M. Sc. Dissertation, Centre for Tropical Veterinary Medicine, University of Edinburgh, Edinburgh, UK.

DOLD, A. P. & M. L. COCKS 2001. Traditional veterinary medicine in the Alice district of the Eastern Cape Province, South Africa. *South African Journal of Science* 97 : 375-379.

DE, J. N. 1967. Further observations on the Ethnobotany of Purulia district in West Bengal. *Indian Forester* 95 (8) : 551-559.

GEERLINGS, ELLEN 200 I. *Sheep husbandry and ethnoveterinaty knowledge of Raika sheep pastoralists in Rajasthan, India.* M. Sc. Thesis, Environmental Sciences, Wageningen University, Wageningen, Netherlands. [Full text available at www. pastoralpeoples.org].

JOSHI, P. 1995. Ethnobotany of the primitive tribes in Rajasthan. Rupa Books Pvt. Ltd., Jaipur:

HOLLAND, B.K. 1994. Prospecting the drugs in ancient texts. *Nature* 1994 : 369-702.

REDDY, K.J. & G. SUDERSANAM 1987. Plants used as veterinary medicine in Chittoor district of Andhra Pradesh, India. *International Journal of Crude Drug Research* 25 : 145-152.

KATEWA, S.S. & B.L. CHAUDHARY 2000. Ethnoveterinary survey of plants of Rajsamand district, Rajasthan. *Vasundhara* 5 : 95-99.

MAHESHWARI, J.K., B.S. KALAKOTI & L. BRIJ 1986. Ethnomedicines of Bhil tribe of Jhabua district, M. P. *Ancient Science of Life* 5 (4) : 255-261.

PRIYADARSAN, S.S. 1991. Herbal veterinary medicines in an Ancient Sanskrit work – The Garuda Purana. *Ethnobotany* 3 : 83-87.

SEBASTINE, M.K. 1984. Plants used as veterinary medicines, galactogogues and fodder in forest areas of Rajasthan. *Journal Economic & Taxonomic Botany* 5 : 785-788.

———— & M.M. BHANDARI 1984. Some plants used as veterinary medicines by Bhils. *International Journal of Tropical Agriculture* 2 : 307-310.

SINGH, V. & R.P. PANDEY 1998. *Ethnobotany of Rajasthan, India.* Scientific Pubilshers, Jodhpur.

# TRADITIONAL MEDICINAL KNOWLEDGE ON PLANTS FROM VELLIANGIRI HILLS OF COIMBATORE DISTRICT, TAMILNADU

M. MURUGESAN, V. BALASUBRAMANIAM & N. NAGARAJAN

*P. G and Research Department of Botany, Kongunadu Arts and Science College, Coimbatore-29*

**ABSTRACT**

The plant-based traditional knowledge has become a recognized tool in the search for new sources of drugs and neutraceuticals. Present paper deals with the uses of 66 plant species belonging to 58 genera and 36 families made by the Muthuva tribals inhabiting in Velliangiri hills of Coimbatore district, Tamilnadu. The data gathered include several new reports to ethnobotanical literature as well as additional information on different uses of plants already known in literature.

## INTRODUCTION

Tribal knowledge of plants is important not just for the tribal people themselves, but for the wider World. Tribals widely practice herbal medicines to cure various diseases. These medicines are comparatively safe and environment friendly than synthetic ones. Attempts are also made to validate ethnobotanical claims by carrying intensive work on the Phytochemistry, Pharmacognosy, Pharmacology and Toxicology of various herbal drugs. Some splendid examples in drug development relate to the species of *Clerodendrum tomentosum* (Muscle relaxant), *Bacopa monnieri* (Psychoactive) *Asclepias curassavica* (Cardiotonic), *Rauvolfia serpentina* (Mental disorder and insomnia) etc.

Tribal societies and cultures are themselves disappearing and with them their good plant lore. The tribal people are not only familiar with the thousands of plant species in their ecosystems, but also understand the ecological interrelations of various components of their resource base better than most modern foresters, biologists, agronomists and ecologists. Much knowledge accumulated by tribal people regarding ethnopharmacology, the identity, location and mode of use of myriad food stuffs and drug plants, cures for specific ailments, contraceptives, abortifacients, arrow poisons, ichthyotoxic substances etc remains unknown to the scientists. Realizing these facts, the authors took up the present work.

Velliangiri hills, one the higher elevation (1850 m above msl) plateaus of Western Ghats is situated in western part of Coimbatore district, Tamilnadu State. It lies between 76°40' and 77°10' E longitude and 10°55' and 11°10' N latitude. Forests of this hill are of moist deciduous type dominated by *Elaeocarpus munronii, Myristica dactyloides, Fagraea ceilanica, Terminalia bellirica, Ligustrum perottetii* etc. It has a segmented society of indige-nous people like Irulas, Kattumalasars, Paniyans and Muthuvars. Among them, Muthuvars are one of the major tribes having about 40% of population. The language used by them is akin to Tamil. Living close to the nature they have acquired unique knowledge about the use of flora and have developed medicinal pract-ices of their own. Many health hazards like jaundice, fever, stomach pain, snake bite, kidney stone, rye disease and diabetes etc are being treated by using the plant materials.

## METHODOLOGY

During the period of floristic study from 2003-2004, the ethnobotanical details were collected and documented from the Muthuvar tribals. Documentation consists of interviews with traditional healers (medicine men, vaid, elder people) based on the methods given by Jain (1987) and Martin (1995). Ethnobotanical information was noted on data sheets and in the field books. For collection of plant material, local informer accompanied the authors. Medicinal uses of plants were compared with major published literature and several research papers (Anonymous, 1940-76; Jain, 1981, 1996; Chopra *et al.,* 1956; Kritikar & Basu 1935; Nadkarni, 1954; Warrier *et al.,* 1994; Maheshwari, 2000 etc). The uses, which are not mentioned in literature are considered as new or less known and are enumerated in present paper. Very limited work has been done on this subject from this area (Balasubramaniam & Murugesan, 2004).

## ENUMERATION

The plants are arranged alphabetically by botanical name, followed by local name (in parenthesis), family name (in capitals) and uses (parts used, mode of preparation, doses and method of administration).

**Abutilon indicum** (L.) Sweet, (Thuthi), MALVACEAE

*Use :* The leaves and fruits are used as paste to cure skin diseases like ringworm and itches.

**Acacia nilotica** (L.) Willd. subsp. **Indica** (Benth.) Brenan, (Karuvel), MIMOSACEAE

*Use :* Bark, seed and flowers are grind and made into the powder. It is used as tooth powder against toothache. Powder mixed with water is used as an ingredient in the preparation for treating impotency.

**Acalypha fruticosa** Forsskal, (Chinnithazhai), EUPHORBIACEAE

*Use :* 50-100 gm of fresh leaves with one teaspoon of salt are taken orally to get relief from stomach pain.

**Achyranthes aspera** L., (Nayurivi), AMARANTHACEAE

*Use :* Root-paste is administered orally to reduce body temperature during fever.

**Aerva lanata** (L.) Juss. ex Schult., (Poolapoo), AMARANTHACEAE

*Use :* Leaf-decoction with 5 pepper is taken orally as an antidote for scorpion-sting.

**Alangium salvifolium** (L. f.) Wargerin, (Alangi), ALANGIACEAE

*Use :* Seed-oil is applied externally to cure leprosy.

**Albizia amara** (Roxb.) Boivin., (Arappu), MIMOSACEAE

*Use :* Leaf powder is used as hair shampoo.

**Andrographis paniculata** (Burm. f.) Wallich ex Nees, (Syrianangai), ACANTHACEAE

*Use :* One teaspoon of fresh or dried leaf-powder is mixed with water and taken internally in morning and evening as an antidote for snake-bite.

**Artimesia parviflora** Buch.-Ham. ex D. Don, (Masipathini), ASTERACEAE

*Use :* Whole plant is boiled with coconut oil and made in to a paste and given orally to children to cure fever.

**Asparagus racemosus** Willd., (Thanneer Vittan Kizhangu), ASPARAGACEAE

*Use :* Fresh root-juice is taken orally to cure dysentery. Tuber is crushed and made into a paste and administered orally against urinary problems.

**Azadirachta indica** A. Juss., (Vembu), MELIACEAE

*Use :* The bark-decoction is given for fever. Leaves, flowers, fruits, stem-bark and root are taken in equal amount, air dried, powered and administered orally as "Kayakalpam".

**Begonia malabarica** Lam., (Rathachoori), BEGONIACEAE

*Use :* Leaf-juice is administered orally to purify the blood and prevent blood cancer. Stem and leaves soaked in cow milk over night are administered orally in morning as a "Kayakalpam".

**Bridelia retusa** (L.) Spreng., (Mul-vengai), EUPHORBIACEAE

*Use :* Bark of the plant is mixed with the bark of *Anogeissus latifolia* and put to boil in water and that water is used for bath to get relief from body pain.

**Capparis zeylanica** L., (Suduthoratti), CAPPARACEAE

*Use :* Root-paste is taken orally to cure dyspepsia, and it also prevents the incidence of toothache.

**Caralluma adscendens** (Roxb.) R. Br., (Pachai-puli), ASCLEPIADACEAE

*Use :* Young stem is grind with onion and tamarind and made into a paste. It is used to cure digestive problems.

**Cardiospermun halicacabum** L., (Mudakkathan), SAPINDACEAE

*Use :* Leaves and seeds are boiled with coconut oil and poured in to ear to reduce pain.

**Centella asiatica** (L.) Urban, (Vallarai), APIACEAE

*Use :* Boiled leaves are used as vegetable to cure ulcer and digestive problems.

**Cissus quadrangularis** L., (Pirandai), VITACEAE

*Use :* Young stem and leaves are grind with onion, coriander, salt and curry leaves and made in to a paste. This paste is taken orally with food for dyspepsia.

**Cleome monophylla** L., (Ellukku Sakkalathi), CLEOMACEAE

*Use :* Leaves and seeds are boiled with coconut oil and poured in to ear to reduce pain.

**Coccinia grandis** (L.) J. O. Voigt, (Kovai), CUCURBITACEAE

*Use :* Boiled leaves are eaten as vegetable to cure diabetes.

**Cocculus hirsutus** (L.) Diels, (Kattukodi), MENISPERMACEAE

*Use :* Leaf-juice is taken orally to reduce body temperature.

**Corallocarpus epigaeus** (Rottl. & Willd.) Hook. f., (Kizhimookku Kizhangu), CUCURBITACEAE

*Use :* Fresh tuber is grind with turmeric powder and applied externally to cure small pox and chicken pox.

**Curculigo orchioides** Gaertner, (Nilappanai), HYPOXYDACEAE

*Use :* Tubers are crushed and eaten orally to cure snake-bite. It is also used as aphrodisiac tonic.

**Cyclea peltata** (Lam.) Hook. f. & Thoms., (Patchi), MENISPERMACEAE

*Use :* The leaves are crushed and made in to a paste and taken orally for whooping cough.

**Diplocyclos palmatus** (L.) C. Jeffery, (Iyvirali), CUCURBITACEAE

*Use :* Leaf-paste is used as curative for chronic rheumatism.

**Eclipta prostrata** (L.) L., (Karisalan Kanni), ASTERACEAE

*Use :* Dried leaves are grind with salt, charcoal and made in to a powder. This powder is used as antiscorbutic.

**Euphorbia hirta** L., (Amman-pacharisi), EUPHORBIACEAE

*Use :* Latex is applied externally as astringent.

**Euphorbia thymifolia** L., (Chinnamman-pacharisi), EUPHORBIACEAE

*Use :* Leaf-juice is mixed with buttermilk and taken orally to cure jaundice.

**Evolvulus alsinoides** (L.) L., (Vishnu Kranthi), CONVOLVULACEAE

*Use :* Whole plant is boiled and eaten to improve memory power.

**Fagraea ceilanica** Thunb., (Kattu Kathiri), LOGANIACEAE

*Use :* Fruit-paste is applied externally to cure psoriosis.

**Ficus auriculata** Lour., (Atthi), MORACEAE

*Use :* Two or three fruits are eaten in empty stomach to get relief from anaemic disease.

**Ficus racemosa** L., (Vellai-atthi), MORACEAE

*Use :* Root-paste is taken orally for dysentery. Leaf-decoction is used for ulcers.

**Glinus lotoides** L., (Siruseruppadai), AIZOACEAE

*Use :* Leaves of *Glinus lotoides, Centella asiatica, Eclipta prostrata, Sphearanthus indicus* are powdered and stored for a month in the dried fruit shell of *Luffa cylindrica*, and subsequently taken orally as a stimulant.

**Helicteres isora** L., (Valampuri, Idampuri), STERCULIACEAE

*Use :* Fruit is mixed with pepper and garlic and boiled in coconut oil. One or two drops of this oil are poured in to ear to reduce pain.

**Hemidesmus indicus** (L.) R. Br., (Nannari), PERIPLOCACEAE

*Use :* Whole plant is grind with water and taken orally against fever.

**Indigofera caerulea** Roxb., FABACEAE

*Use :* Fresh leaves are mixed with fresh leaves of *Leucas aspera* and grind with 5 or 6 pepper and made in to a paste which is used as a successful remedy in snake-bite. Leaf-juice used to cure blindness.

**Indigofera tinctoria** L., FABACEAE

*Use :* Leaf-powder is stored in the dried fruit shell of *Luffa cylindrica* for a month and taken orally for 45 days as a blood purifier.

**Indoneesiella echioides** (L.) Sreemadh., (Gopurapoodu), ACANTHACEAE

*Use :* Root-powder with water is administered orally to strengthen the muscles.

**Mimosa pudica** L., (Thottal-surungi), MIMOSACEAE

*Use :* Leaves are crushed and boiled with ginger oil and applied externally to cure psoriosis.

**Myristica dactyloides** Gaertner, (Kattu-jathikai), MYRISTICACEAE

*Use :* Fruit is grind with dried ginger and ocimum seeds and made in to a powder which is consumed orally with milk as sex stimulant.

**Naringi crenulata** (Roxb.) Nicolson, (Vellaivilvam), RUTACEAE

*Use :* Bark, leaf and roots are grind and made in to a paste which is administered orally to increase the fertility among women.

**Pavonia procumbens** (Wall. ex W. & A.) Walp., (Pazhampasi), MALVACEAE

*Use :* Whole plant is grind with 2 or 3 onion and buttermilk and made into a paste which is used to cure the troubles in urinary tract of males.

**Phyllanthus amarus** Schum. & Thonn., (Keelanelli), EUPHORBIACEAE

*Use :* Root-paste is mixed with buttermilk and administered orally to cure jaundice.

**Phyllanthus emblica** L., (Nellikai), EUPHORBIACEAE

*Use :* Fruit decoction is mixed with green gram and used as curative for blood pressure.

**Plectranthus amboinicus** (Lour.) Spreng., (Karpooravalli), LAMIACEAE

*Use :* Whole plant is boiled with oil, made in to paste and administered orally to cure acute bronchitis.

**Plumbago zeylanica** L., (Kodiveli), PLUMBAGINACEAE

*Use :* Root-paste is used as an antidote for insect bite. Bark of the root is made in to a powder and mixed with jaggery and used to cure leucorrhoea

**Pterocarpus marsupium** Roxb., (Vengai), FABACEAE

*Use :* Bark is mixed with *Terminalia chebula* fruit, boiled and sieved. The extract so obtained is applied orally to get relief from tooth-ache.

**Rauvolfia serpentina** (L.) Benth. ex Kurz, (Sarpagantha), APOCYNACEAE

*Use :* Decoction of roots and flowers is given for treatment of fever. Root-juice is used as an antidote for snake-bite.

**Sansevieria roxburghiana** Schultes & Schultes f., (Pambukali), DRACAENACEAE

*Use :* Root is grind with turmeric powder and made in to a paste which is applied orally as an antidote for snake venom.

**Scoparia dulcis** L., (Sakkarai-vembu), SCROPHULARIACEAE

*Use :* Whole plant is air-dried, powdered and taken orally to reduce blood pressure (anti-hypertensive).

**Senna auriculata** (L.) Roxb., (Avarai), CAESALPINIACEAE

*Use :* Tender leaves are crushed with 2 or 3 onion and turmeric powder and applied externally to cure chicken pox.

**Senna italica** Mill., (Nila-avarai), CAESALPINIACEAE

*Use :* Leaf and fruit-decoction is useful to check dysentery

**Senna occidentalis** (L.) Link, (Ponnavarai), CAESALPINIACEAE

*Use :* Leaf and fruit are grind and made into a paste which is mixed with milk and administered orally as purgative.

**Solanum nigrum** L., (Manathakkali), SOLANACEAE

*Use :* Decoction prepared by grinding whole plant with *Euphorbia hirta* is used against eye troubles.

**Solanum trilobatum** L., (Thuthuvalai), SOLANACEAE

*Use :* Flowers and roots are grind and made in to a paste which is mixed with the paste prepared by grinding leaf, fruit and tender leaves of *Mukia maderaspatana*. This paste is mixed with water and taken orally to cure cold.

**Sphaeranthus indicus** L., (Kottai-karanthai), ASTERACEAE

*Use :* Whole young plant is air-dried and powdered and given orally for skin diseases.

**Stachytarpheta jamaicensis** (L.) Vahl, (Ezhuthani-poodu), VERBENACEAE

*Use :* Root-paste is applied over forehead to cure headache. Leaves are mixed with sesame oil and boiled in water. The latter is used to bath to recover from antispasmodic problem.

**Strychnos nux-vomica** L., (Yettikai), LOGANIACEAE

*Use :* Fruit, bark and leaves are mixed with the leaves of *Aloe vera* and grind into a paste which is taken orally with milk to cure piles.

**Terminalia bellirica** (Gaertner) Roxb., (Dhanikai), COMBRETACEAE

*Use :* Fruit-powder is mixed with milk and used against diarrhoea. Smoke created by leaves is used for young children to ward of evil sprits.

**Terminalia cuneata** Roth, (Maruthamaram), COMBRETACEAE

*Use :* Bark-powder is used as cardiotonic and cardiac stimulant.

**Tylophora indica** (Burm. f.) Merr. (Nanjukodi), ASCLEPIADACEAE

*Use :* Leaf-paste (Leaf + 2 or 3 onion + 3 or 4 pepper) is mixed with milk and taken orally in empty stomach for asthma.

**Tribulus terrestris** L., (Neringi), ZYGOPHYLLACEAE

*Use :* Decoction of whole plant is used for urine complaints.

**Wattakaka volubilis** (L. f.) Stapf, (Perunkurinjan), ASCLEPIADACEAE

*Use :* Leaves are grind with pepper and made in to a paste which is applied orally to cure dyspepsia.

**Wedelia chinensis** (Osbeck) Merr., (Manjal-karisalankanni), ASTERACEAE

*Use :* Leaves are grind with the roots of *Phyllanthus amarus* and made in to a paste which is taken orally to cure jaundice.

**Withania somnifera** (L.) Dunal, (Amukkiran), SOLANACEAE

*Use :* Rhizome-powder is mixed with seed powder of *Hygrophila schulli* and *Ionidium suffruticosum,* mixed with water and taken orally as aphrodisiac.

**Wrightia tinctoria** (Roxb.) R. Br., (Veppalai), APOCYNACEAE

*Use :* Leaf-paste and latex are applied externally to cure blisters.

## DISCUSSION

Present ethnobotanical explorations conducted in foot hills of Velliangiri hills resulted in the data collection of traditional plant uses of 66 plant species belonging to 58 genera and 36 families. Among them, majority of species are from families

Euphorbiaceae (6), Asteraceae (4), Asclepiadaceae (3), Caesalpiniaceae (3), Fabaceae (3), Mimosaceae (3), Solanaceae (3) and Cucurbitaceae (3).

## ACKNOWLEDGEMENTS

The authors are thankful to the Joint Director, Botanical Survey of India, Southern Circle, Coimbatore-3 for permitting to consult the Herbarium and to Dr. M. Aruchami, Secretary & Dr. K. Kulandaivelu, Principal, Kongunadu Arts and Science College, Coimbatore-29 for facilities and encouragements. The first author is also thankful to the UGC, Hyderabad for providing necessary financial assistance.

## REFERENCES

ANONYMOUS 1940-1976. The Wealth of India : A dictionary of Indian Raw Materials and Industrial Products, Vol. 1-11. C.S.I.R., New Delhi.

BALASUBRAMANIAN, P. & S.N. PRASAD 1994. Ethnobotany & conservation of medicinal plants by Irulas of Nilgiris Biosphere Reserve. *J. Econ. Taxon. Bot.* Addl. Series 12 : 252-259.

BALASUBRAMANIAM, V. & M. MURUGESAN 2004. A note on the commercially exploited medicinal plants of the Velliangiri Hills, Coimbatore district, Tamilnadu. *Ancient Science of life* 23 (3) : 9-13.

CHOPRA, R.N., S.L. NAYAR & I.C. CHOPRA 1956. *Glossary of Indian Medicinal Plants.* C.S.I.R., New Delhi.

JAIN, S.K. 1981. *Glimpses of Indian Ethnobotany.* Oxford & IBH Publishing Co., New Delhi.

———— 1987. *A Manual of Ethnobotany.* Scientific Publishers, Jodhpur.

———— 1996. *Ethnobiology in relation to human Welfare.* Deep Publications, New Delhi.

KIRTIKAR, K.R. & B.D. BASU 1935. *Indian Medicinal Plants.* International Book Distributors, India.

MAHESHWARI, J. K. 2000. *Ethnobotany and Medicinal Plants of Indian Subcontinent.* Scientific Publishers, Jodhpur.

MARTIN, G. J. 1995. *Ethnobotany : A methods manual.* Chapman & Hall, London.

NADKARNI, A. K. 1954. *Indian Materia Medica.* Popular Book Depot Ltd.

VAIDYARATNAM, P.S. 1990. *Warrier's Arya Vaidya Sala, Kottakal, India : Medicinal Plants,* Vols. 1-5. Orient Longman Ltd., Madras.

WARRIER, P.K., V.P.K. NAMBIAR & C. RAMAKUTTY 1994. *Indian medicinal plants,* Vols. 1-5. Orient Longman Ltd., Madras.

# TRADITIONAL PHYTOTHERAPY FOR DIARRHOEA, DYSENTERY AND LIVER DISORDERS IN CHHATISGARH STATE

S.P. JAIN, NAGENDRA PRASAD MISHRA, JANARDAN SINGH & BINITA KUMARI

*Central Institute of Medicinal and Aromatic Plants (CIMAP), P. O. CIMAP, Lucknow-226 016*

## ABSTRACT

The Chhattisgarh State is predominantly inhabited by tribal populations. These tribals are heavily dependent on forest wealth for meeting their daily requirements including medicines. Diarrhoea, dysentery and liver disorders are most prevalent water born diseases amongst the tribal populations and, therefore, mortality rate of children and adolescent are very high in these areas. For treatment of such diseases, the tribals use a number of plant drugs. In the present study, the ethno-medico-botanical uses of 45 plants in the treatment of diarrhoea, dysentery and liver disorders have been collected from various published sources and authors own survey in the tribal areas of Chhattisgarh.

## INTRODUCTION

The State of Chhatisgarh was carved out of Madhya Pradesh and established as Indias 26th State on 1st November 2000. It lies between 17° to 23° 7' N latitude and 83° 38' E longitude and is spread over an area of 135194 sq km. It is bounded in north by Uttar Pradesh, Jharkhand in the north-east, Orissa in the east, Andhra Pradesh in the west and north-west. The elevation of the State varies from 150-1000 m from sea level and consists of a number of hills and undulated plateau with low hills, narrow valleys, broad catchment area of many important rivers such as Mahanadi, Sheonath, Hadeo, Mand, Eeb, Pairi, Jonk, Kelo Udanti, Indrawati, Arpa and Maniyari.

The entire area supports vegetation composed of dry deciduous and semi-ever-green species of very rich forest resources. It is estimated that about 40% of the total area of the State is under forest cover (Verma *et al.*, 1986).

The State of Chhatisgarh is predominantly inhabited by tribals (30%). There are about 42 tribal communities in

Chhatisgarh, principal among them is the Gond tribe. Besides this, a large population of Kanwar, Brinjhwar, Bhaina, Bhatra, Uraon, Munda, Kamar, Halba, Baiga, Sanwra, Korwa, Bharia, Nageshia, Manghwar, Kharia and Dhanwar tribes also inhabit the State (Russel & Lal, 1916; Srivastava, 1984). The tribal populations are more or less dependent on forest wealth for meeting their various requirement, viz. food, cloth, shelter and medicine. For treatment of various diseases, they are solely dependent upon the plants mostly available within the forests.

Diarrhoea, dysentery and liver disorders are most prevalent water born diseases amongst the tribal populations and, therefore, mortality rate of children and adolescent are very high in these areas. For treatment of such diseases, the tribals use a variety of plant drugs available within the forest.

## MATERIAL AND METHODS

The ethno-medico-botanical uses of plants in the specified diseases, viz. diarrhoea, dysentery and liver disorders, were collected based on authors own observations during field survey (1990-94, 1998-2003) and from scrutiny of published literatures (Jain, 1965; Sahu *et al.,* 1983; Maheshwari *et al.,* 1997; Jain *et al.,* 1997, 2003). These data were scrutinized, compiled and presented at a glance in Table 1.

## ENUMERATION

Table 1. Medicinal plants used for diarrhoea, dysentery and liver diseases.

| Botanical name & Family | Local name | Part used | Mode of Adm. | Use in diseases |
|---|---|---|---|---|
| ***Achyranthes aspera*** L. (Amaranthaceae) | Chiurchitta | RT | Powder | Diarrhoea & dysentery |
| ***Albizia lebbeck*** (L.) Benth. (Mimosaceae) | Khari | SB | Fresh decoction | Diarrhoea & dysentery |
| ***Andrographis paniculata*** (Burm. f.) Wall. ex Nees (Acanthaceae) | Bhuyilim | PL | Decoction | Liver disorders |
| ***Arisaema tortuosum*** (Wall). Schott | - | T | Juice | Liver disorders |
| ***Blepharispermum subsessile*** DC. (Asteraceae) | Rasnajadi | RT | Juice | Diarrhoea & dysentery |
| ***Butea monosperma*** (Lamk.) Taub. (Fabaceae) | Palas | SB | Decoction | Dysentery |
| | | SD | Paste | Diarrhoea & dysentery |
| ***Capparis zeylanica*** L. (Capparaceae) | Aodona | RT | Water extract | Diarrhoea & dysentery |
| ***Careya arborea*** Roxb. (Lecythidaceae) | Kumbhi | SB | Powder | Dysentery |
| ***Cassia fistula*** L. (Caesalpiniaceae) | Bhalumasari | SB | Powder | Dysentery |

| | | | | |
|---|---|---|---|---|
| ***Ceriscoides turgida*** (Roxb.) Tirveng. (Rubiaceae) | Kukkayelka | SD | Extract | Diarrhoea |
| ***Cissampelos pareira*** L. (Menispermaceae) along with *Piper longum* L. & *Zingiber officinale* Rosc. | Guanar | RT | Decoction & powder | Liver disorder, diarrhoea & dysentery |
| ***Clerodendrum indicum*** (L.) Ktze. (Verbenaceae) | Bhabargmug | RT | Juice | Diarrhoea & dysentery |
| ***Costus speciosus*** (Koen.) Smith (Costaceae) | Kevukanda | RH | Chewing | Diarrhoea & dysentery |
| ***Croton roxburghii*** Balak. (Euphorbiaceae) | Putri | RT, SB | Paste | Dysentery |
| ***Curcuma angustifolia*** Roxb. (Zingiberaceae) | Teekur kaanda | RH | Chewing | Liver disorder |
| ***Curculigo orchioides*** Gaertn. (Hypoxidaceae) | Talmule | RT | Decoction | Dysentery & diarrhoea |
| ***Desmodium gangeticum*** (L.) DC. (Fabaceae) | Salparni | RT | Powder | Diarrhoea |
| ***Diospyros peregrina*** (Gaertn.) Gurke (Ebenaceae) | Mokhpal | SB<br>F | Powder | Dysentery & diarrhoea |
| ***Eclipta prostrata*** (L.) L. (Asteraceae) | Balomansi | PL | Decoction | Liver disorders |
| ***Emblica officinalis*** Gaertn. (Euphorbiaceae) | Amla | F<br>B | Powder & Juice | Diarrhoea |
| ***Embliea tsjeriam-cottam*** DC. (Myrsinaceae) | Baibarang | SD | Extract & Decoction | Diarrhoea |
| ***Flemingia strobilifera*** (L.) R. Br. ex Ait. (Fabaceae) | Nappe | RT | Powder | Dysentery |
| ***Helicteres isora*** L. (Sterculiaceae) | Alti | F | Powder | Dysentery |
| ***Hemidesmus indicus*** (L.) R. Br. (Periplocaceae) with *Piper nigrum* L. | Kapuri, Anant mul, Dudhiya & Kalimirch | RT | Infusion & Powder | Dysentery |
| ***Hibiscus sabdariffa*** L. (Malvaceae) | Lalthapa | F | Extract | Dysentery & diarrhoea |
| ***Holarrhena pubescens*** (Buch.-Ham.) Wall. ex G. Don (Apocynaceae) | Korya | RT | Extract | Dysentery & diarrhoea |
| ***Lannea coromandelica*** (Houtt.) Merrill (Anacardiaceae) | Ghighan | G | Chewing | Dysentery & diarrhoea |
| ***Mangifera indica*** L. (Anacardiaceae) | Aam | SB | Extract | Dysentery & diarrhoea |
| ***Oroxylum indicum*** (L.) Vent. (Bignoniaceae) | Dundilamu | RT | Extract | Dysentery |
| ***Ougeinia oojeinensis*** (Roxb.) Hochr. (Fabaceae) | Sadan | SB | Decoction | Dysentery |

| | | | | |
|---|---|---|---|---|
| ***Phyllanthus amarus*** Schum. & Thonn. (Euphorbiaceae) | Bhui-amla | PL | Extract | Liver disorders |
| ***Plumbago zeylanica*** L. (Plumbaginaceae) | Chituar | RT, LF | Decoction | Diarrhoea & dysentery |
| ***Psidium guajava*** L. (Myrtaceae) | Amund | LF | Powder | Liver disorders |
| ***Pterospermum acerifolium*** Willd. (Sterculiaceae) | Chimpan | FR | Juice | Liver disorders |
| ***Diospyros melanoxylon*** Roxb. var. ***tupru*** (Buch.-Ham.) Singh with *Mangifera indica* L. | Kala tendu & Aam | SB & LF | Powder | Liver disorders |
| ***Scoparia dulcis*** L. (Scrophulariaceae) | Cranghae | PL | Extract | Dysentery |
| ***Shorea robusta*** Roxb. ex Gaertn. f. (Dipterocarpaceae) | Sal | SB | Powder | Diarrhoea & dysentery |
| ***Solanum virginianum*** L. (Solanaceae) | Bhejri | F | Vegetable | Diarrhoea & dysentery |
| ***Spilanthes calva*** DC. (Asteraceae) | Akarkara | PL | Juice | Diarrhoea & dysentery |
| ***Tephrosia purpurea*** (L.) Pers. (Fabaceae) | Darkurthe | RT | Water extract | Diarrhoea & dysentery |
| ***Uraria lagopodioides*** (L.) Desv. (Fabaceae) | Chalki | RT | Decoction | Diarrhoea & dysentery |

Abbreviations : RT- Roots, S- Stem, SD- Seeds, SB-Stem Bark, T-Tuber, PL- Whole Plant, RH-Rhizome.

## DISCUSSION AND CONCLUSION

In the present study the folklore uses of 45 plants which are used by different tribes of Chhatisgarh State, living in different pockets of Ambikapur, Raipur, Baster, Naringarh, Raigarh and Bilaspur districts etc have been recorded for the treatment of most prevalent diseases, viz. diarrhoea, dysentery and liver disorders.

The perusal of data indicates that the plants used by different tribals for treating different diseases vary in their identity, content and application for the treatment of a particular ailment. The statistical analysis of plants shows that out of 45 plants, nine plants are used to treat liver disorders while rest of the plant are used in the treatment of diarrhoea and dysentery. Conserversely, it can also be inferred that various plants used as medicine may possess similar chemical constituents or bioactive principals. It is, therefore, suggested that critical validation of various folklore claims should be done, since development of new drugs/formulations is a need of today.

## ACKNOWLEDGEMENTS

The authors are grateful to Dr. S.P.S. Khanuja, Director, Central Institute of Medicinal and Aromatic plants, Lucknow for providing necessary facilities for undertaking this work. We are also thankful to the authority of Botanical Survey of India, Calcutta for Herbarium facilities and the authority of forest department, Chhattisgarh for extending help in survey work.

## REFERENCES


ANONYMOUS 1990. *Glimpses of Medico-botany of Baster District (M.P.),* pp. 97-124. Central Council for Research in Ayurveda and Siddha, New Delhi.

DAKWALA, R.N. 1983. Further contribution towards the ethnobotany of M. P. plants used against diarrhoea and dysentery. *Anc. Sci. Life* 2 (3) : 169-170.

JAIN, S.K. 1965. Medicinal plant lore of the tribals of Baster. *Econ. Bot.* 19 (3) : 236-50.

JAIN, S.P. & S.C. SINGH 1997. An Ethno-medico-botanical survey of Ambikapur District, M.P. *Contribution to Indian Ethnobotany,* 3rd Revised Edition, pp. 83-91. Scientific Publishers, Jodhpur.

————, S.C. SINGH, J. SINGH & SUSHIL KUMAR 2003. Ethno-medico-botanical survey of Raipur District, Chhattisgarh State. In Singh, V. & A.P. Jain (Eds.) *Ethnobotany and Medicinal plants of India & Nepal,* pp. 266-271. Scientific Publishers, Jodhpur.

MAHESHWARI, J.K., R.M. PAINULI & R.P. DWIVDI 1997. Notes on Ethnobotany of the Oraon and Korwa Tribes of Madhya Pradesh. *Contribution to Indian Ethnobotany,* 3rd Revised Edition, pp. 67-82. Scientific Publishers, Jodhpur.

RUSSEL, R.V. & HIRA LAL 1916. *The tribes and castes of the central India,* Vols. 1-4. Cosmo Publication, New Delhi.

SAHU, T.R., I. SAHU & R.N. DAKWALA 1983. Further contribution towards the Ethnobotany of Madhya Pradesh 2. Plants used against diarrhoea and dysentery. *Anc. Sci. Life* 2 (3) : 169-70.

SHAH, N.C. & S.C. SINGH 1990. Ethno-medico-botanical studies in M. P. India-an appraisal. *Current Research Med. & Arom. Plants* 12 (4) : 201-217.

SRIVASTAVA, R. 1984. Tribals of Madhya Pradesh and Forest Bill of 1980. *Man in India* 64 (3) : 320-321.

VERMA, D. M., N. P. BALAKRISHNAN & R. D. DIXIT 1993. *Flora of Madhya Pradesh.* Botanical Survey of India, Calcutta.

# UNCONVENTIONAL WILD EDIBLE FRUITS OF ARUNACHAL PRADESH

H.S. MAHAPATRA & HOMEN MEDHI

*Botanical Survey of India, Arunachal Field Station, Itanagar-791 111*

**ABSTRACT**

The State of Arunachal Pradesh harbours a number of unconventional wild edible fruits plants. In the present paper 56 such plants have been reported which have a great potentiality for common human consumption.

## INTRODUCTION

Arunachal Pradesh, situated in the northeast corner of India lies between 26°28' and 19°30' North latitude and 90°30' and 97°30' East longitude, extending over an area of 83,743 sq km and possesses a hilly terrain with elevation ranging from 200 m to 7750 m.

Phytogeographical position, irregular and undulated topography with wide variation of climate and soil and heavy rainfall have accounted to the rich and fascinating floristic diversity of *ca* 4500 species of flowering plants (Chowdhery, 1999) which includes large number of economic plants being used by the local community as a part of their staple food. Many of them solely depend on these wild plant products in their day-to-day life for sustenance.

Till date there have been very few reports on the unconventional edible plants of Arunachal Pradesh. Though, Pal (1984), Haridasan *et al.* (1990), Chowdhery *et al.* (1996) and Baishya *et al.* (2002) have thrown some light on the subject, but there is no comprehensive account available dealing exclusively with the unconventional wild edible plants of Arunachal Pradesh. The present study deals with 56 species of unconventional wild edible fruit-plants with potential nutritive value occurring in their natural habitat in Arunachal Pradesh. It is the sweetish pulp or the fleshy palatable pericarp of the ripe fruits and often seeds that are commonly consumed. Most of the wild fruits dealt here are eaten raw when ripe.

## MATERIAL AND METHODS

During the routine survey of plant resource of Arunachal Pradesh, a number of unconventional wild edible fruits have been collected along with huge informations from the local communities. The identification of the specimens have been confirmed and the information collected on their edible fruits are checked by consulting available literature. The voucher specimens are preserved in herbarium of Arunachal Field Station, Botanical Survey of India (ARUN). In the present study about 56 species of plants are reported from Arunachal Pradesh in alphabetical sequence along with family name followed by local name, habit and phenology. The local names are abbreviated as : Apa. - Apatani, Adi. - Adi, Hin. - Hindi, Beng. - Bengali, Asm. - Assamese, HM. - Hill Miri.

## ENUMERATION

Table 1. List of unconventional wild edible fruits of Arunachal Pradesh.

| Name of species | Family | Local Name | Habit | Phenology |
|---|---|---|---|---|
| *Actinidia callosa* Lindley | Actinidiaceae | - | Stragglers | May.-Nov. |
| *Ampelocissus barbata* (Wallich) Planch. | Vitaceae | - | Woody climbers | April-Sept. |
| *Aspidocarya uvifera* Hook. f. & Thomson | Menispermaceae | - | Woody climbers | Feb.-Aug. |
| *Bauhinia malabarica* Roxb. | Caesalpiniaceae | Tenga-kotra | Trees | April-Dec. |
| *Bauhinia vahlii* Wight & Arn. | Caesalpiniaceae | Nak-kati-lewa (Asm.) | Woody climbers | April-Dec. |
| *Benthamidia capitata* (Wallich) Hara | Cornaceae | Dieng-sohjaphon (Khasi) | Shrubs | April-Sept. |
| *Cissus repens* Lamk. | Vitaceae | Taru-beku (Apa.) | Climbers | July-Dec. |
| *Citrus limon* (Linn.) Burm. f. | Rutaceae | Jipin (Adi) | Shrubs | Aug.-Nov. |
| *Cleistocalyx operculatus* (Roxb.) Merr. & Perry | Myrtaceae | Godha-jamuk (Asm.) | Trees | March-Aug. |
| *Coffea benghalensis* Heyne ex Roem. & Schult. | Rubiaceae | Akhaji, Kothnaphul (Asm.) | Shrubs | Feb.-Dec. |
| *Cornus controversa* Hemsl. ex Prain | Cornaceae | Dieng-phait (Khasi) | Trees | April-July |
| *Elaeocarpus floribundus* Bl. | Elaeocarpaceae | Jalphai (Asm.) | Trees | May-Dec. |
| *E. lanceifolius* Roxb. | Elaeocarpaceae | - | Trees | March-Nov. |
| *Entada phaseoloides* (Linn.) Merr. | Mimosaceae | Ghila-lewa (Asm.) | Woody climbers | July-Nov. |
| *Fissistigma polyanthum* (Hook. f. & Thomson) Merr. | Annonaceae | Karle-phingnu (Mik.) | Woody climbers | March-Sept. |
| *Flacourtia jangomas* (Lour.) Raeusch. | Flacourtiaceae | Ponial (Asm.) | Trees | March-Jan. |

| | | | | |
|---|---|---|---|---|
| *Garcinia cowa* Roxb. ex DC. | Clusiaceae | Kuji-thekera (Asm.) | Trees | March-Aug. |
| *G. lancaeafolia* (G. Don) Roxb. | Clusiaceae | Rupohithekera (Asm.) | Shrubs | Feb.-Oct. |
| *G. xanthochymus* Hook. f. | Clusiaceae | Tepol-tenga (Asm.) | Trees | Feb.-Oct. |
| *Gardania campanulata* Roxb. | Rubiaceae | Bitmara (Asm.) | Shrubs | March-Dec. |
| *Holboellia latifolia* Wallich | Lardizabalaceae | - | Climbing shrubs | March-Aug. |
| *Lepidium virginicum* Linn. | Brassicaceae | Halim (Hin. & Beng.) | Herbs | Oct.-Jan. |
| *Litchi chinensis* Sonner. | Sapindaceae | Naga-lichi, Takra (Asm.); Togong-asing (Abor.) | Trees | Jan.-June |
| *Limacia oblonga* Miers | Menispermaceae | - | Climbing shrubs | Nov.-Jan. |
| *Limonia acidissima* Linn. | Rutaceae | - | Trees | March.-Aug. |
| *Mahonia acanthifolia* G. Don | Berberidaceae | Taming (Apa.) | Shrubs | April-July |
| *Melastoma normale* D. Don | Melastomataceae | Pudi-raju (Adi); Akysanyi (Apa.) | Shrubs | March-Aug. |
| *Murraya paniculata* (Linn.) Jack | Rutaceae | Nyibumtarum (Adi.); Kamini (Asm.) | Shrubs | Feb.-Jan. |
| *Ophiorrhiza fasiculata* D. Don | Rubiaceae | Akhap (Abor.) | Shrubs | Sept.-March |
| *Osbeckia nepalensis* Hooker | Melastomataceae | Pudirasa (Adi.) | Shrubs | May-Dec. |
| *Parkia roxburghii* G. Don | Mimosaceae | Manipuriurohi (Asm.) | Trees | Oct.-April |
| *Podophyllum hexandrum* Royle | Podophyllaceae | - | Scapigerous herbs | Sept.-Jan. |
| *Pongamia pinnata* (Linn.) Pierre | Fabaceae | Karach, Karchaw (Asm.) | Trees | Feb.-Dec. |
| *Prunus rufa* Steud. ex Hook. f. | Rosaceae | Gonde (Apa.) | Trees | April-May |
| *Pyrus pashia* Buch.-Ham. ex D. Don | Rosaceae | Semo (Apa.) | Trees | March-Oct. |
| *Rubus ellipticus* Smith | Rosaceae | Jillying (Apa.) | Scandent shrubs | Dec.-May |
| *R. insignis* Hook. f. | Rosaceae | Tape-tare (Adi.) | Scandent shrubs | Sept.-April |
| *R. niveus* Thunb. | Rosaceae | Nikhee (Apa.); Kiblupum (HM.) | Scandent shrubs | April-Sept. |
| *R. paniculatus* Smith | Rosaceae | Cheche-nimiri (Nishi); Taptara (Adi.) | Scandent shrubs | Nov.-March |
| *R. rosifolius* Smith | Rosaceae | Hitimbulum (Apa.); Kebo lepum (Adi.); Tashin-ta (Abor.) | Shrubs | Sept.-Feb. |
| *Saurauia armata* Kurz | Actinidiaceae | Himpum (Adi.) | Shrubs | June-Oct. |
| *S. punduana* Wallich | Actinidiaceae | - | Shrubs | March-Feb. |
| *S. roxburghii* Wallich | Actinidiaceae | Chipung-changne (Nishi); Hinche (Adi.) | Trees | May-Sept. |

| | | | | |
|---|---|---|---|---|
| *Silene heterophylla* Lour. | Caryophyllaceae | Jajru (Apa.); Jajra (HM.); Kumumeku (Adi.) | Herbs | April |
| *Spermacocoe hispida* Linn. | Rubiaceae | - | Herbs | Throughout the year |
| *Stixix suaveolens* (Roxb.) Pierre | Capparaceae | Titegille (Ad.); Rokputtutum (Apa.); Madhumalati (Asm.) | Woody climbers | Aug.-Oct. |
| *Syzygium malaccense* (L.) Merr. & Perry | Myrtaceae | Pani-jamuk (Asm.) | Shrubs | March-Aug. |
| *Tetrastigma lanceolarium* (Roxb.) Planch. | Vitaceae | Tadi-ripumriubi (Miri) | Woody climbers | Feb.-Sept. |
| *Toddalia asiatica* (L.) Lamk. | Rutaceae | Tiktaksen (Nishi); Tanoai (Apa.) | Scandent shrubs | Nov.-April |
| *Vangueria madagascariensis* Gmel. | Rubiaceae | - | Trees | April-Oct. |
| *Viburnum corylifolia* Hk. f. | Caprifoliaceae | Soh-lang (Kashi) | Shrubs | March-Sept. |
| *Zanthoxylum acanthopodium* DC. | Rutaceae | Yokhung (Apa.) | Straggling shrubs | Dec.-May |
| *Z. armatum* DC. | Rutaceae | Khangi-shing (Monpa) | Shrubs | April-Oct. |
| *Z. rhetsa* (Roxb.) DC. | Rutaceae | Onger-asing (Abor); Bajarmali (Asm.) | Trees | April-Dec. |
| *Ziziphus oenoplia* (L.) Mill. | Rhamnaceae | Ingsu-dampo (Mikir) | Shrubs | Sept.-Feb. |
| *Z. rugosa* Lamk. | Rhamnaceae | Bon-bogori (Asm.); Khusuguchhi (Aka.) | Trees | March-Aug. |

## DISCUSSION AND CONCLUSION

In fact, today's cultivated fruits plants have their wild ancestors, which have been improved by selection and advance breeding techniques. The wild plant resources of Arunachal Pradesh are rich in *Musa* and *Citrus* species and is considered to be their centre of origin. The State has rich crop plant diversity, viz. barley, maize, buck wheat, finger millet, foxtail millet, French beans, soybean, cow pea, black gram, pea, scarlet bean, pumpkin, cucumber, allium, ginger, chayote, tree tomato, pome and stone fruit (Arora, 1991). Arunachal Pradesh falls under Hindustan Centre of origin of cultivated plants (Vavilov, 1926). No doubt, Arunachal Pradesh is the treasure house of wild edible plants in general and unconventional edible fruits plant in particular. Further, it is interesting that these wild sources of food are very rich in Carbohydrates, Starch, Protein, Sugar and Oil. Moreover, these plants provide great scope as income generating resources for the local communities. Hence, the inventorisation and documentation of such unconventional wild edible fruits will help not only to bring them in cultivation for wide consumption but also to develop effective strategies for their conservation.

## ACKNOWLEDGEMENTS

Authors are thankful to the Director, Botanical Survey of India, Kolkata for facilities and also to Dr. A.K. Baishya, Joint Director, Botanical Survey of India, Arunachal Field Station, for valuable suggestions and encouragement.

## REFERENCES

ARORA, R. K. 1991. Native food plants of the tribals in northeastern India. In Jain, S.K. (Ed.) *Contribution to Ethnobotany of India*, pp. 137-152. Scientific Publishers, Jodhpur, India.

BAISHYA, A.K., S. HAQUE, P.J. BORA & N. KALITA 2001. Flora of Arunachal Pradesh - A overview. *Arunachal Forest News* 19 (1 & 2) : 1-25.

————, B. BALODI & G.D. PAL 2002. *Floristic diversity of Arunachal Pradesh: Environmental Planning and Sustainable Development* 16 : 173-206. G.B. Pant Institute of Himalaya Environment & Development, Almora, Uttaranchal.

CHOWDHERY, H. J. 1999. *Arunachal Pradesh.* In Mudgal, V. & P.K. Hajra (Eds.) *Floristic Diversity and Conservation Strategies in India* 2 : 547-614. Botanical Survey of India, Kolkata-1.

————, G.S. GIRI, G.D. PAL, A. PRAMANIK & S.K. DAS 1996. *Materials for the flora of Arunachal Pradesh.* Botanical Survey of India, Kolkata-1.

HAIDASAN, K., L.R. BHUYAN & M. L. DEORI 1990. Wild edible plants of Arunachal Pradesh. *Arunachal Forest News* 8 (1 & 2) : 1-8.

PAL, G. D. 1984. Observations on Ethnobotany of tribals of Subansiri, Arunachal Pradesh. *Bulletin Bot. Surv. India* 26 (1 & 2) : 26-37.

VAVILOV, N. J. 1926. Studies on the origin of cultivated plants. *Bull. Appl. Bot. Pl. Breed.* 26 (2) : 1-248.

# USEFUL PLANTS OF KABI SACRED GROVE, SIKKIM

S.S. DASH & A.S. CHAUHAN*

*Botanic Garden of Indian Republic, Botanical Survey of India, Noida*
**Botanical Survey of India, Sikkim Himalayan Circle, Gangtok*

## ABSTRACT

The paper provides the detail account regarding utilization of the plants embedded in Kabi sacred grove, Sikkim. The medicinal plants used in traditional treatment of different ailments by the local inhabitants of the area are appended in the tabular form.

## INTRODUCTION

Sacred groves are the outcome of man's spiritual belief and secular appreciation of the values attached with plants. They represent a tradition of conservation, management and even sustainable development of natural resources irrespective of belief, faith and their origin. Though the traditions, beliefs, faith and ethos vary from time to time, place to place and community to community, yet the general theme and concept of conservation and management have not changed over centuries. All the sacred groves in the nature are embedded with multitude human impulses, brotherhood and interaction of different communities staying together.

The studies of sacred groves have become important worldwide today as they are not only embedded with rich floristic diversity amidst high endemism but also serve as natural habitats for many rare species. The importance of sacred groves in biodiversity conservation has long been recognized (Kosambi, 1962; Gadgil & Vartak, 1976, 1996; Hajra, 1979; Gadgil *et al.,* 1993; Lal, 1989; Ramakrishnan *et al.,* 1998). Sacred groves have been reported to exist in many parts of the World (Anonymous, 1995) such as Ghana, Japan, Nigeria, Senegal, Indonesia, Syria, Turkey etc. In India, the sacred groves are mostly found in Bihar, Karnataka, Kerala, Madhya Pradesh, Maharashtra, Manipur, Meghalaya, Orissa, Rajasthan and Sikkim. However, no systematic studies have so far been conducted on the sacred groves of Sikkim, except some sketchy references in gazetteers and Dash & Chauhan (2001).

Sikkim, one of the smallest States of India located in the Eastern Himalaya between 27°5" to 28°19 N and 87° to 88° E, is blessed with a number of ethnic and cultural diversities. The indigenous communities include Nepali, Bhutia, Lepcha, Sherpa etc which coexist amicably thereby complementing each other due to their long interdependence of recognizable reciprocal obligations to participate in social, cultural, linguistic, religious and economic activities. A critical study in all the 4 districts of Sikkim revealed about 35 sacred areas which are conserved through either one or the other spiritual ethos. Out of 35 sacred areas, 24 are terrestrial and 11 are aquatic sacred lakes or water bodies. But, due to many fold urbanization, scientific break - through and fast changing scenario of the rural societies, these sacred areas are shrinking rapidly. Today there is a need to give immediate attention to preserve these fragile ecosystems using successful traditional management practices reinforced with proper modern knowledge. The present paper highlights the useful plants of the grove with an intention to bring the uses to limelight for sustainable utilization and necessary conservation of biodiversity and habitats.

## Study Site

Kabi Sacred grove is situated about 22 km away towards North of Gangtok and just beside the Gangtok-Chungthang Border Road highway at an elevation of 1950 m from mean sea level. The grove occupies about 3 sq km of area and is bounded by households and degraded forests from all sides.

## Climate

The area enjoys a temperate monsoonic climate. Depending upon rainfall and temperature, the year can be broadly divided into 3 distinct well marked seasons, viz. rainy season (June to September), prolonged winter season (October - March) and very short summer season (April - May).

The temperature normally ranges from 6.5° in February to 27.2° in June. The maximum average winter temperature remains 13.6° C and the minimum average summer temperature normally remains below 16° C. The temperature data of Kabi for the years 1999 to 2000 is given in table 1.

South-east monsoon is the main source of rainfall. Maximum rainfall occurs from early June to September and July and August are the wettest months. December is the driest month with minimum rainfall. Pre-monsoon showers or thunder showers, often accompanied by hail, occur in the month of April and May. The winter season remains almost dry with a negligible amount of shower except in October and March. Monthly rainfall recorded in the Kabi grove varies from 10.7 mm to 788.8 mm. The details is given in the table 2.

As the grove is located in the temperate zone of Himalaya and amidst high density of vegetation, it remains adequately humid throughout the year. Maximum relative humidity varies between 85-100% in June to September and minimum 50-65% from November to January.

## Spiritual value and history of the Kabi Sacred Grove

The Kabi Sacred grove, locally known as *Kabi-Lungchuk,* a lepcha word means *blood stained stones* symbolizing the friendship between Lepchas and Bhutias. Etymologically it says *Ka=Hand; Wi= Blood; Lung=Stone; Chuk=Erection.*

The story goes back to the 12th century AD when the childless king of Sikkimese Bhutia tribe *Jo Khye Bhumsa* ruled over Tibet area. Even after several efforts, he failed to have a child and came to Sikkim for treatment. Sikkim was then under the rule of *Thekung Tek Nikung Nel* - an ancestral Lepcha King. After receiving mantras from the Lepcha King of Sikkim he returned back to Tibet giving words that as he will become a father he will handover his child to the Sikkim ruler. When he was blessed with a baby boy, he kept his promise and erected nine stones at this sacred spot and fenced the stones with animal intestine. Mount Kanchendzonga was the symbol of witness to this ritual. He invoked the entire guardian deities of Sikkim by cutting his finger and dropping blood from finger tips on the stones and took a solemn oath of brotherhood with the existence of Mount Kanchendzonga and river flowing through.

Every year on the 15th July, the Sikkimese Lepcha and Bhutia people gather in this sacred grove seeking blessing for their generations to have happiness and prosperity thereby remembering the noble and holy promise made by their fore-fathers on this spot.

## METHODOLOGY

The detail study of Kabi Sacred Grove was carried out from April 1998 to May 2000. The boundary of the area was defined with the help of State revenue map and existing wired fencing. Several exploration tours were undertaken in every season and the detailed floristic survey was done. About 750 plant specimens were collected and brought to herbarium of Botanical Survey of India, Gangtok and identified with existing available literature. All the specimens processed, labeled and deposited in the Herbarium, Botanical Survey of India, Sikkim Himalayan Circle, Gangtok (BSHC).

During exploration work, traditional botanical knowledge about useful plants found inside the grove was also collected. The details regarding the method of preparation of various doses, method of administration of the doses were discussed and documented. Later, the information was cross checked with *local headman, Jakris and Lamas.* The filtered information is presented in this paper.

## RESULTS

The biodiversity of Kabi sacred grove gives a service to the community by providing a spiritual refugee and source for germplasm conservation as well as some economic utilization of plants. During the study, it was observed that in spite of two cordial restriction, i.e. not to cut tree and not to destroy the diversity, beneficiaries do have collect wood, medicinal parts of plants, edible fruits and beautifying orchids from the grove. A good number of plants growing in the Grove have economic potential which may render services and products to the local people as well as commercially.

### Sacred plants

Eight plants or parts of plants are worshiped or used during different rituals by the people; these are *Drepanostachyum intermedium* (Munro) Keng., *Arundo donax* L., *Elaeocarpus lanceifolius* Roxb., *Elaeocarpus sikkimensis* Mast., *Exbucklandia populnea* (R. Br. ex Griff.) R. Br., *Erythrina arborescens* Roxb., *Erianthus rufipilus* (Steud.) Griseb., and *Thunbergia lutea* T. Ander. The sacred plants by and large are protected by the people and this mode of preservation over a large period of time reflects the social and ecological stability of the region and helps to conserve the germplasm.

## Food

Wild fruits and leaves not only support the human diet but also to animal husbandry system as fodder. The study revealed 11 species of plants with edible fruits and 12 species of plants with edible leaves.

## Fuel wood

Local inhabitants around the grove mainly depend on wood biomass for fuel which is usually collected from the outskirt areas. However, occasionally fuel wood is also collected from inside the grove. Eight plants mainly used as fuel wood available in the sacred grove are *Clerodendrum colebrookeanum* Wall., *Clerodendrum viscosum* Vent, *Eurya acuminata* DC., *Myrsine semiserrata* Wall., *Euodia trichotoma* (Lour.) Planch., *Vaccinium vacciniaceum* (Roxb.) Sleumer, *Viburnum erubescens* Wall. ex DC., *Rhododendron arboretum* Sm. etc.

## Timber valued trees

Sixteen types of timber plants are present in the Kabi Sacred Grove. Although these plants are not allowed to cut, but a long range preservation of these tree species may helps to meet demand of timber in future through germplasm conservation.

## Medicinal plants

The recurring theme of Ayurveda is that the body and mind should remain healthy and there must be equilibrium between the two. In this age of synthetic drug it is estimated that about 64% of the total global population dependents on traditional medicine for their health care system (Farnsworth, 1994), while in India 75 - 80% of the population depends upon the traditional folklore medicines (Jain, 1994). The local inhabitants around the Kabi sacred grove also mostly depend upon the herbal resources and collect medicinal plants which are used by them very often to treat against various ailments they suffer from. Table 5 shows the details of the medicinal plants, the parts used against the diseases and the methods of preparation and mode of administrations.

Table 1. Monthly average temperature variation at Kabi.

| Months | April 1999 | May 1999 | June 1999 | July 1999 | August 1999 | Sept. 1999 | Oct. 1999 | Nov. 1999 | Dec. 1999 | Jan. 2000 | Feb. 2000 | Mar. 2000 |
|---|---|---|---|---|---|---|---|---|---|---|---|---|
| Max. | 24.4 | 24.2 | 27.3 | 24.1 | 24.5 | 24.3 | 18.2 | 14.9 | 13.6 | 11.4 | 10.9 | 13.6 |
| Min. | 16.6 | 15.7 | 20.6 | 20.5 | 20.1 | 18.7 | 17.2 | 11.7 | 10.2 | 6.8 | 6.5 | 8.8 |

Table 2. Monthly rainfall variation in the Kabi.

| Months | April 1999 | May 1999 | June 1999 | July 1999 | August 1999 | Sept. 1999 | Oct. 1999 | Nov. 1999 | Dec. 1999 | Jan. 2000 | Feb. 2000 | Mar. 2000 |
|---|---|---|---|---|---|---|---|---|---|---|---|---|
| Rainfall | 747.3 | 579 | 788 | 772 | 724 | 459.7 | 278.6 | 15.8 | 10.7 | 45 | 38.2 | 225.6 |
| Total rainy day | 18 | 23 | 23 | 31 | 29 | 22 | 13 | 5 | 6 | 9 | 6 | 21 |

Table 3. Plants with edible fruits and leaves.

| Plants with edible fruits | Plants with edible leaves |
|---|---|
| *Agapetes serpens* (Wt.) Fleumer | *Aconogonum molle* (D. Don) Hara |
| *Castanopsis indica* (Roxb.) Miq. | *Cardamine circaeoides* Hook. f. & Thoms. |
| *Choerospondias axillaris* (Roxb.) Burtt & Hill | *Houttuynia cordata* Thunb. |
| *Euodia trichotoma* (Lour.) Planch. | *Hovenia dulcis* Thunb. var. *acerba* (Lind.) Sengupta & Safui |
| *Fragaria nubicola* (Lind. ex Hook. f.) Lacaita | *Laportia terminalis* Wt. |
| *Holboellia latifolia* Wall. | *Lysimachia debilis* Wall. |
| *Persea fructibera* Kost. | *Maoutia puya* (Hook.) Wedd. |
| *Prunus cerasoides* D. Don | *Nasturtium officinale* R. Br. ex Aiton |
| *Rubus ellipticus* Smith | *Peperomia heyneana* Miq. |
| *Tetrastigma serrulatum* (Roxb.) Planch. | *Persicaria chinensis* (L.) H. Gross |
| *Trichosanthes lepiniana* (Naud.) Cogn. | *Polystichum* spp. *Urtica diocia* L. |

Table 4. Timber valued plants in Kabi sacred grove.

| Species | Local name | Cost (Rs/c³)* | Used for |
|---|---|---|---|
| *Albizia odoratissima* (L. f.) Benth. | Kakur Siris | 200/- | House construction |
| *Albizia procera* (Roxb.) Benth. | Seto siris | 200/- | Cabinet, planks |
| *Alnus nepalensis* D. Don | Utis | 180/- | House construction |
| *Aphanamixis polystachya* (Wall.) Parker | Lahasune | 250/- | Beams, poles |
| *Castanopsis indica* (Roxb.) Miq. | Dhalne katus | 260/- | Beams, poles |
| *Castanopsis lanceaefolia* (Kurz) Hickel & A. Camus | Patte katus | 260/- | Beams, furniture |
| *Castanopsis tribuloides* (J. E. Sm.) A. DC. | Murse katus | 260/- | Beams, furniture |
| *Engelhardia spicata* Leschen. ex Blume | Mouwa | 300/- | Beams, furniture |
| *Exbucklandia populnea* (R. Br. ex Griff.) R. Br. | Pipli | 200/- | Rarely used for sacred purpose |
| *Lindera pulcherrima* (Nees) Benth. ex Hk. f. | Sissi | -- | Poles |
| *Michelia cathcartii* Hook. f. & Thoms. | Tite champ | 550/- | Furniture |
| *Michelia doltsopa* Buch.-Ham. ex DC. | Ranichamp | 450/- | Furniture |
| *Michelia velutina* DC. | Gugay chanp | 450/- | Furniture |
| *Persea fructibera* Kost. | Lapshi | -- | Poles |

| | | | |
|---|---|---|---|
| *Phoebe attenuata* (Nees) Nees | Lapche Kawla | 250/- | Planks, Cabinets |
| *Schima wallichii* (DC.) Korthals | Aule chilaune | 300/- | Planks, house beams, Poles |

Table 5. Medicinal plant of Kabi sacred grove and their usefulness.

| Name of plants & Family | Parts used | Mode of use & remedy for |
|---|---|---|
| *Acorus calamus* L. (ARACEAE) | Root, rhizome | i) Decoction of the root is taken orally to cure common fever, common cough and cold.<br>ii) Rhizomes paste is made into small tablets of 5 mg each. One tablet twice daily taken orally to cure mental depression, headache and epilepsy. |
| *Ainsliaea latifolia* (D. Don) Schultz-Bip. (ASTERACEAE) | Root | The roots are pounded to a paste and the paste after dilution with warm water is taken orally for stomachache caused due to indigestion. |
| *Ajuga macrosperma* Wall. ex Benth. (LAMIACEAE) | Root | i) Decoction of the root is taken orally to cure urinary disorders and burning sensation during urination.<br>ii) Decoction of the root is also taken orally to cure hangover caused due to over consumption of liquor. |
| *Anaphalis contorta* (D. Don) Hook. f. (ASTERACEAE) | Flower | Flowers paste is applied on bleeding wounds as antiseptic. |
| *Aphanamixis polystachya* (Wall.) Parker (MELIACEAE) | Seed | Seed paste is given orally to children to cure stomachache and expelling intestinal worms. |
| *Artemisia nilagirica* (Clarke) Pamp. (ASTERACEAE) | Leaves, flowering shoot | i) Leaf paste is given orally to cure intestinal worm.<br>ii) Two to three teaspoon of thick decoction of leaf and flowers given orally to cure fever. |
| *Begonia cathcartii* Hk. f. & Thoms. (BEGONIACEAE) | Whole plant | Two teaspoon of the plant paste mixed with warm milk is given as a antipyretic drug (febriguge). |
| *Boenninghausenia albiflora* (Hook.) Reichenb. ex Meissn. (RUTACEAE) | Leaves | Leaf paste is applied in and around the inflammatory part to get a quick relief from pain and swelling. |
| *Brugmansia suaveolens* (Humb. & Bonp. ex Willd.) Berch & Presl (SOLANACEAE) | Leaves | Leaf paste is tied in a clean cloth and squeezed in nasal cavity to clear congestion and sinusitis. |
| *Centella asiatica* (L.) Urban (APIACEAE) | Flowers | Flower paste is applied around wounds of cattle to speedy recovery (antibiotic). |
| *Cinnamomum tamala* (Buch.-Ham.) Nees & Eberm. (LAURACEAE) | Bark | Bark paste is administered orally thrice a day to cure diarrhoea. |
| *Clerodendum viscosum* Vent (VERBENACEAE) | Root | Root paste is applied on the affected area to get relief from pain and swelling (inflammation). |
| *Cuscuta reflexa* Roxb. (CUSCUTACEAE) | Stem | Stem paste is applied around the swollen parts to get relief from pain (rheumatism). |
| *Dichroa febrifuga* Lour. (HYDRANGIACEAE) | Root | Root paste is administered twice daily for fever (febrifuge). |

| | | |
|---|---|---|
| *Dichrocephala integrifolia* (L. f.) Kuntze (ASTERACEAE) | Leaves | Leaf paste is tied around fresh wound to stop bleeding (astringent). |
| *Didymocarpus pedicellatus* R. Br. (GESNERIACEAE) | Leaves | Leaf paste is applied on cuts and wounds as antiseptic. |
| *Drymaria villosa* Cham. & Schlecht. (CARYOPHYLLACEAE) | Whole plant | i) Ten grams of plant paste is taken orally daily for cough and cold.<br>ii) Plant paste is tied in a cloth, one or two drops of juice is squeezed in nasal cavity to cure sinusitis. |
| *Erythrina arborescens* Roxb. (FABACEAE) | Seed | Seed are grounded to a paste and applied around affected parts to alleviate swelling and pain (rheumatism). |
| *Euodia trichotoma* (Lour.) Planch. (RUTACEAE) | Fruit, bark | 10 gms of fruit and bark paste is taken orally to cure dysentery. |
| *Eurya acuminata* DC. (THEACEAE) | Seed, leaf paste | i) Fruit paste is applied around the affected parts to get relief from pain in rheumatism.<br>ii) Leaf paste is applied around the affected area and taken orally as antidote for insect biting. This application reduces the burning sensation of the affected area. |
| *Hedera nepalensis* K. Koch. (ARALIACEAE) | Stem, leaves & fruits | Stem bark is crushed into a paste and about 5 gms of paste is taken orally with warm water for stomachache. ii) Equal amount of leaves and fruit paste is mixed and taken orally as a sexual stimulant for women. |
| *Hedyotis scandens* Roxb. ex D. Don (RUBIACEAE) | Root | About 5 gms of root bark is crushed into a paste and applied around the affected area in inflammation and sprain. |
| *Houttuynia cordata* Thunb. (SAURURACEAE) | Leaves | i) Leaf paste taken orally to cure chronic constipation (purgative).<br>ii) It also cures stomachache and gastritis. |
| *Lyonia villosa* (Hook. f. ex Cl.) Hand.-Mazz. (VACCINIACEAE) | Leaves | Leaf paste prepared at night and applied in the morning to cure skin diseases, particularly itching sensation the epidermal areas. |
| *Maesa chisia* Buch.-Ham. ex D. Don (MYRSINACEAE) | Leaves, bark | Equal amount of leaf and bark paste are mixed and applied all over the body of domestic animals for relief from insects, worms and leeches. |
| *Nasturtium officinale* R. Br. ex Aiton (BRASSICACEAE) | Whole plant | Plant paste is taken orally to cure cold and cough. |
| *Ophiorrhiza treutleri* Hk. f. (RUBIACEAE) | Leaves | Leaf paste is taken orally to cure burning sensation in urine (diuretic). |
| *Osbeckia nepalensis* Hook. (MELASTOMATACEAE) | Flowers | Flowers paste is taken orally and applied on wounds for rapid recovery (antibiotic). |

| | | |
|---|---|---|
| *Persicaria hydropiper* (L.) Spach (POLYGONACEAE) | Leaves | Leaf paste is taken orally to cure irregular menses (emmenagogue). |
| *Plantago erosa* Wall. (PLANTGINACEAE) | Leaves | Leaf paste is applied around the injuries (antiseptic). |
| *Polygala arillata* Buch.-Ham. ex Don (POLYGALACEAE) | Root | i) Root paste is taken orally to cure chronic dysentery.<br>ii) Root bark and stem bark paste in equal amount cure fever when taken orally (febrifuge). |
| *Ranunculus diffusus* DC. (RANUNCULACEAE) | Whole plant | Leaf paste is applied on and around the injuries (antiseptic). |
| *Rhododendron arboreum* Sm. (ERICACEAE) | Flower bud | Paste of the flower bud is taken orally to cure blood dysentery. |
| *Rhus succedanea* L. (ANACARDIACEAE) | Branch gall, leaves & bark | i) The paste of branch galls is made into small tablets of 5 mg each. Two tablets taken orally thrice a day to cure cough (expectorant).<br>ii. Equal amount (10 gm each) of leaf and bark is made into a paste & taken orally to cure blood dysentery. |
| *Rubus ellipticus* Smith (ROSACEAE) | Fruit | Root paste is given orally against jaundice and liver ailments. |
| *Rumex nepalensis* Spreng. (POLYGONACEAE) | Leaves | i) Leaf paste is given orally to cure stomach upset.<br>ii) Leaf decoction is used as antiseptic. |
| *Sanicula elata* Buch.-Ham. ex D. Don (APIACEAE) | Whole plant, flowers | i) About 5 gms of plant paste is taken orally to cure blood dysentery.<br>ii) Flowers paste is used as antiseptic. |
| *Solanum viarum* Dunal (SOLANACEAE) | Fruit, bark | i) Equal amount fruit and bark is mixed and crushed into a paste. One teaspoon of the paste is taken orally twice daily to cure viral cold.<br>ii) The paste is also inhaled to cure nasal congestion. |
| *Stellaria patens* D. Don (CARYOPHYLLACEAE) | Whole plant | Plant paste is applied around the anus to get relief from pain in piles. |
| *Thunbergia grandiflora* (Roxb. ex Rottl.) Roxb. (THUNBERGIACEAE) | Leaves | Leaf decoction is taken orally to cure indigestion. |
| *Thunbergia lutea* T. Ander. (THUNBERGIACEE) | Seeds | Seed paste is taken orally for one week for abortion (abortifacient). |
| *Thysanolaena maxima* Kuntze (POACEAE) | Root | Root paste is applied around boils and carbuncles for easy burst. |

## Miscellaneous uses

Apart from this, people do collect plant specimens from the grove which fetches direct monetary benefits to them. These plants are different kinds of mushroom, *Thysanolaena maxima* Kuntze, *Cinnamomum tamala* (Buch.-Ham.) Nees & Eberm. (Bark), *Rubia cordifolia* L. (Dye), *Engelhardia spicata* Leschen. ex Blume (Bark for tanning), *Smilax lanceaefolia* Roxb. (Drum Stick) etc.

## ACKNOWLEDGMENT

The authors are thankful to Director, Botanical Survey of India, Kolkota for encouragement and facilities.

## REFERENCES

ANONYMOUS 1995. *Sacred places and Vernacular conservation,* pp. 20-21. MAB-23, UNESCO, Man and Biosphere Programme, Paris.

DASH, S.S. & A.S. CHAUHAN 2002. Kabi sacred Grove in Sikkim, relevance to conservation. In Das, A.P. (Ed.) *Prospective of Biodiversity,* pp. 713-731. Bishen Singh Mahendra Pal Singh, Dehra Dun.

FARNSWORTH, N.R. 1994. Ethno-pharmacology and drug development. In Chadwick, D.J. & U. March (Eds.) *Ethnobotany and search for New Drugs,* pp. 42-51. Ciba Foundation Symposium 183, Wiley, Chichester.

JAIN, S.K. 1994. Ethno-pharmacology and drug development. In Chadwick, D.J. & U. March (Eds.) *Ethnobotany and search for New Drugs,* pp. 153-164. Ciba Foundation Symposium 183, Wiley, Chichester.

GADGIL, M. & F. BERKES 1991. Traditional Resources Management systems. *Resource Management & Optimization* 1 (3-4) : 127-141.

————, ———— & C. FOLKE 1993. Indigenous knowledge for biodiversity conservation. *Ambio* 22 (2-3) : 151-156.

———— & V.D. VARTAK 1976. Sacred groves of Western Ghats of India. *Economic Botany* 30 : 152-160.

———— & ———— 1996. Sacred groves in India : A plea of continuous conservation. *Journ. Bombay Nat. Hist. Soc.* 72 : 314-320.

HAJRA, P.K. 1979. *Law Londoh or Sacred grove at Mawphlong.* Visitors guide, Govt. Press, Shillong.

KOSAMBI, D.D. 1962. *Myth and Reality.* Popular Press, Bombay.

LAL, J.B. 1989. *Indian forest Myth and Reality.* Nataraj Publishers, Dehra Dun.

RAMAKRISHNAN, P.S., K.G. SAXENA & U.M. CHANDRASEKHAR (eds.) 1998. Conservation - the sacred for Biodiversity management. Oxford & IBH Publishing Co. Pvt. Ltd., New Delhi.

# UTILIZATION OF MACROPHYTES IN NORTH BIHAR, INDIA

K.B. NANDAN & C.B. SINGH

*Department of Botany, Sundarvati Mahila Mahavidyalaya, Bhagalpur-812 001, Bihar (INDIA)*

**ABSTRACT**

Over sixty macrophytes growing profusely in the wetlands of North Bihar (India) are utilized for food, fodder, fish food, poultry/duck feed, medicine, basketry material, eating plates, worship, handicraft etc. The exploitation of wild macrophytes frames economic backbone of the region and that the aquatic cash crops *(Euryale ferox, Trapa natans* var. *bispinosa)* further boost it up. Their sale for medicine (red flowers of *Nymphaea nouchali)*; fish food (*Ceratophyllum demersum, Hydrilla verticillata)* and handicraft/basketry materials/products consolidate economy of poor native folk whose upliftment calls for encouragement of basketry, handicraft, integrated aquaculture etc.

## INTRODUCTION

North Bihar (India) abounds in varied wetlands as ponds, pools, ditches, chaurs, marshes, lakes etc (Munshi & Munshi, 1985; Munshi, 1988; Ahmad *et al.,* 1990; Dehadrai, 1994; Verma, 1994 a & b; Verma *et al.,* 1996). These water bodies harbour diverse macrophytes as their integral components. The macrophytes which grow profusely therein, as floating, submerged or emerged, become easily available to the native folk for utilization. A perusal of literature related to their economic aspects (Jha *et al.,* 1991 a & b, 1998; Nandan & Singh, 2003) revealed that the works so far done are either fragmentary or confined mostly to their uses as food for human, cattle and fishes. Therefore, a survey has been conducted in North Bihar with an aim to have a comprehensive picture on their utilization, and the present communication is to precisely record its outcome.

## STUDY AREA

The area of present investigation is North Bihar (India) extending over about 56,960 km$^2$, with its 14.45 per cent land under waterlogging. It is located in approximately between the geographical limits of longitudes 83°50'0" E to 88°17'40"

E and latitudes 25°13'46" N to 27°31'15" N (Fig. 1), and at an average altitude round about 61 m above msl. It is subjected to recurrent waterlogging on both permanent and temporary scales due to its bowl-shaped physiography, location near the Himalayan foothills and moderate annual rainfall. The overall consequence witnesses as the regular occurrence of innumerable natural wetlands mainly in the form of lakes, chaurs, marshes etc scattered throughout its surface. The lakes are perennial whereas the chaurs and marshes are seasonal in nature. The chaurs hold shallow water for about eight months in a year and usually get dried during summer (Rai & Munshi, 1982; Munshi & Munshi, 1985), but the marshes contain water saturated sediments with no or little standing water along the vegetation (Munshi & Munshi, 1995).

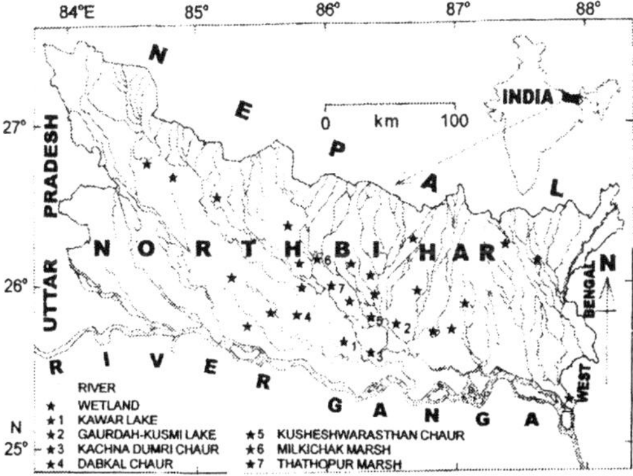


Map 1. Location map of North Bihar, India depicting important wetlends.

The climate of the area is monsoonic. It is characterised by hot-dry summer (March to mid-June), warm-moist rainy (mid-June to October) and cold-sub-humid winter (November to February). The rainfall is uneven with its annual mean value in between 81.8 cm to 143.6 cm (Table 1) and monthly peak value in July/August (Fig. 1). The temperature ranges from 15.1°C (January) to 31.7°C (May/June) around the central place of the region (i.e. around Samastipur). The relative humidity varies between 56.1-80.7 per cent, with August/ September as the most humid months of the year. The pattern of monthly rainfall, temperature and relative humidity for the rest research sites follow the same trend with slight deviations except the magnitude of rainfall.

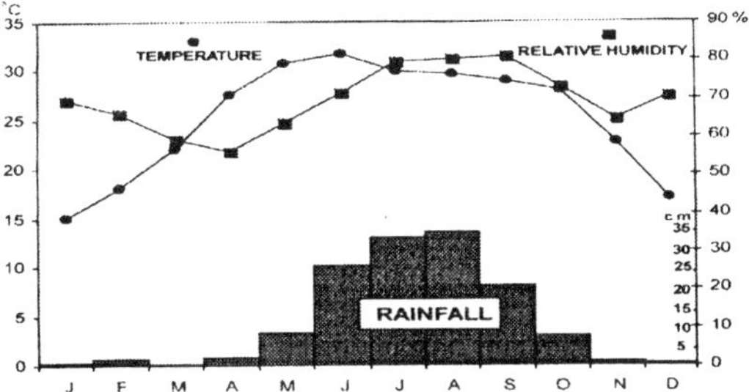


Fig. 1. Monthly rainfall, temperature and relative humidity around Samastipur, North Bihar, India (based on average for the period 1996-2002; data recorded from Meteorological Station, Rajendra Agricultural College, Pusa, Samastipur).

## MATERIALS AND METHODS

A series of explorations were conducted over a period of seven years. The macrophytes were collected frequently in every season from two lakes (Kawar lake, Begusarai and Gaurdah-Kusmi lake, Saharsa), three chaurs (Kachna Dumri chaur, Khagaria; Dabkal chaur, Samastipur and Kusheshwarasthan chaur, Darbhanga) and two marshes (Milkichak marsh, Darbhanga and Thathopur marsh, Darbhanga), and also from ditches, ponds, marshes, lakes, chaurs etc located in the vicinity of the selected main seven wetlands as well as those situated along the road-sides of the study area. The specimens were critically examined and properly identified with the help of available literature. The local people were interviewed/interrogated to gather information about the utilization/ economic aspects of these plants or parts/products thereof in their daily life. In addition to visiting adjacent surrounding villages, the local temporary markets, called '*haats*', were also attended to assess their trade status/prospect from time to time.

Table 1. Characteristics of wetlands under investigation.

| Name of the wetland | Character | | | | |
|---|---|---|---|---|---|
| | Location (NL/EL) | Altitude (m) | Area ($km^2$) | Annual rainfall (cm)* | Mean depth (m) |
| Kawar lake, Begusarai | 25°38'00"/86°08'00" | 35.00 | 24.00 | 143.64 | 1.50 |
| Gaurdah-Kusmi lake, Saharsa | 25°43'30"/86°33'12" | 37.00 | 2.15 | 123.75 | 1.30 |
| Kachna Dumrichaur, Khagaria | 25°39'27"/86°26'11" | 37.00 | 0.36 | 106.48 | 0.71 |
| Dabkal chaur, Samastipur | 25°47'00"/85°47'00" | 44.00 | 2.23 | 138.44 | 0.85 |
| Kusheshwarasthan chaur, Darbhanga | 25°49'19"/86°18'36" | 37.80 | 2.10 | 81.80 | 1.13 |
| Milkichakmarsh, Darbhanga | 26°08'47"/85°57'00" | 45.00 | 1.30 | 81.80 | 0.03 |
| Thathopur marsh, Darbhanga | 25°58'13"/86°02'27" | 42.00 | 1.24 | 81.80 | 0.02 |

*Average for the period 1996-2002 from the nearest meteorological stations

## ENUMERATION

Table 2. Macrophytes used in North Bihar, India.

| Botanical name/Local name (in parenthesis) | Habit | Family | Plant parts/ products | Uses |
|---|---|---|---|---|
| **FERNS:** | | | | |
| *Azolla pinnata* R. Br. (Pani ghaas) | FF | Azollaceae | Whole plant | Used as duck feed |
| *Marsilea minuta* L. (Susnisag) | WE | Marsileaceae | Leaf | Rarely cooked as vegetable |
| **ANGIOSPERMS:** | | | | |
| **A. Dicots :** | | | | |
| *Aeschynomene aspera* L. (Sarkanda, Korhila) | WE | Fabaceae | Tender shoot | Used as green forage |
| | | | Pith | Used in handicraft to make various items mainly headwear of bridegrooms (Vernacular *maur*) and in fishery to make fish-net floats |

| | | | | |
|---|---|---|---|---|
| *A. indica* L. (Sarkanda, Korhila) | WE | --Do-- | --Do-- | --Do-- |
| *Alternanthera paronychioides* St.-Hill. | WE | Amaranthaceae | Shoot | Used as fodder |
| *A. philoxeroides* (Mart.) Griseb. | WE | --Do-- | --Do-- | --Do-- |
| *A. sessilis* (L.) R.Br. ex DC. | WE | --Do-- | --Do-- | --Do-- |
| *Centella asiatica* (L.) Urb. (Brahmi) | WE | Apiaceae | Leaf powder | Mixed in milk and given to cure madness |
| *Ceratophyllum demersum* L. (Kacher, Darah ghaas) | US | Ceratophyllaceae | Whole plant | Used as fish trap; feed for fishes especially grass carp *(Ctenopharyngdon idella)* and rohu *(Labeo rohita)*; duck feed |
| *Euryale ferox* Salisb. (Makhana) | RF | Nymphaeaceae | Perisperm (Seed) | Eaten roasted |
| *Ipomoea aquatica* Forssk. (Karmi) | RF/WE | Convolvulaceae | Dried stem | Used as fuel |
| | | | Young leaf | Cooked as vegetable (Vern. Karmisag) |
| *Nelumbo nucifera* Gaertn. (Kamal, Purain) | RF | Nelumbonaceae | Rhizome powder | Taken to cure piles |
| | | | Leaf | Used as eating plate especially on ceremonial occasions |
| | | | Flower | Used for decoration and offered to deities by Hindus during worship |
| | | | Fruiting torus/Seed | Cooked as vegetable |
| *Nymphaea nouchali* Burm. f. (Bhent, Kumudni, Koka) | RF | Nymphaeaceae | Rhizome/Seed | Eaten cooked |
| | | | Rhizome/ Flower | Mixed with black pepper *(Piper nigrum)*, pasted and taken against menstrual disorder |
| | | | Rhizome powder | Taken in indigestion, dysentery, piles etc |
| | | | Flower | Used in worship |
| *Nymphaea* sp. (Sorkha) | RF | --Do-- | Rhizome | Eaten raw with slight sweet taste; mixed with black pepper (*Piper nigrum)*, made into paste and taken as antibiotic against carbuncle |
| | | | Rhizome paste | Spreaded over carbuncle |
| *Trapa natans* L. var. *bispinosa* (Roxb.) Makino (Singhara) | RF | Trapaceae | Seed | Eaten raw or cooked |

**B. Monocots :**

| | | | | |
|---|---|---|---|---|
| *Aponogeton natans* (L.) Engl. & Krause (Ghechu) | RF | Aponogetonaceae | Root | Eaten cooked |
| *Commelina benghalensis* L. (Kankaua ghaas) | WE | Commelinaceae | Shoot | Used as fodder |
| *Cynodon dactylon* (L.) Pers. (Doobhi) | WE | Poaceae | | Used as an excellent fodder |
| *Cyperus compressus* L. | WE | Cyperaceae | --Do-- | Used as fodder |
| *C. difformis* L. | WE | --Do-- | --Do-- | --Do-- |
| *C. exaltatus* Retz. | WE | --Do-- | --Do-- | --Do-- |
| *C. iria* L. (Motha) | WE | --Do-- | Stem-split | Used in weaving baskets |
| | | | Shoot | Feed to cattle |
| *C. rotundus* L. (Chichorh) | WE | --Do-- | Tuber | Eaten raw |
| | | | Shoot | Used as forage |
| *Echinochloa colona* (L.) Link (Sama) | WE | Poaceae | Shoot | Used as green forage |
| | | | Grain | Eaten roasted or cooked |
| *E. crus-galli* (L.) Beauv. (Barsama) | WE | --Do-- | --Do-- | --Do-- |
| *Eichhornia crassipes* (Mart.) Solms (Jalkumbhi) | FF | Pontederiaceae | Whole plant | Rarely decomposed to prepare manure especially for potato crop for its better yield |
| | | | Leaf | Used as inferior fodder |
| *Eleocharis atropurpurea* (Retz.) Presl | WE | Cyperaceae | Shoot | Used as fodder |
| *E. dulcis* (Burm. f.) Trin. ex. Henschal | WE | --Do-- | --Do-- | --Do-- |
| *E. palustris* (L.) R. Br. | WE | --Do-- | --Do-- | --Do-- |
| *Eragrostis nutans* (Retz.) Nees ex Steud. | WE | Poaceae | --Do-- | --Do-- |
| *Eriochloa procera* (Retz.) Hubb. | WE | --Do-- | --Do-- | --Do-- |
| *Fimbristylis dichotoma* (L) Vahl | WE | Cyperaceae | --Do-- | --Do-- |
| *F. falcata* (Vahl) Kunth | WE | --Do-- | --Do-- | --Do-- |
| *F. littoralis* Gaud. | WE | --Do-- | --Do-- | --Do-- |
| *F. miliacea* (L) Vahl | WE | --Do-- | --Do-- | --Do-- |
| *Fuirena ciliaris* (L) Roxb. | WE | --Do-- | --Do-- | --Do-- |
| *Hydrilla verticillata* (L. f.) Royle (Samar, Sebar, Darah ghaas) | AS | Hydrocharitaceae | Whole plant | Used as fish trap; feed for fishes especially grass carp *(C. idella)* and rohu *(L. rohita);* duck feed |

| | | | | |
|---|---|---|---|---|
| *Hygroryza aristata* (Retz,) Nees ex Wt. & Arn. | FF | Poaceae | Shoot | Used as fodder |
| *Ischaemum rugosum* Salisb. | WE | --Do-- | --Do-- | --Do-- |
| *Kyllinga brevifolia* Rottb. | WE | Cyperaceae | --Do-- | --Do-- |
| *K. tenuifolia* Steud. | WE | --Do-- | --Do-- | --Do-- |
| *Leersia hexandra* Sw. (Garar ghaas) | WE | Poaceae | Shoot | Used as green forage |
| *Lemna minor* L. | FF | Lemnaceae | Whole plant | Feed to poultry/ducks |
| *L. perpusilla* Torr. | FF | --Do-- | --Do-- | --Do-- |
| *Oryza rufipogon* Griff. (Jhauradhan, Janglidhan) | WE | --Do-- | --Do-- | --Do-- |
| *Ottelia alismoides* (L.) Pers. (Kauathuthi) | AS | Hydrocharitaceae | Leaf<br>Seed | Used as fish feed<br>Eaten cooked |
| *Panicum paludosum* Roxb. | WE | Poaceae | Shoot | Used as forage |
| *P. repens* L. | WE | --Do-- | --Do-- | --Do-- |
| *Paspalidium flavidum* (Retz.) A. Camus | WE | --Do-- | --Do-- | --Do-- |
| *P. germinatum* (Forssk.) Stapf | WE | --Do-- | --Do-- | --Do-- |
| *P. punctatum* (Burm. f.) A. Camus | WE | --Do-- | --Do-- | --Do-- |
| *Paspalum scrobiculatum* L. | WE | --Do-- | --Do-- | --Do-- |
| *Phragmites vallatoria* (Pluk. ex L.) Veldk. (Narkat) | WE | --Do-- | Culm | Weaved into chairs, baskets, mats; made into indigenous flute (Vern. *bansuri)* and pens (Vern. *likhni)* for rural school children; dried culms used to prepare hedge |
| | | | Tender shoot | Used as fodder |
| *Potamogeton crispus* L. (Sebar) | AS | Potamogetonaceae | Rhizome<br>Leaf | Cooked as food<br>Used as fish feed |
| *P. nodosus* Poir. | AS | --Do-- | Leaf | Used as fish feed |
| *P. pectinatus* L. | AS | --Do-- | --Do-- | --Do-- |
| *Saccharum spontaneum* L. (Kans, Khar) | WE | Poaceae | Shoot | Used in thatching of roofs; in making baskets/mats etc; as fodder, fuel; to make hedge |
| *Schoenoplectus articulatus* (L.) Palla (Patpati, Khobhi) | WE | Cyperaceae | Shoot<br>Seed | Used as fodder<br>Consumed as food |

| | | | | |
|---|---|---|---|---|
| *S. lateriflorus* (Gmel.) Lye (Patpati) | WE | --Do-- | Shoot | Feed to cattle |
| *S. supinus* (L.) Palla | WE | --Do-- | --Do-- | --Do-- |
| *Spirodela polyrhiza* (L.) Schleid. | FF | Lemnaceae | Whole plant | Feed to poultry/ducks |
| *Typha angustifolia* L. (Pater) | WE | Typhaceae | Leaf | Weaved into baskets etc |

Symbols : FF = Free Floating; RF = Rooted Floating; US = Unanchored Submerged; AS = Anchored Submerged; WE = Wetlant Emergent.

## RESULTS AND DISCUSSION

The survey reveals that over sixty macrophytes which grow profusely in the different wetlands of North Bihar, are under exploitation (Table 2). The native folk of the region collect the aquaphytes/their parts or products and put them for their diverse household purposes or sell them for cash in the nearby haats/markets. The utilization includes as food (roots of *Aponogeton natans;* rhizomes of *Nymphaea nouchali, Nymphaea* sp. - a probable hybrid propagating by vegetative means only and not flowering at all, *Potamogeton crispus;* tubers of *Cyperus rotundus;* leaves of *Marsilea minuta, Ipomoea aquatica;* fruiting torus - vernacular kamalgatta of *Nelumbo nucifera;* grains of *Echinochloa colona, E. crus-galli;* seeds of *Euryale ferox, Nymphaea nouchali, Nelumbo nucifera, Trapa natans* var. *bispinosa, Ottelia alismoides, Schoenoplectus articulatus*), fodder (shoots of *Aeschynomene aspera, A. indica, Alternanthera paronychioides, A. philoxeroides, A. sessilis, Commelina benghalensis, Cynodon dactylon, Cyperus compressus, C. difformis, C. exaltatus, C. iria, C. rotundus, Echinochloa colona, E. crus-galli, Eleocharis atropurpurea, E. dulcis, E. palustris, Eragrostis nutans, Eriochloa procera, Fimbristylis dichotoma, F. miliacea, Fuirena ciliaris, Hygroryza aristata, Ischaemum rugosum, Kyllinga brevifolia, K. tenuifolia, Leersia hexandra, Oryza rufipogon, Panicum paludosum, P. repens, Paspalidium flavidum, P. germinatum, P. punctatum, Paspalum scrobiculatum, Phragmites vallatoria, Saccharum spontaneum, Schoenopletus articulatus, S. lateriflorus, S. supinus;* leaves of *Eichhornia crassipes)*, fish food (whole plants of *Ceratophyllum demersum, Hydrilla verticillata;* leaves of *Ottelia alismoides, Potamogeton crispus, P. nodosus, P. pectinatus)*, poultry/duck feed (whole plants of *Azolla pinnata, Lemna minor, L. perpusilla, Spirodela polyrhiza)*, medicine (rhizomes of *Nelumbo nucifera, Nymphaea nouchali, Nymphaea* sp.; leaves of *Centella asiatica;* red flowers of *Nymphaea nouchali)*, basketry material (shoots of *Saccharum spontaneum,* stems of *Cyperus iria,* culms of *Phragmites vallatoria,* leaves of *Typha angustifolia)*, eating plate (leaves of *Nelumbo nucifera)*, decoration (flowers of *Nelumbo nucifera)*, worship (flowers of *Nymphaea nouchali, Nelumbo nucifera)*, miscellaneous purposes (pith of *Aeschynomene* sp.; shoots of *Saccharum spontaneum, Phragmites vallatoria)* etc. Makhana *(Euryale ferox)* and Sighara (*Trapa natans* var. *bispinosa)* are grown as aquatic cash crops, the former as the foremost and latter ranked next to it in the region.

In general the exploitation of wild macrophytes frames the economic backbone of the area and the aquatic cash crops further boost it up. Besides their household utilization, the commercial sale of wild macrophytes/their parts or products for medicine (red flowers of *Nymphaea nouchali);* fish food (entire plants of *Ceratophyllum demersum* and *Hydrilla verticillata* under the trade name 'darah ghaas'); handicraft items of *Aeschynomene* sp.; basketry materials/products made from stems of *Cyperus iria,* shoots of *Saccharum spontaneum,* leaves of *Typha angustifolia,* culms of *Phragmites vallatoria* etc pays too much in the economy of common people of the area. The cultivation of aquatic cash crops on commercial scale in the diverse wetlands also plays a vital role in their economic life, the main beneficiaries are the landlords. The integrated aquaculture involving the macrophyte(s) especially makhana or/and singhara, and air-breathing fishes like Garai *(Channa punctatus),* Kawai *(Anabas testudines),* Singhi *(Heteropneustes fossilis),* Mangur *(Clarias batrachus)* etc enhance their economy (Laal, 1981; Munshi, 1988; Munshi & Munshi, 1995; Verma *et al.,* 1996).

The efficient utilization of available resources to an optimum level is supposed to bring prosperity in the concerned region. The macrophytes are emerging as one of the most potential resources due to their high productivity and easy availability as food, fodder, medicine, fibre, ornament, paper pulp, basketry material etc. (Kresovich *et al.,* 1981; Munshi & Munshi, 1985; Jha *et al.,* 1998; Nandan & Singh, 2003). They are growing luxuriantly in the wetlands of North Bihar and are looking forward as a very potential candidate for their optimum utilization in the face of poor economy of the region. Therefore, it is imperative to encourage basketry, handicraft, integrated aquaculture etc to consume the resource potential of macrophytes of the region with the objective of economic upliftment of the native population which live below the poverty line.

**ACKNOWLEDGEMENTS**

The authors are thankful to all those who rendered their co-operation during the course of this exhaustive investigation.

**REFERENCES**

AHMAD, S.H., A.A. FAZAL & A.K. SINGH 1990. Retrospect and prospect of fisheries of Bihar. *Fishing Chimes* 9 (11) : 46-49.

DEHADRAI, P.V. 1994. Swamps of North Bihar. *Bull. Nat. Inst. Ecol.* 7 : 17-21.

JHA, V., G.K. BARAT & U.N. JHA 1991a. Nutritional evaluation of *Euryale ferox* Salisb. (Makhana). *J. Food Sci. Technol.* 28 (5) : 326-328.

————, A.N. KARGUPTA, R.N. DUTTA, U. . JHA, R.K. MISHRA & K.C. SARASWATI 1991b. Utilization and conservation of *Euryale ferox* Salisb. in Mithila (North Bihar). *Aqua. Bot.* 39 : 295-314.

————, Y.P. THAKUR & A.M. VERMA 1998. Human welfare in relation to diversity of edible aquaphytes in North Bihar. In Roy, A.K., J.V.V. Dogra & S.K. Varma (eds.) *Phytodiversification and Human Welfare,* pp. 159-174. M. D. Publications Pvt. Ltd., New Delhi.

KRESOVICH, S., C.K. WAGNER, D.A. SCANTLAND & W.T. LAWHON 1981. Emergent aquatic plants : Biological and economic perspectives. *Biomass* 1 : 127-144.

LAAL, A.K. 1981. Studies on the ecology and productivity of swamps in North Bihar in relation to production of fishes and other agricultural commodities. Ph.D. Thesis, Bhagalpur University, Bhagalpur (India).

MUNSHI, J. D. 1988. Ecology of gangetic plain of Bihar. *Int. J. Ecol. Environ. Sci.* 14 : 87-101.

——— & J.S.D. MUNSHI 1995. *Fundamentals of freshwater biology.* Narendra Publishing House, Delhi.

MUNSHI, J.S.D. & J.D. MUNSHI 1985. Nature of freshwater resources of Bihar. In Srivastava, C.B.L. & C.G. Suresh (eds.) *Recent Advances in Zoology,* pp. 127-139. Rastogi & Company, Meerut.

NANDAN, K.B. & C.B. SINGH 2003. Useful macrophytes in Kawar lake, North Bihar, India. *Indian J. Forestry* 26 : (In press).

RAI, D.N. & J.S.D. MUNSHI 1982. Ecological characteristics of Chaurs of North Bihar. In Gopal, B., R.E. Turner, R.G. Wetzel & D.F. Whigham (eds.) *Wetland Ecology and Management,* Part II, pp. 89-95. International Scientific Publications, Jaipur, India.

VERMA, A.M. 1994a. Integrated fish farming with Makhana *(Euryale ferox). Fishing Chimes* 14 (2) : 13.

——— 1994b. Integrated fish farming with Singhara *(Trapa bispinosa). Fishing Chimes* 14 (9) : 29.

———, S.H. AHMAD & V. JHA 1996. Integrated culture of air-breathing carnivorous fishes with makhana *(Euryale ferox* Salisb.) in a derelict wetland of North Bihar, India. *J. Freshwater Biol.* 8 (2) : 117-120.

# WILD EDIBLE PLANTS OF MAYURBHANJ DISTRICT, ORISSA, INDIA

K.K. BEHERA**, N.M. MISHRA*, N.K. DHAL & N.C. ROUT

*Regional Research Laboratory, Bhubaneswar-751 013, Orissa, India*
**Parija Post-Graduate Department of Botany, Ravenshaw College, Cuttack-753 003, Orissa*
***Email : kambaska@yahoo.co.in*

## ABSTRACT

Six thousand medicinally important plant species have been reported from India (Warrier *et al.*, 1993). The source of poly herbal drugs from plant is very essential to meet the challenges of recent allopathic therapy. Not only the therapy but also the source of food from plant origin is the need of the hour to meet the supplementary food requirements of the over growing global population. The present paper deals with 124 edible plant species under 49 families belonging to 88 genera consumed as food supplement by the tribals of Mayurbhanj district apart from their seasonal crops and vegetables.

## INTRODUCTION

Mayurbhanj district is located between 21°17' and 22°34' North longitudes and between 85°40' and 87°10' East latitude. The district is bounded by the Singhbhum district of Jharkhand and Midanapore district of West Bengal in the north, Balasore and Keonjhar in the south, Midanapore and Balasore districts in the east, Keonjhar and Singhbhum in the west. The total area of the district is 10,418.00 sq km and about 50% of the area is covered by forests. The predominant species in the forest is Sal (*Shorea robusta* Gaertn. f.), where as the Similipal range shows diversified vegetation pattern, i.e from tropical deciduous to a virgin semi-evergreen forest which express the climatic climax type of vegetation (Brahmam & Saxena, 1990).

Agriculture is the main occupation of tribal people but forest and its products play important role to meet their multifarious requirements such as food, medicine, fibres etc (Bal, 1942; Saxena & Brahmam, 1989). For food, they are mainly dependent on agriculture but also collect roots, tubers, leaves, flowers, fruits as supplementary food. The total population of the district is 1,884,580, out of which scheduled caste is 131,765 and scheduled tribe is 1,090,626. There are 46 tribes inhabiting in this district which is 64.86% of the total population of the district (Tiwari *et al.,* 1999). The dominated tribes of the district are Santal, Munda , Kolha, Bhumija, Bathudi, Lodha, Gond, Kondha, Saunti etc. Although, the ethnomedicinal investigation have been done by various workers (Bal, 1982; Brahmam & Saxena, 1990, 1994-1996; Haines, 1921-25; Nayak *et al.,* 2003a, b; Jain 1991; Saxena & Dutta, 1975; Tiwari *et al.,* 1999 etc), the information on edible plants is meager in these publications. The present work concentrates in the alternative food habits of the tribals and local folks.

## MATERIALS AND METHODS

Ethnobotanical survey have been carried out in various tribal pockets of the district during 2001-2003. First hand information on food plants were collected in the field and confirmed through the tribal healers located in various forest patches. The plants are arranged alphabetically by botanical names followed by family and local names, uses of plants as food are given in briefs with specific tribes (S-Santal, Sa-Saunti, Ba-Bathudi, Bh-Bhumija, Mu-Munda, Lo-Lodha, K-Kondha, Ko-Kolha, G-Gond). Place of collection and collection number for each plants are given in the Table 1. The voucher specimens have been deposited in the herbarium of Parija Post-Graduate Department of Botany, Ravenshaw College, Cuttack.

## OBSERVATIONS

The paper records 124 plant species consumed by the tribals either raw or after cooking. This vast resource of edible plants would give a platform to the nutritionist, food scientist to work with. The nutritional value of leafy vegetables and wild nuts are yet to be worked out. The population of rhizomatous plants in this area is alarmingly depleting, hence warrants measures for conservation.

Table 1. List of wild edible plants of Mayurbhanj district.

| Name of the plant/ Family | Local name | Place of collection/ collection number | Plant parts and Uses |
|---|---|---|---|
| *Abelmoschus crinitus* Wall. Malvaceae | Bhedi-kandda | Asanabani/1058 | Tender fruits are eaten as raw. (Ba, Mu, Lo) |
| *Abelmoschus moschatus* Medic. Malvaceae | Jangali-bhindi | Gadia pal/1204 | Tender fruits mixed with *Tamarindus indica* juice are cooked and eaten with Handia. |
| *Acacia nilotica* (L.) Delile subsp. *indica* (Benth.) Brenan Mimosaceae | Babul | Chuinpushi/1103 | Gum is edible for stomach disorder. (S, G, Ba) |
| *Achyranthes aspera* L. Amaranthaceae | Apamaranga | Ghagaradi/1256 | Seeds paste taken as chatani both in dinner and lunch. (S, Ba, Bh, Ko, K) |

| | | | |
|---|---|---|---|
| *Acorus calamus* L. Araceae | Bacha | Kaptipada/1257 | Fresh rhizome is eaten as raw against throat infection. (Ba, Mu, Lo) |
| *Aegle marmelos* (L.) Corr. Rutaceae | Bel, Belo | Kathapal/1200 | Ripen fruit pulp is edible. (S, Mu, Bh, K) |
| *Agave angustifolia* Haw. Agavaceae | Baramasi | Salamundli/1240 | Young leaves are cooked and used as vegetable. (S, Sa, Mu, Ko, Ba) |
| *Alangium salvifolium* (L. f.) Wang. Alangiaceae | Dhalanga, Ankulo | Pathar khani/1252 | Fruits are edible. (S, Mu, Bh, K) |
| *Alternanthera sessilis* (L.) R. Br. ex DC. Amaranthaceae | Madaranga-sago | Petasal/1250 | Young leaves are cooked and used as vegetable. (S, Sa, Mu, Ko, Ba) |
| **Amaranthus caudatus* L. Amaranthaceae | Khadasago, Dhemanasago | Kuliana/1234 | Leaves and tender shoots are cooked and used as vegetable. (S, Sa, Mu, Ko, Ba) |
| **Amaranthus spinosus* L. Amaranthaceae | Kantaneutia, Kanta-khara | Poda astia/1230 | Leaves and tender shoots are cooked and used as vegetable. (Ba, Mu, Lo) |
| **Amaranthus tricolor* L. Amaranthaceae | Nautia | Lulungu/1232 | Leaves and tender shoots are cooked and used as vegetable. (S, Mu, Bh, K) |
| **Amaranthus viridis* L. Amaranthaceae | Jangali-chauli | Badasahi/1235 | Leaves and tender shoots are cooked and used as vegetable. (S, Sa, Mu, Ko, Ba) |
| *Amorphophalus bulbifer* (Roxb.) Bl. Araceae | Dhara-oal | Khaladi/1243 | Corm and young leaves are edible. (S, Sa, Mu, Ko, Ba) |
| *Amorphophalus paeoniifolius* (Dennst.) Nicolson Araceae | Olua, Ol | Chhanua/1245 | Corm and young leaves are edible. (S, Mu, Bh, K) |
| *Ampelocissus latifolia* (Roxb.) Planch. Vitaceae | Paninohi, Kanjinoi | Kulidiha/1238 | Ripen fruits are edible. (Ba, Mu, Lo) |
| *Ampelocissus tomentosa* (Rəth) Planch. Vitaceae | Katabhanganoi | Sidamchandrapur/1239 | Ripen fruits are edible. (S, Sa, Mu, Ko, Ba) |
| *Anogeissus latifolia* (Roxb. ex DC.) Wall. ex Guill. & Perr. Combretaceae | Dhaura | Gandidhar/1244 | Boiled gum is eaten for stomach disorder. (Ba, Mu, Lo) |
| *Antidesma acidium* Retz. Euphorbiaceae | Nunnunia | Kaladam/1054 | Leaves are cooked and used as vegetable. (S, Ko, Ba, G) |
| *Ardisia solanacea* Roxb. Myrsinaceae | Tinkoli, Bat-gouri | Sarabasa/1056 | Young leaves are cooked and used as vegetable. (S, Sa, Mu, Ko, Ba) |
| *Arisaema tortuosum* (Wall.) Schott Araceae | Kushnopheni | Handiphuta/1057 | Young leaves are cooked and used as vegetable. (S, Ko, Ba, G) |

| | | | |
|---|---|---|---|
| *Artocarpus heterophyllus* Lam. Moraceae | Panaso | Kantagadi/558 | Fruits are edible as vegetable. (S, Sa, Mu, Ko, Ba) |
| *Asparagus racemosus* Willd. Liliaceae | Gaichero, Satawari | Thakurmunda/1554 | Tuberous roots are edible. (S, Mu, B, K) |
| *Basella rubra* L. Basellaceae | Poi, Pui | Mahuldiha/755 | Leaves and young shoots are cooked and eaten as vegetable. (Ba, Mu, Lo) |
| *Bauhinia semla* Wunderl. Caesalpiniaceae | Amthi | Mankadapada/773 | Tender leaves are used as green vegetable. (S, Ko, Ba, G) |
| *Bauhinia vahlii* Wight & Arn. Caesalpiniaceae | Siadilata, Siali | Mankadapada/756 | Ripen fruits are edible. (S, Sa, Mu, Ko, Ba) |
| *Bauhinia variegata* L. Caesalpiniaceae | Kanchan | Mankadapada/777 | Young leaves are cooked and used as vegetables. (S, Ko, Ba, G) |
| *Boerhavia diffusa* L. Nyctaginaceae | Purunisag | Padmapokhori/1260 | Young leaves are cooked and used as vegetables. (S, Sa, Mu, Ko, Ba) |
| *Bombax ceiba* L. Bombacaceae | Simuli, Se-mulo | Jadida/778 | Young fruits are cooked and eaten as vegetables. (S, Ko, Ba, G) |
| *Borassus flabellifer* L. Arecaceae | Tala | Kotoria/1050 | Fruits are edible. (S, Su, Mu, Ko, Ba) |
| *Bridelia retusa* (L) Spr. Euphorbiaceae | Pani-kasi, Khais | Sarat/1403 | Fruits are edible. (S, Mu, B, K) |
| *Buchanania cochinchinensis* (Lour.) Almeida Anacardiaceae | Chara | Jadida/779 | Ripe fruits and fleshy part of seeds are edible. (S, Ko, Ba, G) |
| **Cajanus cajan* (L.) Millsp. Fabaceae | Harada | Jadida/780 | Young fruits and seeds are edible. (Ba, Mu, Lo) |
| *Cajanus scarabaeoides* (L.) du Petit-Thou. Fabaceae | Kandulo | Kotoria/1040 | Young fruits and seeds are edible. (S, Sa, Mu, Ko, Ba) |
| *Canavalia verosa* (Roxb.) W. & A. Fabaceae | Kalasima | Noto/1401 | Young fruits and seeds are edible. (S, Ko, Ba, G) |
| *Capparis zeylanica* L. Capparaceae | Asadua | Kotoria/1033 | Fruits are cooked and used as vegetables. (S, Ko, Ba, G) |
| *Carissa carandas* L. Apocynaceae | Ankukoli | Kotoria/1066 | Fruits are edible. (S, Sa, Mu, Ko, Ba) |
| *Carissa spinarum* L. Apocynaceae | Dudkoli, Ankhukoli | Kotoria/1074 | Fruits are edible. (Ba, Mu, Lo, G, Su) |
| *Cassia fistula* L. Caesalpiniaceae | Sunari | Gandidhar/1212 | Flowers are edible. (S, Sa, Mu, Ko, Ba) |
| *Catunaregam spinosa* (Thunb.) Tirving. Rubiaceae | Kalai-kanta | Jadida/782 | Young fruits are cooked and eaten as vegetable. (S, Ko, Ba, G) |

| | | | |
|---|---|---|---|
| *Cayratia auriculata* (Wall.) Gamble Vitaceae | Luduri, Luduri-danka | Phulgadia/1290 | Young shoots are cooked and eaten as vegetable. (S, Ko, Ba, G) |
| *Centella asiatica* (L.) Urban Apiaceae | Thalkudi | Kotoria/1075 | Fresh leaves are edible. (S, Su, Mu, Ko, Ba) |
| *Ceropegia hirsuta* Wight & Arn. Asclepiadaceae | Pharbaghaga | Phulgadia/1291 | Tubers are edible. (Ba, Mu, Lo, G, Su) |
| *Chenopodium album* L. Chenopodiaceae | Bathua-saga | Kotoria/1067 | Leaves are cooked and eaten as vegetable. (S, Mu, Bh, K) |
| **Coccinia grandis* (L.)Voigt Cucurbitaceae | Kunduri | Sidamchandrapur/939 | Fruits and young leaves are eaten as vegetables. (S, Mu, B, K) |
| **Colocasia esculanta* (L.) Schott Araceae | Saru | Kuldiha/1211 | Root tuber and spadix are edible. (S, Ko, Ba, G) |
| *Commelina bengalensis* L. Commelinaceae | Kanisiri | Dukura/1281 | Fresh leaves are edible. (S, Sa, Mu, Ko, Ba) |
| *Commelina erecta* L. Commelinaceae | Panikanisiri | Handiphuta/1268 | Fresh leaves are edible. (S, Bh, Mu, Lo, Ko) |
| *Corchorus capsularis* L. Tiliaceae | Nalita | Phulgadia/1292 | Leaves are cooked and eaten as vegetable. (Ba, Mu, Lo, G, Su) |
| *Cucumis melo* L. Cucurbitaceae | Kharabhuja | Mahulabani/938 | Fruits are eaten. (S, Mu, B, K) |
| **Cucurbita maxima* Duch. ex Lam. Cucurbitaceae | Baitikakharu | Sidamchandrapur/933 | Fruits, young leaves and shoots are eaten as vegetable. (Ba, Mu, Lo, G, Sa) |
| **Cucurbita pepo* L. Cucurbitaceae | Kakharu | Jarada/944 | Fruits, young leaves and shoots are eaten as vegetable. (S, Ko, Ba, G) |
| *Curcuma aromatica* Salisb. Zingiberaceae | Parbati-palua | Phulgadia/1293 | Whitish powdered extracted from rhizome is edible. (S, Mu, B, K) |
| *Dendrocalamus strictus* (Roxb.) Nees Poaceae | Banso | Jaranata/1296 | Young shoots are eaten as vegetables. (S, Ko, Ba, G) |
| *Dendrophthoe falcata* (L. f.) Etting. Loranthaceae | Madanga | Ambadiha/1196 | Flower juice and fruits are edible. (Ba, Mu, Lo, G, Su) |
| **Dillenia aurea* J. E. Smith Dilleniaceae | Awoo, Oao | Pathrakhani/1215 | Fruits are used as vegetable. (Ba, Mu, Lo, G, Sa) |
| *Dillenia indica* L. Dilleniaceae | Rai | Siadia/977 | Fruits are cooked and used as vegetable. (S, Ko, Ba, G) |
| *Dioscorea alata* L. Dioscoreaceae | Khamba-alu | Sarabasa/930 | Root tubers are cooked and used as vegetable. (S, Ko, Ba, G) |

| | | | |
|---|---|---|---|
| *Dioscorea belophylla* Voigt ex Haines<br>Dioscoreaceae | Tunga-alu | Kantagadi/939 | Root tubers are cooked and used as vegetable. (S, B, Mu, Lo, Ko) |
| *Dioscorea bulbifera* L.<br>Dioscoreaceae | Pita-alu | Phulgadia/1294 | Root tubers are cooked and used as vegetable. (S, Ko, Ba, G) |
| *Dioscorea esculanta* (Lour.) Burkill<br>Dioscoreaceae | Susni-alu | Monabhanga/1352 | Root tubers are cooked and used as vegetable. (Ba, Mu, Lo, G, Sa) |
| *Dioscorea glabra* Roxb.<br>Dioscoreaceae | Kanta-alu | Monabhanga/1353 | Root tubers are cooked and used as vegetable. (S, Bh, Mu, Lo, Ko) |
| *Dioscorea hamiltonii* Hook. f.<br>Dioscoreaceae | Suta-alu | Phulgadia/1396 | Root tubers are cooked and used as vegetable. (S, Ko, Ba, G) |
| *Dioscorea hispida* Dennst.<br>Dioscoreaceae | Bainya-alu | Kaliasahi/796 | Root tubers are cooked and used as vegetable. (Ba, Mu, Lo, G, Sa) |
| *Dioscorea oppositifolia* L.<br>Dioscoreaceae | Pani-alu | Champajhar/815 | Root tubers are cooked and used as vegetable. (S, Ko, Ba, G) |
| *Dioscorea puber* Bl.<br>Dioscoreaceae | Danga-alu | Champajhar/816 | Root tubers are cooked and used as vegetable. (S, B, Mu, Lo, Ko) |
| *Diospyros melanoxylon* Roxb.<br>Ebenaceae | Kendu | Noto/1400 | Ripen fruits are edible. (S, Mu, Bh, K) |
| *Diospyros peregrina* (Gaertn.) Gurke<br>Ebenaceae | Mankara kendu | Monabhanga/1399 | Ripen fruits and young leaves are edible. (S, Mu, Bh, K) |
| *Diospyros sylvatica* Roxb.<br>Ebenaceae | Kalidukha | Dunduria/1401 | Young fruits are edible. (Ba, Mu, Lo, G, Su) |
| *Eryngium foetidium* L<br>Apiaceae | Jangali-dhania | Monabhanga/1369 | Leaves are cooked and eaten as vegetable. (S, Ko, Ba, G) |
| *Eulophia herbacea* Lindl.<br>Orchidaceae | Masina | Sankhaiadi/1370 | Tubers are edible. (S, Bh, Mu, Lo, Ko) |
| *Eulophia nuda* Lindl.<br>Orchidaceae | Gutti-kamal | Noto/1444 | Tubers are edible. (Ba, Mu, Lo, G, Sa) |
| *Ficus auriculata* Lour.<br>Moraceae | Raja-dimiri | Kotoria/1175 | Raw fruits are cooked and eaten as vegetable. (S, Ko, Ba, G) |
| **Ficus benghalensis* L.<br>Moraceae | Bara | Kutting/932 | Ripen fruits are edible. (S, Lo, B, Mu) |

| | | | |
|---|---|---|---|
| **Ficus hispida* L. Moraceae | Panidimiri | Nuasahi/919 | Ripen fruits are edible. (S, Bh, Mu, Lo, Ko) |
| **Ficus racemosa* L. Moraceae | Dimiri | Podadiha/920 | Ripen fruits are edible. (S, Ko, Ba, G) |
| *Gardenia gummifera* L. f. Rubiaceae | Bhurudukuli | Taldiha/924 | Ripen fruits are edible. (Ba, Mu, Lo, G, Su) |
| *Hibiscus sabdariffa* L. Malvaceae | Khatta palanga | Monabhanga/1333 | Leaves and flowers are cooked and eaten as vegetable. (S, Ko, Ba, G) |
| *Ipomoea aquatica* Forssk. Convolvulaceae | Kalama-saga | Nagapal/912 | Leaves and young shoots are cooked and used as vegetable. (S, B, Mu, Lo, Ko) |
| **Ipomoea batatas* (L.) Lam. Convolvulaceae | Kanda-Mula | Kotoria/1171 | Root tubers are edible. (S, Bh, Mu, Lo, Ko) |
| *Ipomoea mauritiana* Jacq. Convolvulaceae | Katha kalama | Debakunda/1199 | Root tubers are edible. (Ba, Mu, Lo, G, Sa) |
| *Lablab purpureus* (L.) Sweet Fabaceae | Simbo | Monabhanga/1398 | Young fruits are cooked and used as vegetable. (S, Bh, Mu, Lo, Ko) |
| **Lagenaria siceraria* (Molina) Standley Cucurbitaceae | Lau | Monabhanga/1334 | Fruit and young shoots are cooked and eaten as vegetable. (S, Ko, Ba, G) |
| *Luffa acutangula* (L.) Roxb. Cucurbitaceae | Jahni | Kotoria/1134 | Young fruits are cooked and eaten as vegetable. (S, B, Mu, Lo, Ko) |
| *Luffa cylindrica* (L.) M. Roem. Cucurbitaceae | Tadari | Pagadadiha/930 | Young fruits are cooked and eaten as vegetable. (S, Mu, B, K) |
| *Macrotyloma unifloram* (Lam.) Verdc. Fabaceae | Kolatha | Kotoria/1138 | Seeds are edible. (S, Bh, Mu, Lo, Ko) |
| **Madhuca indica* Gmel. Sapotaceae | Mahula | Udala/1262 | Raw fruits are used as chatani. (S, Lo, Bh, Mu) |
| **Mangifera indica* L. Anacardiaceae | Amba | Pagadadiha/945 | Fruits are edible. (S, Ko, Ba, G) |
| *Marsilea minuta* L. Marsiliaceae | Sun-suniya | Pagadadiha/966 | Leaves are cooked and used as vegetable. (S, Ko, Ba, G) |
| *Martynia annua* L. Martyniaceae | Bagha-nekhi | Bhimitali/1226 | Ripe fruits are edible. (Ba, Mu, Lo, G, Su) |
| **Mentha spicata* L. Lamiaceae | Podina | Jidra/1225 | Leaves are used for the preparation of chatani. (S, Bh, Mu, Lo, Ko) |

| | | | |
|---|---|---|---|
| *Mimusops elengi* L. Sapotaceae | Baulo | Kotoria/1141 | Ripe fruits are edible. (S, Bh, Mu, Lo, Ko) |
| **Momordica dioica* Roxb. ex Willd. Cucurbitaceae | Kaankada | Pagadadiha/975 | Fruits are cooked and eaten as vegetable. (S, Ko, Ba, G) |
| **Moringa oleifera* Lamk. Moringaceae | Sajana | Nagapal/1221 | Young leaves and young fruits are cooked and eaten as vegetable. (S, Ko, Ba, G) |
| **Murraya koenigii* (L.) Spreng. Rutaceae | Bhursunga | Kotoria/1137 | Leaves are used in curry as a flavour. (Ba, Mu, Lo, G, Su) |
| **Naringi crenulata* (Roxb.) Nicols. Rutaceae | Kaitha | Monabhanga/1388 | Fruit pulp is edible. (Ba, Mu, Lo, G, Sa) |
| *Nelumbo nucifera* Gaertn. Nelumbonaceae | Kanii | Udala/1236 | Tubers are edible. (S, Ko, Ba, G) |
| *Neolamarkia cadamba* (Roxb.) Bosser Rubiaceae | Kadamba | Sagadi/1246 | Fruits are edible. (S, Mu, Bh, K) |
| *Pachyrhizus erosus* (L.) Urb. Fabaceae | Chini-alu | Khunta/1246 | Root tubers are edible. (S, B, Mu, Lo, Ko) |
| **Phoenix sylvestris* (L.) Roxb. Arecaceae | Khajuri | Khunta/1249 | Ripen fruits are edible. (S, Bh, Mu, Lo, Ko) |
| *Polygonum plebeium* R. Br. Polygonaceae | Muthi-sago | Podadiha/1010 | Leaves are cooked and eaten as vegetable. (S, Ko, Ba, G) |
| *Pueraria tuberosa* (Willd.) DC. Fabaceae | Bhuni-kakharu | Pagadadiha/972 | Tuber-roots are edible. (S, Mu, Bh, K) |
| **Punica granatum* L. Punicaceae | Dalimbo | Kotoria/1149 | Ripen fruits are edible. (S, Bh, Mu, Lo, Ko) |
| **Saccharum officinarum* L. Poaceae | Akhu | Jaranata/737 | Stem is edible. (S, Mu, Bh, K) |
| *Saccharum spontaneum* L. Poaceae | Kasatandi | Jamadiha/892 | Stem is edible. (S, Ko, Ba, G) |
| *Schleichera oleosa* (Lour.) Oken. Sapindaceae | Kusuma | Pagadadiha/1001 | Ripe fruit pulps are edible. (S, B, Mu, Lo, Ko) |
| *Solena amplexicaulis* (Lam.) Gandhi Cucurbitaceae | Bana-kundari | Pagadadiha/872 | Fruits are edible. (S, Bh, Mu, Lo, Ko) |
| *Sesamum orientale* L. Pedaliaceae | Khasa, Rasi | Pagadadiha/931 | Seeds are edible. (S, Bh, Mu, Lo, Ko) |
| *Shorea robusta* Gaertn. f. Dipterocarpaceae | Sal, Kudia | Bhalughudura/785 | Boiled seeds are eaten with salt. (S, Ko, Ba, G) |
| **Spinacia oleracea* L. Chenopodiaceae | Palanga | Jadida/783 | Leaves are cooked and eaten as vegetable. (S, B, Mu, Lo, Ko) |

| | | | |
|---|---|---|---|
| **Syzygium cuminii* (L.) Skeels Myrtaceae | Bada-jamukoli | Uthanisahi/1015 | Ripen fruits are edible. (S, Ko, Ba, G) |
| *Syzygium heyneanum* (Duthie) Wall. ex Gamble Myrtaceae | Chhata-jamukoli | Nuasahi/1215 | Ripen fruits are edible. (Ba, Mu, Lo, G, Su) |
| **Tamarindus indica* L. Caesalpiniaceae | Tentuli | Nachipur/611 | Leaves, fruits and seeds are edible. (S, B, Mu, Lo, Ko) |
| *Terminalia alata* Heyne ex Roth Combretaceae | Asana | Petasal/917 | Gum is edible for stomach disorder. (S, Ko, Ba, G) |
| *Terminalia bellirica* (Gaertn.) Roxb. Combretaceae | Bahada | Bhudurbani/1217 | Fleshy part of the seed is edible. (S, Bh, Mu, Lo, Ko) |
| **Trapa natans* L. var. *bispinosa* (L.) Makino Trapaceae | Kantaphala, Panisingira | Pagadadiha/1006 | Fruits are edible. (S, Mu, Bh, K) |
| *Trichosanthes cucumerina* L. Cucurbitaceae | Ban-potolo | Pagadadiha/1023 | Fruits are cooked and used as vegetable. (S, Ko, Ba, G) |
| **Vigna mungo* (L.) Hepper Fabaceae | Biri | Pagadadiha/978 | Seeds and tender fruits are edible. (S, B, Mu, Lo, Ko) |
| **Vigna radiata* (L.) Wilczek Fabaceae | Munga | Kotoria/1138 | Seeds and tender fruits are edible. (Ba, Mu, Lo, G, Sa) |
| *Vigna ricciardiana* (Tenore) Babu & Sharma Fabaceae | Banamunga | Pagadadiha/1007 | Seeds and tender fruits are edible. (Ba, Mu, Lo, G, Sa) |
| **Vigna unguiculata* (L.) Walp. Fabaceae | Jhuranga | Pagadadiha/1009 | Seeds are edible. (S, Bh, Mu, Lo, Ko) |
| *Woodfordia fruticosa* (L.) Kurz Lythraceae | Dhatki | Champajhar1703 | Flower juice is edible. (S, Ko, Ba, G) |
| *Ziziphus mauritiana* Lamk. Rhamnaceae | Barokoli | Baunsadiha/601 | Fruits are edible. (S, Bh, Mu, Lo, Ko) |
| *Ziziphus nummularia* (Burm. f.) Wt. & Arn. Rhamnaceae | Chhota-barokoli | Bandiraposhi/645 | Fruits are edible. (S, Bh, Mu, Lo, Ko) |
| *Ziziphus oenoplia* (L.) Mill. Rhamnaceae | Kontei-koli | Pagadadiha/1033 | Fruits are edible. (S, Mu, Bh, K) |

* = Cultivated species.

## DISCUSSION

Agriculture and hunting is the main occupation of tribal people of the district. Besides the crops and cultivated seasonal vegetables, they depend on 124 plant species used for human consumption. Out of these, 47 species are used as vegetables, 32 species for fruits and 41 species for other purposes. The plants marked with an asterisk (*) are cultivated and included in this paper to make the work a bit wholesome.

It is observed that the maximum plant part used by tribals is fruit (57 species), followed by leaves (32 species) and underground parts (26 species), seeds (15 species), stem (14 species), flowers (6 species) and gum (3 species) respectively. Figure 1 indicates that the Santals are the aboriginal of this geographical region where as Kondhas are the migrants from the adjoining areas. The large number of species being used by the santals prove their inclination with the vegetation at higher level than other tribals. To over come the Global population it is the high time to conserve the natural genetic resources both *in-vitro* and *in-vivo*.

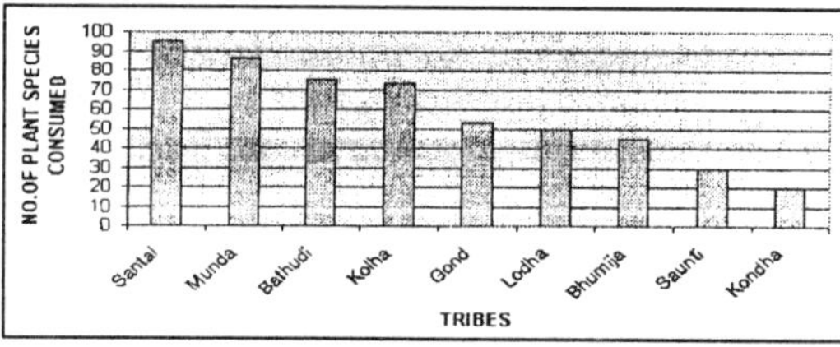


Fig. 1. Histogram showing the no. of species consumed by tribals.

## ACKNOWLEDGEMENTS

The authors wish to acknowledge the Head, Post-Graduate Department of Botany, Ravenshaw College, Cuttack to provide the laboratory facility and Divisional Forest Officer, Similipal range, Mayurbhanj for extending co-operation during the field survey.

## REFERENCES

BAL, S. N. 1942. Useful plants of Mayurbhanj district in Orissa. *Rec. Bot. Surv. India* 6 : 1-119.

BRAHMAM, M. & H.O. SAXENA 1990. Ethnobotany of Gandha mardhan Hills - some noteworthy folk medicinal uses. *Ethnobotany* 2 : 71-77.

HAINES, H. H. 1921-25. *The Botany of Bihar and Orissa,* 6 parts. London.

NAYAK, P.K., R.K. NAYAK & B.P. CHOUDHURAY 2003a. A check list of medicinal plants of Kalahandi district in Orissa. *J. Econ. Taxon. Bot.* 27 (3) : 519-532.

————, ———— & ———— 2003b. Some medicinal weeds of Mahanadi delta. *J. Econ. Taxon. Bot.* 27 (3) : 533-538.

JAIN, S.K. 1991. Some magico-religious beliefs about plants among Adibasis of Orissa. *Adibasi* 12 : 38-44.

SAXENA, H.O. & M. BRAHMAM 1994-1996. *Flora of Orissa,* Vols. I-IV. Bhubaneshwar.

———— & ———— 1989. *The Flora of Similipahar (Similipal) Orissa with particular reference to the Potential Economic Plants.*

———— & P. K. DUTTA 1975. Studies on ethnobotany of Orissa. *Bull. Bot. Surv. India* 17 : 124-131.

TIAWARI *et al.* 1999. *References of Orissa.* Enterprising Publisher, Bhubaneswar, Orissa.

WARRIER, P.K. *et al.* 1993. *Indian medicinal plants.* Orient Longman Publisher, London.

# ETHNOBOTANY OF SOME HEALING PLANTS OF KUNIHAR FOREST DIVISION, DISTRICT SOLAN (HIMACHAL PRADESH), INDIA

SAROJ VERMA & N.S. CHAUHAN

*Department of Forest Products, Dr. Y.S. Parmar University of Horticulture and Forestry, Nauni-Solan 173 230 (HP) India*

## ABSTRACT

Our knowledge of the intimate relationship between early man and plants has come to us mainly through surviving traditions. This relationship which forms the core of inter-disciplinary science, known as Ethnobotany, has attracted much attention. Kunihar forest division falling under the administrative jurisdiction of district Solan boasts of rich plant diversity. The present course of study was undertaken for making preliminary survey of plant wealth of Kunihar forest division, district Solan of Himachal Pradesh. An analysis of data in the present communication reveals the uses of 98 species belonging to 47 families for health care among the people of the area.

## INTRODUCTION

The history of plant exploration is concomitant with the evolution of human beings. Our ancestors studied the plants around them to meet their basic requirements, which laid the foundation of civilization. The enquiry of early man into nearby plant wealth was primarily to cater for basic needs like food, clothing, shelter etc. Incidentally some of them might have proved to be curative against certain ailments, which took wider range and scope by the experimentation on human beings with different plant parts through trial and error (Pandey & Verma, 2005). In recent times, Ayurveda and Chinese system of medicine has hit global market becoming a part of alternative system of medicine in many western and developed countries. India has been one of the major exporters of herbal products as well as raw drugs (Lange, 1996). The source of these plants is the forest wealth of India with diversity of plants, which are found in various edapho-climatic areas from Himalayas to Western Ghats. Ethnobotanical studies provide helpful clues of resources to meet local needs, but there have been only limited surveys to reveal this fact even in ethnically, physio-graphically and floristically rich regions such as India (Arora, 1987). Himachal Pradesh, a hilly State situated in the North-West Himalayas, boasts of a wide diversity of plant forms of European, Tibetan and Chinese origin in its sub-tropical, temperate, alpine and cold desert regions (Chauhan & Thakur, 1995).

Geographically, Kunihar forest division is situated between 39°52′ to 31°22′ longitude and 76°36′ to 77°09′ latitudes. This area constitutes a part of Solan district of Himachal Pradesh. The elevation generally varies from 300m to 2200m above mean sea level. This covers an area of 572.37 sq kms. Kunihar forest division has four ranges, viz. Kunihar, Arki, Darlaghat and Kutharh with head quarters at respective areas. In the past plant explorations more emphasis was laid on their systematics rather than their ethnobotanical uses. The present attempt has been designed to explore the ethnobotanical significance of some healing plants of Kunihar Forest division.

The dialect of this area is called 'Paharhi Boli'. Cast system dominates among the people. Brahmins are considered as the top class, who act as priests and 'Kul Purohits' of people. People have great faith in God and Deities. They celebrate many festivals and fairs, viz. Sharad-purnima fair in the month of January-February, Baisakhi fair, Barhi Dhar fair, Sair fair at Arki etc.

## MATERIALS AND METHODS

Extensive field survey of entire area was carried out during the year 1999-2001. The collection of voucher specimens was done during flowering/ fruiting period to facilitate the process of identification. Only angiospermic plants including seasonal, annual, perennials and woody plant species were collected and no lower plants were collected. Identification of plant species was done according to the field characteristics by consulting various floras and comparison in the Herbarium of the Department of Forest Products, Dr. Y.S. Parmar University of Horticulture and Forestry, Nauni-Solan.

Inventory of the entire floristic elements was prepared. The documented information was used to delineate medicinal and aromatic plants. The information regarding the traditional knowledge, local uses of the plants of the study area containing the information about the local names of the plant, parts used, purpose for which used, mode of administration and curative properties were recorded through interviews and discussion with elderly people, herbal healers, local vaids and rural women and the same have been included in the results.

## GENERAL VEGETATION

Vegetation of the area is mainly dominated by broad leaved mixed vegetation. *Acacia catechu, Albizzia chinensis, Anogeissus latifolia, Cassia fistula, Pinus roxburghii, Terminalia bellirica, Terminalia chebula, Toona hexandra, Pistacea khinjuk* are the main tree species of the area. Shrubby elements constitute chiefly *Punica granatum, Berberis lycium, Tecoma stans, Himalrandia tetrasperma, Flacaurtia indica* etc. Bauhinia vahlii, a woody climber is quite conspicuous in ravines and slopes. Other climbers include *Smilax aspera, Rosa moschata, Cryptolepis buchananii, Celastrus paniculata, Jasminum arborescens J. dispermum, J. humile, Mellettia extensa* etc. Herbaceous element is constituted by mixture of grasses, like species of Phragmitis, Paspalum, Dichanthium, Heteropogon, Bothriochloa, Chrysopogon, Chloris etc and species of Cassia, Rhynchosia, Lathyrus, Bidens, Achyranthes, Solanum and various weeds of agriculture fields like *Cynodon dactylon, Cyperus rotundus, Malvastrum coromandelianum* etc.

## RESULTS

The details of healing plant species of the area have been arranged family wise alongwith their latin name, local names and parts used.

### FABACEAE

***Abrus precatorius*** Linn.

*Local names :* Rati, Lal Gunja

*Uses :* Decoction of the leaves is given in cough and cold. The roots are crushed in water and filtered. This filtrate is given to reduce the effect of poison. The paste of the leaves is applied on the swellings. Bruised seeds are used as poison. Seeds are put in eyes for bringing out foreign particles by making it move in circles under the eyelids.

***Indigofera cassioides*** Rottl. ex DC.

*Local name :* Kathi

*Uses :* Powdered stem bark is used in toothache. Flowers are used in *Triphala Churan.* Seeds are dried and applied on hairs to check the hair fall and breaking of hair.

***Pueraria tuberosa*** (Roxb. ex Willd.) DC.

*Local names :* Soral, Vidarikand

*Uses :* Flowers are cooling, aphrodisiac. Roots are used as demulcent and refrigerant in fever.

### CAESALPINIACEAE

***Bauhinia variegata*** Linn.

*Local names :* Karyalo, Kachnar

*Uses :* Bark decoction used as blood purifier and 25 ml of it also given twice a day for 15 days in tuberculosis. The flower powder is used for haemorrhage. Decoction of the leaves is used in headache caused by malarial fever.

***Cassia fistula*** Linn.

*Local name :* Amaltas

*Uses :* The fruit pulp is used as laxative. The bark of pod is crushed with kesar (*Crocus sativus*), Mishri and rose water. This paste is given for easy and painless delivery. The paste of leaves is applied on ringworm and in few days, it removes even the scars of ringworm. Ash of the pod is given with honey in cough to the children

### MIMOSACEAE

***Mimosa rubicaulis*** Lam.

*Local name :* Kandyari

*Uses :* Leaves are crushed and applied on wounds. The paste of roots with black pepper is used on boils.

### BIGNONIACEAE

***Tecoma stans*** (L.) Juss. ex Kunth

*Local name :* Yellow Elder

*Uses :* Roots are used as antidote to snake or scorpion venom. The decoction of roots is used in urinary problems.

### LAMIACEAE

***Ajuga bracteosa*** Wall. ex Benth.

*Local name :* Nilkanthi

*Uses :* It is used as a bitter tonic. Leaf extract is applied on hairs to kill the lices. Leaf juice is also used in earache. Infusion of leaves is given in pneumonia, bronchitis, gastric pain and typhoid fever.

***Colebrookea oppositifolia*** Smith

*Local names :* Dushno, Pansra, Binda

*Uses :* The leaves are chewed in case of mouth ulcers. The paste of leaves is also applied on wounds and bruises.

***Micromeria biflora*** (Buch.-Ham. ex D.Don) Benth.

*Local name :* Jungli Ajwain

*Uses :* Plant is ground and used on Kachhrali (Ulcer in armpit). This paste is also used on wounds in cattle. The juice of leaves is used in stomachache.

***Ocimum tenuiflorum*** Linn.

*Local name :* Tulsi

*Uses :* The leaves are used as mosquito repellent and kept in a corner in the room. Plant is used in cough and cold. Decoction of the plant is used in 'vata rog'.

***Origanum vulgare*** Linn.

*Local names :* Sathra, Jakhmi-buti

*Uses :* Seed are used in stomachache. The paste of leaves and terminal shoots along with 2-3 black pepper is applied on boils, cuts, wounds and ulcer. A piece of the root bounded in a cloth is tied to the neck of small kids as a protective measure against conjunctivitis.

***Plectranthus mollis*** (Aiton) Spreng.

*Local name :* Chhichhdi

*Uses :* Crushed leaves are used to apply on wounds and cuts. Leaves are applied on septic between toes.

***Roylea cinerea*** (D. Don) Baill.

*Local name :* Kadwo

*Uses :* Tablets are made of the leaves. These tablets with gur are taken in body allergy. These tablets are also given empty stomach to the diabetic patients.

VIOLACAE

***Viola betonicifolia*** J. Smith

*Local name :* Banaksha

*Uses :* Dried flowers are used for cough and cold.

SAPINDACEAE

***Dodonaea viscosa*** (Linn.) Jacq.

*Local name :* Mehndru

*Uses :* Leaves are boiled in water and this water is applied on stiff areas. The paste of leaves is also used on burns. Seeds are edible. About 20ml decoction of leaves is taken once a day for two days to stop excess menstrual flow.

***Sapindus mukorosi*** Gaertn.

*Local name :* Ritha

*Uses :* Unripe fruits are ground and taken with honey in gout.

AMARANTHACEAE

***Achryanthes aspera*** L.

*Local name :* Puthkanda

*Uses :* In painful scorpion stings, the paste of the roots is applied. Ash of the plant is used to cure asthma.

ARACEAE

***Acorus calamus*** Linn.

*Local names :* Vach, Bare

*Uses :* Dry rhizomes are ground and mixed in mustard oil and massaged over the aching joints. Rhizomes are burnt in earthen pot and are made into ash and in this ash Kuth (*Saussurea costus)* and Sendha salt are mixed. This power is given in malaria.

***Scindapsus officinalis*** (Roxb.) Schott

*Local name :* Gathiala

*Uses :* The 'Halwa' of the plant is used in gout. Leaves are given in *Bsoun* (a disease in which teeth start falling and there is smell in the milk) to cattle. The decoction of the fruits is given as expectorant in asthma.

ADIANTACEAE

***Adiantum capillus-veneris*** Linn.

*Local name :* Hansraj

*Uses :* Tea made from the leaves is widely used in cough, especially from cold. It is also used in fever.

***Adiantum incisum*** Forsk.

*Local name :* Mayurshikha

*Uses :* Fronds are mixed in butter and used for controlling the internal heat of the body. These are also used in cough and fever.

AGAVACEAE

***Agave americana*** Linn.

*Local name :* Ramban

*Uses :* Leaves are used to increase the menstrual discharge in ladies. Young shoots are used as vegetable and are also pickled. This pickle is considered 'Niroga' (keeping far from diseases).

VERBENACEAE

***Lantana camara*** Linn.

*Local name :* Phul-lakri

*Uses :* Crushed leaves are applied on swellings. The paste of the leaves is also used in snake bite to reduce the effect of poison.

***Premna latifolia*** Roxb.

*Local name :* Bakar

*Uses :* Leaves are boiled in water and bath is taken in that water in case of body allergy.

***Vitex negundo*** Linn.

*Local name :* Bana

*Uses :* Flowering tips are chewed in case of mouth ulcer. The leaves are used in gastric problems and also in leprosy. Leaves of Bana and Chuli (*Albizia chinensis*) are ground and used as paste on snake and scorpion bites.

POLYGONACEAE

***Persicaria hydropiper*** (Linn.) Spach

*Local name :* Jalpippli

*Uses :* The decoction of plant is given with long pepper and salt to women for irregular periods. The paste of roots is used in toothache. The decoction of roots after periods is given as contraceptive to women.

MALVACEAE

***Malvastrum coromandelianum*** (L.) Garcke

*Local name :* Bala-ka-bhed

*Uses :* Root bark is given with milk to increase the strength. The paste of leaves is applied on the cuts for soon healing. The decoction of leaves is also given in dysentery.

CACTACEAE

***Opuntia dillenii*** (Ker-Gaw.) Haw.

*Local name :* Nagphani

*Uses :* Sharbat (juice) of fruits is used in asthma. Decoction of the fruit with ginger is also given in whooping cough to children.

SANTALACEAE

***Osyris quadripartita*** Salz. ex Decne.

*Local names :* Jangli Chai, Kaile

*Uses :* Leaves and bark of the stem are used to make tea. Crushed leaves are applied on cuts and other wounds.

PLUMBAGINACEAE

***Plumbago zeylanica*** Linn.

*Local names :* Chitrak, Chicha

*Uses :* The paste of root bark after warming is used in treatment of leucoderma and decoction of the roots with cow milk is given in fever. The paste of the root is applied on ulcers. Chitrak, Hard, (*Terminalia chebula*), Aonla (*Phyllanthus emblica*) and long pepper (*Piper longum*) are taken in quantities of 6 masha and are ground.

This powder is boiled in half litre water. When it remains half, filter it and let it be cool. This decoction is given in morning and evening in all type of fevers.

PAPAVERACEAE

***Argemone mexicana*** Linn.

*Local names :* Satyanashi, Swarnkshiri

*Uses :* Plant juice is used to cure leprosy. It is also used as blood purifier. Seeds are used to remove constipation. Paste of roots is applied on skin diseases. Plant juice is also used in ophthalmia.

SCROPHULARIACEAE

***Bacopa monnieri*** (L.) Wettst.

*Local name :* Brahmi

*Uses :* Plant is used in urinary troubles and in constipation. To increase memory fresh leaves should be eaten before sun rise.

ACANTHACEAE

***Justicia adhatoda*** Linn.

*Local names :* Basuti, Basti

*Uses :* One table spoon of leaf juice is given thrice a day to cure cough. The juice of leaves provides relief in problems related to nose, throat and urinary infection. In case of tuberculosis, leaf juice with honey is consumed in the morning and evening. Root powder is taken with water in malarial fever.

GRAMINAE

***Cynodon dactylon*** (L.) Pers.

*Local name :* Doob

*Uses :* The juice of plant is put in nose to check the bleeding. This juice is used to reduce the acidity and constipation.

DIOSCOREACEAE

***Dioscorea esculenta*** (Lour) Burkill

*Local name :* Ratalu

*Uses :* Seeds are crushed and applied on the boils. Roots are used as vegetable.

SOLANACEAE

***Datura stramonium*** Linn.

*Local name :* Dhatura

*Uses :* Leaves of Dhatura, Opium poppy (*Papaver somniferum)* and dried ginger are ground and applied on swellings. Smoking of the leaves is good for asthma patients.

***Nicotiana tabacum*** Linn.

*Local name :* Tambaku

*Uses :* Leaves are used in smoking. The paste of leaves is applied on scorpion stings and in skin diseases. Oil is extracted from seeds. Plant is given in stomachache in cattle.

***Solanum myriacanthum*** Dunal

*Local name :* Chhakmoli

*Uses :* Fruits are given in toothache to cattle. Fruits are crushed and applied on the wounds for soon healing.

***Solanum nigrum*** Linn.

*Local names :* Makoi, Pahatkoin

*Uses :* The juice of the plant is given in leprosy. The vegetable of this is also used in stomachache. Fruits are edible.

***Solanum virginianum*** Linn.

*Local name :* Kateri

*Uses :* Roots are used in *Dashmula*. Root decoction used in cough, asthma and pain in chest. The juice of flowers is given to infants in stomachache.

***Withania somnifera*** (L.) Dunal

*Local name :* Ashwagandha

*Uses :* 1 masha 'Sattu', 1 tola butter and 1 tola Mishri are mixed and used in 'Viray Vikar'. Green plant is crushed and filtered with cloth. This filtrate is put in earthen pot and filled with water. This is kept for 3-4 hours. After that water is filtered and the material which get settled down is taken out and dried. This material is known as 'Sattu' of Asgandh.

EUPHORBIACEAE

***Euphorbia hirta*** Linn.

*Local names :* Dughdika, Laldudhi

*Uses :* Juice of the plant is given to stop bleeding in piles.

***Euphorbia royleana*** Boiss.

*Local name :* Chhuro

*Uses :* The latex is applied on the painful parts. Care should be taken that water should not touch on that part and milk should be kept away from the eyes. The latex is filled in the hollow cavities of decayed teeth.

***Falconeria insignis*** Royle

*Local name :* Chhirna

*Uses :* Milky latex is applied on warts and also effective in alopecia.

***Mallotus philippensis*** (Lamk.) Muell.-Arg.

*Local names :* Kamal, Kamela

*Uses :* Seeds are dried in shade and then ground. Small tablets are made of this powder and are given with cold water in jaundice. Decoction of leaves is good for the 'Vat Rog'. In Polio (myelitis) small and thin roots are tied on the neck. Bark and leaf powder are mixed with butter and applied on sores and boils.

***Ricinus communis*** Linn.

*Local name :* Erand

*Uses :* Leaves are heated and tied over aching joints and swellings caused by Vat Rog. This is also useful in backache. In stomach swelling (Afara) in children, leaves with warm ghee are tied on belly. Juice of leaves mixed in mustard oil is used on burns. About 10g paste of the roots is taken in the morning for three days for abortion.

MENISPERMACEAE

***Cissampelos pareira*** Linn.

*Local names :* Patindu, Patha

*Uses :* Root is laxative, useful in dysentery and urinary troubles. Crushed leaves are applied on septic ulcers. Its roots, black pepper, asafoetida and ginger are crushed and mixed with honey and this is used is colic and indigestion.

***Tinospora cordifolia*** (Willd.) Miers ex Hook. f. & Thoms.

*Local name :* Giloe

*Uses :* Giloe stem, dried ginger, black pepper and long pepper are powdered and taken 2 masha with honey in cough. Powder of root bark is used with curd in asthma. This plant is also used in fever.

RANUNCULACEAE

***Clematis gouriana*** Roxb. ex DC.

*Local name :* Belkum

*Uses :* Decoction of the roots is given thrice a day for three days in stomachache. If the juice of leaf touches the skin, it forms boils.

***Delphinium denudatum*** Wall. ex Hk. f. & Th.

*Local name :* Nirbisi

*Uses :* Roots are dried and powdered and this powder is massaged on head to kill lices. It also increases length of hair.

***Ranunculus sceleratus*** Linn.

*Local name :* Minkali

*Uses :* Plant is crushed and tied on ulcers on feet. The paste of roots with mustard oil is used on swellings.

***Thalictrum foliolosum*** DC.

*Local name :* Jhalamala

*Uses :* Root powder is used in ophthalmia. The paste of root is applied around the forehead in headache. The root powder in the dose of 15g is used thrice a day for two weeks in jaundice and in fever.

CUSCUTACEAE

***Cuscuta reflexa*** Roxb.

*Local name :* Akashbel

*Uses :* The plant is crushed with Munnaka and half glass juice is taken to remove the worms is intestine. Ladies apply the juice of plant on hairs to increase the length of the hairs.

CONVOLVULACEAE

***Evolvulus alsinoides*** (Linn.) Linn.

*Local name :* Shankhpushpi

*Uses :* The plant is ground with cold water and is used to increase age, strength, beauty, appetite and intelligence.

ANACARDIACEAE

***Mangifera indica*** Linn.

*Local name :* Aam

*Uses:* The ash of the leaves along with water or oil is applied on the burns. It gives much relief. Powder of young leaves of mango is useful in diabetes. Decoction of the bark or the leaves is used to wash the wounds to heal them soon. For bleeding gums, the leaves of mango are given to chew.

***Pistacia khinjuk*** Stocks

*Local name :* Kakar-singhi

*Uses :* Powder of galls, pipramul (*Piper longum*) and dried ginger are mixed in equal quantity and are given in fever and Vat Rog. Galls, Kaphal (*Myrica esculenta*), dried ginger, black pepper, Kalunji is mixed in equal quantities and is ground. This powder is given with honey in hiccups, fever, cough and other throat diseases.

ROSACEAE

***Prinsepia utilis*** Royle

*Local names :* Bhekhlo, Bhekhal

*Uses :* The paste of roots is used on wounds for soon healing.

BALSAMINACEAE

***Impatiens balsamina*** Linn.

*Local name :* Binchi

*Uses :* The paste of the plant is considered cooling and is applied on head. Crushed leaves are applied on septic between toes.

LYTHRACEAE

***Woodfordia fruticosa*** (L.) Kurz

*Local name :* Dhawi

*Uses :* Powder of the flowers is given in dysentery and cough. The paste of leaves is used in skin diseases. The juice of leaves is applied three times a day for one week in conjunctivitis.

PUNICACEAE

***Punica granatum*** Linn.

*Local names :* Daru, Anar

*Uses :* The juice of seeds, ghee and syrup of sugar is given in case of vomiting. The bark 50-100 gm is cooked in the water. When the water remains half, let it be cool. This water kills all worms in intestine. Seeds are cardiac and good for digestion. Take 50-60 gm juice of sweet seeds in iron pot and keep it in open for a night. In the morning mix *mishri* in this juice. This juice is given in jaundice.

OLEACEAE

***Jasminum arborescens*** Roxb.

*Local name :* Chameli

*Uses :* The juice of leaves with black pepper and garlic acts as an emetic and also used in obstructions of the bronchial tubes by viscid phlegm.

***Jasminum humile*** Linn.

*Local name :* Pili-chameli.

*Uses :* The paste of leaves is applied on skin diseases, wounds and ulcers. Leaves are chewed in toothache.

JUGLANDACEAE

***Juglans regia*** Linn.

*Local name :* Akhrot

*Uses :* Nut is roasted on fire and is given in cough. Nut is kept in water and then ground. This paste is applied on navel to cure dysentery. Walnut oil is put in the nose to cure paralysis. In kidney stone, one spoon powder of shell of the nut or powder of the nut is given in the morning and evening. The stem bark and leaves are used as Datun (Tooth stick). It prevents the cavities in teeth. The shell of the nut is crushed and applied in the laceration of toes.

COMPOSITAE

***Blumea mollis*** (D. Don) Merr.

*Local names :* Jangli-Tambaku, Kukkuradru

*Uses :* Decoction of roots is taken twice a day for 5 days in urinary problems. The decoction of leaves is given in stomachache. Oil is extracted from seeds.

***Crisium wallichii*** DC.

*Local name :* Chiul

*Uses :* Roots are cooling. In fever the paste of roots is applied on the forehead.

***Gerbera gossypina*** (Royle) P. Beauv.

*Local name :* Kupheru

*Uses :* The white cotton like coating on the underside of the leaf is used as tincture for staunching wounds.

***Senecio desfontainei*** Druce

*Local name :* Jaari

*Uses :* The decoction of roots of this plant and flowers of Bahera (*Terminalia bellirica*) is given in cholera and periodic fever.

LILIACEAE

***Aloe vera*** (L.) Burm. f.

*Local names :* Ghikawar, Gwarpatha, Ghritkumari

*Uses :* Leaves are used as vegetable. The mucilaginous inner portion is warmed and applied on boils to induce puss formation. Juice of the leaves is used on burns.

***Asparagus adscendens*** Roxb.

*Local names :* Sansarmuli, Safed Muli

*Uses :* Roots are used as tonic, demulcent and in dysentery with milk. In case of tonsils, its root is tied in the neck. In headache, the paste of roots is applied on forehead.

PRIMULACEAE

***Androsace rotundifolia*** Hardw.

*Local names :* Nirodhak buti, Golpatti phool

*Uses :* Leaves are eaten with gur and ghee in correcting menstrual flow, thus helping avoiding conception in women.

COMBRITACEAE

***Anogeissus latifolia*** (Roxb. ex DC.) Wall. ex Guill. & Perr.

*Local names :* Chhal, Dhaw

*Uses :* Decoction of the root bark is used in the treatment of biliousness. In 'Munadi fever' (periodic fever), decoction of leaves, long pepper, black salt and harad are given.

***Terminalia bellirica*** (Gaertn.) Roxb.

*Local name :* Bahera

*Uses :* Fruits are used in bronchitis. It is one of major constituents of 'Triphla Churan'. Bark is diuretic, useful in anaemia and leucoderma. The powder of fruits in the dose of 20 gm is used as tonic and laxative. It is also used thrice a day for one week in piles and dyspepsia.

***Terminalia chebula*** Retz.

*Local name :* Harad

*Uses :* If fruit is eaten as such it cures dyspepsia (*Ajiran*). If eaten as powder, it is laxative and if used after boiling cures dysentery and after roasting it cures *Tridosha* (Vat, Pitt and Kapha). Harad cures stomachache and other disease of stomach. In excess sweating, take powder of harad and massage it on body. After sometime take bath with cold water. It reduces the excess sweating.

NYCTAGINACEAE

***Boerhavia repens*** Linn.

*Local name :* Punarnava

*Uses :* Plant is *shothahara, mootrala, hridya, raktavardhaka* i.e. reducing swelling, diuretic, cardiac and blood purifier. Tender shoots are eaten as vegetable. Plant is smoked to treat asthma, used in dropsy and jaundice. Plant paste is applied on scorpion sting. Powder of root with sugar is taken in whooping cough. Plant powder is rubbed on skin to reduce the swelling.

RUBIACEAE

***Catunaregam spinosa*** (Thunb.) Tirveng.

*Local name :* Madanphal

*Uses :* One full hand seeds with long pepper (*Piper longum*) are used in sinus. It cures sinus in one day, but the quantity of seeds should not be more, otherwise it causes bleeding from nose.

***Rubia cordifolia*** Linn.

*Local names :* Manjishtha, Manjith

*Uses :* The decoction of roots with sugar is given to women to cure irregular menstruation. Infusion of the plant with mustard oil is used in eyes and ear diseases.

BOMBACACEAE

***Bombax ceiba*** Linn.

*Local names :* Simbal, Semal

*Uses :* Thorns are crushed in cow milk and paste is applied on pimples. Paste of roots is also taken with unboiled cow milk to regulate the menstruation and with black pepper, it is used to control the white discharge. Cotton is used in pillow stuffing.

CELASTRACEAE

***Celastrus paniculatus*** Willd.

*Local name :* Malkangni

*Uses :* The bark of root is crushed and paste is applied on wounds for soon healing. Paste of roots is given to women as abortificient. Seeds are laxative, stimulant and used in leprosy. It is said that if the seeds are eaten for 62 days, it increase the memory. For this, start from one seed first day, two seeds on second day and so on up to 31 seeds on 31$^{st}$ day. Thereafter decrease the number of seeds one by one so that on 62nd day there is one seed to eat.

CANNABACEAE

***Cannabis sativa*** Linn.

*Local names :* Bhang, Sulfa

*Uses :* Leaves are rubbed with hands to make bhang. It is narcotic and sedative. This bhang is applied on the boils. Charas is prepared from the resinous exudates from leaves. Ganja is prepared from the dried flowering or fruiting tops of female plant. Bhang is used as an appetizer. Leaves are boiled in water and that water is applied on sprain or foot is kept in that water.

APOCYNACEAE

***Carissa carandas*** Linn.

*Local name :* Karaunda

*Uses :* Fruits are used in cough, but if fruits are eaten without having cough, they may cause cough. Leaves are given at the commencement of remittent fever. The leaf juice with honey is used in whooping cough.

***Holarrhena pubescens*** (Buch.-Ham.) Wall. ex G. Don

*Local names :* Indrajau, Kurchi

*Uses :* The root powder is used as snuff (Naswar) to check the bleeding from nose. Seeds and bark of the stem are given in dysentery. In case of dropsy, the powder of the bark is rubbed over the body.

***Nerium oleander*** Linn.

*Local name :* Lal Kaner

*Uses :* Leaves are used in Hriday Rog (Cardiac problems). Leaves are also used in snake bite. Leaves cooked in mustard oil are rubbed on body in case of allergy. Snuff of the dry leaves is used for sneezing and it relieves headache.

***Quirivelia frutescens*** (L.) M.R. & S.M. Almeida

*Local names :* Dudhi bel, Belkamu

*Uses :* Roots are boiled in water and when it remains 1/4$^{th}$, it is given with the paste of black pepper as a blood purifier

***Rauvolfia serpentina*** (Linn.) Benth. ex Kurz

*Local name :* Sarapgandha

*Uses :* The bark of root is chewed to check the vomiting. 10 gm fine paste of roots is used as tranquilizer in hypertension (high BP). The paste of leaves with decoction of black pepper is used in pneumonia.

CONIFERAE

***Cedrus deodara*** (Roxb. ex Lambert.) D. Don

*Local name :* Deodar

*Uses :* The decoction of bark is used in diarrhoea and dysentery. From wood, oil is extracted which is used in skin diseases.

APIACEAE

***Centella asiatica*** (L.) Urban

*Local name :* Brahmi

*Uses :* Leaves are used as brain tonic, in skin diseases, leprosy and as a blood purifier.

***Ferula jaeschkeana*** Vatke

*Local names :* Ghora ghass, Jangli Heeng.

*Uses :* An essential oil is extracted from the roots and seeds. This plant is given to horses to increase their strength. Hence called ghora ghass.

CHEILANTHACEAE

***Cheilanthes farinosa*** (Forsk.) Kaulf.

*Local name :* Hansraj

*Uses :* Decoction of the leaves is used in fever. The juice of the plant is mixed with honey and is taken after meal by the weak people suffering from tuberculosis.

BERBERIDACEAE

***Berberis lycium*** Royle

*Local name :* Kashmal

*Uses :* Roots are boiled in water and concentrated to make *Rasount* which is used in ophthalmia, earache and in gum swellings.

ERICACEAE

***Rhododendron arboreum*** Sm.

*Local name :* Burans

*Uses :* Young leaves are poisonous. Crushed leaves are applied on forehead to cure headache. *Sharbat* (Juice) is prepared form flowers. The dried flowers are fried in ghee and used to check the bloody dysentery.

RHAMNACEAE

***Helinus lanceolatus*** Brandis

*Local name :* Gur-mar-buti

*Uses :* If one eats the root bark, it makes the sweet buds of tongue inert. After this any sweet thing does not give any sweet taste and one can eat more amounts of sweets.

***Ziziphus xylopyrus*** (Retz.) Willd.

*Local names :* Jangli Ber, Chulna

*Uses :* The decoction of stem bark alongwith paste of long pepper and ginger is used in hysteria treatment. Leaf paste is applied on forehead in headache. In case of urinary troubles, two and a half leaves are chewed thrice a day for 15 days.

MORACEAE

***Ficus racemosa*** Linn.

*Local name :* Umbro

*Uses :* The juice of roots, sugar and black zeera. are mixed together. This juice is given to treat the 'Sujak Rog' (gonorrhoea). Green and dried fruits are ground in the water and mixed with Mishri. This juice is given in case of piles and diarrhoea.

***Ficus palmata*** Forsk.

*Local names :* Fegra, Daugla

*Uses :* Dip two dry fruit in water for a night and eat them in the morning. Similarly 2 dry fruits are dipped in the water in the morning and eaten in the evening. It removes pile in one week only. Latex is used on ringworm. Fruits are also used in constipation and diabetes. For skin diseases and purification of blood, 5 figs and 11 munakkas are boiled in 250 ml milk and are given to the patient.

***Ficus religiosa*** Linn.

*Local name :* Pipal

*Uses :* In urinary troubles, 10 gm bark is boiled in half litre water and when it remains 1/4$^{th}$ it is filtered. This filtrate should be taken twice a day. If in this decoction honey is mixed, it is used to purify the blood. Powder of bark is sprinkled over the wounds for soon healing.

RUTACEAE

***Murraya koenigii*** (L.) Spreng.

*Local name :* Gandhela

*Uses :* Leaves are eaten raw in dysentery. Fruits are edible. Leaves are used to flavour the *Curries.* To cure diabetes, 10 leaves are given daily to eat for one week.

***Zanthoxylum armatum*** DC.

*Local names :* Timbro, Tirmir

*Uses :* Fruits are used in toothache. The leaves and fruits are chewed in the teeth enamel disease. The twigs are used as tooth brush and mouth fresheners.

## CONCLUSION

Results show that the practice of ethnobotany is still a matter of cultural heritage of the area. People are making maximum use of the plant species for meeting their day to day household demands. The ethnobotanical knowledge about use of plants against diseases like rheumatism, cardiac disorders, headache, diarrhoea, dysentery, stomachache, fever etc is still practiced in majority of households. Of course, scientific investigations through bio-chemical, clinical and pharmacological screening are required for verification of the folklore and bring out newer and potential sources of medicament for the welfare of the society.

## ACKNOWLEDGEMENTS

Authors are thankful to Department of Forest Products, College of Forestry, Nauni for providing essential assistance for conducting the research and also grateful to the informers of the area who provided valuable information to complete the research work.

## REFERENCES

ARORA, R. K. 1987. Ethnobotany and its role in the domestication and conservation of native plant genetic resources. In Jain, S. K. (Ed.) *A Manual of Ethnobotany,* pp. 94-102. Oxford and IBH, New Delhi.

CHAUHAN, N. S. & SANJEEV THAKUR 1995. Conservation of Forest Genetic Resources of HP. In Press with *Indian J. For. Res.,* Dehra Dun.

LANGE, D. 1996. Medicinal plant market study in Germany-State of project. *Medicinal Plant Conservation* 2 : 9-10.

PANDEY, H.P. & B.K. VERMA 2005. Phytoremedial Wreath : A traditional excellence of healing. *Indian Forester* 131 : 437-441.

# STUDY OF NON-CONVENTIONAL EDIBLE PLANTS USED BY DIFFERENT COMMUNITIES OF BARAK VALLEY, ASSAM (N.E. INDIA)

AJIT KUMAR DAS, B.K. DUTTA* & G.D. SHARMA**

*S.S. College, Hailakandi, Assam – 788151 (India)*
**Department of Ecology & Environmental Sciences, Assam University, Silchar-788011 (India)*
***Nagaland University, Kohima-797011 (India)*

## ABSTRACT

A survey in different areas of Barak Valley has been conducted in different seasons of the year to identify the non-conventional edible plants. The present paper deals with 70 edible plants belonging to 45 families of the flowering and non-flowering plants. Out of which some of the plants are used as edible green vegetable and edible fruits, and for other purposes. The use of these plants varies from place to place. Different parts of the plants are used for different purposes and by adopting different methods. The plant species are arranged alphabetically and for each species scientific name, vernacular name and their uses are given.

## INTRODUCTION

Barak Valley is surrounded by North Cachar Hills and Jaintia Hills in the North, in the South by Mizoram, in the East by the State of Manipur and in the Western side by Tripura and Sylhet district of Bangladesh. The area has an altitude of 26-27 m above MSL and this falls under 24°8' and 25°8' N latitude and 92°15' East longitude. Geologically, Barak Valley is younger than Brahmaputra Valley. It is entirely alluvial in origin and is composed of pebbles, sand, silt, clay and sometimes a mixture of sand and clay containing decomposed vegetable matter. The pH of the soil ranges from 3.5 – 5.6. The temperature regime of the area has a mean maximum temperature of 30.2°C & a minimum temperature of 19.6°C accompanied by fairly high relative humidity. The area experiences a humid subtropical climate. The warmest period in this area is May – September while the December – January is the coldest and dry time of the year. The rainfall starts in the month of March and it reaches its maximum in June – August.

In N.E. India reports on the ethnobotanical work are available (Borthakur, 1996; Gangwar & Ramkrishna, 1990; Rao & Neogi, 1980; Rao & Jamir, 1990 etc). In Barak Valley also some reports are available on ethnobotany : ethnomedicinal plants used by Barman & Manipuri communities (Das & Sharma, 2002), Dwelling Courtyard plants of Karimganj district (Das *et al.*, 2003) etc.

The present work has been carried out to explore the edible plants of Barak Valley. Local markets of the different areas of Barak Valley have been surveyed to identify the species. In some cases information has been gathered from the local people. Plants were collected from the different parts of the valley covering all the seasons of the year. The herbarium of the plants have been prepared as per the methods described by Jain & Rao (1977). For authentic identification of the plants, a number of floras and monographs have been consulted, especially the flora of British India Vol. 1-7 (Hooker 1872 –1897), Flora of Assam Vol. 1-4 (Kanjilal *et al.*, 1934 – 40), Flora of India, Vol. 1-3 (Sharma *et al.*, 1993) and finally the identification was confirmed by consulting the herbarium of the Botanical Ssurvey of India, Eastern Circle, Shillong. Voucher specimens are deposited in the Deptt. of Life Science. The species are arranged in alphabetical order with the name of the family, vernacular name, as well as by mentioning whether it is cultivated or wild. The abbreviations for the communities are used as : Bengali (Beng), Manipuri (M), Reang (R), Kuki (K), Barman (B) H'mar (H) & T.G.C (Tea Garden Community).

Fig. 1. H'mar tribes women selling wild/ cultivated edible plants in the local market.

## ENUMERATION OF EDIBLE PLANT SPECIES

| Name of the species | Vern. name | Habit | Parts used |
|---|---|---|---|
| *Alternanthera philoxeroides* (Mart.) Griseb. (Amaranthaceae) | Sanche (Beng.); Phakchet (M) | Wild | Leaf |
| *Aegle marmelos* (L.) Correa (Rutaceae) | Bel (Beng.); Heirikhagsk (M) | Planted | Fruits & leaf |
| *Alternanthera sessilis* (L.) R. Br. ex DC. (Amaranthaceae) | Salincha, Allencha (Beng); Phakchet (M) | Wild | Leaf |
| *Amaranthus spinosus* L. (Amaranthaceae) | Katanota (Beng); Chengkue (M) | Wild | Leaf |
| *Andrographis paniculata* (Burm. f.) | Kalomegh (Beng); Vubati (M); | Cultivated | Leaf |

| | | | |
|---|---|---|---|
| Wall. ex Nees (Acanthaceae) | Kairate (R) | | |
| *Angiopteris evecta* (Forst.) Hoffm. (Angiopteridaceae) | Gakur (Beng & D); Shamiomamidas (R) | Wild | Rhizome |
| *Anisomeles indica* (Linn.) O. Kuntze (Lamiaceae) | Rujanta (Beng); Thoiding-anguoba (M) | Cultivated | Leaf |
| *Areca catechu* Linn. (Arecaceae) | Supari (Beng); Kwa (M) | Cultivated | Fruits |
| *Artocarpus heterophyllus* Lamk. (Moraceae) | Katal (Beng) | Cultivated | Fruits |
| *Hygrophila schullii* (Buch.-Ham.) Almeida & Almeida (Acanthaceae) | Khula-khara (Beng) | Planted | Leaf |
| *Averrhoa carambola* Linn. (Averrhoaceae) | Kamranga (Beng); Heinijom (M) | Cultivated | Fruits |
| *Azadirachta indica* A. Juss. (Meliaceae) | Nim (Beng); Seizrak (M); Neel-fung (R) | Cultivated | Fruits |
| *Bacopa monnieri* (L.) Wettst. (Scrophulariacae) | Brahmi-sak (Beng) | Cultivated | Leaf |
| *Basella rubra* L. (Basellaceae) | Poi (Beng); Urrkesumban (M) | Cultivated | Leaf |
| *Benincasa hispida* (Thunb.) Cogn. (Cucurbitaceae) | Chalkumra (Beng); Torobot (M) | Cultivated | Fruits & leaf |
| *Boerhavia diffusa* L. (Nyctaginaceae) | Punarnava (Beng); Sanr (T.G.C) | Planted | Leaf |
| *Brassica oleraceae* L. var. *botrytis* L. (Brassicaceae) | Phulkapi (Beng) | Cultivated | Flower |
| *Brassica oleracea* L. var. capitata L. (Brassicaceae) | Bandhakapi (Beng) | Cultivated | Leaf |
| *Cajanus cajan* (Linn.) Millsp. (Papilionaceae) | Arhar (Beng); Mairongi (M) | Cultivated | Leaf, fruits |
| *Carica papaya* Linn. (Caricaceae) | Papaya (Beng); Mongkol (K); Keing Kokia (R) | Cultivated | Fruits |
| *Centella asiatica* (L.) Urban (Apiaceae) | Tankuni (Beng); Peruk (M); Shymasatta(R) | Wild | Leaf |
| *Chenopodium ambrosioides* L. (Chenopodiaceae) | Bethua (Beng); Mangshobi (M) | Wild | Leaf |
| *Cinnamomum tamala* (Buch.-Ham.) Nees & Eberm. (Lauraceae) | Tejpat (Beng, M); Kashpa(R) | Cultivated | Leaf |
| *Coccinia grandis* (L.) Voigt (Cucurbitaceae) | Telakucha (Beng) | Wild | Leaf |
| *Cocos nucifera* Linn. | Narikel (Beng, T.G.C); Pubi (M) | Cultivated | Fruits |

| | | | |
|---|---|---|---|
| (Arecaceae) | | | |
| *Colocasia esculenta* (L.) Schott (Araceae) | Kachu (Beng); Pan (M) | Wild | Leaf, rhizome |
| *Corchorus capsularis* L. (Tiliaceae) | Pat, Naliya (Beng); Parit(R) | Cultivated | Leaf |
| *Coriandrum sativum* L. (Apiaceae) | Danya (Beng) | Cultivated | Leaf & fruits |
| *Costus speciosus* (Koen.) Sm. (Zingiberaceae) | Khongbamtakhei (M); Keri (Beng) | Wild | Root |
| *Curcuma aromatica* Salisb. (Zingiberaceae) | Lam-yaingang (M) | Wild | Rhizome |
| *Curcuma domestica* Valeton (Zingiberaceae) | Haldi (Beng); Yaingang (M) | Cultivated | Rhizome |
| *Cucumis sativus* Linn. (Cucurbitaceae) | Khira (Beng) | Cultivated | Leaf, fruit |
| *Cucurbita maxima* Duchesne ex Lam. (Curcurbitaceae) | Kumra (Beng) | Cultivated | Leaf & fruits |
| *Dillenia indica* L. (Dilleniaceae) | Chalta (Beng); Thaiploy (R) | Wild | Fruits |
| *Dioscorea alata* Linn. (Dioscoreaceae) | Khamoalu (Beng); Rangrangl (R) | Wild | Tuber |
| *Diplazium esculentum* (Retz.) Sw. (Athyriaceae) | Paloisak (Beng); Maihando (R) | Wild | Leaf |
| *Enhydra fluctuans* Lour. (Asteraceae) | Hingcha (Beng) | Wild | Whole plant except root |
| *Hibiscus mutabilis* Linn. (Malvaceae) | Sthalpadma (Beng); Sitalkamal (M) | Ornamental garden plant | Flower |
| *Homalomena aromatica* (Roxb.) Schott (Araceae) | Gandhikachu (Beng); Gandai(B); Kamantri(R) | Wild | Petiole of the leaf |
| *Houttuynia cordata* Thunb. (Saururaceae) | Toningkhok (M) | Wild | Leaf |
| *Ipomoea aquatica* Forsk. (Convolvulaceae) | Kalmisag (Beng) ; Kolamni (M) | Wild | Leaf & stem |
| *Ipomoea batatas* (L.) Lam. (Convolvulaceae) | Rangaialu (Beng) ; Mangra (M) | Cultivated | Whole plant except root. |
| *Leucas aspera* (Willd.) Link (Lamiaceae) | Drunkalosh (Beng) ; Mayang (M) | Wild | Leaf |
| *Litchi chinensis* (Gaertn.) Sonn. (Sapindaceae) | Lichu (M, Beng) | Cultivated | Fruits |
| *Mangifera indica* Linn. (Anacardiaceae) | Am (Beng); Heinou (M); Ami (R) | Cultivated | Fruits |

| | | | |
|---|---|---|---|
| *Marsilea minuta* L. (Marsileaceae) | Susni (Beng); Ishing-yen slun (M) | Wild | Leaf |
| *Mentha arvensis* L. (Lamiaceae) | Pudina (Beng); Nungshi-hidak (M); Kumbaw (R) | Cultivated | Leaf |
| *Mollugo cerviana* (L.) Seringe (Molluginaceae) | Gimasag (Beng) | Wild | Leaf & stem |
| *Momordica charantia* L. (Cucurbitaceae) | Kerela (Beng); Kangla(R)) | Cultivated | Fruits |
| *Monochoria hastata* (L.) Solms (Pontederiaceae) | Chicluri (R) | Wild | Petiole |
| *Moringa oleifera* Lam. (Moringaceae) | Sajina (Beng, M & T.G.C) | Cultivated | Leaf & fruits |
| *Musa paradsiaca* L. (Musaceae) | Kala (Beng); Hali (R) | Cultivated | Fruits |
| *Nyctanthes arbor-tristis* L. (Oleaceae) | Safli (Beng); Singarli (M) | Cultivated | Leaf |
| *Oxalis corniculata* L. | Amrul (Beng); Yensil (M) | Wild | Leaf & stem |
| *Phyllanthus emblica* L. (Euphorbiaceae) | Amlaki (Beng & B); Amlaki (R) | Planted/Wild | Fruits |
| *Piper betle* Linn. (Piperaceae) | Pan (Beng,T.G.C); Panaman (M); Fathai (R) | Cultivated | Leaf |
| *Pisum sativum* L. (Papilionaceae) | Motar (Beng); Houwaitharak (M) | Cultivated | Fruits |
| *Portulaca oleracea* Linn. (Portulacaeae) | Boraloniya (Beng) | Wild | Young stem & leaf |
| *Pouzolzia hirta* (Bl.) Hassk. (Urticaceae) | Chirsag (Beng) | Wild | Young stem & leaf |
| *Psidium guajava* Linn. (Myrtaceae) | Pungdol (M); Peyara (Beng); Gaiang(R) | Cultivated | Leaf & fruits |
| *Punica granatum* Linn. (Punicaceae) | Anar (T.G.C.); Anaras (Beng) | Cultivated | Fruits |
| *Senna tora* (L.) Roxb. (Caesalpiniaceae) | Kalkasenda (Beng) | Wild | Leaf |
| *Solanum nigrum* L. (Solanaceae) | Kaikor (Beng); Leipung-khanga (M) | Wild | Fruits |
| *Solanum tuberosum* L. (Solanaceae) | Alu (Beng, M & T.G.C.) | Cultivated | Tuber |
| *Spilanthes paniculata* Wall. ex DC. (Asterceae) | Ankahcha (Hmar) | Wild | Stem |
| *Syzygium cumini* (Linn.) Skeels (Myrtaceae) | Jam (Beng); Gulamchat (M); Chambu (R) | Introduced | Fruits |
| *Tamarindus indica* L. (Caesalpiniaceae) | Tetul (Beng); Nangge (M) | Cultivated | Fruits |

| | | | |
|---|---|---|---|
| *Terminalia chebula* (Gaertn.) Retz. (Combretaceae) | Hartaki (Beng); Haritok (M) | Wild | Fruits |
| *Wedelia chinensis* (Osb.) Merr. (Asteraceae) | Bhimral (B); Bhingaraj (Beng) | Wild | Leaf |

## DISCUSSION

Survey work done during three years (1999 – 2002) revealed that the local inhabitants are consuming at least 70 edible plant species in different forms, i.e. raw, ripe or cooked. A list of these plants along with their family, botanical name, parts used has been given, where it can be seen that out of the 70 plant species, 24 species provide edible ripe fruits, 42 spp. edible green leaves or edible as a whole, 2 spp. flowers and 3 spp. tubers. The nutritional value of the identified wild edible plants should be worked out so that their contribution to human health in terms of nutrition can be assessed and recommended for the benefit of the people at large. The medicinal value of the edible plants which can contribute to the development of immunity against some diseases should also be assessed to find out their probable contribution in relation to human health. Barak Valley is floristically very rich and information on the edible plants from this area would contribute to our knowledge on the potential uses of plants for the benefit of society.

## ACKNOWLEDGEMENTS

The authors are highly grateful to Dr. T.M. Hynniewta, Head, Eastern Circle, Botanical Survey of India for providing facilities in Herbarium and Library. We are also thankful to local people who shared knowledge with us.

## REFERENCES

BORTHAKUR, S. K. 1996. Postnatal care of women in traditional system in Assam. *Ethnobotany* 8 : 51-53.

DAS, AJIT KUMAR & G. D. SHARMA 2002. Ethnomedicinal plants used by Barman and Manipuri community, Cachar District, Assam. *J. Econ. Taxon. Bot.* 27 (2) : 421-429.

————, K. C. DAS, B. K. DUTTA & G. D. SHARMA 2003. Study of the dwelling courtyard plants and their uses in Karimganj District, Assam, N.E. India. Communicated.

GANGWAR, A. K. & P. S. RAMKRISHNA 1990. Ethnobotanical notes on some tribes of Arunachal Pradesh, North Eastern India. *J. Econ. Taxon. Bot.* 44 (1) : 94-105.

HOOKER, J. D. 1872-1887. *The Flora of British India,* Vol. 1-7. London.

JAIN, S. K. & R. R. RAO 1977. *Hand book of field and herbarium methods.* New Delhi.

KANJILAL, U. N., P. C. KANJILAL, A. DAS. & C. PURKAYSTHA 1934. *Flora of Assam,* Vol. I. Govt. of Assam, Shillong.

————, ———— & A. DAS. 1936. Flora of Assam, Vol. II. Govt. of Assam, Shillong.

————, ————, ———— & R. N. DE 1938. *Flora of Assam,* Vol. III. Govt. of Assam, Shillong.

————, ————, ———— & ———— 1940. *Flora of Assam,* Vol. IV. Govt. of Assam, Shillong.

PARTHA SARATHI DAS & M. DUTTA CHOUDHURY 2003. A survey on non-conventional food plants of Southern Assam. *J. Econ. Taxon. Bot.* 27 (2) : 416-420.

RAO, R.R. & N.C. JAMIR 1990. Ethnobotany of the Angami Nagas of Nagaland. *J. Econ. Taxon. Bot.* 14 (3) : 593-604.

———— & B. NEOGI 1980. Observation on the Ethnobotany of the Khasi and Garo tribes in Meghalaya. *J. Econ. Taxon. Bot.* 1 : 157-162.

SHARMA, B.D., N.P. BALAKRISHNAN, R.R. RAO & P.K. HAJRA (Eds.) 1993. *Flora of India*, Vol. I. (Ranunculaceae-Barclayaceae). B.S.I., Calcutta.

————, ————, & M. SANJAPPA (Eds.) 1993. *Flora of India*, Vol. II. B.S.I. Calcutta.

————, M. SANJAPPA & N.P. BALAKRISHNAN (Eds.) 1993. *Flora of India*, Vol. III. B.S.I., Calcutta.

# ETHNOMEDICINAL SURVEY OF ALAGAR KOVIL HILLS (RESERVED FOREST), TAMILNADU, INDIA

S. GANESAN, N. RAMAR PANDI & N. BANUMATHY

*Centre for Research and P.G. Department of Botany, Thiagarajar College (Autonomous), Madurai – 625 009, Tamil Nadu, India*
*E-mail : sganesan76@yahoo.com*

## ABSTRACT

Ethnomedicinal uses of 111 plant species belonging to 100 genera and 49 families used by the ethnic group, Valaiyans of Alagar Hills of Madurai district, Tamil Nadu, India and their traditional modes of treatment of diseases and various ailments like skin diseases, cold and cough, reducing body heat, ulcer, stomach related problems, fever, piles, jaundice, diabetes etc are presented in this paper.

## INTRODUCTION

The study of Ethnobotany assumes great importance in enhancing our knowledge about the plants used by the ethnic and the tribal communities, the diversity produced and their own sustenance and different traditional technologies and methods adopted by them for conservation of the plant diversity. These days greater emphasis is being laid on the traditional knowledge and the social and cultural aspects of the traditional and ethnic people imparting that knowledge and use that knowledge in "bioprospecting" of biological resource as a new source of food and medicine and other important industrial raw materials.

The ancient man through his own observation and experiences developed knowledge about the usage of plants for edible, medicinal and other purposes. Although in the modern era medicinal facilities and enough food are available to most of the people, still in several under-developed and inaccessible areas of the country food deficiency and lack of medicinal facilities have been prevalent. Plant parts like fruits, tubers, flowers, leaves etc are consumed as principal or supplementary food and used as medicine (Sharma & Singh, 2001).

India is one of the 12 megadiversity countries in the World and has 17,000 flowering plants. Of the designated 25 hotspots in the World, the Eastern Himalaya and the Western Ghats are the 2 hotspots in India. Indian subcontinent represents one of the greatest emporia of ethnobotanical wealth. The country is represented by 400 different tribals and other ethnic groups with about 67.8 million total population (Viswanathan, 2004).

Tamil Nadu is situated on the eastern side of the Indian Peninsula. The State has 29 districts and includes 37 tribal communities. They are distributed in various districts. Three major primitive tribal groups, viz. the Valaya or Valaiyans show that they were the inhabitants of the hill tracks of Madurai, Dindigul and Tanjore districts (Ganesan & Kesavan, 2003); Paliyan and Pulayan tribals inhabit the various places of Western Ghats of Tamil Nadu (Ganesan *et al.*, 2001, 2004).

The available literature is pertaining to the ethnomedicinal aspects of many tribals in different regions. The particular tribal group is used to inhabit specific areas and lead a life depending on the existence of plants in their surroundings. In this regard, information about ethnobotanical studies of Valaiyans staying in two places, Vellimalli hills and Seithur hills (Ganesan & Kesavan, 2003) have already been undertaken. But, there is no such kind of study regarding the Valaiyans of Alagar Hills, Tamil Nadu, India. So an attempt is made in order to know about the ethnomedicinal aspects of Valaiyans dwelling in this area.

## STUDY AREA

The area of investigation lies approximately between 77°30′ and 78°20′ longitude and latitude 10°05′ – 10°09′. The elevation of the area of investigation ranges from 1000 to 3000 feet above mean sea level. Variations in the altitude and rainfall always have a bearing on the vegetation in general. The floristic division of the area includes the vegetation at the foot of the hills of scrub jungle type and at slightly higher levels open mixed deciduous forests, at still higher elevations, the evergreen and grassland formations are noticed.

The area lies between two hills named as "Vellaimalai" in southern side of the area of investigation and "Sennamalai" in northern side of the area. These two hills are situated in the western part of the Alagar hills. Valaiyans have short cut route to reach Alagar kovil through the hills. The areas of tribal inhabitation in Alagar hills are Vembarali, Barali, Alagapury, Thethampatty, Podugampattu and Pattanampatty. In these area Valaiyans live in isolated pockets or in small hamlets in the forms of labourers and small landowners. The investigation was carried out for 10 months from August 2004 to May 2005.

## METHODOLOGY

Ethnomedicinal information was gathered from both the categories of Valaiyans by contacting the medicine man, headman and the persons having thorough knowledge of plants. The information gathered was conformed by cross check between different groups of Valaiyans dwelling in different places. The methodology adopted by previous workers was followed (Jains & Goel, 1995). The information was meticulously entered in the field notebook and the voucher specimens were collected and identified with the help of standard flora (Gamble & Fischer, 1957;

Matthew, 1981, 1991; Nair & Henry, 1983; Henry *et al.*, 1987, 1989). All the voucher specimens preserved in the form of herbarium were deposited in the Centre for Research and PG Department of Botany, Thiagarajar College (Autonomous), Madurai, Tamil Nadu-625009.

## OBSERVATIONS

Valaiyans are good herbalists. The utilization of plants may be in different forms like juice extract, decoction, paste, infusion etc. **Juice extract:** Cleaned plant or plant parts are ground with water. The extract is filtered and filtrate is used as juice extract. **Decoction:** It is generally obtained by boiling the plant parts /pieces in water. **Paste:** Small parts of plant are crushed with water and made into a soft mass. **Infusion:** The cleaned plant or plant parts are soaked in water for few hours or days, then filtered and the filtrate is the infusion. The alphabetical list of plant binomial, family, local name (Tamil) and their mode of utilization are followed.

***Abrus precatorius*** L. Fabaceae, "Kundumani". Seeds paste is used to cure eczema.

***Abutilon indicum*** (L.) Sweet Malvaceae, "Thuththi". The leaf paste is taken orally to cure piles. Leaves also relieve leg pain.

***Acalypha indica*** L. Euphorbiaceae, "Kuppaimeni". The leaf paste mixed with common salt is used to cure eczema and chest pain.

***Achyranthes aspera*** L. Amaranthaceáe, "Nayuruvi". The cooked leaves are consumed to relieve internal piles and roots are used as brush to relieve pain and clean the teeth.

***Aerva lanata*** (L.) Juss. ex Schult. Amaranthaceae, " Kannupeelai". The decoction or juice of the whole plant is taken for urinary problems.

***Alangium salvifolium*** (L.f.) Wangerin Alangiaceae, "Alingil ". The stem is used for brushing the teeth and one or two drops of fruit juice are poured into the eye to cure eye diseases in summer.

***Albizia amara*** ( Roxb.) Boivin Mimosaceae, "Usilai". Shade dried powder is mixed with water and used as a shampoo for cleaning the hairs and to reduce body heat.

***Albizia lebbeck*** (L.) Benth. Mimosaceae, "Vagai". Leaf paste is applied for eczema.

***Aloe vera*** (L.) Burm.f. Liliaceae, "Chotthukatthalai". Leaf paste is applied over the body before bath for reducing body heat.

***Alpinia calcarata*** (Haw.) Roscoe Zingiberaceae, "Chitharathi". Powder of dried rhizome is mixed with water and used to cure cough, cold and digestion.

***Alstonia venenata*** R. Br. Apocynaceae, "Paalai". Milky latex is used to cure wound and cuts.

***Ammannia baccifera*** L. Lythraceae, "Neermalneruppu". Leaf paste is applied to relieve edema/swelling.

***Andrographis paniculata*** (Burm.f.) Wallich ex Nees Acanthaceae, "Nilavaembu, Chirianangai". A handful of leaves are taken and extract is made and mixed with milk to be taken internally to cure snakebite.

***Anisomeles malabarica*** (L.) R. Br. ex Sims Lamiaceae, "Peithumbai". Leaf paste is applied to cure eczema.

***Annona squamosa*** L. Annonaceae, "Sita". The young fruits are dried and made into powder. A spoon of powder is mixed with water and taken internally to cure dysentery.

***Argemone mexicana*** L. Papaveraceae, "Narimirati". The yellow latex is used to cure lips ulcer, pimples and wound healing.

***Argyreia kleiniana*** (Roemer & Schultes) Raiz. Convolvulaceae, "Onankodi". The plant extract is internally used to cure stomach upset.

***Aristolochia bracteolata*** Lam. Aristolochiaceae, "Aaduthinnaapalai". The leaf paste is applied on the head before bath to relieve dandruff and other infection.

***Asparagus racemosus*** Willd. Liliaceae, "Thanneervittan". The bulb extract is used to cure urinary troubles.

***Boerhavia diffusa*** L. Nyctaginaceae, "Mukurattai". Decoction of the roots is used for gas trouble.

***Borassus flabellifer*** L. Arecaceae, "Panai". The pectin layer on the leaves is tied around the cut or wounds for blood clotting.

***Cadaba trifoliata*** (Roxb.) Wight & Arn. Capparaceae, "Purana". Boiled leaves vapours are inhaled to relieve cough and cold.

***Caesalpinia pulcherrima*** (L.) Swartz. Caesalpiniaceae, "Mailkondarai". Seed paste is applied on the infected teeth to cure toothache.

***Calotropis gigantea*** (L.) R.Br. Asclepiadaceae, "Erukku". The leaves are tied around the spot to cure thorn stings.

***Cardiospermum halicacabum*** L. Sapindaceae, "Mudakkaththan". The leaf paste is applied for joint pain, or leaves are prepared in the form of soup and consumed to cure rhuematic pain.

***Carica papaya*** L. Caricaceae, "Pappaali". Milky latex of the plant is applied on teeth in order to relieve teeth infection.

***Carissa carandas*** L. Apocynaceae, "Kalakai". Latex is used for mouth ulcer.

***Catharanthus pusillus*** (Murr.) G. Don Apocynaceae, "Mukkuthipoo". Leaf paste is applied externally for tumour (kalalai).

***Centella asiatica*** (L.) Urban Apiaceae, "Vallarai". The dried plant materials are powdered and mixed with hot water and taken for gas trouble and fried plants are used in the diet of children for improving memory power.

***Cissus quadrangularis*** L. Vitaceae, "Pirandai". The plant paste is taken for improving digestion and inducing appetite.

***Citrus medica*** L. Rutaceae, "Elumichi". Fruit juice mixed with milk is taken two times a day for dysentery.

***Cleome aspera*** J. Koeing ex DC. Cleomaceae, "Karumpoondu". Leaf paste is used to cure eczema.

***Cleome viscosa*** L. Cleomaceae, "Naikadugu". Juice of leaves is poured into ear to relieve earache.

***Clerodendrum inerme*** (L.) Gaertn. Verbenaceae, "Peenaari changu" The leaf paste is applied for psoriasis and other skin infections.

***Coccinia grandis*** (L.) Voigt Cucurbitaceae, "Kovai". Leaf extract is mixed with milk and used in the case of jaundice.

***Cocos nucifera*** L. Arecaceae, "Thennai". Young fruit is used for dysentery.

***Coleus aromaticus*** Benth. Lamiaceae, "Omavali". The leaf extract is used to cure cold and leaf juice is used for digestion.

***Commiphora caudata*** (Wight & Arn.) Engl. Burseraceae, "Kiluvai". Stem exudation (gum) is mixed with water and used as mouth wash to cure mouth ulcer. Gum is also used for wound healing.

***Corallocarpus epigaeus*** (Rottler ex Willd.) Hook. f. Cucurbitaceae, "Kollan kova killangu". The tuber paste is applied on the poisonous bite area in order to remove the poison.

***Cuscuta chinensis*** Lam. Cuscutaceae, "Manjapulluruvi". Leafless, rootless twiner is made in the form of paste and applied on the broken parts of the bone for joining the fracture.

***Cyanotis arachnoidea*** Clarke Commelinaceae, "Vallukkai". Leaf paste is applied on swelling to reduce itching.

***Cymbidium aloifolium*** (L.) Sw. Orchidaceae, "Panai pulluruvi". The succulent leaves are warmed on the flame and the juice is extracted, two or three drops of juice are poured into ear to cure earache.

***Cynodon dactylon*** (L.) Pers. Poaceae, "Arugampul". The whole plant juice is used for reducing the body heat and blood pressure.

***Dichrostachys cinerea*** (L.) W. & A . Mimosaceae, "Vedathalai". Plant extract ia used for rhuematic pain.

***Dodonaea angustifolia*** L.f. Sapindaceae, "Virali". The leaf paste is applied on the surface of blood clot (Haematoma).

***Eclipta prostrata*** (L.) L. Asteraceae, "Karisalankanni". Leaf extracts is applied on the head to relieve dandruff and for blackening the gray hairs.

***Eleusine coracana*** (L.) Gaertn. Poaceae, "Kaelvaragu". The seed powder mixed in hot water and in the form of gel is applied on the forehead to relieve viral fever and cold.

***Enicostema axillare*** (Lam.) A. Raynal Gentianaceae, "Vellaragu". Leaf powder is taken two times a day for white discharge (leucorrhoea) in women. Excess leaf paste is used for antifertility in man.

***Erythrina variegata*** L. Fabaceae, "Mullumurungai". A handful of leaves are ground with rice and made into roasted thosa and eaten for cold and cough.

***Eucalyptus tereticornis*** Smith Myrtaceae, "Thailamaram". Boiled vapours of leaves are inhaled for cough and cold. Plant oil mixed with coconut oil is applied on chest to releive dry cough and chest pain.

***Euphorbia hirta*** L. Euphorbiaceae, "Amman patcharisi". Entire plant paste is mixed with goat milk and taken internally for all stomach upset in children.

***Evolvulus alsinoides*** (L.) L. Convolvulaceae, "Vishnukiranthi". Plant decoction is used for fever.

***Gloriosa superba*** L. Liliaceae, "Kanuvalikodi". The paste of the tuber is used as an abortifacient.

***Gymnema sylvestre*** (Retz.) R. Br. ex Roem. & Schultes Asclepiadaceae, "Chirukurunjan".

The dried leaf powder is used to cure diabetes and to reduce the sugar level of blood.

***Hedyotis puberula*** (G. Don) Arn Rubiaceae, "Impural". A teaspoon of dried plant powder is mixed with water and taken internally for asthma.

***Hemidesmus indicus*** (L.) R. Br. Asclepiadaceae, "Nannaari". The extract of entire plant is given for fever.

***Hibiscus rosa-sinensis*** L. Malvaceae, "Semparuththi". Shade dried powdered flower is used for cleaning hair and to prevent falling of hairs (Alopecia).

***Hybanthus enneaspermus*** (L.) F. Muell. Violaceae, "Orithalthamarai". The plant extract is used to reduce body heat and also used to increase male fertility.

***Hygrophila schulli*** (Buch.-Ham.) Almeida & Almeida Acanthaceae, "Neermulli". Seeds are soaked in water and the water is taken two times per day for relieving menstrual problem in women.

***Hyptis suaveolens*** (L.) Poit. Lamiaceae, "Karunchsatachi". Leaf paste is used to cure eczema.

***Indigofera tinctoria*** L.Fabaceae, "Avuri". Leaf decoction is used to cure skin diseases.

***Jatropha curcas*** L. Euphorbiaceae, "Chanimuthu & Kaattamanakku". Plant is soaked in rice water and flamed on fire and in warm condition, the leaves are tied around the painful area.

***Jatropha gossypifolia*** L. Euphorbiaceae, "Adalai". Latex of the plant is applied on lips to cure ulcer in the mouth and stem is used for cleaning teeth and to strengthen the gum.

***Justicia adhatoda*** L. Acanthaceae, "Adatodai". Leaf decoction is used to relieve fever.

***Kleinia grandiflora*** (Wall. ex DC.) Rani Asteraceae, "Muyalkathu". The succulent leaf juice is poured into the ear to cure earache and leaf paste is used for eczema.

***Lawsonia inermis*** L. Lythraceae, "Maruthani". The leaf paste is applied on cracks of the foot ("Piththa vedippu") and also it provides cooling effect to the body.

***Leonotis nepetiifolia*** (L.) R. Br. Lamiaceae, "Theanthumpai". Leaf paste is used for eczema.

***Leucas aspera*** (Willd.) Link Lamiaceae, "Thumbai". Boiled vapours of leaves are inhaled to relieve cough and cold.

***Mangifera indica*** L. Anacardiaceae, "Maa". Latex of the plant cures wounds on the legs of livestock.

***Marsilea minuta*** L. Marsileaceae, "Aarakkerai". Dried leaf powder mixed with hot water is taken for diabetes.

***Melia azedarach*** L. Meliaceae, "Malaivembu". The juice of the bark is taken internally in early morning for three days to relieve stomach pain and also used as vermifuge.

***Melothria maderaspatana*** (L.) Cong. Cucurbitaceae, "Mosumosukkai". The leaf extract is taken internally to cure piles and applied on hair to blacken the gray hair.

***Mollugo nudicaulis*** Lam. Molluginaceae, "Parpaadagam". The entire plant is used to cure cold, and leaf paste is used for skin diseases.

***Morinda pubescens*** J. E. Smith Rubiaceae, "Manjanathi". The leaf extract is used to cure dysentery.

***Moringa oleifera*** Lam. Moringaceae, "Murungai". Cooked leaves and flowers are eaten to increase fertility in man. A handful of bark juice of the plant is used to cure stomach pain.

***Musa paradisiaca*** L. Musaceae, "Vazhai". Plant extract is given for snake bite and also used for burnt wounds.

***Ocimum canum*** Sims Lamiaceae, "Naithulasi". Boiled vapours of leaves are inhaled to relieve headache and fever.

***Ocimum tenuiflorum*** L. Lamiaceae, "Nallathulasi". The infusion of leaves is consumed for cold and cough.

***Pavetta indica*** L. Rubiaceae, "Therani". Leaf paste is used for snakebite.

***Pedalium murex*** L. Pedaliaceae, "Yaanainerunji". Leaf paste is applied on swellings to reduce itching.

***Pergularia daemia*** (Forsskal) Chiov. Asclepiadaceae, "Vaeliparuththi". The leaf paste is consumed to relieve gas trouble.

***Phyllanthus amarus*** Schum. & Thonn. Euphorbiaceae, "Kizhaanelli". The whole plant paste is mixed with goat milk and taken internally for three days to cure jaundice.

***Polygala arvensis*** Willd. Polygalaceae. Root paste is used for anti-inflammation.

***Pongamia pinnata*** (L.) Pierre Fabaceae, "Pungam". The seed oil is used to cure rheumatic pain and swelling.

***Senna auriculata*** (L.) Roxb. Caesalpiniaceae, "Aavaram poo". Dried powdered flowers are used for cleaning hairs and reducing body heat.

***Senna italica*** Mill. Caesalpiniaceae, "Nilavagai". The leaf powder is mixed with hot water and taken for throat pain.

***Senna occidentalis*** (L.) Link Caesalpiniaceae, "Thakarai". Leaf paste is used for skin diseases.

***Sesbania grandiflora*** (L.) Poiret Fabaceae, "Agaththi". The leaves are prepared in the form of a soup and taken as vermifuge and also to cure peptic ulcer.

***Solanum nigrum*** L. Solanaceae, "Milaguthakkali". The plant prepared in the form of soup is taken for stomach ulcer. The leaf paste is also used to cure rabies.

***Solanum torvum*** Sw. Solanaceae, "Sundai". The leaf paste is used to bath the newly born baby to protect from any infection.

***Solanum trilobatum*** L. Solanaceae, "Thudhuvaelai". The leaf juice is taken in the form of soup for cough and cold.

***Solanum virginianum*** L. Solanaceae, "Kandankathiri". The whole plant powder is mixed with water and taken internally to cure cold and wheezing trouble.

***Sphaeranthus indicus*** L. Asteraceae, "Kottaikaranthi". A spoon of plant powder is mixed with water and taken internally for blacken the gray hairs.

***Syzygium cumini*** (L.) Skeels Myrtaceae, "Navel". The dried seed powder mixed with hot water is taken for reducing blood sugar level.

***Tamarindus indica*** L. Caesalpiniaceae, "Puli". The seed coat paste is applied on the spot of scorpion bite to relieve pain, or scratched seed in warm condition is placed on the area of scorpion bite to relieve pain.

***Tephrosia purpurea*** (L.) Pers. Fabaceae, "Kozhunji". Root bark paste is taken for stomach pain.

***Tragia involucrata*** L. Euphorbiaceae, "Senthatti". The paste of the fruit is applied on the forehead to relieve one side headache and a handful of leaf paste is mixed with water and taken internally to relieve stomach pain at the time of menstruation period.

***Tribulus terrestris*** L. Zygophyllaceae, "Nerunji". The plant extract is used to remove the kidney stones.

***Vernonia cinerea*** (L.) Less. Asteraceae, "Mukuttipoondu". The plant paste is used to cure paralysis.

***Vitex negundo*** L. Verbenaceae, "Karunochchi". Boiled vapours of leaves are inhaled to relieve running nose and headache.

***Wrightia tinctoria*** (Roxb.) R. Br. Apocynaceae, "Vetpalai". Leaf paste mixed with neem oil is applied for eczema.

***Zaleya decandra*** (L.) Burm. f. Aizoaceae, "Sathasaaranetti". The leaves are cooked and consumed for removing gas trouble.

***Zingiber roseum*** (Roxb.) Rosc. Zingiberaceae, "Inji". Juice of the rhizome mixed with honey is taken internally to improve digestion and to relieve giddiness (piththamayakam).

Apart from the medicinal use of individual plants, a combination of two or more plants may be utilized in medicine.

I. A. *Cuminum cyminum* L. Apiaceae, "Seeragam"

B. *Murraya koenigii* (L.) Spreng. Rutaceae, "Karuveppilai"

Leaf of *Murraya koenigii* ground with fruits of *Cuminum cyminum* and the extract is taken with buttermilk for dysentery.

II. A. *Allium sativum* L. Liliaceae, "Vellai poodu"

B. *Cissus quadrangularis* L. Vitaceae, "Pirandai"

C. *Capsicum frutescens* L. Solanaceae, "Milakai"

Above plants are made into green salad and taken for gas trouble.

III. A. *Cleome gynandra* L. Cleomaceae, "Thaivelai"

B. *Abutilon indicum* (L.) Sweet Malvaceae, "Thuuthi"

Extract of above plant roots is used to cure stomach upset in livestock's.

IV. A. *Cocos nucifera* L. Arecaceae, "Thennai"

B. *Trigonella foenum-graecum* L. Fabaceae, "Vendayam"

A spoon seeds of *Trigonella foenum-graecum* are soaked in tender coconut water and left in open fullmoon day for 12hrs. In the early morning the coconut water along with seeds are taken internally to get relief from white discharge in women and to reduce body heat.

V. A. *Piper betle* L. Piperaceae, "Vetrilai"

B. *Zingiber roseum* (Roxb.) Rosc. Zingiberaceae, "Inji"

Leaf extract of *Piper betle* is mixed with rhizome extract of *Zingiber roseum* and taken internally to get relief from liver infection.

VI. A. *Tephrosia purpurea* (L.) Pers. Fabaceae, "Kozhunji"

B. *Piper nigrum* L. Piperaceae, "Milagu"

Leaves of *Tephrosia purpurea* are ground with fruits of *Piper nigrum* and taken with buttermilk to get relief from white discharge in women.

VII. A. *Ocimum tenuiflorum* L. Lamiaceae, "Nallathulasi"

B. *Piper nigrum* L. Piperaceae, "Milagu"

C. *Solanum virginianum* L. Solanaceae, "Kandankaththari"

Leaves of *Ocimum tenuiflorum* and *Solanum virginianum* are ground with fruits of *Piper nigrum* and the extract is taken internally to relieve cold and cough.

VIII. A. *Coriandrum sativum* L. Apiaceae, "Koththamalli"

B. *Zingiber roseum* (Roxb.) Rosc. Zingiberaceae, "Inji"

The seeds of *Coriandrum sativum* and rhizome of *Zingiber roseum* are ground with water and the extract is taken internally for giddiness.

## DISCUSSION

The tribal and rural population of India is highly dependent on medicinal plant therapy for meeting their healthcare needs. This attracted the attention of several botanists and plant scientists who directed vigorous researches towards the rediscovery of several medicinal plants and there was a spurt of scientific literatures. In the present investigation, about 111 plant species belonging to 100 genera and 49 families are recorded which are employed for medicinal purposes. Out of which, 110 plant species represented Angiosperms and one plant is assigned to Pteritophytes (*Marselia minuta*).

About 20 plants are applied for skin related diseases, viz. *Abrus precatorius, Acalypha indica, Albizia lebbeck, Ammannia baccifera, Anisomeles malabarica, Argemone mexicana, Aristolochia bracteolata, Senna occidentalis, Cleome aspera, Clerodendrum inerme, Cyanotis arachnoidea, Eclipta prostrata, Hyptis suaveolens, Indigofera tinctoria, Lawsonia inermis, Leonotis nepetiifolia, Mollugo nudicaulis, Notonia grandiflora, Pedalium murex* and *Wrightia tinctoria.*

Further, about 10 plants, viz. *Alpinia calcarata, Cadaba trifoliata, Eleusine coracana, Erythrina variegata, Eucalyptus tereticornis, Leucas aspera, Mollugo nudicaulis, Ocimum tenuiflorum, Solanum virginianum, Solanum trilobatum* are used for cold and cough.

About 8 plants, viz. *Albizia amara, Aloe vera, Senna auriculata, Cocos nucifera, Cyanodon dactylon, Hybanthus enneaspermus, Lawsonia inermis* and *Trigonella foenum-graecum* are consumed for reducing body heat.

*Abutilon indicum, Acalypha indica, Cardiospermum halicacabum, Senna italica, Dichrostachys cinerea, Eucalyptus tereticornis* and *Pongamia pinnata* are employed in the treatment of various pains related to leg, joint, chest, throat etc.

*Allium sativum, Boerhavia diffusa, Capsicum frutescens, Centella asiatica, Cissus quadrangularis, Pergularia daemia* and *Zaleya decandra* are taken for gas trouble.

*Justicia adhatoda, Eleusine corocana, Evolvulus alsinoides, Hemidesmus indicus* and *Ocimum canum* are utilized for fever.

*Argemone mexicana, Carissa carandas, Commiphora caudata, Sesbania grandiflora* and *Solanum nigrum* are given for the treatment of various ulcers.

Some plants are administered for stomach related problems such as stomach upset (*Abutilon indicum, Argyreia kleiniana, Cleome gynandra* and *Euphorbia hirta*), dysentery (*Annona squamosa, Citrus medica, Cocos nucifera, Cuminum cyminum*), stomach pain (*Melia azedarach, Moringa oleifera, Tephrosia purpurea*).

The plants given for women related problems include : *Hygrophila schulli, Piper nigrum, Tephrosia purpurea, Tragia involucrata* and *Trigonella foenum-graecum.* For male fertility *Hybanthus enneaspermus* and *Moringa oleifora* are used.

Three plants are used to cure piles (*Abutilon indicum, Achyranthes aspera* and *Melothria maderaspatana*), to cure diabetes (*Gymnema sylvestre, Marsilea minuta, Syzygium cumini*), for earache (*Cleome viscosa, Cymbidium aloifolium, Notonia grandiflora*).

The following ailments are cured by employing two plants or one plant only.

| | |
|---|---|
| Jaundice | *Coccinia grandis* and *Phyllanthus amarus* |
| Urinary problems | *Aerva lanata* and *Asparagus racemosus* |
| Headache | *Ocimum canum* & *Vitex negundo* |
| Toothache | *Carica papaya* and *Caesalpinia pulcherrima* |
| Asthma | *Hedyotis umbellata* and *Solanum virginianum* |
| Giddiness | *Coriandrum sativum* and *Zingiber roseum* |
| Paralysis | *Vernonia cinerea* |
| Kidney stone | *Tribulus terrestris* |
| Tumour | *Catharanthus pusillus* |

The studies on ethnomedicinal lore of Seithur hills revealed the uses of 36 plant species belonging to 33 genera represented in 24 families of flowering plants by Valaya tribals of Virudhunagar. Out of which, about 7 species were used for scabies and other skin diseases, 2 species for jaundice, 4 species for headache, 3 species for fever and one species for diabetes.

The studies conducted on the ethnomedicinal plants used by Valaiyans of Vellimalai hills (Ganesan & Kesavan, 2003) indexed 84 angiospermic plants species belonging to 28 genera distributed in 40 families for the treatment of various disorders such as wounds, cuts, stomach pain, diabetes, fever, eczema, dandruff, cold, body heat, poisonous bites.

Ethnobotanical aspects of 161 species of Angiospermic plants distributed in 139 genera representing 60 families were observed from Thottianiackans of Semmalai hills, Tiruchirapalli district (Venkatesan *et al.*, 2005). He recorded 119 plants for medicinal purposes.

Ganesan *et al.* (2004) conducted ethnobotanical survey of Lower Palani hills of Tamil Nadu. They enumerated 45 species of plants distributed among 42 genera belonging to 26 families as ethnomedicinal value from Paliyans and Pulayans for the treatment of various ailments.

Valaiyans utilized a large number of plants for medicinal and other purposes. Some of the uses are common and already reported from other places (Karuppusamy *et al.*, 2000; Ganesan & Kesavan, 2003; Ganesan *et al.*, 2004 and Venkatesan *et al.*, 2005). The uses of certain species gained importance due to their manifold properties. For example, *Abutilon indicum* is used for piles and leg pain; *Acalypha indica* for eczema and chest pain; *Mollugo nudicaulis* for cold and skin diseases; *Notonia grandiflora* for earache and eczema. Sharma & Singh (2001) and Jain & Patole (2001) have also reported such activities in certain plant species (*Cassia fistula* and *Gloriosa superba)* in their studies on ethnobotanical uses of plants.

Information on some very useful medicines known to the tribal or ethnic communities through experiences of age is usually passed on from generation to generation. The tribal ways of life are rapidly breaking of under the impact of improved communication facilities and still more under the various tribal welfare schemes of the government (Matthew, 1981). As the tribal population is gradually adapting modern ways of life, their heritage of traditional knowledge in plants is being lost for ever.

For the ethnobotanist, the impressive and urgent task is to record for posterity what ever is valuable in the traditions of the tribes, their way of life, their knowledge of plants before all these disappear before the close of century. At present, the valuable diversity of plants is under serious threats due to their habitat destruction, overexploitation, shifting of cultivation and several anthropogenic and natural interference.

## REFERENCES


GAMBLE, J.S. & C.E.C. FISCHER 1959. *The flora of the Presidency of Madras*. Rep. Ed. I-III. Botanical Survey of India, Calcutta.

GANESAN, S. & L. KESAVAN 2003. Ethnomedicinal plants used by the ethnic group Valaiyans of Vellimalai hills (Reserved Forest), Tamil Nadu, India. *J. Econ. Taxon. Bot.* 27 (3) : 754-760.

————, T. ELUVAKKAL & L. KESAVAN 2001. Medicinal plants used by Paliyans of Sirumalai hills, Sathuragiri hills and Kumily hills of Tamilnadu. India. Presented in: *State level seminar on conservation of biodiversity present and future scenario.* Sponsored by UGC, Ayyanadar College, Sivakasi, Tamilnadu, March 23 & 24.

————, N. SURESH & L. KESAVAN 2004. Ethnomedicinal survey of Lower Palni Hills of Tamil Nadu. *Indian Journal Traditional Knowledge* 3 (3) : 299-304.

HENRY, A.N., G.R. KUMARI & V. CHITRA 1987. *Flora of Tamil Nadu*, Series–I, Vol. II. Botanical Survey of India, Coimbatore.

————, V. CHITHRA & N.P. BALAKRISHNAN 1989. *Flora of Tamil Nadu*, Series–I, Vol. III. Botanical Survey of India, Coimbatore.

JAIN, A.K. & S.N. PATOLE 2001. Less-known medicinal uses of plants among some tribal and rural communities of Pachmarchi forest (M.P.). *Ethnobotany* 13 : 96-100.

JAIN, S.K. & A.K. GOEL 1995. *A manual of Ethnobotany*. Scientific Publishers, Jodhpur.

KARUPPUSAMY, S., K.M. RAJASEKARAN & R. KUMUTHAKALAVALLI 2000. Needs for diversity conservation of traditional medicinal plant resources of Dindigul district, Tamil Nadu. *Ecol. Env. & Cons.* 6 (3) : 273-278.

MATTHEW, K.M. 1981. Materials for a flora of the Tamilnadu Carnatic. Rapinat Herbarium, India.

———— 1991. *An Excursion flora of Central Tamil Nadu, India.* Oxford and IBH Publishing Co., New Delhi.

NAIR, N. C & A. N. HENRY 1983. *Flora of Tamil Nadu*, Series–I, Vol. I. Botanical Survey of India, Coimbatore.

SHARMA, P.P. & N.P. SINGH 2001. Ethnomedicinal uses of some edible parts in Dadra, Nagar Haveli and Daman (U.T.). *Ethnobotany* 13 : 121-125.

VENKATESAN. G., S. GANESAN, L. KESAVAN & N. BANUMATHY 2005. Ethnobotany of Ethnic group Thottianaickan of Tiruchirapalli District, Tamilnadu, India. *Advances in Medicinal Plants* 1 : 59-75. Asian Medicinal Plants and Health Care Trust, Jodhpur.

VISWANATHAN, M. B. 2004. Ethnobotanically important plants. *Tamil Nadu Biodiversity strategy and action plan of wild plant biodiversity*, pp. 172-231. Tamil Nadu Forest Department, Chennai.

# OBSERVATION ON ETHNOBOTANY OF DARJEELING DISTRICT, WEST BENGAL

A.M. SAREN, A.C. HALDER & P.R. SUR

*Central Botanical Laboratory, Botanical Survey of India, B. Garden, Howrah-711103*

**ABSTRACT**

The present paper deals with 71 species of vascular plants belonging to 69 genera used traditionally by the tribals of Darjeeling district of West Bengal for curing various ailments of man and cattle.

## INTRODUCTION

Darjeeling district lies on the extreme northern boundary of West Bengal. It is situated between the 28°31'-27°13' North and 87°59'-88°53' East in the Eastern Himalayan region of India. Out of four sub-divisions of Darjeeling district, Kalimpong, Kerseong and Darjeeling Sadar comprise the hilly areas covering an area of 2417 sq km. Rest one Siliguri subdivision comprises almost plain region. The hilly region is further divided in 8 blocks. A number of mountain ranges and deep river valleys are unique features in the region. The Siliguri range stands on the western border of Darjeeling district. Sandakphu (3630 m) is the highest mountain peak of West Bengal.

The temperature during summer varies usually between 15°-20° C, but the winter is very cold with temperature falling even 5°- 1° C. Snowfall also occurs in winter season. Average annual rainfall is 300 cm.

The population of the hilly areas of Darjeeling district is above 1 million, of which 78.42% population lives in rural areas and 21.51% in urban areas. The major ethnic groups of this region are Lepcha, Bhutia and Nepalese. Nepalese is a conglomerate of Mongoloid tribals like Sherpa, Limbu, Tamag, Yolmo, Kagalay etc, and the Aryan non-tribals like Chhetri, Bahun, Sanyasi, Kami, Damai etc. Each of this tribal group has its own distinct culture, language and traditions. But, Nepali language is the medium of communication between the tribes. These tribals also have rich ethnobotanical traditions.

Since ancient times the Himalayan mountains, particularly the Eastern

Himalayan mountains in India, are known to be the home of vast number of medicinal plants. There is no doub: that Darjeeling district is very rich in ethnobotanical resources. Though, the discovery and manufacture of synthetic drugs and antibiotics took place rapidly and the primitive systems of medicine have become gradually overshadowed, yet, in some of the far-flung villages of the district inhabited by the tribal people, the primitive systems can still be found to be strongly in usage.

## METHODOLOGY

Three field trips covering different remote areas of the district of Darjeeling were undertaken during December 2001 to February 2004. Important plants of ethnobotanical importance were collected and preserved in the form of herbarium sheets. Ethnobotanical informations were recorded from knowledgeable sources. All the specimens have been deposited in the Ethnobotanical Herbarium, Central Botanical Laboratory, Botanical Survey of India, Howrah.

## RESULTS

The different plants, available in the locality, which are used by the tribal people, are summarised in an alphabetical order by the scientific names with respective families, collection number (No.) and local names (L.N.).These are followed by a brief note on uses as revealed by informants during field trips. Pharmacological and clinical trials will help confirmation of the efficacy of the reported herbs.

Abbreviation used in the enumeraton of taxa are : Loc-Locality; L.N-Local name; B-Bengali; Nep-Nepali.

***Achyrenthes aspera*** Linn. (Amaranthaceae)

*L.N.:* Ulto-kuro (Nep.)

*Loc.:* Sukhna forest (Darjeeling), No. 5544, 5457

*Use :* Flower paste is given as a drink prepared with sugar candy to cure stomach ulcer.

***Acorus calamus*** Linn. (Araceae)

*L.N.:* Bozo (Nep.)

*Loc.:* Dolopchan (Kalimpong), No. 5521

*Use :* The plantation of this plant around boundaries of dwellings keeps house free from entering of snakes.

***Ageratina adenophora*** (Spreng.) R. King & Robinson (Asteraceae)

*L.N.:* Ban-mara (Nep.)

*Loc.:* Suknha forest, Tagda tea garden, Darjeeling, No. 5543, 5414

*Use :* Leaf paste is applied to fresh cuts to stop bleeding.

**Ageratum conyzoides** Linn. (Asteraceae)

*L.N.:* Ilami (Nep.)

*Loc.:* Dolopchan Kalimpong, No. 5509

*Use :* Pasted of leaf or juice is applied on fresh cuts to stop bleeding.

***Aloe vera*** (L.) Burm. f. (Liliaceae)

*L.N.:* Ghritakumari (B.)

*Loc.:* Kalijhora, Darjeeling, No. 5461

*Use :* Pasted of stem is applied to relieve forehead pain immediately.

***Alstonia scholoris*** (Linn.) R.Br. (Apocynaceae)

*L.N.:* Chatian (Nep.)

*Loc.:* Champa & Khokli forest basti (Sukhna), No.5564

*Use :* Bark juice is given as health tonic and for curing cough.

***Amomum subulatum*** Roxb. (Zingiberaceae)

*L.N.:* Elachi (Nep.)

*Loc.:* Sukhna, Darjeeling, No. *s.n.*

*Use :* Root and berries are given for improving appetite and for recovery of bitter test.

***Andrographis paniculata*** (Burm. f.) Wall. ex Nees (Acanthaceae)

*L.N.:* Kalmegh (Nep., B.)

*Loc.:* Kalijhora, Darjeeling, No. 5464

*Use :* The whole plant paste is given in stomach trouble and deworming the stomach, specially for the children.

***Annona reticulata*** Linn. (Annonaceae)

*L.N.:* Nona-phal, Sarifa (Nep.)

*Loc.:* Kalijhora, Darjeeling, No. 5467

*Use :* Ripen fruits are edible, sweet in taste.

***Artemisia dubia*** Wall. ex Besser (Asteraceae)

*L.N.:* Tita-apti (Nep.)

*Loc.:* Dolopchan, Kalimpong, Kamath, Kalijhora, Darjeeling, No. 5517, 5434

*Use :* Infusion of young leaf is taken in headache and for high blood pressure. Smell of pasted leaf is inhaled to stop vomiting and bleeding from nose. Powdered plant is used for making incense.

***Berginia ciliata*** (Haw.) Sternb. forma ***ligulata*** Yeo (Saxifragaceae)

*L.N.:* Raya sag (Nep.)

*Loc.:* Tagda tea garden, Darjeeling, No. 5429

*Use :* The plant is taken as vegetable.

***Boehmeria macrophylla*** Hornem. var. ***scabrella*** (Roxb.) Long. (Urticaeae)

*L.N.:* Khasre (Nep.)

*Loc.:* Sukhna, Darjeeling, No. *s.n.*

*Use :* Leaf is a good fodder and ripe fruits are edible.

***Boerhavia diffusa*** Linn. (Nyctaginaceae)

*L.N.:* Punar-nova (B.)

*Loc.:* Sukhna forest area, Darjeeling, No. *s.n.*

*Use :* The whole plant is given in eye diseases and in jaundice. It is also prescribed for falling of hairs.

***Cajanus cajan*** (Linn.) Millsp. (Fabaceae)

*L.N.:* Lohori-dal (Nep.)

*Loc.:* Sukhna, Darjeeling, No. 5558

*Use :* Stem and leaf decoction is taken as 'dal' to cure jaundice. Dose should be continued till recovery, once or twice a day.

***Calotropis gigantea*** (Linn.) R. Br. (Asclepiadaceae)

*L.N.:* Dhotra, Akand (Nep.)

*Loc.:* Sukhna, Darjeeling, No. 5570

*Use :* Latex is used to cure bone fracture. Leaf is used to cure wounds/septic ulcers.

***Cardamine hirsuta*** Linn. (Brassicaceae)

*L.N.:* Simraya (Nep.)

*Loc.:* Tagda, Darjeeling, No. 5417

*Use :* Juice is given to cure high blood pressure.

***Cardiospermum halicacabum*** Linn. (Sapindaceae)

*L.N.:* Fuk-fuke-jhar (Nep.)

*Loc.:* Dolopchan, Kalimpong, Darjeeling, No. 5529

*Use :* The whole plant is given to cows which prevents attacks of worm.

***Catharanthus roseus*** (Linn.) G. Don (Apocynaceae)

*L.N.:* Nayan tara (Nep.)

*Loc.:* Kalijhora, Darjeeling, No. 5462

*Use :* Young leaf is given for high blood pressure. Two to five young leaves or white flowers are chewed raw in empty stomach and the juice is taken four to five days to control high blood sugar.

***Centella asiatica*** (Linn.) Urban (Apiaceae)

*L.N.:* Dollejor,Gol-pata (Nep.)

*Loc.:* Tagda, Dolopchan, Kalimpong, Darjeeling, No. 5409, 5560

*Use :* The leaf juice is given to cure pneumonia and cold and cough. It is also prescribed in curing tonsillitis. Leaf paste is given with sugar in tonsillitis. It is also a good medicine for curing stomach disorders, amoebiosis etc.

***Cinchona officinalis*** Linn. (Rubiaceae)

*L.N.:* Cinchona (Nep.)

*Loc.:* Latpanchar, Kalijhora, Darjeeling, No. 5446

*Use :* Root-bark and stem-bark juice is given to cure fever, especially the malarial fever.

***Cinnamomum tamala*** (Buch.-Ham.) Nees & Eberm. (Lauraceae)

*L.N.:* Tejpata (Nep.)

*Loc.:* Karmatrh, Kalijhora, Darjeeling, No. 5437, 5560

*Use :* Leaves are used as spices.

***Cissus quadrangularis*** Linn. (Vitaceae)

*L.N.:* Harjora, Haddichur (Nep.)

*Loc.:* Sukhna, Darjeeling, No. 5554

*Use :* Whole plant is prescribed for curing ulcer and removing constipation. Stem paste is applied as poultice on bone-fracture to set the fracture smoothly and firmly

***Cleome gynandra*** Linn. (Cleomaceae)

*L.N.:* Banmasi (Nep.)

*Loc.:* Latpanchar, Kalijhora, Darjeeling, No. 5445

*Use :* The juice of the whole plant is given to cure blood sugar.

***Clerodendrum viscosum*** Vent. (Verbenaceae)

*L.N.:* Ghata-pata (Nep.)

*Loc.:* Sukhna, Darjeeling, No. 5545

*Use :* Young leaf-paste or juice is effective in removing worms from stomach.

***Curcuma longa*** Linn. (Zingiberaceae)

*L.N.:* Haledo, Hardi (Nep.)

*Loc.:* Dolopchan, Kalimpong, Darjeeling, No. 5509

*Use :* Pasted rhizome is given in appetite and in any type of stomach trouble.

***Cuscuta reflexa*** Roxb. (Convolvulaceae)

*L.N.:* Sarlaka-tarul (Nep.)

*Loc.:* Sukhna, Darjeeling, No. 5557

*Use :* Plant paste or juice is given in jaundice. Juice should be kept in a bottle after extracting from the plant, and teaspoonful is given to the patient every morning in empty stomach for 10-12 days or till recovery of jaundice.

***Cymbopogon citratus*** (DC.) Stapf (Poaceae)

*L.N.:* Citol (Nep.)

*Loc.:* Sukhna, Darjeeling, No. 5551

*Use :* Oil obtained from the grass is antiseptic and also used as insect repellant.

***Cynodon dactylon*** (Linn.) Pers. (Poaceae)

*L.N.:* Dubo (Nep.)

*Loc.:* Karmath, Darjeeling, No. 5438

*Use :* The whole plant juice extracted is given in pneumonial fever twice/thrice daily till recovery.

***Daphne bholua*** Buch.-Ham. ex D. Don (Thymelaeaceae)

*L.N.:* Loktaa (Nep.)

*Loc.:* Rishap hills, Kalimpong, Darjeeling, No. 5530

*Use :* Stem is used for making paper.

***Datura arborea*** Linn. (Solanaceae)

*L.N.:* Dhukra, Pampha (Nep.)

*Loc.:* Dolopchan, Kalimpong, Darjeeling, No. 5501

*Use :* Seed paste is given in diarrhoea. Leaf is used as good fodder.

***Datura metel*** Linn. (Solanaceae)

*L.N.:* Datura (Nep.)

*Loc.:* Dolpochan, Kalimpong, Darjeeling, No.5524

*Use :* Fruit juice is given for three days in madness. High dose is dangerous for human being.

***Dendrobium moschatum*** Swartz (Orchidaceae)

*L.N.:* Jaring (Nep)

*Loc.:* Sukhna, Darjeeling, No. 5578

*Use :* Leaves are prescribed in jaundice after cooking. The root juice is prepared after rubbing it on wooden surface and given to jaundice patient once or twice a day.

***Dioscorea bulbifera*** Linn. (Dioscoreaceae)

*L.N.:* Khamalu (B.)

*Loc.:* Sukhna, Darjeeling, No. *s.n.*

*Use :* Tuber is given in urinary track infection, also taken as vegetable.

***Drymaria cordata*** Willd. (Caryophyllaceae)

*L.N.:* Abhijal (Nep.)

*Loc.:* Tagda tea garden, Darjeeling, No. 5408, 5503, 5552

*Use :* Green leaves after rubbing is given to inhale to cure cough and cold with headache. The plant paste is covered with cloth and boiled, its vapours coming out are inhaled by the patient suffering from cold and cough. The pasted leaf is also inhaled to relieve sinus.

***Elettaria cardamomum*** (L.) Manton (Zingiberaceae)

*L.N.:* Elachi (Nep.)

*Loc.:* Latpanchar, Kalijhora, Darjeeling, No. 5447

*Use :* Fruits are used as spices.

***Erycibe paniculata*** Roxb. (Convolvulaceae)

*L.N.:* Bhuin champa (Nep.)

*Loc.:* Tagda tea garden, Darjeeling, No. 5407

*Use :* The tuber paste is given as a poultice to cure bone-fracture and in rheumatic pain.

***Eupatorium nodiflorum*** DC. (Asteraceae)

*L.N.:* Kalijhar (Nep.)

*Loc.:* Karmath, Kalijhora, Darjeeling, No. 5431

*Use :* Leaf paste is applied on fresh cuts and sores to protect infection.

***Gonostegia hirta*** (Blume) Miq. (Urticaceae)

*L.N.:* Chiplay (Nep.)

*Loc.:* Dolopchan, Kalimpong, Darjeeling, No. 5519

*Use :* The whole plant is given in bone-fracture along with the rhizome of 'Bhui-champa' (*Erycibe paniculata* Roxb.), which works quickly.

***Hoya lanceolata*** Wall. ex D. Don (Euphorbiaceae)

*L.N.:* Kurkuriya jhar (Nep.)

*Loc.:* Dolpchan, Kalimpong, Darjeeling, No. 5506

*Use :* The drink is made from pasted plant is given in diarrhoea and stomach trouble.

***Jatropha curcas*** Linn. (Euphorbiaceae)

*L.N.:* Kadam (Nep.)

*Loc.:* Sukhna, Darjeeling, No. 5555, 5432

*Use :* Stem is used as tooth brush which cures tooth problem like pyorrhoea, bleeding from teeth etc.

***Justicia adhatoda*** Linn. (Acanthaceae)

*L.N.:* Asru (Nep.), Basak (Beng.)

*Loc.:* Sukhna, Champa & Khokli Basti, Darjeeling, No. 5458, 5565

*Use :* Leaf juice with honey is given in cold and cough. Leaf paste or juice is given in jaundice, twice daily till cure. It has repellant power to protect any cultivated plant or nursery plant from the attack of insects or worms when a fencing is given around the garden; specially it is used in tea-gardens for this purposes.

***Lagerstroemia parviflora*** Roxb. (Lythraceae)

*L.N.:* Bura-Dhamro (Nep.)

*Loc.:* Sukhna, Darjeeling, No. 5549

*Use :* Timber yielding plant.

***Leucas cephalotes*** (Koen. ex Roth) Spreng. (Lamiaceae)

*L.N.:* Sarifa (Nep.)

*Loc.:* Sukhna, Darjeeling, No. 5465

*Use :* Leaf juice is very effective in scabies.

***Litsea cubeba*** (Lour.) Pers. (Lauraceae)

*L.N.:* Sil-timber (Nep.)

*Loc.:* Dolpchan, Kalimpong, Darjeeling, No. 5522

*Use :* Seed oil is prescribed in typhoid fever.

***Lyonia ovalifolia*** (Wall.) Drude (Ericaceae)

*L.N.:* Angeri (Nep.)

*Loc.:* Rishap, Lava, Kalimpong, Darjeeling, No. 5536

*Use :* Young bud juice is applied to cure scabies or other skin diseases.

***Malvaviscus arboreus*** Cav. (Malvaceae)

*L.N.:* Chusniful (Nep.)

*Loc.:* Tagda tea garden, Darjeeling, No. 5406

*Use :* Pasted flowers are given in diarrhoea, also it can be sucked directly for this purpose.

***Mentha longifolia*** (Linn.) Hudson (Lamiaceae)

*L.N.:* Tulsi pata (Nep.)

*Loc.:* Dolopchan, Kalimpong, Darjeeling, No. 5520

*Use :* Leaf paste or juice is prescribed in pneumonial fever.

***Moringa oleifera*** Lamk. (Moringaceae)

*L.N.:* Sajna (Nep.)

*Loc.:* Sukhna, Darjeeling, No. 5559

*Use :* Leaf is boiled with 'dal' and the soup is given for controlling high blood pressure.

***Nephrolepis auriculata*** (Linn.) Trimen (Nephrolepidaceae)

*L.N.:* Jamin-amla (Nep.)

*Loc.:* Tagda tera garden, Darjeeling, No. 5412

*Use :* Root bulb is taken during stomach trouble and in gas formation.

***Nephrolepis cordifolia*** (L.) Presl (Nephrolepidaceae)

*L.N.:* Ban-ukhu (Nep.)

*Loc.:* Karmath, Kalijhora, Darjeeling, No. *s.n.*

*Use :* Stem juice is given to cure blood sugar or diabetes.

***Ocimum canum*** Sims. (Lamiaceae)

*L.N.:* Silam (Nep.)

*Loc.:* Tagda tea garden, Darjeeling, No. 5404

*Use :* Wild edible fruit taken when ripe.

***Oxalis corniculata*** Linn. (Oxalidaceae)

*L.N.:* Chari-amito (Nep.)

*Loc.:* Dolopchan, Kalimpong, Darjeeling, No. 5514

*Use :* The whole plant is used for preparing 'chatny'

***Paedaria scandens*** (Lour.) Merr. (Rubiaceae)

*L.N.:* Padli-larang (Oraon)

*Loc.:* Sukhna, Darjeeling, No. 5563

*Use :* Pasted leaves are prescribed for increasing digestive power and also for curing amoebiosis. Stem is used as rope for making fencing.

***Phyla nodiflora*** (Linn.) Greene (Verbenaceae)

*L.N.:* Phullijhar, Porpore (Nep.)

*Loc.:* Champa & Rhokli basti Sukhna, Darjeeling, No. 5567

*Use :* Leaf paste is applied on fresh cuts to stop bleeding. Whole plant is a good fodder.

***Phyllanthus emblica*** Linn. (Euphorbiaceae)

*L.N.:* Amla (Nep.)

*Loc.:* Sukhna, Kurseong, Darjeeling, No. 5553

*Use :* Fruits are given in gastric disorders.

***Psidium guajava*** Linn. (Myrtaceae)

*L.N.:* Ambak (Nep.)

*Loc.:* Sukhna, Darjeeling, No. 5562

*Use :* Stem bark juice is given two spoonfuls twice daily for 2-3 days to cure white dysentery.

***Pteridium aquilinum*** (L.) Kuhn (Pteridiaceae)

*L.N.:* Unio (Nep.)

*Loc.:* Karmath, Kalijhora, Darjeeling, No. *s.n.*

*Use :* Leaf paste is applied to cows affected with worm to deworm quickly. Leaves are placed in the hatchery during hatching time to keep the chicks unharmed from worms or any virus etc.

***Pterygota alata*** (Roxb.) R. Br. (Sterculiaceae)

*L.N.:* Lator (Nep.)

*Loc.:* Sukhna, Darjeeling, No. 5548.

*Use :* Ripe seeds are eaten both by man and monkey.

***Rhododendron barbatum*** Wall. ex G. Don (Ericaeae)

*L.N.:* Guras (Nep.)

*Loc.:* Rishap, Lava, Darjeeling, No. *s.n.*

*Use :* Leaf paste or juice is very effective for taking out the fish bone from throat.

***Roylea cinerea*** (G. Don) Baill. (Labiateae)

*L.N.:* Tita-pati (Nep.)

*Loc.:* Rishap, Lava, Darjeeling, No. 5535

*Use :* Leaf paste is given to smell to stop bleeding from nostrils.

***Rubus ellipticus*** Sm. (Rosaceae)

*L.N.:* Aaisilo (Nep.)

*Loc.:* Dolpochan, Kalimpong, Darjeeling, No. 5528

*Use :* Bark juice is taken as a drink in strong diarrhoea. Ripe fruits are edible.

***Saurauia nepaulensis*** DC. (Actinidiaceae)

*L.N.:* Gogun (Nep.)

*Loc.:* Dolopchan, Kalimpong, Darjeeling, No. 5512

*Use :* Leaf is a good fodder. The young plant twig is cut into pieces and boiled and decoction is given to cows for increasing milk.

***Scoparia dulcis*** Linn. (Scrophulariceae)

*L.N.:* Chini-jhar, Jharco (Nep.)

*Loc.:* Champa & Khokli basti, Sukhna, Darjeeling, No. 5547, 5566

*Use :* Pasted plant juice is given in diabetes. Root paste is given in pneumonial fever. Brooms are also made from this plant.

***Smilax aspera*** Linn. (Smilacaceae)

*L.N.:* Kukurdaaino (Nep.)

*Loc.:* Mairung village, Kalimpong, Darjeeling, No. 5526.

*Use :* Leaf paste is applied on ulcerous wounds of cattle.

***Solanum ferox*** Linn. (Solanaceae)

*L.N.:* Ruk tomator (Nep.)

*Loc.:* Musikola, Darjeeling, No. 5405.

*Use :* Seed paste is given for curing gastric trouble. Fruits are cooked as vegetable.

***Swertia angustifolia*** Buch.-Ham. (Gentianaceae)

*L.N.:* Chiraaito(Nep.)

*Loc.:* Lava, Kalimpong, Darjeeling, No.5531

*Use :* Stem is cut into pieces and boiled in water and the decoction is prescribed 3 times per day till recovery for curing malaria or other kind of fever.

***Terminalia bellirica*** (Gaertn.) Roxb. (Combretaceae)

*L.N.:* Borra (Nep.)

*Loc.:* Way to Latpanchar, Karmath, Darjeeling, No. 5449, 5550

*Use :* Fruit's pulp is given for chewing to stop cough and cold. Root bark and fruit paste together are given in piles and in dysentery. Fruit decoction is given to cure gastric problem.

***Thysanolaena maxima*** (Roxb.) O. Ktze. (Poaceae)

*L.N.:* Amelesu (Nep.)

*Loc.:* Musikola, Dolopchan, Kalimpong, Darjeeling, No. 5403, 5508

*Use :* Root paste is prescribed by the Sherpa as a poultice on a boil which burst it and brings out the pus from the boil. Brooms are also made from this plant.

***Trichosanthes bracteata*** (Lam.) Voigt (Cucurbitaceae)

*L.N.:* Makal (Nep.)

*Loc.:* Risap hills, Lava, Kalimpong, Darjeeling, No. 5553

*Use :* Fried seeds are eaten.

***Urtica dioica*** Linn. (Urticaceae )

*L.N.:* Garia, Sisno (Nep.)

*Loc.:* Dolpchan, Kalimpong, Darjeeling, No. 5507

*Use :* Flower is taken with 'dal' for controlling high blood pressure.

***Vitex negundo*** Linn. (Verbenaceae)

*L.N.:* Nishinda (Nep.)

*Loc.:* Sukhna, Darjeeling, No. 5571

*Use :* The whole plant is given for baldness and for curing boils. Also used as insect repellant in the paddy fields. The young twigs are used as good tooth brush, effective in pyorrhoea.

## ACKNOWLEDGEMENTS

We are thankful to Director, Botanical Survey of India, Kolkata and Deputy Director, Central Botanical Laboratory, Howrah for giving facilities and encouragement.

## REFERENCES

CHHETRI, D.R. 2002. Medicinal plants used as antiseptic agents by the traditional healers of Darjeeling Himalyas. *J. Traditional Knowledge* 3 (3) : 271.

DIXIT *et al.* 1999. Fern of edible and medicinal value used by tribals of Darjeeling. *Jour. Ind. Bot. Soc.* 80 : 289-295.

JAIN, S. K. 1984. Bibliography of ethnobotany. New Delhi.

LAMA, P. C. 1989. A preliminary report on the ethnobotanical importance of the Sukia Pokhari region of Darjeeling Himalayas. *Journ. Bengal Nat. Hist. Soc.* 8 (1) : 56.

RAI, S. K. 1999. Notes on some less known plants of ethonobotanical importance from Darjeeling Himalayas. *Journ. Hill. Res.* 12 (2) : 160.

SRIVASTAVA, S.C. *et al.* 2003. Some plants of Ethnobotanical importance from Darjeeling Dist., West Bengal. *J. Econ. Taxon. Bot.* 27 (4) : 974.

YOUZONE, G.S. *et al.* 1984. Medicinal Plants of Darjeeling Dist. *J. Econ. Taxon. Bot.* 5 (3) : 605-616.

# ENUMERATION OF SOME POTENTIAL AND TRIBAL MEDICINAL PLANTS OF BODAMALAI HILLS IN NAMAKKAL DISTRICT OF TAMIL NADU

S.P. ANAND & R. JEYACHANDRAN

*Department of Plant Biology and Plant Biotechnology, St. Joseph's College (Autonomous with Five Stars & CPE), Tiruchirappalli – 620 002. Tamil Nadu, South India.*
*E Mail : anand_srs@yahoo.co.in*

## ABSTRACT

The Bodamalai hill tribe of Tamil Nadu is endowed with vast knowledge of medicinal plants. This research article deals with the study of some ethnomedicinal importance of 142 plants belonging to 56 families. Different medicinal plants are available in the locality used by the tribals for various diseases such as jaundice, diarrhoea, fever, skin diseases etc, and are also used for the treatment of animals. The botanical names, family, local name (Tamil name) and their uses are listed in this article.

## INTRODUCTION

According to a World Health Organization estimate, approximately 80% of indigenous population in developing countries depend on traditional medicine for primary health care needs, major portion of these involves the use of medicinal plants. (Benneman *et al.,* 1983). At present, around 70% drugs used as medicines are modeled on natural compounds, yet these are derived from only plant species. They traditionally occupy an important position in the socio-cultural, spiritual and medicinal arena. Recognizing the World's increasing concern on herbal medicines and India's rich traditional heritage and folklore in utilizing plants as medicine (Gupta *et al.,* 2004), most of the modern researchers on the herbal medicines have hinged round traditional folklore medicines namely ethnomedicine (Jain, 1980).

Same kind of research work carried out in some other regions of Tamil Nadu include Vallan-Kadu and Pollachi of Coimbatore district (Rajaram, 2004; Balasubramanium *et al.,* 1997) and Palani hills area of Dindugal district (Ganesan *et al.*, 2004).

Bodamalai hills is situated in Namakkal district of Tamil Nadu. The hill region comes under Eastern Ghats of Tamil Nadu. It covers a total area of 111 sq km, geographic location at 11° 29′ 52″ to 11° 35′ 12″ N and 78° 20′ 49″ to 78° 20′ 49″ E, and altitude of 200 m to 1194 m above MSL. The hill contains Southern thorn and Southern thorn scrub forest types. At the top hill of Bodamalai, there are four villages, namely Kadamalai, Kilur, Melur and Jambuttu malai (Plate 1).

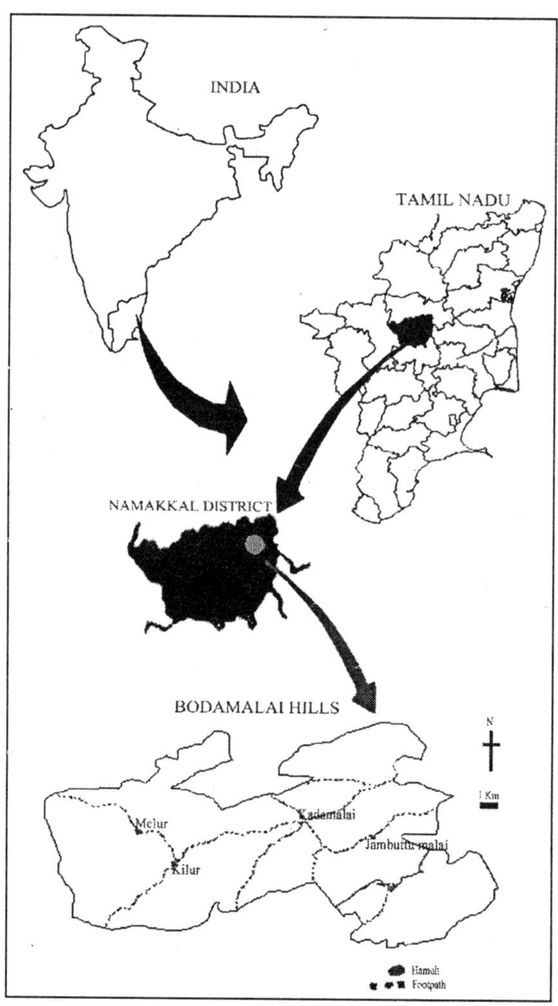


Plate 1. Location map of Bodamalai hill.

## METHODOLOGY

The survey of medicinal plants was done with the help of five field trips and with the aid of the people of the foot hills of Puduppatti, Pattanam, Vadugam and Mangalapuram villages. The old knowledgeable tribal people were interviewed about the medicinal plants and their usage, and medicinal plants were collected and notes were gathered about their usage. The collected sample specimens have been preserved for future references. After every field visit, the collected specimens were identified in Rapinet Herbarium (St. Joseph's College (Autonomous), Tiruchirappalli – 620 002, Tamil Nadu, India).

## OBSERVATIONS

The different kinds of plants used by tribal people and some medicinal plants available on that hill area are summarized in an alphabetical order by their botanical names, with respective family, vernacular names (Tamil name) and a brief notes on uses collected from tribals of Bodamalai hill. Some leading literature on medicinal plants, such as Kirtikar & Basu (1935), Chopra *et al.* (1956), Warriear *et al.* (1994-96), Chatterjee & Pakrashi (1994-97), Trivedi (2004) and Trivedi & Sharma (2004) was consulted to verify the folk medicinal uses.

**Abrus precatorius** L. (Fabaceae); Kundumani

Seeds cure skin diseases. Stem-paste cures jaundice.

**Abutilon indicum** (L.) Sweet (Malvaceae); Thutti

Seeds act as cooling agent for human body.

**Acacia nilotica** (L.) Willd. ex Del. ssp. **indica** (Benth.) Brenan (Mimosaceae); Karuvelam

Pods are used in impotency and are effective for skin diseases and curing mouth ulcers.

**Acalypha indica** L. (Euphorbiaceae); Kuppaimani

Paste of leaves is applied on burns with juice of lime and to cure eye inflammation.

**Aegle marmelos** (L.) Corr. (Rutaceae); Vilvam

Leaf-juice cures dysentery and diarrhoea. Fruit acts as a cooling agent.

**Alangium salvifolium** (L.f.) Wang. (Alangiaceae); Alangil

Leaf-juice is used for dog bite and to heal up feet wound. Root-bark is used for fever and skin diseases.

**Aerva lanata** (L.) Juss. ex Schult. (Amaranthaceae); Pollapoo

It cures urinary problems.

**Albizia lebbeck** (L.) Benth. (Mimosaceae); Vagaimaram

Stem, leaves and flowers are used for snake bite and scorpion sting.

**Anisochilus carnosus** Wall. ex Benth. (Lamiaceae); Satrupoonthalai

Leaf-paste cures the wounds on the legs. It is also used as a remedy for cold and cough.

**Anisomeles indica** (L.) Ktze. (Lamiaceae); Paaiemirati

Plant is used to cure digestive problems. Leaf-juice cures fever.

**Annona squamosa** L. (Annonaceae); Setha

Leaf-paste is applied to cure body pain.

**Aristolochia indica** L. (Aristolochiaceae); Peerumarunthukodi

Root is used in toothache. Fruit acts as a tonic and restorative.

**Artocarpus heterophyllus** Lam. (Moraceae); Palamaram

Leaves are useful in healing ulcers, curing fever, wounds and skin diseases.

**Asparagus racemosus** Willd. (Liliaceae); Thannervittankizanku

It is a powerful drug capable of improving memory power. Used to cure diarrhoea, dysentery and cough.

**Atalantia monophylla** (L.) Corr. (Rutaceae); Kattuezhumichi

Leaves are used for snake bite. Root acts as antiseptic agent.

**Azadirachta indica** A. Juss. (Meliaceae); Vambu

It is a grand remedy for fever. Bark acts as an anticancer drug and pesticide and insecticide.

**Bauhinia variegata** L. (Caesalpiniaceae); Maliyatthi

Dried buds are used to cure dysentery, piles, diarrhoea, worms and against tumors.

**Blepharis maderaspatensis** (L.) Heyne ex Roth (Acanthaceae); Nanthrapundu

Leaves are ground with egg and onion and applied for bone fracture.

**Boswellia serrata** Roxb. ex Colebr. (Burseraceae); Vellaikunkilium

It used to cure skin diseases, asthma, dysentery, piles, fever and cough.

**Breynia vitis-idaea** (Burm. f.) Fischer (Euphorbiaceae); Pavallapundu

Plant is used to cure diarrhoea.

**Buchanania cochinchinensis** (Lour.) Almeida (Anacardiaceae); Murrukamaram

Extract of leaves is used as anticancer. Root is used as cordial tonic and for skin diseases.

**Butea parviflora** Roxb. (Fabaceae); Poonaikali

Stem is used in dysentery. Bark is used in toothache.

**Cadaba fruticosa** (L.) Druce (Capparaceae); Vizhuthi

Leaf-paste with oil is used for sores.

**Calotropis gigantea** (L.) R. Br. (Asclepiadaceae); Erukku

Leaves are used to cure joint pain and swellings.

**Canarium strictum** Roxb. (Burseraceae); Karunkungiliam

Gum is used with gingili oil in rheumatic pains & chronic skin diseases such as psoriasis.

**Canavalia virosa** (Roxb.) Wt. & Arn. (Fabaceae); Kattuthampattai

Seeds stimulate sleep.

**Canthium parviflorum** Lam.(Rubiaceae); Karaicceti

Roots and leaves are used as diuretic.

**Canthium dicoccum** (Gaertn.) Teijsm. & Binn. var. **umbellatum** (Wight) Sant. & Merch. (Rubiaceae); Varapoola

It cures diarrhoea and dysentery.

**Capparis sepiaria** L. (Capparaceae); Thurati

Seed acts as an antidote to snake bite.

**Capparis zeylanica** L. (Capparaceae); Athattai

Leaf paste is applied for small pox, to control cholera, piles and swellings.

**Caralluma umbellata** Haw. (Asclelpiadaceae); Kallimullian

Plant acts as water substitute for thirst quenching, and used to cure paralysis.

**Cardiospermum halicacabum** L. (Sapindaceae); Mudakatthan

Leaves are used in paralysis.

**Carissa spinarum** L. (Apocynaceae); Perunkilla

Plant prevents growth of intestinal worms, cure burning wounds and skin diseases.

**Cassine glauca** (Rottb.) O. Ktze. (Celastraceae); Sillupamaram

It acts as a good medicine for all poison bites.

**Cayratia pedata** (Lam.) Juss. ex Gagnep. (Vitaceae); Iynthilaikodi

Useful to cure burning wounds and leaf is used for ulcers.

**Centella asiatica** (L.) Urban (Apiaceae); Vallarai

Leaves are used in nervous problems; whole plant acts as a brain tonic.

**Chamaecrista absus** (L.) Irwin & Barneby (Caesalpiniaceae); Kattukollu

Seed is used in skin diseases. Leaves are used in cough and poisonous bites.

**Cipadessa baccifera** (Roth) Miq. (Meliaceae); Pullipuchedi

Leaf-paste is taken orally to control diarrhoea.

**Cissampelos pareira** L. (Menispermaceae); Appatta

Whole plant is used in diarrhoea.

**Cissus quadrangularis** L. (Vitaceae); Pirantai

Stem and leaf paste is used to cure bone fracture.

**Cissus vitiginea** L. (Vitaceae); Nilapirantai

Plant-paste cures piles and ear-ache.

**Cleistanthus collinus** (Roxb.) Benth. ex Hook. f. (Euphorbiaceae); Oduvan

Leaves are used as insect repellant in agricultural fields.

**Clitoria ternatea** L. (Fabaceae); Sankupookodi

Plant is used to cure snake poison. It gives strength and has more digestive power.

**Coccinia grandis** (L) Voigt (Cucurbitaceae); Koovaikodi

It cures skin diseases, mouth ulcer and diabetes.

**Cocculus hirsutus** (L) Diels (Menispermaceae); Sirukattukodi

Leaf-juice is used for gonorrhoea.

**Commiphora berryi** (Arn.) Engler (Burseraceae); Mullukiluvan

Gum resin is used for snake bite and scorpion sting.

**Croton bonplandianum** Baill. (Euphorbiaceae); Vinaipondu

Leaves are used as antibacterial agent.

**Cryptolepis buchananii** Roem. & Schult. (Asclepiadaceae); Palkodi

Plants are used for body pain and dysentery.

**Curculigo orchioides** Gaertn. (Hypoxidaceae); Nilapanai

Roots are used in toothache and to control diarrhoea.

**Dalbergia paniculata** Roxb. (Fabaceae); Achamaram

Leaf-paste is used to cure rheumatic pain.

**Datura stramonium** L. (Solanaceae); Umathai

It is used to cure fever, complications of diarrhoea and skin diseases.

**Derris scandens** (Roxb.) Benth. (Fabaceae); Podanikodi

Young leaf-paste is used to cure diarrhoea.

**Dioscorea oppositifolia** L. (Dioscoreaceae); Kavalikizanku

Tubers are used for external application to reduce swellings.

**Diospyros montana** Roxb. (Ebenaceae); Vakkanai

Fruits are applied on external boils.

**Dolichandrone falcata** (Wall. ex DC.) Seem. (Bignoniaceae); Kadalathi

Bark is used for fish poison and as abortive.

**Erythrina variegata** L. (Fabaceae); Kalyana, Murungai

Bark is used in leprosy, skin diseases and fever.

**Erythroxylum monogynum** Roxb. (Erythroxylaceae); Devadaru

Wood and bark are used for dyspepsia and continued fever.

**Eucalyptus globulus** Labill. (Myrtaceae); Thilamaram

Roots are used as antiseptic and diaphoretic.

**Euphorbia hirta** L. (Euphorbiaceae); Ammaanpacharishi

A decoction is used for asthma and chronic bronchial infections.

**Euphorbia nivulia** Buch.-Ham. (Euphorbiaceae); Illaikalli

Stem mixed with green ginger is given to patients of dog bites.

**Euphorbia tirucalli** L. (Euphorbiaceae); Kodikalli

Milk acts as a poison controlling agent.

**Evolvulus alsinoides** (L.) L. (Convolvulaceae); Vishnukiranthi

Leaves are used to make cigarettes smoked in chronic bronchial infection and asthma.

**Feronia limonia** (L.) Swingle (Rutaceae); Vilangaimaram

Leaves are used to cure diarrhoea, vomiting and cough. Ripe fruits are used to cure dysentery.

**Flacourtia indica** (Burm. f.) Merr. (Flacourtiaceae); Sottaikalla

Fruits are used for jaundice, enlarged spleen and cholera.

**Gardenia resinifera** Roth (Rubiaceae); Kattukoyia

Fruits are used for digestion and to cure stomach pain.

**Garuga pinnata** Roxb. (Burseraceae); Yannaikkarai

Juice of the leaves is used for asthma. Decoction of the roots is used in pulmonary affections.

**Gliricidia sepium** (Jacq.) Kunth ex Walp. (Fabaceae); Vivazyathagari

It is used for stomach pain.

**Gloriosa superba** L. (Liliaceae); Kannvazhipookodi

Roots are used to cure leprosy and parasitical infection of skin piles.

**Grewia rhamnifolia** Heyne ex Roth (Tiliaceae); Udmbaamaram

Paste of crushed leaves is applied on wounds and cuts.

**Gymnema sylvestre** (Retz.) R. Br. ex Schult. (Asclepiadaceae); Sirukurinja

Leaves are used for diabetes and to control urinary problems.

**Hardwickia pinnata** Roxb. (Caesalpiniaceae); Kodapalai

Paste of this plant is used for gonorrhoea.

**Hedyotis puberula** (G. Don) Arn. (Rubiaceae); Sayavar

Leaves and roots cure asthma. Roots are a good medicine against snake-bite.

**Heliotropium indicum** L. (Boraginaceae); Thelkoudukupundu

Plant is used to cure wounds and fever. Root decoction is used for cough and fever.

**Hemidesmus indicus** (L.) R. Br. (Asclepiadaceae); Nannari

Plant is used to cure diarrhoea and inflammed eyes.

**Holarrhena pubescens** (Buch.-Ham.) Wall. ex G. Don (Apocynaceae); Kodapallai

Plant acts against dysentery, used to expel worms from the body, cure diseases of the bloods and leucoderma.

**Hugonia mystax** L. (Linaceae); Mothorudan

Roots are used for externally reducing inflammatory swellings.

**Hybanthus enneaspermus** (L.) F. Muell. (Violaceae); Orethilthamarai

Plant juice acts as a tonic and roots are used for bowel complaints in children.

**Indigofera tinctoria** L. (Fabaceae); Avuri

Leaf-juice is used as prophylactic against hydrophobia.

**Ipomoea pes-caprae** (L.) R. Br. (Convolvulaceae); Attukallkilangu

Root is used to cure joint pain and foot pain.

**Ipomoea sepiaria** Koen. ex Roxb. (Convolvulaceae); Talikkodi

Plant juice is used as deobstruent and diuretic.

**Ixora arborea** Roxb. ex J. E. Sm. (Rubiaceae); Perunthutti

Bark decoction is given as a tonic for anaemia and general debility. Flowers are used for whooping cough.

**Jasminum angustifolium** (L.) Willd. (Oleaceae); Kattumullai

Roots are used for external application for ringworm and leprosy.

**Jatropha curcas** L. (Euphorbiaceae); Pacchai Kataamanuku

Plant juice is used for scabies and ring worm. It gives strength to teeth and cures toothache.

**Jatropha glandulifera** Roxb. (Euphorbiaceae); Sikapu Kataamanuku

Plant juice is used to clean the eyes. Roots are used for reducing glandular swellings.

**Kleinia grandiflora** (Wall. ex DC.) N. Rani (Astraceae); Musakadhukalli

This plant cures hydrophobia disease.

**Lannea coromandelica** (Houtt.) Merrill (Anacardiaceae); Uthian

Bark is used for cuts, wounds, ulcers, diarrhoea and dysentery. Leaves are used for elephantiasis.

**Lantana camara** L. (Verbenaceae); Unnichedi

It is used for cough, cold and fever.

**Lawsonia inermis** L. (Lythraceae); Maruthondri

Bark is used to cure jaundice and skin diseases.

**Leucas aspera** (Willd.) Link (Labiatae); Thumbai

Flowers are used to cure cold. Leaves juice is used to cure skin diseases.

**Macaranga peltata** (Roxb.) Muell.-Arg. (Euphorbiaceae); Vattakunni

Bark is used to cure backbone pain. It removes kidney stones.

**Madhuca longifolia** (Koen.) Macbr. (Sapotaceae); Eluppai

Bark is used for ulcers, itches and bleeding.

**Mallotus philippensis** (Lam.) Muell.-Arg. (Euphorbiaceae); Nallathumbai

Cures ringworm and skin brown spots.

**Mangifera indica** L. (Anacardiaceae); Mangaai

Seed is used to cure asthma. Dried flowers are used to cure diarrhoea.

**Melia azedarach** L. (Meliaceae); Malaivambu

Flowers and leaves paste is used to cure headache. Bark is used to cure neck bulging.

**Melothria maderaspatana** (L.) Cong. (Cucurbitaceae); Musumusukkai

Tender shoots are used for toothache. Leaves are used to cure body pain.

**Mimosa pudica** L. (Mimosaceae); Thottarsinunge

Leaves are used to cure dysentery, kidney and liver problems.

**Morinda tinctoria** Roxb. (Rubiaceae); Nunna

Roots are used to cure dysentery. Leaves are useful in curing fever and diarrhoea.

**Moringa concanensis** Nimmo ex Dalz. & Gibs. (Moringaceae); Kattumurungai

Roots are used to cure fever. Fruits cure liver diseases, spleen and paralysis.

**Naravelia zeylanica** (L.) DC. (Ranunculaceae); Vatamkolli

Plants cure paralysis, skin diseases and ulcers.

**Ocimum canum** Sims. (Labiatae); Naaitulasi

Seed paste cures lung and chest pains.

**Ocimum tenuiflorum** L. (Labiatae); Nalltulasi

It is used to cure cough and cold. Leaf juice prevents earaches.

**Oxalis corniculata** L. (Oxalidaceae); Cenkondai

Plant is used as digestive agent. It cures anaemia and dysentery.

**Pachygone ovata** (Poir.) Miers ex Hook. f. & Thom. (Menispermaceae); Peerukattukodi

Dried fruits are used to control insect bites.

**Pavonia odorata** Willd. (Malvaceae); Malaiccentotti

Roots are used to cure dysentery. It cures joint and muscle pain.

**Pergularia daemia** (Forsk.) Choiv. (Asclepiadaceae); Uttamanikodi, Vealiparuthi

Plants cure urinal problems, fever and asthma. Leaves are used to cure leprosy.

**Phyllanthus amarus** Schum. & Thonn. (Euphorbiaceae); Kizhanelli

It is one of the best medicines for jaundice. Powder of leaves is used to cure ulcers.

**Pongamia pinnata** (L.) Pierre (Fabaceae); Pungan

Flowers cure diabetes. They also prevent stomach problem.

**Pouzolzia zeylanica** (L.) Benn. (Urticaceae); Kalluruki

This plant cures sexually transmitted diseases. It controls snake poisoning.

**Premna tomentosa** Willd. (Verbenaceae); Vakkanai, Malaithakku

Roots cure stomach disorders.

**Quirivelia frutescens** (L.) M. R. & S. M. Almeida (Apocynaceae); Udargodi

Used for vitiated conditions of pitta, burning sensation and leprosy

**Rhynchosia minima** (L.) DC. (Fabaceae); Kattukollu

Used to describe a drug that causes abortion.

**Sansevieria roxburghiana** Schult. & Schult. f. (Liliaceae); Marul

The leaf is introduced into heat flame and made into juice that cures earache.

**Santalum album** L. (Santalaceae); Sandanam

Powder of wood is used to cure pimples and skin diseases.

**Sapindus emarginatus** Vahl (Sapindaceae); Poovathikottai

Fruits are used to cure asthma, cholera and diarrhoea.

*Boswellia serrata* Roxb.

*Cissus vitiginea* L.

*Euphorbia nivulia* Buch.-Ham.

*Gardenia resinifera* Roth

*Gymnema sylvestre* (Retz.) R. Br. ex Schult

*Holarrhena pubescens* (Buch.-Ham.) Wall. ex G. Don

Plate 2. Some medicinal plants of Bodamalai Hills.

*Hugonia mystax* L.

*Scilla hyacinthina* (Roth) Macbr.

*Sterculia urens* Roxb.

*Terminalia chebula* (Gaertn.) Retz.

*Vitex altissima* L.f.

*Zehneria scabra* (L.f.) Sond.

Plate 3. Some medicinal plants of Bodamalai Hills.

**Sarcostemma intermedium** Decne. (Asclepiadaceae); Kodikalli

In case of poisoning, the plant milk stimulates vomitting.

**Scilla hyacinthina** (Roth) Macbride (Liliaceae); Narivengayam

It is used to cure dysentery and acts against poison.

**Securinega leucopyrus** (Willd.) Muell.-Arg. (Euphorbiaceae); Pulanji

Leaves are made into a paste with tobacco and used to destroy worms in sores.

**Senna auriculata** (L.) Roxb. (Caesalpiniaceae); Avarampoo

Leaf and flowers are used for hair improving. Seeds are used in diabetes.

**Senna montana** (Heyne ex Roth) Singh (Caesalpiniaceae); Ponnavarai

It is a cooling agent and cures diabetics.

**Senna tora** (L.) Roxb. (Caesalpiniaceae); Thakarai

It prevents recurring diseases such as malaria.

**Sida acuta** Burm. f. (Malvaceae); Pazampasi

Leaves juice is used with coconut oil to prevent hair fall.

**Sida cordifolia** L. (Malvaceae); Arivalmaniponndu

Juice is used for healing wounds, and juice with water is given to the spermatorrhoea patients.

**Solanum erianthum** D. Don (Solanaceae); Kattushundai

The fruits are tied around the neck for relief in cough.

**Sterculia urens** Roxb. (Sterculiaceae); Senthanugai

Leaves are applied to wounds, fractures and cracked skin. They are also used to cure throat infection.

**Strychnos nux-vomica** L. (Loganiaceae); Ettikottai

Bark juice is used for cure cholera and dysentery. Roots are used for fever.

**Strychnos potatorum** L. f. (Loganiaceae); Thatthan

Fruits are used to control dysentery. Bulbs stimulate vomitting.

**Syzygium jambos** (L.) Alston (Myrtaceae); Naval

Seeds are used to cure diarrhoea and dysentery and fruits are used in liver complaints.

**Tectona grandis** L. f. (Verbenaceae); Thakku

Flowers and seeds are used to promote the flow of urine.

**Terminalia cuneata** Roth (Combretaceae); Neermaruthu

The leaf-juice is a good remedy for earache. Bark is used for ulcers.

**Terminalia chebula** (Gaertn.) Retz. (Combretaceae); Kadukkai

Fruits are used for digestive, antiseptic and diuretic purposes.

**Tinospora cordifolia** (Willd.) Miers ex Hk. f. & Th. (Menispermaceae); Centhilkodi

Plant juice is used for diuretic and urinal problems.

**Toddalia asiatica** (L.) Lam. (Rutaceae); Milakaranai

Fruits are useful to cure fever, dyspepsia, diarrhoea, cough, wound and ulcers.

**Tribulus terrestris** L. (Zygophyllaceae); Sirunirungil

Leaves and fruits dissolve the stone formation in kidney and cure various kidney problems.

**Tridax procumbens** L. (Astraceae); Vettukayapoondu

Plant juice is applied on cuts and wounds. Leaves are used to cure dysentery and diarrhoea.

**Ventilago madraspatana** Gaertn. (Rhamnaceae); Surulpattai

Root bark is used to relive stomach disorders and fevers.

**Vitex altissima** L. f. (Verbenaceae); Mayiladi

Plant is used to promote growth of hairs. Flowers are used to cure cholera, diarrhoea and fever.

**Vitex negundo** L. (Verbenaceae); Nochee

Leaf extract is applied to the swelling of the joints.

**Wattakaka volubilis** (L.f.) Stapf (Asclepiadaceae); Perunkurinja

Plant juice is used for cooling, curing cold and for eye diseases. It is also used against snake bite.

**Wrightia tinctoria** (Roxb.) R. Br. (Apocynaceae); Veppalai

Leaf-juice is used to cure jaundice and stomach pain. It relives toothache and acts as antidiuretic.

**Zehneria scabra** (L.f.) Sonder (Cucurbitaceae); Akazakarudan kodi

The plants are used for treating stomach pain, fever and skin diseases, both for the human as well as animals.

**Ziziphus mauritiana** Lam. (Rhamnaceae); Elanthai

Seeds control diarrhoea & are used against poison. Root decoction cures fever and wounds.

**Ziziphus oenoplia** (L.) Mill. (Rhamnaceae); Soorai

Root-bark is used to cure and heal wounds. Fruit acts as digestive agent.

## ACKNOWLEDGEMENT

The authors express their deep sense of gratitude to Tamil Nadu State Government for financial assistance. They are also grateful to the management of St. Joseph's College (Autonomous), Tiruchirappalli-2, Tamil Nadu, South India, for providing necessary facilities to carry out research.

## REFERENCES

BALASUBRAMANIAN, P., A. RAJASEKARAN & S. N. PRASAD 1997. Folk medicine of the Irules of Coimbatore forests. *Anci. Sci. Life* 16 (3) : 222-226.

BENNEMAN, R., J. BURTON & W. C. CHEN 1983. *Traditional medicine and health care.* World Health Organization, Geneva, Switzerland.

CHATTERJEE, A. & S.C. PAKRASHI 1994-97. *The treatise of Indian medicinal plants.* National Institute of Science communication, New Delhi.

CHOPRA, R. N., S. L. NAYAR & I. C. CHOPRA 1956. *Glossary of Indian medicinal plants.* CSIR, New Delhi.

GUPTA, S., M. C. PORWAL & P. S. ROY 2004. Indigenous knowledge on some medicinal plants curing the Nicobari Tribe of Car Nicobar Island. *Indian Journal Traditional Knowledge* 3 (3) : 287-293.

JAIN, S.K. 1980. Ethnobotanical research unfolds near visits of traditional medicine. In Jain, S.K. (Ed.) *Glimpses of Indian ethnobotany,* pp. 191-198. Oxford & IBI Pub., New Delhi.

KIRTIKAR, K.R. & B.D. BASU 1935. *Indian medicinal plants,* Vols. 1-4. Jayyed Press, New Delhi.

RAJARAM, N. 2004. Tribal medicines of Vellian-kadu and Paalamali of Comibatore district,

Tamil Nadu. *Advances Plant Sciences* 17(2): 389-393.

TRIVEDI, P.C. 2004. *Medicinal plants – Utilization and Conservation.* Aavishkar Publishers, The Diamond Printing Press, Jaipur, India.

——— & N.K. SHARMA 2004. *Ethnomedicinal plants.* Pointer Publishers, the Diamond Printing Press, Jaipur, Rajasthan, India.

WARRIER, P.K., V.P.K. NAMBIAR & C. RAMANKUTTY 1994-96. *Indian Medicinal Plants*, Vols. 1-5. Orient Longman Ltd., Madras.

# WILD PLANTS USED BY THE TRIBES OF NOKREK BIOSPHERE RESERVE, MEGHALAYA AND THEIR CONSERVATION

BASANT KUMAR SINGH & HIMADRI SEKHAR DEBNATH

*Botanical Survey of India, Industrial Section, Indian Museum, 1-Sudder Street, Kolkata – 700 016, West Bengal*

**ABSTRACT**

Systematic survey was conducted on some wild plants used by the tribal people living in Nokrek Biosphere Reserve of Meghalaya. Although 21 plants were recorded to have medicinal, edible and constructional values and widely used by the tribes of Nokrek Biosphere Reserve. At the same time, it also throws light on need of conservation of these plants.

## INTRODUCTION

Since time immemorial human being has made use of plants in different ways. Particularly the rural people and tribes living in remote villages and forest areas still depend to a great extent on the wild plants to fulfill their needs in the form of food, medicine and construction material. At the same time, the tribal people are very much aware about the conservation of these important plants for future use. They have their own way or method to protect these plants. Some of the tribal medicines and their indigenous knowledge for using the plant resources are also incorporated in our civilized society. However, much larger number of such indigenous knowledge has been confined to certain pockets of North-east India. It is estimated that one fifth of the total 17,000 angiosperms found in India constitute useful plants (Rawat & Choudhury, 1988).

Nokrek Biosphere Reserve is located in between 25°20′ to 25°29′ N latitude and 90°13′ to 90°35′ E longitude in the North-Western part of Meghalaya having the parts of East, West and South Garo Hill districts of Meghalaya. The total area of this Biosphere Reserve is 820 sq. km distributed in three zones, viz. Core Zone of 47.48 sq km, Buffer Zone of 227.92 sq km and a Transition Zone of 544.60 sq km.

The main tribes living inside Nokrek Biosphere Reserve are Garo, Hajongs, Roches, Banias etc. According to the 2001 census the human population in the

Biosphere Reserve is 22,084, inhibiting the buffer and transition zones. They are living in the remote villages and in the vicinity of the forests. They are very much dependent on forest resources for their survival. Since modern medicinal facilities are yet to reach in many remote parts of the Biosphere Reserve, ethnic people believe in the traditional systems of folk medicines and rely extensively on their own herbal cures. They use the plant products in the form of food, medicines and also construction purposes. Though the biodiversity of this area is under extreme pressure due to regular practice of shifting cultivation by the local people as it is the only way of survival for them, they take immense care to conserve the plants they use in either way.

Study of plants having medicinal and other values have been carried out from time to time in different parts of India by several workers, such as Nayar *et al.* (1989), Rawat & Choudhury (1988), Jain (1965), Majumdar *et al.* (1978) and several others. Although in Meghalaya, the study of economical important plants and their ethnobotanical importance have already been emphasized by several workers (Rao & Sanpru, 1981; Kumar *et al.*, 1982; Biswas & Bhuyan, 1983 etc), yet not much work has been done in Nokrek Biosphere Reserve with reference to the tribes living inside the Reserve. It is very important to know the use of plants growing in wild. Moreover, the existence of a Biosphere Reserve totally depends upon the man and nature relationships. So, it is also necessary to protect the Biosphere Reserve with proper management implementation and sustainable use.

## METHODOLOGY

During present study, few villages, namely Daribokgre, Asagre, Chokpot, Chandigre, Sakalgre, Mandalgre and Budhigre located inside the Nokrek Biosphere Reserve, were surveyed during 2006-2007, so as to gather detailed information on each and every plant species found useful for the tribes. The information and data has been mainly collected from *Oja* (Medicine Man) and other people who have the knowledge of the use of these plants. Routine methods of botanical collection and herbarium technique were followed as suggested by Jain & Rao (1977). Specimens have been collected & deposited in the in the herbarium of Botanical Survey of India, Industrial Section, Indian Museum (BSIS), Kolkata.

## ENUMERATION

| Plant species | Family | Vernacular name | Use | Plants part used |
|---|---|---|---|---|
| **Plants used as medicines** | | | | |
| *Ageratum conyzoides* L. | Asteraceae | Sazadri | Cuts and wounds | Leaf juice |
| *Bambusa tulda* Roxb. | Poaceae | Wada | Stop bleeding | Peel of young stem |
| *Centella asiatica* (L.) Urban | Apiaceae | Misnachel | Eye infection, dysentery and stomachache | Leaf juice |
| *Cinnamomum tamala* Nees | Lauraceae | Tezpatta | Toothache | Leaf |
| *Citrus indica* Tanaka | Rutaceae | Miming-narang | Blood dysentery | Leaf and fruit |

| | | | | |
|---|---|---|---|---|
| *Curcuma angustifolia* Roxb. | Zingiberaceae | Boldam | Headache | Leaf juice |
| *Melastoma malabathricum* L. | Melastoma-taceae | Khakhuchi | Dysentery and worm infection | Leaf and young twigs |
| *Oxalis corniculata* L. | Oxalidaceae | Munamuni | Indigestion, gastric problems | Whole plant |
| *Solanum sisymbrifolium* Lam. | Solanaceae | Barankuni | Toothache | Fruits |
| *Spilanthus paniculata* Wall. ex DC. | Asteraceae | Kimka | Toothache and worm infection | Flowers and leaves |
| **Plants used as food** | | | | |
| *Bambusa tulda* Roxb. | Poaceae | Wada | Eaten as vegetable and soda | Tender stem, the ash of mature stem and leaf known as *Kharchi* is used to cook food. |
| *Calamus erectus* Roxb. | Arecaceae | Sukmal | Eaten as chatni | Stem sap and tender stem |
| *Colocasia esculenta* (L.) Schott | Araceae | Tha | Eaten as vegetables | Underground stem |
| *Eryngium foetidum* L. | Apiaceae | Samskal | Eaten as chatni and Curry | Young leaf and tender stem |
| *Ficus hispida* L. | Moraceae | Khatap | Eaten as vegetable | Fruits |
| *Hydrocera triflora* W. & A. | Balsaminaceae | Wathpu | Eaten as vegetables | Young leaves and tender stem |
| *Passiflora quadrangularis* L. | Passifloraceae | Alagoti | Eaten fruit juice | Ripen fruit |
| *Rubus ellipticus* Sm. | Rosaceae | Tikesambak | Eaten as fruit | Fruit |
| **Plants used for construction** | | | | |
| *Calamus erectus* Roxb. | Arecaceae | Sukmal | Used for constructing different articles for domestic purposes | Stem |
| *Dendrocalamus hamiltonii* Nees & Arn. ex Munro | Poaceae | Wada | Used for constructing houses & other articles of domestic use | Stem |
| *Saccharum procerum* Roxb. | Poaceae | Ampang | Used as roofing material | Leaves |

## DISCUSSION & CONCLUSION

The study revealed that Nokrek Biosphere Reserve in Meghalaya is rich in various plants with different economical values. The way by which the tribal people of Nokrek Biosphere Reserve are using the plants has some similarity with some other parts of India also. At the same time, they are also very keen in the preservation of these plants for future. Their consumption is very much resource based. It is seen during the field visits that the local people are using the forest products sustainably by collecting only the needed amount of plant part and do not destroy whole plant. Moreover, they use the more abundant plants in large amount than the less available plant. The availability of the plants decided by them is totally based on their observations and experience. But, in case of bamboo species, they use it in different ways, as it is an abundant species in Nokrek Biosphere Reserve. During one field trip they directed one of the authors not to take the specimen in much amount as that was a very important medicinal plant and was less frequent in the forest. This example shows their understanding of the forest resources and its importance, which is the sign of healthy development of the Biosphere Reserve. But, the regular practice of shifting cultivation and rapid deforestation in the biosphere reserve are leading to the fast disappearance of these plants. The indigenous knowledge of the utilization of plants passes on through oral communication from one generation to another. It has also been observed that the younger generation is no more interested to learn the uses of these plants from the elders. Therefore, it is important to preserve the indigenous knowledge of our tribal communities. It is also important to preserve these plants species before they become endangered. This study will serve as a basic information for any further study in these regards.

## ACKNOWLEDGEMENTS

The authors are thankful to Ministry of Environment and Forests, New Delhi for the financial assistance to undertake the field tours. The authors are also thankful to Dr. M. Sanjappa, Director, Botanical Survey of India and Mr. P.R. Marak, DFO, East and West Garo Hill Districts, Wildlife Division, Tura for their kind cooperation.

## REFERENCES

BISWAS, S. & T.C. BHUYAN 1983. On the identity of some food plants of Garo hills, Meghalaya. *Indian Journal Forestry* 6 (3) : 208-213.

JAIN, S.K. 1965. On the prospects of some new or less known medicinal plant resources. *Ind. Med. J.* 59 : 270-272.

———— & R.R. RAO 1977. *Handbook of Field and Herbarium methods.* New Delhi.

KUMAR, Y., K. HARIDASAN & R.R. RAO 1982. Ethnobotanical notes on certain medicinal plants among some Garo people around Balpakhram Sanctuary in Meghalaya. *Bull. Bot. Surv. India* 22(1-4) : 161-165.

MAJUMDAR, R., K.C. TIWARI, S. BHATTACHARJEE & A.R. NAIR 1978. Some folklore medicines from Assam and Meghalaya. *Quart. J. Crude Drug Res.* 16 : 4.

NAYAR, M.P., K. RAMAMURTHY & V.S. AGRAWAL 1989. *Economic Plants of India,* Vol. I. BSI, Calcutta.

RAO, M.K. & R. SANPRU 1981. Some plants in the Life of Garos. In Jain, S.K. (Ed.) *Glimpses of Indian Ethnobotany,* pp. 153-160. Oxford and IBH Publishing Co., New Delhi.

RAWAT, M.S. & S. CHOUDHURY 1988. *Ethnomedico botany of Arunachal Pradesh (Nishi and Apatani tribe).* Bishen Singh Mahendra Pal Singh, Dehra Dun.

# MEDICINAL PROPERTIES OF SOME LOCALLY AVAILABLE GREEN LEAFY VEGETABLES OF PONDICHERRY REGION

A. PRAGASAM, K. KADAVUL & S. SIVARANJINI

*Department of Plant Science, K.M. Centre for P.G Studies, Lawspet, Pondicherry – 605 008, India*

## ABSTRACT

The nutritive and medicinal properties of green leafy vegetables are high because of the presence of indispensable minerals, vitamins, antioxidants and alkaloids. They are cheap and the natural sources of growth promoting and health protecting functions. They have a special place in daily diet. They are used as alternative, antidote, antipyretic, anthelmintic, aphrodisiac, appetizer, carminative, demulcent, diuretic, expectorant, laxative, refrigerant and tonic. Their regular use prevents and cures the diseases of ear, eye, liver, lungs, nerves, skin and other vital organs. The medicinal properties of 30 species belonging to 25 genera and 21 families are discussed.

## INTRODUCTION

The green leafy vegetables (GLV) store a small amount of carbohydrates, more protein and very little amount of fat. The food value of leaf vegetables is low due to the presence of large amount of water (70 to 95 percent). Their nutritive values are high because of the presence of the indispensable minerals (iron, calcium, potassium, magnesium) and vitamins (vitamin A, B, C, E and K) that aid our digestion, metabolism and growth. About 280 gram fresh weight of vegetables per head as daily minimum consumption is required by an adult for normal health. However, about 25 gram per head is the average consumption in India. They are very cheap and natural sources of growth promoting and health protecting functions. According to Gopalan *et al.* (1989) GLV contain all important nutrients required for daily consumption in one's diet. They have reported the nutritive value of 67 GLV. Awasthi & Tandon (1987) have identified 30 GLV as commonly consumed types.

It is well known that the prevalence of the degenerative diseases such as cancer, cataract and heart disease is lower in Asians due to their predominant vegetarian habits with ample use of fresh vegetables. These are rich in nutrients and other phytochemicals as compared with cereals and legumes (Tarwadi & Agte, 2003). The recent studies have emerged with health attributes for these food components, with their antioxidant action and have been implicated to have a protective role in the majority of oxidation stress-related disorders (Reddy & Bhat, 1999). The intake of antioxidant components present in food is now considered as important as vitamins for health promotion and protection against damage due to oxidative stress (Fogliano *et al.*, 1999). According to Susan Aldridge (2004) the antioxidants, such as lutein and zeaxanthin found in GLV, protect against cataract clouding of eye lens linked to oxidative stress. He further stressed that these antioxidants are better than vitamin E. Meke Faber *et al.* (2002) found that GLV increase serum retinol concentration. According to Maharishi (1999) dark green leafy vegetables have a special place in daily diet and they help to balance pitta and kapha.

Since the GLV offer a cheap but rich source of a number of nutrients and vitamins having antioxidant properties, there is a reason to emphasis their importance as cheap and natural sources that have protective functions. So, the present work has been taken up to make a survey of the leafy vegetables cultivated or collected and sold in the markets of Pondicherry and their medicinal importances.

## MATERIALS AND METHODS

Pondicherry is an enclave of India surrounded by Villupuram district of Tamilnadu. It is at the north between 11°46′ and 12°3′ latitude and between 76°36′ and 79°53′ of eastern latitude. The area of Pondicherry is 290 square kilometers, of which 25832-hectare land is under cultivation. The total population of the enclave is 735000, of which the urban population constitutes 506000 (68.84 per cent) and the rural population is 229000 (31.16 per cent). The literary rate is 80.90 per cent.

There are many well organized vegetable markets in and around Pondicherry. Many visits were made to the Grand Bazaar at Nehru Street, Uzhavar Sandai at Lawspet and near Botanical Garden. Apart from ordinary vegetables, the green leafy vegetables are also sold in these markets. Some of the greens are cultivated (*Allium cepa, Alternanthrera sessilis, Amaranthus roxburghianus, A. tricolor, A. tristis, Basella rubra, Centella asiatica, Coriandrum sativum, Hibiscus cannabinus, Mentha arvensis, Moringa oleifera, Murraya koenigii, Pisonia grandis, Portulaca oleracea, Raphanus sativus, Sesbania grandiflora, Solanum nigrum, Spinacia oleracea, Talinum portulacifolium, Trianthema portulacastrum, Trigonella foenum-graecum, Weddelia chinensis)* and others are collected from wild condition *(Amaranthus spinosus, A. viridis, Cardiospermum halicacabum, Cissus quadrangularis, Cleome gynandra, Ipomoea aquatica, Marsilea quadrifolia, Solanum trilobatum)* and sold in the market. The greens were bought to the laboratory and identified with the help of the Flora of the Presidency of Madras (Gamble, 1967), The Flora of Tamilnadu Carnatic (Matthew, 1991) and Materia Medica (Nadkarni,

1976). The vernacular names were ascertained by enquiring the vendors and consulting the standard books such as Wealth of India (1972), Economic Botany (Sambamurthy & Subramaniyan, 1982), Dictionary of Economic Plants of India (Singh *et al.*, 1983), Economic Plants of India (Yoganarasimhan, 2000).

## OBSERVATIONS

Thirty species of green leafy vegetables belonging to 25 genera and 21 families were collected, identified and described. The binomial, family (in parentheses), English name (E), Hindi name (H) and Tamil name (T) of each species is given.

***Allium cepa*** L., (Liliaceae), Onion (E), Pyaz (H), Vengaya thal (T).

Leaves exhibit medicinal property and are used in cooking. This could be taken as raw or boiled. Thirst, piles, internal heat could be cured with its leaves. It increases the flow of urine and regulates menstruation.

***Alternathera sessilis*** (L.) R. Br. ex DC., (Amaranthaceae), Saranti sag (H), Kanhonagone soppu (H), Ponnanganni (T).

If this plant is taken regularly with butter for 40 days, it improves eyesight and cures cataract. It also cures liver disorders, stomatitis, bad breath, gonorrhoea and heart burn. To cure piles, this plant is consumed regularly along with garlic. This also gives complexion and strength to body.

***Amaranthus roxburghianus*** Nevski, (Amaranthaceae), Sirukeerai (T).

It is used as vegetable. If consumed daily along with food it reduces iron deficiency. It is found effective in tuberculosis, liver disorders and urinary infections. It is also used as an antidote in poisonous bite.

***Amaranthus spinosus*** L., (Amaranthaceae), Amnaranth (Pickly) (E), Kantewali chaulai, Goja (H), Mullu Keerai (T).

The leaves are eaten as vegetable. It is supposed to improve appetite. It is used as an antidote for snake-bite. Paste of the leaves reduces inflammation and edema. The mixture of celery, garlic and this plant is useful to check stomachache. It regulates the flow of urine. It is used to arrest spermatorrhoea.

***Amaranthus tricolor*** L., (Amaranthaceae), Chinese spinach (E), Bari chaulai, Lal sag (H), Thandukeerai (T).

It is used in salads and also provides a good vegetable. It is used as cooling spring vegetable. It is used to cure gonorrhoea , eczema, piles and diarrhoea. It is also used as demulcent, diuretic, ascaricide and as antidote for snake-bite.

***Amaranthus viridis*** L., (Amaranthaceae), Kuppaikeerai (T).

It gives luster and smoothness to the skin. It is normally given along with other pulses to cattle to obtain more milk. It is a good appetizer and acts as diuretic.

***Basella rubra*** L., (Basellaceae), Indian spinach (E), Poi (H),Sivappu salakeerai (T).

Leaves are used in catarrhal affections and to hasten suppuration. They are especially useful as laxative in children and pregnant women.

***Cardiospermum halicacabum*** L., (Sapindaceae), Kanphuti (H), Modkathan Keerai (T).

It cures joint pains, herpes, eczema, rheumatism, arthritis, rigidity of the muscles, rigidity of joints, nervous debility and skin diseases. Roasted leaves are squeezed and the juice is used as ear drops, checks otorrhoea. Roasted leaves are used to treat inflammation of the testicles also.

***Centella asiatica*** (L.) Urban, (Apiaceae), Asiatic pennywort (E), Brahmi (H), Vallarai (T).

Leaves are cooked and eaten as vegetable. It is a tonic, diuretic and local stimulant. It is used for eczema, psoriasis, ulceration, enlargement of glands, chronic rheumatism, epilepsy and indigestion, lack of appetite, diarrhoea and leucorrhoea. It is also given to improve memory. The paste of the leaves with cloves and garlic is given for liver disorder.

***Cissus quadrangularis*** L., (Vitaceae), Hadjora (H),Perandai (T).

Leaves are used as vegetable and thovayal is also made. Bleeding piles, indigestion, dyspepsia, diarrhoea, fistula, urinary tract disorders and asthma are treated with this. It is an appetizer, strengthens bones and expels intestinal worms.

***Cleome gynandra*** L., (Cleomaceae), Hulul (H), Velai keerai(T).

The leaves are eaten as vegetable, used as flavoring agent and also pickled. They are bitter but cooking removes the bitterness. Leaves are used as counter irritant in headache, neuralgia, rheumatism and other local pains. They are also applied to boils to prevent formation of pus.

***Coriandrum sativum*** L., (Apiaceae), Coriander leaves (E), Hara Dhania (H), Kothamalli (T).

The leaves are pungent and aromatic, used in flavouring curries and soups. Leaves when green are eaten raw as well as used for preparing a sauce or chutney. They are carminative and antibilious.

***Hibiscus cannabinus*** L., (Malvaceae), Kenaf (E), Patsan, Ambari (H), Pulichai keerai (T).

It is used as vegetable. Pickles and chutney are also prepared from this. It is laxative, refrigerant and gives digestive power. It is also used to treat intestinal ulcers, cardiac problems, high blood pressure, rh umatism, anorexia and syphilis.

***Ipomoea aquatica*** Forsk., (Convolvulaceae), Water spinach (E), Kalmi sag (H), Vellai Keerai (T).

It is used as vegetable and as an antidote to opium and arsenol poisoning. Used as emetic and purgative.

***Marsilea quadrifolia*** L., (Marsileaceae), Aarakeerai (T).

This is used as diuretic, refrigerant, tonic and aphrodisiac. It cures diabetes and bile disorders if consumed regularly with food. Infusion of dried leaves with milk and sugar candy, if taken twice daily, can give excellent results in controlling blood in urine.

***Mentha arvensis*** L., (Lamiaceae), Field mint (E), Pudina (H), Pudina (T).

It is used in flavouring food and thovayal is also prepared . Infusion of the plant with lemon grass is used in treating headache, vomiting and fever. Oil extracted form this plant is similar to peppermint and used in treating headache. If taken internally along with water, it acts as an appetizer.

***Moringa oleifera*** Lamk., (Moringaceae), Drum stick tree, Horse-radish tree (E), Sahinjna (H), Murungai (T).

It is eaten as vegetable. It is antispasmodic, stimulant, expectorant, diuretic, anthelmintic and abortifacient. It is given for high blood pressure, cough, asthma, anaemia and conjunctivitis.

***Murraya koenigii*** (L.) Sprengel, (Rutaceae), Curry leaves (E), Gandhela (H), Kariveppilai (T).

It is used in all preparation as flavouring agent. Chutney is prepared form the leaves. Curry leaf powder is prepared from the dried leaves and eaten with rice. Infusion of the leaves is useful in vomiting. Green tender leaves are eaten raw for curing dysentery. When boiled in milk and ground they form a good application to poisonous bites and to eruptions. Decoction of the leaves is given with butter as a febrifuge.

***Pisonia grandis*** R. Br., (Nyctaginaceae), Lettuce tree (E), Bhaggachura (H), Nachukootai keerai (T).

Fresh leaves moistened with eau-de-cologne are used to subdue inflammation of an elephant nature of legs and other parts. Leaves are diuretic and purgative.

***Portulaca oleracea*** L., (Portulacaceae), Common purslane (E), Khursa (H), Paruppukeerai (T).

Leaves are eaten as soup, salad and curry. Plants are used in diseases of kidney, liver and spleen. Used as external application in burns and various forms of skin diseases. Leaves are astringent, refrigerant, diuretic and emollient.

***Rephanus sativus*** L., (Brassicaceae), Raddish (E), Muli (H), Mullangi (T).

Young leaves are used as vegetable. Radish preparations are useful in liver and gall bladder troubles. Also used for neuralgic, headache, sleeplessness and chronic diarrhoea. Juice of fresh leaves is used as diuretic and laxative.

***Sesbania grandiflora*** (L.) Poir., (Fabaceae), Agati sesbania (E), Basna (H), Agati keerai (T).

The leaves have a sharp bitter and sweet taste. They are indigestible, anthelmintic, tonic and antipyretic. Cure kapha, itching, leprosy, night blindness, epilepsy and gout. Useful in ophthalmia.

***Solanum nigrum*** L. (Solanaceae), Black nightshade (E), Makoi (H), Manathakali keerai (T).

The leave are cooked and eaten as vegetable. An infusion of plant is used as an enema in infants having abdominal upsets. It is also a household remedy for anthrax pustules. Freshly prepared extract of plant is considered useful in treating cirrhosis of liver. Boiled leaves and tender shoots are recommended to patients suffering from dropsy. This act as refrigerant when the body has excess of heat. It cures stomatitis, burns, intestinal ulcers, bronchitis, cough and tuberculosis.

***Solanum trilobatum*** L., (Solanaceae), Thooduvalai keerai (T).

The leaves are roasted with onion in gingili oil and used as cardiac tonic.It is carminative, stimulant, expectorant, nutritive, diaphoretic and aphrodisiac. If taken as vegetable, it relieves diseases of ear due to phlegm, asthma, indigestion and spermatorrhoea. The leaf juice cures dullness of hearing.

***Spinacia oleracea*** L., (Chenopodiaceae), Spinach (E), Palak (H), Pasalaikeerai (T).

Leaves are eaten as vegetable. It is used in soups and salads. It is one of nature's most perfect foods. It is good for heart and boosts energy. It has anti-inflammatory and anticancer properties. It cures arthritis and osteoporosis.

***Talinum portulacifolium*** (Forsk.) Asch. & Schweinf., (Portulacaceae), Kozhikeerai (T).

Leaves are eaten as vegetable. They are nutritive and aphrodisiac.

***Trianthema portulacastum*** L., (Aizoaceae), Horse purslane (E), Santhi, Lalsabuni (H), Shaaranaj (T).

Decoction of the leaves is used as vermifuge and is useful in rheumatism. It is also used as an antidote to alcoholic poison.

***Trigonella foenum-graecum*** L., (Fabaceae), Fenugreek (E), Methi (H), Venthiya keerai (T).

Regular use of the plant cures asthma, eczema, indigestion, eye problems, stomachache , fatigue and back pain. It also gives strength and vitality.

***Weddelia chinensis*** (Osbeck) Merrill, (Asteraceae), Bhangra (H), Manjal Karisalankanni (T).

Leaves are used for flavouring foods, given for jaundice, regarded as tonic and alterative, useful in cough, cephalagia and diseases of skin. A decoction of the herb is used in uterine hemorrhage and menorrhagia.

## DISCUSSION

Nutritional anaemia is a major public health problem in all the developing countries of the World. The surveys conducted by the National Research Laboratory have shown that about 20 per cent of the Indian population suffers from iron deficiency or anaemia. This is due to the predominant use of cereals and poor intake of dietary iron. The antioxidants such as β-carotene, luetein, tocophoerol (Vitamin E), zeaxanthin and phenols check autoxidation. The intake of antioxidant compound is as important as vitamins for health promotion and protection against damage due to oxidative stress (Fogliano *et al.,* 1999).

Considering the values of green leafy vegetables as double fortification of both iron and antioxidant, there is a good reason to emphasize their importance among people as cheap and natural sources having protective factors (Tarwadi & Agte, 2003). The GLV contain more vitamins, minerals and fewer calories than any other vegetables. They are rich in vitamins A, B, C, E and K and the antioxidants tocopherol, β-carotene and zeaxanthin. They can be cultivated easily, available most of the time in a year and can be eaten at any stage of maturity. They are used as alternative, antidote, antipyretic, anthelmintic, aphrodisiac, appetizer, carminative, demulcent, diuretic, expectorant, laxative, refregerant and tonic. Their regular use prevents and cures the diseases of ear, eye, liver, lungs, nerves, skin and other vital organs.

The green leafy vegetables are inexpensive and it is advisable to include at least 50 gram of GLV daily in ones diet. They contain all important nutrients required for growth and maintenance of health. Hence, GLV must be consumed by children, pregnant and nursing women to obtain much needed β-carotene, calcium, magnesium and iron. This is particularly so on a predominantly cereal based diets of the poor who suffer from the dietary deficiency of these nutrients. Hence, steps must be taken to encourage cultivation of GLV in kitchen gardens so that they are available all through the year. Planting trees such as drumstick, agathi, curry leaf and tree- lettuce in the backyard must be encouraged.

Thus, the present study reveals that the greens discussed here are consumed by people daily or twice in a week or once in a week or occasionally for nutrition, better health and for curing diseases.

## REFERENCES

AWASTHI, C. P. & P. K. TANDON 1987. Quality studies on some common green leafy vegetable in India. *Progr. Hort.* 19 : 207-212.

FOGLIANO, V., V. VERDE, G. RANDAZZO & A. RITIENI 1999. Method for measuring antioxidant activity and its application to monitoring the antioxidant capacity of wines. *J. Agric. Food Chem.* 4 : 1035-1040.

GAMBLE, J. S. 1967. *Flora of the Presidency of Madras.* Botanical Survey of India, Calcutta, India. (Reprinted Ed.)

GOPALAN, C., B.V. RAMA SASTRI & S.C. BALASUBRAMANIAN 1994. *Nutritive value of Indian foods.* National Institute of Nutrition, I.C.M.R., Hyderabad, India.

MAHARISHI, 1999. *Ayurveda.* Newsletter.

MATTHEW, K.M. 1991. *An Excursion Flora of Central Tamilnadu, India.* Oxford & IBH Publishing Company, New Delhi, India.

MEKE FABER, A.A.S., PHUNGULA, L. SONJA, A. MUHAMMAD, & A.J. SPINNLER BENADE 2002. Home gardens focusing as the production of yellow and dark green leafy vegetables increase the serum retinal concentration of 2-5 year old children in south Africa. *American J. Nut.* 76 (5) : 1048-1054.

NADKARNI, A.K. 1976. *Indian Materia Medica.* Popular Prakashan, Bombay, India.

REDDY, G.B. & K.C. BHAT 1999. Protection against UVB inactivation (*in vitro*) of rat lens enzymes by natural antioxidants. *Mol. Cell. Biochem.* 194 : 41-45.

SAMBAMURTHY, A.V.V.S. & N.S. SUBRAHMANIYAM 1989. *A Text Book of Economic Botany.* Wiley Eastern Limited, New Delhi, India.

SUSAN ALDRIDGE 2004. Leafy vegetables prevent cataract. *J. Nut.* 134 : 3225-3232.

TARWADI, K. & V. AGTE 2003. Potential of commonly consumed green leafy vegetables for their antioxidant capacity and its linkage with the micronutrient profile. *Int. J. Food Sci. Nut.* 54 : 417-431.

SINGH, UMRAO, A. M. WADHWANI & B. M. JOHRI 1983. *Dictionary of Economic Plants of India.* ICAR, New Delhi, India.

WEALTH OF INDIA, 1970-1992. *Raw Materials,* Vols. 1-11. Council of Scientific and Industrial Research, New Delhi.

YOGANARASIMMAN, S. N. 2000. *Economic Plants of India.* Regional Research Institute, Bangalore, India.

# MEDICINAL PLANTS OF FAMILY LAMIACEAE OF KASHMIR – I

GOWHER GUNA, M.Y. SHAH & A.R. NAQSHI

*Department of Pharmaceutical Sciences, University of Kashmir, Srinagar – 190 006, India*

## ABSTRACT

The present account provides a glimpse of enthnobotanical studies in Kashmir Valley of family Lamiaceae. As many as 19 plant species mostly of wild nature are found in common use – medicinal treatment as well as for other purposes.

## INTRODUCTION

In recent decades interest in herbal medicine and folk-lore remedies has increased in India and other developing countries as the modern allopathic medicines have proved not only to be beyond the reach of a common man but many of them are also known to have side effects. The flora of Kashmir Himalaya comprises about 2104 species (Dar & Naqshi, 2001) and many of them are used medicinally. Numerous plant species of medicinal/pharmacological interest constitute an integral component of the flora of Kashmir. Despite a good account of flora being known now, it is unfortunate that ethnobotanically only few stray cases are available. To fill up this lacuna, the authors have initially undertaken the detailed study of medicinal plants of family Lamiaceae of Kashmir.

Kashmir lies between 33°20′ and 34° 54′ N latitudes and 73°55′ and 75°35′ E longitudes, covering an area of 15,948 sq km. Topographically, it is a deep elliptical bowl-shaped valley bounded by lofty mountains of the Pir Panjal Range in the south and south west and the Great Himalayan Range in the north and east, with 64% of the total area being mountainous. The climate of the area is temperate with July as the hottest month and January as the coldest. The annual rainfall is more than 700 mm.

## MATERIALS AND METHODS

Several field trips in and around the study area were undertaken during 2002. Plants were collected and then properly identified. Voucher herbarium specimens have been deposited at Department of Pharmaceutical Sciences, University of Kashmir. Ethnobotanical information was collected from knowledgeable persons, local Hakims, elder men, women, shepherds, Gujjars and Bakerwalls belonging to different sects and ethnic groups. Repeated queries were made to get the data verified and confirmed.

## OBSERVATIONS

The observations made during the course of this study are given in the following enumeration. A, H, K, S, E and U after vernaculars stand for Arabic, Hindi, Kashmiri, Sanskrit, English and Urdu names, respectively. Varnacular names are followed by collection number and place of collection.

**Ajuga bracetosa** Wall. ex Benth. (*Janiadam* –K, *Faniadaur*–U, *Nilkanthi*-S); *GAG 123; Narbal.*

*Uses :* Herb astringent, febrifuge, aperient, toxic and diuretic, used in gout, rheumatism, palsy and amenorrhoea.

**Lamium album** Linn. : (White Dead nettle – E); GAG 180; Kangan

*Uses :* Astringent; decoction used in hemorrhages of uterus and nose. Plant yields an essential oil. Roots resolvent and vulnerary. Flowers mild astringent, haemostatic, hypnotic, depurative & toxic. Leaves edible, a good source of carotene.

**Lavandula angustifolia** Mill. : (Lavender – K, True Lavender/ Common Lavender – E); GAG 78; Srinagar.

*Uses :* Flowers and flowering tops yield an essential oil – lavender oil, used in perfumery, production of lavender water and soap industry. Used medicinally as a carminative and mild stimulant, and for flavouring pharmaceutical preparations.

**Leonurus cardiaca** Linn. : (*Guma*–K, *Common Motherwort* – E); *GAG 73; Aharbal.*

*Uses :* Dried leaves and flowering tops diuretic, analgesic, emmenagogue, prescribed in hysteria and heart palpitation. They also tone up uterine membrane and generative organs, allay nervous irritability and afford relief from pain in stomach and gall bladder. Leaves may cause dermatitis.

**Majorana hortensis** Moench Syn. *Origanum majorana* Linn. : (*Murwa*–H, *Sweet Marjoram* – E); *GAG 126; Harwan.*

*Uses :* Leaves are used as condiment. Fresh leaves are employed for garnishing and in salads, also used for flavouring vinegar. Oil applied to sprain, stiff and paralytic limbs and toothache. Infusion of plant is used as a stimulant, sudorific, emmenagogue and galactagogue; also used in hysteria and paralysis.

**Marrubium vulgare** Linn. (Troper – K, Paharigandana- H, Hore hound/ Hoarhound – E); GAG 23; Rainawari.

*Uses :* Expectorant and diuretic; laxative in large doses. Dried leaves and flowering tops constitute the crude drug. Used against colds and coughs and pulmonary affections. Extract useful for cardiac extrasytoles. Yields an essential oil used in liquors. Sap is used in eye infections and leaf in skin infections.

**Mentha arvensis** Linn. : (Pudina – K, Pudina – H, Field Mint/ Corn Mint – E); GAG 200; Gulmarg.

*Uses :* Stimulant and carminative. Infusion of leaves used as a digestive and in rheumatism. Dried plant is antiseptic, refrigerant, emmenagogue and diuretic.

**Mentha longifolia** (Linn.) Huds. Syn. *M. sylvestris* Linn. : (*Jangli Pudina* – K, *Horsemint* – E); *GAG 201; Tangmarg.*

*Uses :* Carminative, antiseptic and stimulant; decoction used in fever and heat apoplexy. Eaten in the form of chutney. Leaves and flowering tops yield an essential oil which is used as a substitute for imported peppermint oil for flavoring confectionery,

**Mentha piperata** Linn. emend. Huds. : (*Pudina* – K, *Paparaminta* – H, *Peppermint* – E); *GAG 209; Srinagar.*

*Uses :* Stimulant, stomachic and carminative, used for allaying nausea and flatulence. Hot infusion given in stomachache and colicky diarrrhoea. Source of peppermint oil, extensively used as a flavouring and in pharmacy. For internal use preferred to menthol because of agreeable taste. Externally applied in rheumatism, neuralgia, congestive headache & toothache.

**Micromaria biflora** (Buch.-Ham. ex D. Don) Benth. : (*Indian Wild Thyme* – E); *GAG 99; Ferozpur.*

*Uses :* Used for worm infested wounds of cattle. Four physiological types (all yielding essential oils) camphorate, citrate, menthata and pulegata are known; the principal constituent of the oil from the first one is camphor; citral in the second; d-menthone in the third and pulegone in the last one.

**Nepeta cataria** Linn. : (Gandhsoi – K, Catnip/ Catmint – E); GAG 210; Bijbehara.

*Uses :* Leaves and shoots are used as flavoring agent; dried leaves employed in soups and stews. Leaves and flowering tops carminative, tonic, diaphoretic, refrigerant, soporific. Leaves chewed in toothache. A volatile oil, oil of Catnip, obtained form the herb is used as a lure for trapping wild animals of cat family.

**Ocimum tenuiflorum** Linn. : (*Babreboel* - K, *Tulsi* – H, *Sacred Basil* or *Holy Basil* – E); *GAG 227; Jawahar Nagar.*

*Uses :* Leaves yield oil with an appreciable note of cloves. It possesses insecticidal and antibacterial properties and is mosquito repellant. Seeds yield a fixed oil. Leaves stimulant, diaphoretic, antiperiodic and expectorant, used in catarrh and bronchitis, ringworm and other cutaneous diseases; infusion used as a stomachic. Decoction of roots is given as a diaphoretic in malarial fevers. Seeds mucilaginous and demulcent and used in genitourinary disorders.

**Origanum vulgare** Linn. : (Wan Baber/ Marzanjosh – K, Sathra – H, Common or Wild Marjoram – E, Sathra/ Mridu- marurama- U); GAG 77; Kangan.

*Uses :* Contains a volatile oil and tannin. Origanum oil is carminative, stomachic, diuretic, diaphoretic and emmenagogue, used as a stimulant and tonic in diarrhoea and colic, also applied in chronic rheumatism, toothache and earache. Given in whooping cough and bronchitis because of its spasmolytic action.

**Perovskia abrotanoides** Karel. : *GAG 151; Srinagar.*

*Uses :* Used as an antiacid and in tenesmus. Flowers soaked in water applied to the body of patient suffering from fever as a cooling medicine.

**Plectranthus rugous** Wall. ex Benth. : (*Soli / Thaper* - L); *GAG 133; Aharbal.*

*Uses :* Plants used to keep off fleas. Leaf used in abdominal pain and round worm expulsion.

**Prunella vulgaris** Linn. : (*Kalweoth / Syangara* – K, *Dharu* – H); *GAG 179; Gulmarg.*

*Uses :* Flowers/ mixture of flowers and leaves is used in headache, cerebral and brain disorders, cold, headache and gastric disturbances. Herb is antiseptic, expectorant, used for fevers and coughs; considered antirheumatic, alterative and toxic. Green leaves smeared with castor oil and warmed over fire applied externally to painful piles.

**Rosmarinus officinalis** Linn. : (*Rosmary* – K, *Rasmari* – H, *Rosemary* – E); *GAG 186; Aharbal.*

*Uses :* Rosemary oil is mildly irritant and used as a carminative and cardiac stimulant.

**Salvia moorcroftiana** Wall. ex Benth. : (*Sholar* –K); *GAG 52; Srinagar.*

*Uses :* Leaves used for relief from cough and employed for extraction of guinea–worms; also applied as a poultice to boils, wounds and chronic skin affections. Seeds emetic; also used for colic, dysentery and hemorrhoids; stalks peeled and eaten; roots used in high fever.

**Thymus linearis** Benth. ex Benth. : (*Javind*- K, *Banajwain* – H, *Wild thyme* – E, *Masho/ Washa*-U); *GAG 23; Ferozpur.*

*Uses :* Leaves and floral shoots employed for suppression of wine and menstruation and for convulsive and whooping coughs; also used as a sedative in radiculagia and epilepsy. Seeds vermifuge. Oil used in toothache. Extract of whole herb used by pregnant ladies for smooth delivery.

## ACKNOWLEDGMENTS

The senior author is highly thankful to CSIR, New Delhi for financial assistance.

## REFERENCES

ANONYMOUS 1976. *Medicinal plants of India*, Vol. I. Indian Coun. Med. Res., New Delhi.

BISWAS, K. 1956. Common Medicinal plants of Darjeeling and the Sikkim Himalayas. Govt. Press, Calcutta.

CHOPRA, R.N., S.L. NAYAR & I.C. CHOPRA 1956. *Glossary of Indian Medicinal plants.* CSIR, New Delhi.

————, ———— & ———— 1969. *Suppliment to Glossary of Indian Medical plants.* Coun. Sci. Indl. Res., New Delhi.

DAR, G.H., VIR JEE, P. KACHROO & G.M. BUTH 1984 Ethnobotany of Kashmir – I. Sind Valley. *J. Econ. Taxon. Bot.* 5 (3) : 668-675.

———— & A.R. NAQSHI 2001. Plant Resources of Kashmir: Diversity, Utilization and conservation In Ashok Pandita (Ed.) *Natural Resources of W. Himalaya.* Srinagar, Kashmir.

DASTUR, J.F. 1951. *Useful plants of India and Pakistan.* Tarapore Wala Sons & Co., Bombay.

HILL, A.F. 1952. *Economic botany.* McGraw Hill Book Co., New York.

JAIN, S.K. 1963. Studies in Indian ethnobotany - II. Plants used in medicine by the tribals of M. P. *R. R. L. Bull.* 1 (2) : 126-128.

———— 1975. *Medicinal plants.* 2nd ed. NBT, New Delhi, (India).

———— 1980. *Glimpses in Indian Ethnobotany.* Oxford & IBH Publishing Co., New Delhi.

KACHROO, P. 1973. Ethnobotany of Kashmiris - II. Glossary of Kashmiri Plant names. *J. Sc. Univ. Kash.* 1 (1-2) : 1-10.

———— & I.M. NAHVI 1976. Ethnobotany of Kashmiris - I, In Singh & Kachroo (Eds.) *Forest Flora of Srinagar.* Dehra Dun.

KANGO, G.H. & B. DHAR 1981. Studies in Trans-human pastoralism in Northwest Himalayas. *Tech. Bull. RMP.* 1.

KIRTIKAR, K.R. & B.D. BASU 1935. *Indian Medicinal Plants.* Allahabad.

KOUL, A.K. *et al.* 1982. Some novel tea substitutes for Kashmir. *Ind. Forester* 108 (4) : 310-313.

KUMAR, G.M. & A.R. NAQSHI 1990. Ethnobotany of Jammu – II. *J. Econ. Taxon. Bot.* 14 (1) : 67-74.

MAHESHWARI, P. & U. SINGH, 1965. *Dictionary of Economic Plants in India.* ICAR, Delhi.

MUKHERJEE, S. K. 1969. Survey of plant resources of India & its use. *Bull. Bot. Surv. India* 11 (3-4) : 217-223.

NAQSHI, A. R., M. Y. BABA & S. ARA 1992. Ethnobotanical studies of Kashmir – Jhelum valley. *Recent Advances in Medicinal, Aromatic and Spice Crops* 2 : 371-379. New Delhi.

SCHERY, R.W. 1952. *Plants for man*. Prentice-Hall, Inc., U.S.A.

STEWART, R.R. 1972. *An annotated Catalogue of the vascular plants of W. Pakistan and Kashmir*. Fakhri Press, Karachi.

UHLIG, H. 1973. *Wanderhirten in westlichen Himalaya Chopans, Gujars, Bakerwals, Gaddi. Transaction & papers 1969.* Publication No. 47

UPHOF, J.C. TH. 1968. *Dictionary of Economic Plants*. 2nd ed. Verlag von J. Cramer, Germany.

USHER, G. 1974. *A Dictionary of the plants used by man*. London.

WATT, G. 1889-1893. *A dictionary of the Economic Products of India,* Vol. I – VI. Govt. Press, Calcutta.

# MEDICINAL FLORA OF KASHMIR VALLEY

GOWHER AYOOB, MUDASIR AHMAD, M.Y. SHAH & A. R. NAQSHI

*Department of Pharmaceutical Sciences, University of Kashmir, Srinagar-190 006*

## ABSTRACT

The diversity of medicinal plants in the Kashmir valley region is very vast. Kashmir valley with its lofty mountains abound a great variety of plants in general and medicinal ones in particular. These plants are being used both locally as well as at industrial level for alleviating diverse type of disorders. Some of the plants have very high medicinal/ therapeutic value, while some of them are used ethnobotanically in Kashmir valley as well as by ethnic communities world over. The present detailed article highlights the variety of medicinal plants and their use in curing various ailments.

## INTRODUCTION

Kashmir, commonly called as the Kashmir Valley, is an integral but geologically younger part of the main Himalayan range, lying between 33°20' and 34°54'N latitudes and 73°55' and 75°35'E longitudes, covering an area of 15,948 sq km. Topographically, it is a deep, elliptical, bowl-shaped Valley bounded by lofty mountains of the Pir Panjal Range in the south and southwest and the Great Himalayan Range in the north and east, with 64% of the total area being mountainous.

Fig. 1. Sketch map of Jammu and Kashmir in India with its various areas.

The Valley is an asymmetrical fertile basin stretching from southeast to north-westernly direction. Its diagonal length (from southeast to northwest corner) is 187 km, while the breadth varies considerably, being 115.6 km along the latitude of Srinagar (Kaul, 1977).

The raw material, particularly of botanical origin, is available in plenty in our country, particularly in Kashmir Valley. Numerous plant species of medicinal/pharmacological interest constitute an integral component of flora of Kashmir Valley. Despite a good account of flora is now known, it is unfortunate that therapeutic value of only few plants is available. The present study has been envisaged in the context of above mentioned facts.

## METHODOLOGY

Periodic surveys of different areas of Kashmir Valley were conducted during 2004-2006. Information regarding the medicinal uses of the herbs was gathered through daily interviews with knowledgeable old native people who showed familiarity with traditional local remedies.

First information on medicaments was gained through casual conversation with many persons whose age ranged from 50 to 92 years. Then, for each species referred to as having medicinal use, precise questions were asked about the therapeutic applications and the modes of preparation and administration.

All the plants mentioned were recognized by the informants, then collected and taxonomically identified with the help of available literature. Voucher herbarium specimens are now preserved in the Herbarium of Department of Pharmaceutical Sciences of Kashmir University.

Table 1. List of medicinal plant species of Kashmir Valley.

| *Species* | *Family* | *Part used** | *Pharmacological activity*** | *Flowering period* | *Altitudinal Range* |
|---|---|---|---|---|---|
| *Abelmoschus esculentus* (L.) Moench | Malvaceae | Wp | 1, 7, .8, 14 | May- Jul | Plains (Cultivated) |
| *Abutilon theophrastii* Medik | Malvaceae | Wp | 2,3, 6, 8, 21, 28, 30, 33, 56, 57, 64, 76 | Jun- Sep | Plains |
| *Achillea millefolium* L. | Asteraceae | Fr, L, S, Wp | 2, 3, 4, 5, 6, 9, 10, 11, 12, 15, 16 | May- Oct | Plains-Subalpines-Alpines |
| *Achyranthes aspera* L. | Amaranthaceae | L, Sd, Wp | 1, 2, 5, 12, 18, 19, 20, 23, 50, 51 | Jun- Aug | Plains-Subalpines |
| *Aconitum heterophyllum* Wall. ex Royle | Ranunculaceae | R | 6, 7, 8, 18, 37 | Jul-Aug | Subalpines |
| *Aconitum moschatum* (Bruehl) Stapf | Ranunculaceae | Wp | 52 | Aug-Sep | Alpines |
| *Acorus calamus* L. | Araceae | Us | 5, 14, 18, 21, 22, 24, 25, 26, 37, 52, 53 | Jun-Jul | Plains |

| | | | | | |
|---|---|---|---|---|---|
| *Adiantum capillus-veneris* L. | Adiantaceae | L, Us | 2, 3, 5, 6, 17, 21, 27, 28, 29, 37 | Jun- Aug | Plains-Subalpines |
| *Aesculus indica* (Wall. ex Camb.) Hk. f. | Hippocastanaceae | Fr, R, Sd, Sb | 6, 8, 21, 29, 37 | Jul- Sep | Plains (Cultivated) |
| *Ailanthus altissima* (Mill.) Swingle | Simaroubaceae | R, Sb, Wp | 5, 15, 29, 60, 61, 62 | Jul-Aug | Plains (Cultivated) |
| *Ajuga bracteosa* Wall. ex Benth. | Lamiaceae | Wp | 1, 5, 14, 21, 49, 57, 59, 63 | Jun-Jul | Subalpines |
| *Alcea rosea* L. | Malvaceae | Wp | 5, 15, 16, 28, 30, 53, 55, 57, 64 | Jun-Jul | Plains (Ornamental) |
| *Allium cepa* L. | Liliaceae | Fr, L, Us | 1, 12, 14, 26, 27, 65, 66 | Jun- Aug | Plains (Cultivated) |
| *Allium sativum* L. | Liliaceae | Us | 1, 2, 3, 5, 7, 12, 15, 21, 24, 31, 37, 55, 61, 88 | Jun-Jul | Plains (Cultivated) |
| *Amaranthus caudatus* L. | Amaranthaceae | Sd, Wp | 1, 3, 4, 8, 69 | Jul- Sep | Plains |
| *Amaranthus spinosus* L. | Amaranthaceae | Wp | 15, 18, 53 | Jul-Aug | Plains |
| *Anagallis arvensis* L. | Primulaceae | Wp | 4, 23, 29, 37, 52, 62, 68, 70, 71 | Jun- Aug | Plains |
| *Androsace mucronifolia* Watt. | Primulaceae | Wp | 43 | Jun- Aug | Alpines |
| *Androsace rotundifolia* Hardw. | Primulaceae | Wp | 11, 61 | May- Jul | Subalpines-Alpines |
| *Anemone obtusiloba* D. Don | Ranunculaceae | R, Sd | 16, 59, 72 | Jun-Jul | Subalpines-Alpines |
| *Anemone tetrasepala* Royle | Ranunculaceae | Wp | 11, 21 | Jul-Aug | Alpines |
| *Angelica roylei* (Lindl.) Mukh. & Constance | Apiaceae | R, Wp | 2, 3, 5, 9, 21, 33, 34, 35, 36, 49 | Jun- Aug | Subalpines |
| *Anthemis cotula* L. | Asteraceae | Fl, R | 3, 8, 15, 17, 21, 61, 72 | May- Oct | Plains |
| *Aralia cachemirica* Decne. | Araliaceae | R | 27 | Jul-Aug | Subalpines |
| *Arctium lappa* L. | Asteraceae | L, R, Sd | 2, 3, 5, 9, 15, 16, 20, 21, 25, 26, 27, 29, 30, 39, 40, 41, 55, 58, 61, 73 | Jul- Sep | Plains |
| *Arenaria serpyllifolia* L. | Caryophyllaceae | Wp | 74 | Mar-Sep | Plains |
| *Argemone mexicana* L. | Papaveraceae | Wp | 5, 15, 16, 21, 25, 28, 29, 37, 49, 52,59, 62, 70, 75 | Feb- May | Plains-Subalpines |

| | | | | | |
|---|---|---|---|---|---|
| *Arisaema jacquemontii* Blume | Araceae | Us, Wp | 16, 43, 62, 76 | Jul-Aug | Alpines-Subalpines |
| *Arnebia benthamii* (Wall. ex D. Don) Jonhston | Boraginaceae | Ap, Ft, R | 6, 21, 31, 58, 61 | May- Jul | Alpines-Subalpines |
| *Artemisia absinthium* L. | Asteraceae | Wp | 2, 5, 6, 9, 10, 17, 19, 21, 25, 41, 43, 49, 50, 55 | Jul- Sep | Plains |
| *Artemisia capillaris* Thunb. | Asteraceae | Wp | 1, 2, 3, 5, 15, 18, 24, 25, 34, 38, 49 | Jul-Dec | Subalpines |
| *Artemisia dubia* Wall. ex Bess. | Asteraceae | Ap | 61, 68, 76 | Jul-Aug | Subalpines |
| *Artemisia maritima* L. | Asteraceae | Ap, Ft | 6, 16, 43, 54 | Aug | Subalpines-Alpines |
| *Artemisia wallichiana* Bess. | Asteraceae | Wp | 7, 9, 21 | Jul- Sep | Subalpines-Alpines |
| *Asparagus filicinus* Buch.-Ham. ex D. Don | Liliaceae | R, Sd | 8, 30, 43, 55, 57, 64 | May-Jun | Plains |
| *Asparagus officinalis* L. | Liliaceae | Ap | 15, 18, 21, 28, 30, 33, 40, 49, 52, 59, 70, 77, 78 | Jul-Aug | Plains (Cultivated) |
| *Astragalus polyacanthus* Royle ex Benth. | Papilionaceae | R | 9, 18, 21, 39, 67 | Jul | Plains-Subalpines |
| *Astragalus rhizanthus* Royle ex Benth. | Papilionaceae | R | 1, 11, 16, 18, 27, 31, 39,45 | Jun-Jul | Alpines |
| *Atropa acuminata* Royle | Solanaceae | L, R | 2, 3, 21, 25, 27, 30, 37, 39, 50, 55, 56, 57, 76 | Jun-Jul | Subalpines-Alpines |
| *Avena fatua* L. | Poaceae | Fl | 15 | May- Aug | Plains |
| *Barbaraea vulgaris* R. Br. | Brassicaceae | L, Wp | 8, 41, 68 | Apr-Jun | Subalpines |
| *Berberis lyceum* Royle | Berberidaceae | Fr, L, Rb, S | 2, 3, 5, 6, 8, 12, 14, 21, 23, 25, 31, 50, 56, 59, 60, 67 | Jun-Jul | Subalpines- Plains |
| *Bergenia ciliata* (Haw.) Sternb. | Saxifragaceae | R, Wp | 6, 28, 31, 40, 41, 53, 69, 76 | Jun-Jul | Subalpines-Alpines |
| *Bergenia ligulata* (Wall.) Engl. | Saxifragaceae | Us | 29, 40, 41 | Mar- Jul | Subalpines-Alpines |
| *Bistorta amplexicaulis* (D. Don) Greene | Polygonaceae | L, R | 6, 8, 55 | Jul-Sep | Subalpines-Alpines |
| *Bunium persicum* (Boiss.) Fedtsch. | Apiaceae | Wp | 36 | Jun-Aug | Subalpines |
| *Bupleurum falcatum* L. var. *marginatum* (DC.) Cl. | Apiaceae | R | 49, 71 | Jul-Sep | Subalpines-Alpines |

| | | | | | |
|---|---|---|---|---|---|
| *Bupleurum longicaule* Wall. ex DC. | Apiaceae | R | 5, 29, 49 | Jul- Sep | Subalpines-Alpines |
| *Calendula officinalis* L. | Asteraceae | Wp | 2, 3, 7, 15, 16, 29, 41, 52 | Mar-May | Plains (Ornamental) |
| *Caltha palustris* L. var. *alba* (Camb.) Hk. f. & Th. | Ranunculaceae | R, Wp | 52, 55 | Jun-Aug | Subalpines- Plains |
| *Cannabis sativa* L. | Cannabaceae | Ft, Sd | 2, 6, 12, 14, 21, 22, 31,44, 46, 50, 52, 55, 64, 78, 79 | Apr- Sep | Subalpines- Plains |
| *Capsella bursa-pastoris* (L.) Medik. | Brassicaceae | Ap | 2, 5, 6, 7, 10, 11, 17, 21, 25, 30, 31, 52, 54, 64, 80 | Mar- Jun | Plains |
| *Carpesium cernuum* L. | Asteraceae | R | 3 | Sep-Nov | Subalpines- Plains |
| *Carthamus lanatus* L. | Asteraceae | Ap | 1, 2, 3, 5, 42, 65, 71 | Jul-Sep | Plains |
| *Carum carvi* L. | Apiaceae | Fr | 1,2,3, 5, 7, 8, 15, 27, 32,42 | Jul-Aug | Subalpines-Alpines |
| *Castanea sativa* Mill. | Fagaceae | Wp | 1, 3, 15, 57, 61, 76 | Apr-May | Plains |
| *Celosia argentea* L. | Amaranthaceae | L, Sd, Wp | 3, 4, 5, 18, 27, 36, 41, 49 | Jul-Sep | Plains (Cultivated) |
| *Celtis australis* L. | Ulmaceae | Fr, Sd | 8, 21 | Mar -May | Plains- Subalpines (rare) |
| *Centaurea iberica* Trev. ex Spreng. | Asteraceae | Ft | 16 | Jul-Aug | Plains |
| *Cerastium glomeratum* Thuill. | Caryophyllaceae | Wp | 8 | Jul- Oct | Plains |
| *Chenopodium album* L. | Chenopodiaceae | Sd, Wp | 43, 49, 55, 59, 64 | Jun-Aug | Plains-Subalpines |
| *Chenopodium ambrosioides* L. | Chenopodiaceae | Wp | 15, 21, 43, 52, 62, 66, 70 | Jul-Aug | Plains |
| *Chenopodium botrys* L. | Chenopodiaceae | Wp | 3, 4, 62 | Jun-Aug | Plains-Subalpines |
| *Cichorium intybus* L. | Asteraceae | Wp | 1, 2, 3, 5, 8, 15, 18, 27, 34, 36, 47, 60 | Mar-Sep | Plains |
| *Cicuta virosa* L. | Apiaceae | Wp | 18, 25 | Jun-Aug | Subalpines |
| *Codonopsis clematidea* Cl. | Campanulaceae | Sd | 2 | Jul-Aug | Subalpines-Alpines |
| *Codonopsis obtusa* Nannf. | Campanulaceae | R, Wp | 75 | Jul-Aug | Subalpines-Alpines |
| *Colchicum luteum* Baker | Liliaceae | Us | 2, 5, 15, 21, 23, 56, 78 | Jun-Aug | Subalpines |
| *Conium maculatum* L. | Apiaceae | R, Sd, Wp | 10, 12, 15, 21, 25, 31, 33, 52, 72, 76 | Jun-Aug | Plains |

| | | | | | |
|---|---|---|---|---|---|
| *Convolvulus arvensis* L. | Convolvulaceae | L, Wp | 5, 52, 53, 74 | Mar-Dec | Plains |
| *Conyza canadensis* (L.) Cronq. | Asteraceae | Ft | 2, 11, 16, 21, 27, 40, 41, 55, 57, 67, 68, 76 | Jun- Aug | Plains |
| *Coriaria nepalensis* Wall. | Coriariaceae | R, Wp | 5, 21, 76, 80 | May-Jun | Plains |
| *Corydalis govaniana* Wall. | Papaveraceae | Us,Wp | 3, 12, 25 | Jun-Aug | Subalpines |
| *Cotula anthemoides* L. | Asteraceae | Wp | 21, 37, 55, 64, 80 | Apr- Oct | Plains |
| *Cuscuta chinensis* Lam. | Cuscutaceae | Sd | 1, 8, 18, 25, 28, 71, 82 | Jun-Aug | Plains-Subalpines |
| *Cuscuta reflexa* Roxb. | Cuscutaceae | Wp | 2, 14, 16, 31, 37, 43, 49, 55, 59 | Jul-Aug | Subalpines |
| *Cydonia oblonga* Mill. | Rosaceae | Fr | 1, 3, 5, 6, 8, 28, 29, 31, 37, 37, 41, 53, 57, 58, 76 | Apr-May | Plains (Cultivated) |
| *Cynoglossum glochidiatum* Wall. ex Benth. | Boraginaceae | Fl, R, Sd | 2, 7, 33, 45 | Jun- Sep | Subalpines- Plains |
| *Cyperus rotundus* L. | Cyperaceae | Wp | 4, 5, 6, 13, 14, 16, 24, 25, 26, 29, 30, 31, 34, 36, 44, 46, 49, 52, 60, 62, 64, 74 | May-Sep | Plains |
| *Daphne mucronata* Royle | Thymellaeaceae | L, R | 16, 23, 36, 58, 59, 67 | Apr- Sep | Plains-Subalpines |
| *Datura stramonium* L. | Solanaceae | Fl, Fr, L, Sd | 14, 16, 21, 25, 31, 56, 58, 67, 76 | Jun-Jul | Plains |
| *Daucus carota* L. | Apiaceae | R, Sd, Wp | 15, 16, 19, 21, 25, 27, 30, 33, 39, 40, 68 | Jun-Aug | Plains |
| *Descurainia sophia* (L.) Webb. | Brassicaceae | Fl, L, Sd | 2, 6, 8, 29, 37, 40, 41, 43, 53, 59, 64, 76 | Apr-Jul | Plains-Subalpines |
| *Dianthus caryophyllus* L. | Caryophyllaceae | Fl, Wp | 25, 31, 43, 67, 71, 73 | Nov- Jan | Plains (Ornamental) |
| *Dictamnus albus* L. | Rutaceae | R, Wp | 5, 6, 15, 16, 17, 19, 25, 30, 43, 55, 60, 71 | Jun-Jul | Subalpines |
| *Digitalis lanata* Ehrh. | Scrophulariaceae | L | 25, 31 | Jun-Aug | Subalpines |
| *Digitalis purpurea* L. | Scrophulariaceae | L | 5, 10, 11, 25, 30, 31, 51, 72 | May- Aug | Subalpines- Plains |
| *Dioscorea deltoidea* Wall. ex Griseb. | Dioscoreaceae | Us | 5, 9, 10, 21, 23, 25, 50, 76 | Jun-Jul | Subalpines |
| *Epilobium angustifolium* L. | Onagraceae | L, Wp | 2, 3, 4, 5, 25, 37, 45, 59 | Jul-Aug | Subalpines-Alpines |
| *Eriobotrya japonica* (Thunb.) Lindl. | Rosaceae | L, Sd | 2, 5, 8, 18, 26, 27, 29, 49 | Dec- Feb | Plains(Cultivated) |

| | | | | | |
|---|---|---|---|---|---|
| *Erodium cicutarium* (L.) L'Herit. | Geraniaceae | Wp | 2, 11, 53 | May-Jun | Plains |
| *Euphorbia hispida* Boiss. | Euphorbiaceae | Wp | 5, 8, 16, 43, 67 | Jun- Nov | Plains-Subalpines |
| *Euphorbia thomsoniana* Boiss. | Euphorbiaceae | R, Wp | 58, 59 | Jun- Sep | Subalpines-Alpines |
| *Ferula jaeschkeana* Vatke | Apiaceae | Ap | 10, 75 | Apr- Jun | Subalpines-Alpines |
| *Ficus carica* L. | Moraceae | Fr | 5, 8, 16, 41, 76 | Apr- Dec | Plains(Cultivated) |
| *Fragaria nubicola* Lindl. ex Lacaita | Rosaceae | Fr, L | 8, 30, 39, 53, 57 | Apr- Jun | Subalpines |
| *Fritillaria cirrhosa* D. Don | Liliaceae | Us,Wp | 3, 26, 37, 76 | May- Jul | Subalpines-Alpines |
| *Fritillaria imperialis* L. | Liliaceae | Wp | 15, 30, 51, 52 | Apr-Jun | Plains-Subalpines |
| *Fritillaria roylei* Hook. | Liliaceae | Us | 3, 5, 6, 31, 37, 42, 73, 76 | Jun-Jul | Subalpines |
| *Fumaria indica* (Haussk.) Pugsley | Fumariaceae | Wp | 6, 41, 16, 21, 22, 25, 30, 43, 49, 71 | Mar- Jun | Plains |
| *Galium aparine* L. | Rubiaceae | Wp | 30, 59 | Mar- Jul | Plains |
| *Gentiana carinata* (D.Don) Griseb. | Gentianaceae | Wp | 7, 77 | May- Sep | Plains-Subalpines |
| *Gentiana tianshanica* Rupr. | Gentianaceae | R | 6, 21, 67, 71 | Aug- Oct | Alpines |
| *Geranium pratense* L. | Geraniaceae | R, Wp | 41 | Jun- Aug | Subalpines-Alpines |
| *Geranium wallichianum* D. Don ex Sweet | Geraniaceae | R, Wp | 2, 21, 44, 55, 57, 80 | Jul- Sep | Subalpines |
| *Geum elatum* Wall. ex G. Don | Rosaceae | R | 7, 8, 53, 54, 57 | Jun-Jul | Subalpines |
| *Geum roylei* Wall. ex Bolle | Rosaceae | R, Wp | 2, 57, 67, 76 | Jun- Aug | Subalpines |
| *Glycyrrhiza glabra* L. | Papilionaceae | R | 1, 2, 3, 4, 5, 6, 7, 13, 16, 18, 21, 29, 49, 52, 64, 66 | May-Jun | Plains-Subalpines |
| *Hemerocallis fulva* L. | Liliaceae | Ap, L, R | 1, 5, 8, 35, 64 | May--Jul | Plains-Subalpines |
| *Heracleum candicans* DC. | Apiaceae | R | 7, 16, 25, 33 | Jul-Aug | Subalpines-Alpines |
| *Hernaria hirsuta* L. | Illecebraceae | Wp | 16, 30, 77 | Jun- Aug | Plains-Subalpines |
| *Hibiscus syriacus* L. | Malvaceae | L, S, Sb | 1, 4, 5, 21, 53, 64 | Jul- Sep | Plains |

| | | | | | |
|---|---|---|---|---|---|
| *Humulus lupulus* L. | Cannabaceae (Urticaceae) | Fl,Wp | 1, 3, 5, 7, 15, 29, 48, 60, 64, 68, 74 | Jun-Jul | Plains-Subalpines |
| *Hyoscyamus niger* L. | Solanaceae | L | 15, 25, 50, 76 | Jun-Jul | Plains-Subalpines |
| *Hypericum perforatum* L. | Hypericaceae | Wp | 2, 3, 5, 16, 21, 22, 30, 37, 41, 53, 55, 56, 58, 59 | Jun- Oct | Plains |
| *Indigofera heterantha* Wall. ex Brandis | Papilionaceae | Fl | 41 | May- Jul | Plains-Subalpines |
| *Inula racemosa* Hook. f. | Asteraceae | R, Wp | 2, 3, 4, 6, 14, 16, 33, 36, 37, 68 | Jun- Sep | Subalpines-Alpines |
| *Inula royleana* DC. | Asteraceae | Ap, R | 12, 16, 52, 62 | Jun- Sep | Subalpines |
| *Iris decora* Wall. | Iridaceae | Us | 7, 21, 30, 41, 55 | Jul | Plains |
| *Iris kashmiriana* Baker | Iridaceae | Us | 55 | May-Jun | Plains |
| *Iris lactea* Pall. | Iridaceae | Us | 36, 49, 68, 77 | Jun- Aug | Plains-Subalpines |
| *Juglans regia* L. | Juglandaceae | Fr, R, Sb | 2, 3, 4, 8, 16, 21, 27, 29, 41, 43, 55, 56, 57, 59, 62, 67, 73 | Feb-Apr | Plains-Subalpines |
| *Jurinea dolomiaea* Boiss. | Asteraceae | R | 6, 7, 16, 64, 68 | Jun- Sep | Subalpines |
| *Lamium album* L. | Lamiaceae | Fl, Wp | 5, 11, 25, 56, 59, 73 | Jun- Aug | Subalpines-Alpines |
| *Lavatera kashemiriana* Camb. | Malvaceae | R | 39 | Jun-Aug | Subalpines |
| *Linum usitatissimum* L. | Linaceae | Sd | 5, 8, 18, 23, 25, 31, 39, 42, 49, 64 | Feb- May | Plains (Cultivated)-Subalpines |
| *Logfia arvensis* (L.) Holub. | Asteraceae | Wp | 1, 6, 73, 76 | Jul- Sep | Plains-Subalpines |
| *Lomatogonium carinthiacum* Reichb. | Gentianaceae | Wp | 16, 49, 76, 80 | Aug-Oct | Subalpines |
| *Lychnis coronaria* (L.) Desr. | Caryophyllaceae | R | 16, 49, 76, 80 | Jul-Aug | Subalpines |
| *Malva neglecta* Wallr. | Malvaceae | Fl, L | 8, 37, 76 | Apr-Sep | Plains |
| *Malva parviflora* L. | Malvaceae | R, Sd, Wp | 8, 16, 37, 41, 58 | Jun-Jul | Plains |
| *Malva sylvestris* L. | Malvaceae | Fl, Sd | 2, 8, 16, 28, 37, 40, 76, 83 | Apr- Jul | Plains |
| *Marrubium vulgare* L. | Lamiaceae | Ap | 3, 7, 12, 21, 30, 31, 37, 49, 59 | May-Jun | Plains |
| *Medicago sativa* L. | Papilionaceae | Wp | 4, 18 | May- Sep | Plains-Subalpines |
| *Melilotus alba* Medik. ex Desr. | Papilionaceae | Wp | 16 | Jul-Aug | Plains-Subalpines |

| | | | | | |
|---|---|---|---|---|---|
| *Mentha arvensis* L. | Lamiaceae | L, Wp | 1, 3, 4, 10, 18, 21, 25, 47, 52, 62, 75 | Jul-Aug | Plains |
| *Mentha piperita* L. | Lamiaceae | Wp | 7, 9, 14, 29, 36, 76 | Jun-Jul | Plains |
| *Mentha spicata* L. var. *longifolia* | Lamiaceae | Wp | 1, 3, 4, 29 | Jul-Aug | Plains-Subalpines |
| *Morus nigra* L. | Moraceae | Fr, L, Sb | 8, 27, 31, 41, 43, 59 | Mar- Jun | Plains |
| *Narcissus tazetta* L. | Amaryllidaceae | Us | 8, 16, 21, 29, 30, 52, 59, 72 | Dec- May | Plains |
| *Nepeta cataria* L. | Lamiaceae | Ap | 8, 17, 21, 25, 33, 41, 43, 45, 53, 64, 67, 68, 71 | Jul-Aug | Plains |
| *Nerium oleander* L. | Apocynaceae | L, Sb, Wp | 10, 15, 25, 52, 62, 70, 74 | Jun-Jul | Plains |
| *Ocimum basilicum* L. | Lamiaceae | Fl, Sd, Wp | 1, 2, 3, 4, 5, 29, 30, 53, 68, 76 | Jul-Aug | Plains |
| *Ocimum tenuiflorum* L. | Lamiaceae | L, R, Sd | 6, 7, 16, 27, 28, 37, 43, 49, 62, 71, 73, 76 | Jun-Jul | Plains |
| *Origanum vulgare* L. | Lamiaceae | Wp | 1, 3, 4, 5, 18, 26, 27 | May- Sep | Subalpines-Alpines |
| *Oxalis acetosella* L. | Oxalidaceae | Wp | 2, 6, 8, 11, 30, 67, 68, 80 | Apr- Sep | Subalpines |
| *Oxalis corniculata* L. | Oxalidaceae | L | 2, 6, 11, 30 | Mar-Sep | Plains- Subalpines |
| *Oxyria digyna* (L.) Hill. | Polygonaceae | Wp | 8 | Jun- Aug | Subalpines-Alpines |
| *Paeonia emodi* Wall. ex Royle | Paeoniaceae | Wp | 7, 21, 25, 30, 44, 53, 76, 77 | May-Jun | Subalpines |
| *Papaver rhoeas* L. | Papaveraceae | Fr, L, Sd | 6, 8, 25 | Apr- Sep | Plains |
| *Papaver somniferum* L. | Papaveraceae | L, Sd | 5, 15, 19, 21, 33, 76 | Apr- Jun | Plains |
| *Peganum harmala* L. | Zygophyllaceae | Sd, Wp | 3, 5, 16, 17, 19, 25, 27, 31, 34, 36, 43, 45, 46, 47, 52, 55 | Jun- Oct | Plains- Subalpines |
| *Phytolacca latbenia* H. Walter | Phytolaccaceae | Wp | 5, 15, 23, 25 | Jun- Aug | Subalpines |
| *Picrorhiza kurrooa* Royle ex Benth. | Scrophulariaceae | Us | 1, 2, 3, 6, 18, 36, 37, 49, 52, 66, 73 | Jul-Aug | Subalpines-Alpines |
| *Plantago lanceolata* L. | Plantaginaceae | L, Sd | 2, 3, 5, 7, 11, 14, 18, 29, 31, 37, 67, 73, 76 | Jul-Aug | Plains- Subalpines |
| *Pleuropteropyrum alpinum* Koidz. | Polygonaceae | L, R | 8, 23, 53, 57 | Jun-Jul | Plains- Subalpines |

| | | | | | |
|---|---|---|---|---|---|
| *Podophyllum hexandrum* Royle | Podophyllaceae | Us | 5, 6, 8, 15, 16, 18, 19, 23, 38, 47, 49, 52,59,60,62, 72 | Apr- Jun | Subalpines-Alpines |
| *Polemonium caeruleum* L. ssp. *himalayanum* (Baker) Hara | Polemoniaceae | Wp | 25, 57 | May-Aug | Subalpines |
| *Polygala sibirica* L. | Polygalaceae | R | 2, 33, 37, 45, 76 | May- Jul | Plains- Subalpines |
| *Portulaca oleracea* L. | Portulacaceae | Sd,Wp | 3, 4, 5, 12, 13, 14, 25, 27, 30, 31, 41, 49, 50, 62, 67, 83 | Mar- Aug | Plains |
| *Potentilla supina* L. | Rosaceae | R | 6, 8, 57 | Jul-Aug | Plains- Subalpines |
| *Primula denticulata* Sm. | Primulaceae | L, Wp | 7, 21, 49, 76 | Apr- Jul | Subalpines-Alpines |
| *Prunella vulgaris* L. | Lamiaceae | Wp | 1, 5, 15, 23, 26, 29 | Jun-Oct | Plains- Subalpines |
| *Punica granatum* L. | Punicaceae | Wp | 1, 2, 3, 4, 5, 10, 27, 29, 30, 37, 43, 46, 52, 53, 62, 64, 67, 83 | May- Jul | Plains |
| *Rabdosia rugosa* (Wall. ex Benth.) Hara | Lamiaceae | L | 21, 43, 62, 73 | May- Sep | Plains- Subalpines |
| *Ranunculus laetus* Wall. ex G. Don | Ranunculaceae | Wp | 3, 4, 60 | Apr-Aug | Subalpines- Plains |
| *Ranunculus muricatus* L. | Ranunculaceae | Wp | 6, 55, 76 | Mar- Apr | Plains |
| *Rheum australe* D. Don | Polygonaceae | R | 2, 3, 5, 7, 11, 21, 29, 31, 39, 41, 52, 55, 70, 72, 83 | Aug-Sep | Subalpines-Alpines |
| *Rhus succedanea* L. | Anacardiaceae | Fr, L | 15, 53, 57, 76 | May-Jun | Plains |
| *Ribes nigrum* L. | Grossulariaceae | Fr | 1, 2, 3, 29, 31, 34, 74, 83 | Jun-Jul | Plains- Subalpines |
| *Ricinus communis* L. | Euphorbiaceae | L, Sd | 6, 7, 16, 36, 39, 42, 43, 49, 52, 53, 59, 64, 69 | Jun-Aug | Plains (Cultivated) |
| *Robinia pseudoacacia* L. | Papilionaceae | Fl, Wp | 4, 5, 7, 8, 14, 16, 30, 31, 38, 52, 57, 59, 72, 83 | Mar- Apr | Plains (Cultivated) |
| *Rosa damascena* Herrm. | Rosaceae | Fl | 2, 8, 21, 25, 31, 37, 49, 57, 59 | Jun-Jul | Plains (Cultivated) |
| *Rosmarinus officinalis* L. | Lamiaceae | Ap | 1, 2, 3, 4, 5, 7, 15, 18, 19, 21, 25, 29, 31, 38, 45, 49, 62, 76, 79, 83 | Jun-Jul | Plains |
| *Rubia cordifolia* L. | Rubiaceae | Wp | 1, 2, 3, 5, 14, 25, 26, 29, 41, 44 | Jun- Nov | Plains- Subalpines |
| *Rumex dentatus* L. | Polygonaceae | L | 2, 6, 7, 8, 12, 16, 23, 30, 57, 67 | Jun- Oct | Plains- Subalpines |
| *Rumex nepalensis* Spreng. | Polygonaceae | R | 59, 67 | Jun-Aug | Plains- Subalpines |

| | | | | | |
|---|---|---|---|---|---|
| *Rumex patientia* Gaud. ex Meissn. | Polygonaceae | R | 8, 59 | Jun- Sep | Plains- Subalpines |
| *Sagittaria trifolia* L. | Alismataceae | L, Us | 2, 16, 21, 30, 36 | Apr- Sep | Plains |
| *Salix alba* L. | Salicaceae | L, Sb | 2, 6, 8, 21, 23, 53, 55, 57, 64, 67, 78 | Apr-May | Plains |
| *Salix babylonica* L. | Salicaceae | L, Sb | 6, 8, 43, 55, 75 | Mar- Apr | Plains |
| *Salvia moorcroftiana* Wall. ex Benth. | Lamiaceae | L, R, Sd | 6, 16, 21, 36, 41, 56, 73, 77 | Jun-Aug | Plains- Subalpines |
| *Sambucus wightiana* Wall. ex Wt. & Arn. | Sambucaceae | Wp | 2, 6, 30, 37, 42, 52, 59, 71 | Jun-Jul | Subalpines |
| *Saussurea costus* (Falc.) Lipsch. | Asteraceae | R, Wp | 1, 2, 5, 7, 18, 31, 49, 62, 67, 82 | Jun- Sep | Subalpines- Alpines |
| *Saussurea simpson-iana* (Field & Gard.) Lipsch. | Asteraceae | Wp | 8, 16, 25, 41, 57, 68, 73 | Jul-Oct | Alpines |
| *Saxifraga moorcrofti-ana* (Wall. ex Ser.) Sternb. | Saxifragaceae | Wp | 40 | Aug-Sep | Alpines |
| *Sedum ewersii* Ledeb. | Crassulaceae | Wp | 57, 68 | Jun-Sep | Subalpines- Alpines |
| *Senecio jacquemontianus* | Asteraceae | R | 16, 25, 41, 52 | Jun-Sep | Alpines |
| *Senecio laetus* Edgew. | Asteraceae | Wp | 5 | May- Jul | Subalpines- Alpines |
| *Senecio vulgaris* L. | Asteraceae | Wp | 2, 4, 7, 8, 30, 31, 43, 52, 56, 64, 71, 78 | Jan- Dec | Plains- Subalpines |
| *Siegesbeckia orientalis* L. | Asteraceae | Wp | 2, 3, 4, 8, 21, 23, 25, 27, 29, 31, 39,41,55,63,73, 76 | Apr-Dec | Plains |
| *Sisymbrium irio* L. | Brassicaceae | L, Sd | 6, 8, 37, 68, 76 | Mar-May | Plains |
| *Skimmia anquitilia* Taylor & Airy Shaw | Rutaceae | L | 52, 53 | Apr- Jun | Subalpines |
| *Solanum nigrum* L. | Solanaceae | Ap, Fr | 2, 6, 14, 16, 21, 25, 30, 31, 36, 46, 49, 56, 59, 67, 80, 81 | Jan- Dec | Plains- Subalpines |
| *Solidago decurrens* Lour. | Asteraceae | Wp | 11, 30, 39, 40, 67 | Jul- Sep | Subalpines |
| *Sonchus oleraceus* L. | Asteraceae | L, R, Wp | 6, 8, 59 | Mar-Nov | Plains |
| *Stachys floccosa* Benth. | Lamiaceae | Wp | 8, 17, 30 | Jun-Jul | Subalpines- Plains |

| | | | | | |
|---|---|---|---|---|---|
| *Strobilanthes urticifolia* Kuntze | Acanthaceae | L | 6 | Jun- Oct | Subalpines |
| *Swertia petiolata* D. Don | Gentianaceae | Wp | 7, 14, 31 | Jul- Sep | Subalpines-Alpines |
| *Tagetes minuta* L. | Asteraceae | Wp | 3, 4, 5, 15, 25, 62 | Sep- Nov | Plains |
| *Tanacetum cinerariaefolium* (Trev.) Schultz-Bip. | Asteraceae | Fl | 5, 15, 25, 62, 81 | Jul-Aug | Plains-Subalpines |
| *Tanacetum dolichophyllum* (Kitam.) Kitam. | Asteraceae | L, R | 2, 67 | Jul-Aug | Subalpines-Alpines |
| *Taraxacum officinale* Weber | Asteraceae | L, R | 5, 6, 8, 15, 18, 21, 23, 30, 31, 38, 39, 40, 49, 55, 56, 59 | Mar-Nov | Plains- Subalpines-Alpines |
| *Thalictrum minus* L. | Ranunculaceae | Wp | 6, 12, 14, 15, 30, 31, 67, 73, 80 | Jul | Subalpines |
| *Thymus linearis* Benth. | Lamiaceae | Wp | 3, 5, 7, 8, 17, 21, 25, 41, 43, 44, 55, 57, 58, 63, 76 | Jul-Aug | Subalpines- Plains |
| *Trachyspermum ammi* (L.) Sprague | Lamiaceae | Fr,Wp | 4, 8, 29, 67, 68, 70 | Jun-Jul | Plains(cultivated) |
| *Tribulus terrestris* L. | Zygophyllaceae | Ap,Fr, R | 2, 3, 5, 12, 18, 21, 30, 31, 33, 40, 45, 52, 70, 83 | Mar-Dec | Plains |
| *Tussilago farfara* L. | Asteraceae | Fl,Wp | 1, 3, 25, 31, 36, 52, 76 | Mar-Oct | Subalpines-Alpines |
| *Urtica dioica* L. | Urticaceae | L, R | 2, 5, 7, 15, 16, 18, 25, 29, 36, 50 | May- Sep | Plains- Subalpines |
| *Valeriana jatamansii* Jones | Valerianaceae | Us | 12, 14, 21, 22, 25, 31, 67, 68 | Apr- Jul | Subalpines |
| *Verbascum thapsus* L. | Scrophulariaceae | Wp | 2, 3, 6, 16, 25, 28, 29, 33, 57, 76, 83 | May-Sep | Plains- Subalpines |
| *Verbena officinalis* L. | Verbenaceae | Wp | 6, 23, 25, 42, 55, 63, 64, 73 | Jul-Aug | Plains |
| *Veronica beccabunga* L. | Scrophulariaceae | Wp | 8, 16, 30, 31, 41, 56 | May- Aug | Subalpines |
| *Vincetoxicum auriculatum* (Royle ex Wt.) Jagtap & Singh | Asclepiadaceae | L, R, Wp | 62, 72 | Jun | Subalpines |
| *Viola indica* W. Beck. | Violaceae | Wp | 6, 16, 21, 28, 59, 76 | May- Jul | Subalpines |
| *Viola odorata* L. | Violaceae | Wp | 2, 4, 5, 6, 21, 37, 59, 78 | Mar- May | Subalpines |
| *Viscum album* L. | Loranthaceae | Wp | 1, 2, 5, 15, 18, 27, 34, 49 | Mar- Jun | Plains- Subalpines |

| | | | | | |
|---|---|---|---|---|---|
| *Xanthium indicum* Koenig. | Asteraceae | Fr, L, Wp | 2, 3, 5, 25, 27, 52, 62, 67 | Mar-Dec | Plains |
| *Zanthoxylum armatum* DC. | Rutaceae | Wp | 7, 16, 53, 62 | Mar- Apr | Plains |
| *Ziziphus mauritiana* Lam. | Rhamnaceae | Wp | 1, 2, 22, 25, 36, 76 | Jun-Jul | Plains |

Symbols : *Ap-Aerial parts; Fl-Flower; Ft-Flowering tops; Fr-Fruit; L-Leaves; R-Root; Rb-Root bark; Sd-Seeds; S-Stem; Sb-Stem bark; Us-Underground stem; Wp-Whole plant.

**1-Antioxidant; 2-Anti-inflammatory; 3-Antibacterial; 4-Antifungal; 5-Antitumor; 6-Antipyretic; 7-Gastroprotective; 8-Nutritive/Tonic; 9-Digestive; 10-Antifertility; 11-Antihaemmorhagic; 12-Hypotensive; 13-Hypertensive; 14-Antispasmodic/ Spasmolytic; 15-Allergic; 16-Dermaprotective; 17-Emmenagogue; 18-Immunomodulator; 19-Abortifacient; 20-Growth stimulating; 21-Analgesic; 22-Antidepressant; 23-Antiarthritic; 24-Hypolipidemic; 25-CNS activity; 26-Anticoagulant; 27-Hypoglycaemic; 28-Demulcent; 29-Antiviral; 30-Diuretic; 31-Cardiovascular activity; 32-Chemopreventive; 33-Aphrodisiac; 34-Vasorelaxant; 35-Endocrine regulatory; 36-Antiallergic; 37-Expectorant; 38-Choleretic/Cholagogue; 39-Renoprotective; 40-Litholytic; 41-Wound healing; 42-Lactation promoting /Lactogogue; 43-Antih elmenthic; 44-Anticonvulsant; 45-Sexual character promoting; 46-Hypothermic; 47-Radioprotective; 48-Cell signalling; 49-Hepatoprotective; 50-Antiandrogenic; 51-Cardiotoxic; 52-Poisonous; 53-Antidiarrhoeal; 54-Leucorrhoea; 55-Antirheumatic; 56-Antihaemmorhoidal; 57-Astringent; 58-Hair promoting; 59-Laxative; 60-Antiplasmodial; 61-Herbicidal; 62-Insecticidal/ Pesticidal; 63-Depurative; 64-Gynogenic; 65-Osteoclastogenic activity; 66-Antileishmenial activity; 67-Antinfectious; 68-Stimulative; 69-Biomarker; 70-Molluscicidal; 71-Diaphoretic; 72-Emetic; 73-Antitoxic; 74-Antenzymatic; 75-Anti-implantation; 76-Anti-respiratory troubles; 77-Antidropsy; 78-Antigout; 79-Hyperglycaemic; 80-Opthalmia; 81-Teratogenic; 82-Tumerogenic; 83-Muscle activity.

## RESULTS

In the present studies, floristic diversity of medicinal plants of the Kashmir Valley has been dealt with. Mention has been made of 220 species belonging to 77 families. The results of the study are reported in Table 1 (Gowher, 2006), where plants are arranged in alphabetical order. In the respective columns, the botanical name, the family to which it belongs, the part used as a medicine, the pharmacological activity, flowering period and altitudinal range are listed.

## DISCUSSION AND CONCLUSIONS

During present investigation of the study area, 220 medicinal plant species, belonging to 178 genera distributed over 77 families have been dealt with. This forms about 10.45% of the total 2104 vascular plants reported from Kashmir Valley (Stewart, 1972; Dar *et al.*, 1995; Singh *et al.,* 2002). On the basis of the pharmacological, phytochemical and other related data some medicinal plants have been designated as of 'High Medicinal Value'. These plants can be brought under cultivation and as such cultivation protocols need to be developed. Simultaneously present study revealed that some plant species, though used frequently by locals of the study area and ethnic communities in other parts of the world for certain ailments, have not been investigated on the modern scientific lines. This is also corroborated by the scarce pharmacological literature available. Ethnobotany and ethnomedicine can provide information that is useful as a 'prescreen' to select plants for experimental pharmacological studies. As such, this data from ethnomedicine can be analysed with the aim of selecting a reasonable number of plants to be tested in bioassay systems that are believed to predict the action of these drugs in humans.

## ACKNOWLEDGEMENT

The senior author is highly thankful to CSIR, New Delhi for financial assistance.

## REFERENCES

ANONYMOUS 1976. *Medicinal plants of India*, Vol. I. Indian Conc. Med. Res., New Delhi.

CHOPRA, R.N., S.L. NAYAR & I.C. CHOPRA 1956. *Glossary of Indian Medicinal plants*. CSIR, New Delhi.

DAR, G.H., A.R. NAQSHI & S. ARA 1995. New records and new taxa of flowering plants from Jammu & Kashmir State. 1970-1992. *Oriental Science* Special publication : 33-44.

GOWHER, GUNA 2006. *Medicinal Flora of Kashmir Valley*. Ph.D. Thesis, Department of Pharmaceutical Sciences, University of Kashmir.

KAUL, A.K. 1977. Kashmir Physiography. *Inquiry* 4 (7) : 1-20.

NAQSHI, A.R., M.Y. BABA & S. ARA 1992. Ethnobotanical studies of Kashmir-Jhelum Valley. *Recent Advances in Medicinal, Aromatic and Spice Crops* 2 : 371-379. New Delhi.

NASIR, E. & S.I. ALI (Eds.) 1970. *Flora of Pakistan*. Karachi, Islamabad.

SINGH, N.P., D.K. SINGH & B.P. UNIYAL 2002. *Flora of Jammu & Kashmir*, Vol. I. BSI., Kolkata.

STEWART, R.R. 1972. Flora of West Pakistan. An Annotated Catalogue of Vascular Plants of West Pakistan and Kashmir. Karachi, Pakistan.

# ETHNOMEDICINAL PLANTS USED IN SKIN DISEASES BY SOME INDO-MONGOLOID COMMUNITIES OF ASSAM

JITU BURAGOHAIN* & B.K. KONWAR

**Dept. of Botany, Namrup College, Parbatpur-786 623, Dist. Dibrugarh, Assam*
*E-mail : jitu_buragohain@yahoo.co.uk*
*Dept. of Molecular Biology & Biotechnology, Tezpur University, Napaam-784 028, Dist. Sonitpur*

**ABSTRACT**

An ethnobotanical study was carried out among five Indo-Mongoloid communities distributed in five districts of Upper Assam, India to document plants used in various skin diseases. A total of 68 plant species belonging to 40 families were described. All these plants and plant parts need to be evaluated through phyto and pharmaco-chemical investigations to discover their potentiality as drugs.

## INTRODUCTION

Skin diseases are of common occurrence among the rural masses due to poor hygienic conditions, poor sanitation facility and contaminated water. Traditional herbal medicines used by different communities play an important role in alleviating such skin diseases. They are safe, effective and inexpensive and in many case, the only method of medication. The ethnobotanical study was undertaken among some Indo-Mongoloid communities, namely Mishing, Sonowal Kachari, Motok, Deori and Ahom, largely distributed in five districts of Upper Assam to document plants used in various skin diseases. Although several studies on ethnomedicinal plants of Assam have been carried out, the ethnobotany of the said communities concerning the use in skin diseases is poorly known.

Assam is a botanically rich State in India, which is situated between 24°2′-27°6′ N latitudes and 89°8′-96° E longitudes and covers an area of 78,523 sq kilometers [Fig. 1]. The State is bordered by the nations Bhutan and Bangladesh. Assam has a humid subtropical climate with extremely heavy rainfall. A sizeable area of the State is covered with dense tropical forests of bamboo, and at large elevation evergreens. Assam experiences heavy rainfall between March and September with very high humidity in the summer months. Winter sets in from the month of November and lasts till the end of February. The total number of districts in Assam is 23 and the total population as per 2001 census is 26,638,407. Assam is inhabited by a number of ethnic tribes belonging to the Indo-Mongoloid race. The dominant Indo-Mongoloid communities of Assam are Bodo, Mishing, Karbi, Dimasha, Rabha, Tiwa, Sonowal Kachari, Ahom, Tai Turung, Tai Khamyang, Deori, Chutia, Koch, Motok and Moran. Plants form an integral part of their culture, which is evident from their food habit, customs, marriage system and household practices. The present study was carried out among five Indo-Mongoloid communities such as Mishing, Sonowal Kachari, Motok, Ahom and Deori distributed largely in the five districts of upper Assam. The people of these communities, particularly the rural folk have long been using plants for their various ailments. Paddy cultivation is their main agricultural practice and, therefore, rice is the main staple food. They speak Assamese which is the official language of Assam. Although they have accepted modern health care facilities, the rural folk at large still prefer to stay with their traditional herbal medication.

The aim of the study was to explore and document the plants used in different skin diseases by the said Indo-Mongoloid communities, which might be of considerable interest to the development of potential antimicrobial drugs.

## METHODOLOGY

Regular field trips were conducted during the period from January 2002 – April 2004 in the said districts of Upper Assam. The investigation was carried out where the population distribution of the investigated communities was dense. During the trips, contacts were made with the village heads, herbal practitioners as well as elderly men and women of the respective communities. Prior informed consent was obtained from the village heads and from the participants in the study. Information on herbal medication was gathered through conversations, interviews and discussions. For this, a questionnaire was designed for interviewing the participants of different communities. During the fieldwork, repeated verification of data by different informants and at different places was carried out. Only the specific and reliable information crosschecked with different informants and at different places was incorporated in the present study. The information gathered was also crosschecked with some published literature (Bora & Kumar, 2003; Dutta, 1983; Kanjilal *et al.*, 1940; Purkayastha *et al.*, 2005; Chopra *et al.*, 1956; Jain, 1991; Kirtikar & Basu, 1935). The collected plant species were authentically identified with the help of local floras and Botanical Survey of India, Shillong. Voucher specimens were deposited to the herbarium of the Department of Molecular Biology and Biotechnology, Tezpur University, Assam, India. The recorded plant species were alphabetically arranged followed by voucher specimen number in Table1 along with their families, local names, English names, flowering and fruiting periods, life forms and medicinal uses.

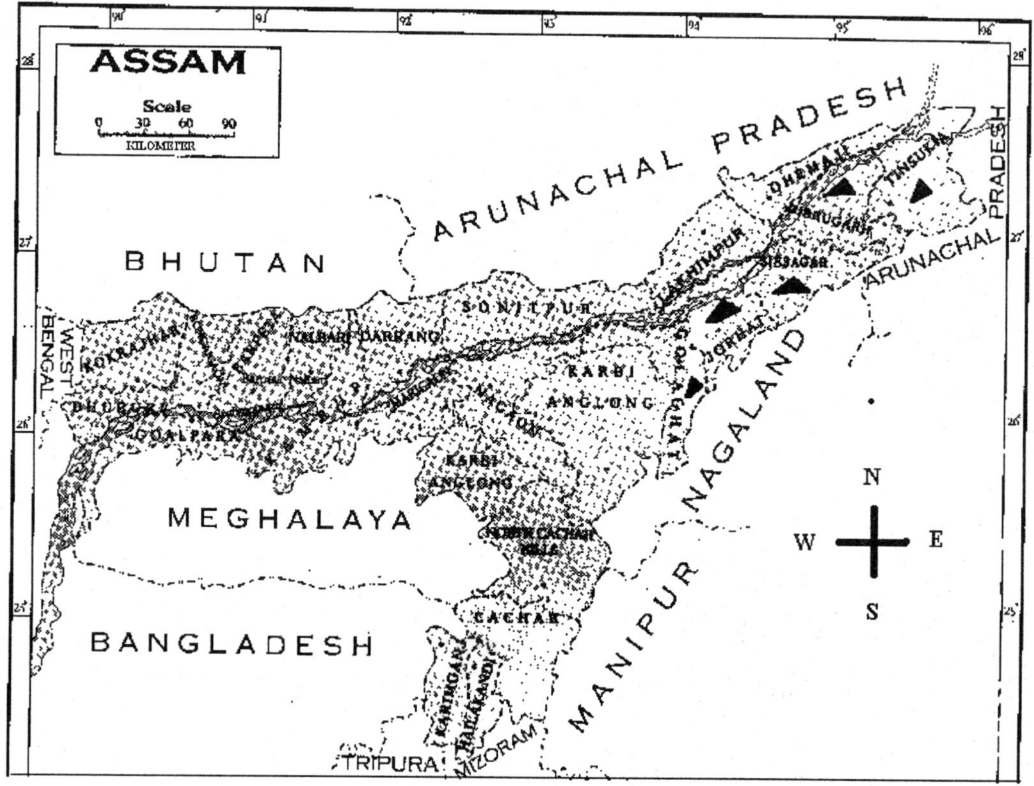


Fig. 1. Geographical map of Assam
(Black shaded triangles represent the five districts where study was carried out).

## RESULTS AND DISCUSSION

In the present investigation, information on 68 plant species belonging to 40 families was collected, verified and authenticated [Table 1]. Among them, 18 species are shrubs and trees, 17 herbs, 6 climbers, 5 undershrubs and 1 each of creeper, grass, trailer and palm. These plants are used in the treatment of abscesses, septic ulcers, scabies, ringworms, allergy and pimples.

The most cited plant family is Fabaceae (containing 6 species), followed by Caesalpiniaceae (5 species), Zingiberaceae (4 species), Amaranthaceae, Rubiaceae, Rutaceae, Euphorbiaceae (3 species each). Leaf is the most widely used plant part accounting for 32 plant species in a total of 68 reported plants. This is followed by latex (8 species), fruit (7 species), root (7 species), bark (5 species), seed and rhizome (3 species). Most of the herbal remedies are taken externally in the form of paste. The plant parts are crushed and made into paste for application over the area of the disease. Only in five occasions oral administration is prescribed. In some cases, along with the plant parts a little amount of salt or milk or molasses or honey is used. This addition might be to enhance the efficacy of herbal remedies or to make the remedy more palatable masking the undesirable taste when taken orally. The medicinal plants are usually collected from wild habitat as and when there is a need. Some of the information reported in this study concerning *Amaranthus spinosus*, *Areca catachu*, *Flemingia strobilifera*, *Meyna spinosa* etc are found to be new in the literature of Indian medicinal plants and deserves further study.

Table 1. List of recorded ethnomedicinal plants of Assam used in skin diseases.

| ***Name of the species*** | ***Family*** | ***Local name*** | ***English name*** | ***Flowe-ring & fruiting*** | ***Life form*** | ***Part used & mode of administration and uses*** |
|---|---|---|---|---|---|---|
| *Abrus precatorius* L. MBBT/0029 | Fabaceae | Latumoni | Crab's eye vine | July-Nov | Climber | Root-paste is applied to cure leucoderma. |
| *Achyranthes aspera* L. MBBT/0047 | Amaranthaceae | Bioni-hakuta, Ubhota-bonsosh | Prickly chaff flower | Oct-April | Herb | Seed-paste with little salt is applied on carbuncles. |
| *Albizzia lebbeck* (L.) Benth. MBBT/0200 | Mimosaceae | Sirish | Silk flowered sau | Mar-Sept | Tree | Juice of the bark is applied on the ringworm, scabies and septic ulcer. Fruit-paste is applied on sore eyes. |
| *Allium sativum* L. MBBT/0249 | Liliaceae | Nohoru | Garlic | Nov-Dec | Herb | Bulbs are crushed and mixed with coconut oil and applied on ringworm and scabies. |
| *Alpinia nigra* (Gaertn.) Burtt. MBBT/0268 | Zingiberaceae | Tora | — | May-Aug | Herb | Tubers are eaten fresh which is said to cure scabies and other skin diseases. |
| *Alstonia scholaris* (L.) R.Br. MBBT/0120 | Apocynaceae | Chationa | Devil's tree | Oct-Mar | Tree | Fresh latex is applied on abscess and septic ulcer for quick healing. |
| *Amaranthus spinosus* L. MBBT/0054 | Amaranthaceae | Hati-khutura | Spiny amaranth | Jan-Dec | Herb | The paste of whole plant is applied on eczema. |
| *Amaranthus tricolor* L. var. | Amaranthaceae | Bishalyakarani | — | Jun-Nov | Herb | Leaf-paste is applied on abscess and septic |

| | | | | | | |
|---|---|---|---|---|---|---|
| *tristis* (Prain) Nayar., MBBT/0231 | | | | | | ulcer for quick healing |
| *Androgaphis paniculata* (Burm.f.) Wall. ex Nees MBBT/0176 | Acanthaceae | Chirota, Kalmegh | — | Nov-Jun | Herb | Leaf-paste is applied on abscess. |
| *Areca catechu* L. | Arecaceae | Tamul | Betelnut | Jan-Dec | Palm | The nuts are dried in the sun and crushed into powder, which is applied on septic ulcer for quick healing. |
| *Azadirachta indica* A. Juss. MBBT/0124 | Meliaceae | Neem | Margosa tree | Mar-July | Tree | Leaves are crushed and applied on ringworm. Crushed bark is applied to cure abscess. Leaves are boiled and the boiled water is used in bathing, which is said to cure scabies and other skin diseases. |
| *Basella rubra* L. MBBT/0079 | Basellaceae | Puroi | Indian spinach | Aug-Dec | Climber | Leaf-paste is applied on allergy for quick relief. |
| *Bauhinia variegata* L. MBBT/0260 | Caesalpiniaceae | Kanchan | — | Oct-July | Tree | Leaf-paste is applied on septic ulcer for quick healing. |
| *Boerhavia diffusa* L. MBBT/0270 | Nyctaginaceae | Purnonowa | Hog weed | Jun-Nov | Herb | Leaf-paste is applied on abscess. |
| *Bombax ceiba* L. | Bombacaceae | Simolu | Silk cotton tree | Jan-May | Tree | Thorns are crushed and mixed with milk and applied on pimples. |
| *Butea monosporma* (Lamk.) Taub. MBBT/0251 | Fabaceae | Polash | Flame of the forest | April-Sept | Tree | Latex is applied on ringworm and other skin diseases. |
| *Calotropis procera* (Ait) Ait. | Asclepiadaceae | Akon | — | April-Sept | Shrub | Latex is applied on carbuncles. |

| | | | | | | |
|---|---|---|---|---|---|---|
| f. MBBT/0253 | | | | | | |
| *Carica papaya* L. MBBT/0193 | Caricaceae | Amita | Papaya | April-Oct | Shrub | Fresh latex is applied on ringworm, pimples and sore eyes. |
| *Cassia fistula* L. MBBT/0182 | Caesalpiniaceae | Sonaru | Golden flower, Indian laburnum | April-Sept | Tree | Leaf-paste is applied on septic ulcer for quick healing. |
| *Centalla asiatica* (L.) Urban MBBT/0010 | Apiaceae | Bor-manimuni | Asiatic penn-ywort, Indian pennywort | April-Sept | Herb | Leaf-paste is applied on abscess & carbunc-les for quick healing. |
| *Cinnamomum tamala* (Buch.-Ham.) Nees & Eberm. MBBT/0135 | Lauraceae | Tezpat | Bay leaf | April-Aug | Tree | Fresh leaf-paste is applied on allergy for quick relief. |
| *Citrus auran-tifolia* (Christm. & Panz.) Swingle MBBT/0068 | Rutaceae | Gol-nemu | Common lime | July-Dec | Shrub | Fruit-juice with little coconut oil is applied on scabies. |
| *Citrus limon* (L.) Burm. f. MBBT/0138 | Rutaceae | Kaji-nemu | Lemon | Mar-July | Shrub | Fruit-juice is applied on the body to get relief from prickly heat. |
| *Colocasia fontanesii* Schott MBBT/0234 | Araceae | Kola kochu | Black taro | May-Aug | Herb | Petiole juice is applied to remove warts. |
| *Crotalaria pallida* Ait. MBBT/0042 | Fabaceae | Ghonta karna | ___ | Jun-Dec | Under-shrub | Leaf-paste is applied to cure scabies and ringworm. |
| *Curcuma aromatica* Salisb. | Zingiberaceae | Bon-halodhi, Keturi halodhi | Wild turm-eric, Yellow zedoary | April-Jun | Herb | Fresh rhizome-paste is applied on the ringworm & scabies. |
| *Curcuma longa* L. | Zingiberaceae | Halodhi | Turmeric | April-Jun | Herb | Dry rhizome along with a little lime is applied to remove warts. |
| *Cynodon dactylon* (L.) Pers. MBBT/0184 | Poaceae | Dubori-bon | Dhub grass, Barmuda grass | Jan-Dec | Grass | Leaf-juice is applied on allergy and prickly heat to get quick relief. |
| *Datura stramonium* L. MBBT/0076 | Solanaceae | Dhatura | ___ | Aug-April | Shrub | Leaf-paste is applied on eczema. |

| | | | | | | |
|---|---|---|---|---|---|---|
| *Daucus carota* L. MBBT/0235 | Apiaceae | Gajor | Carrot | | Herb | Root-juice is applied on scabies. |
| *Dendrcnide sinuata* (Bl.) Chew. | Urticaceae | Surat | ___ | Aug-Jan | Shrub | Roots with the rhizome of turmeric are crushed into a paste & applied on septic ulcer for quick healing. |
| *Enhydra fluctuans* Lour. MBBT/0241 | Asteraceae | Helonchi sak | Water cress | Nov-Jan | Herb | Leaves are takèn as vegetable to cure leucoderma. Leaf-juice is applied on prickly heat to get quick relief. |
| *Erythrina stricta* Roxb. MBBT/0237 | Fabaceae | Ronga-modar | ___ | Mar-Jun | Tree | Leaf-paste is applied on abscess. |
| *Euphorbia neriifolia* L. | Euphorbiaceae | Siju | ___ | Dec-May | Shrub | Stem-paste is applied on abscess. |
| *Ficus benghalensis* L. MBBT/0276 | Moraceae | Bor goch, Bot goch | Banyan tree | April-Oct | Tree | Latex is applied on abscess, septic ulcer and cracked heels for quick healing. |
| *Ficus religiosa* L. MBBT/0172 | Moraceae | Ahot goch | Peepul tree | April-Sept | Tree | Bark is crushed and mixed with little milk and applied on scabies. Latex is applied on cracked heels for quick healing. |
| *Flemingia strobilifera* (L.) R. Br. ex Ait. MBBT/0001 | Fabaceae | Makhioti | ___ | Mar-July | Shrub | Pounded roots are applied on ringworm. |
| *Holarrhena pubescens* (Buch.-Ham.) Wall. ex G. Don MBBT/0174 | Apocynaceae | Kutoj | ___ | | Tree | The juice of the bark is taken to cure leprosy. |
| *Ipomoea aquatica* Forsk. MBBT/0045 | Convolvulaceae | Pani-kolmow | Swamp cabbage | Aug-Feb | Trailer | Leaf-juice is applied on prickly heat to get quick relief. |
| *Justicia adhatoda* L. MBBT/0098 | Acanthaceae | Bahok tita | ___ | Oct-April | Shrub | Leaf-paste with little sugar is applied on abscess. |
| *Kalanchoe pinnata* (Lamk.) Pers. MBBT/0247 | Crassulaceae | Dupor tenga, Pategaja | Sprout leaf plant | Nov-Mar | Herb | Leaf-paste is applied on abscess. |

| | | | | | | |
|---|---|---|---|---|---|---|
| *Lawsonia inermis* L. MBBT/0273 | Lythraceae | Jetuka | Henna plant | Mar-Sept | Shrub | Leaf-paste is applied on abscess and septic ulcer for quick healing. |
| *Mangifera indica* L. MBBT/0120 | Anacardiaceae | Am | Mango | Mar-Jun | Tree | Latex is applied on sore eyes. |
| *Mesua ferrea* L. MBBT/0136 | Clusiaceae | Nahor | Iron wood tree | Mar-Jun | Tree | Leaf-paste is applied to cure pimples. |
| *Meyna spinosa* Roxb. ex Link MBBT/0250 | Rubiaceae | Kutkura | ___ | Mar-Nov | Shrub | Ripe fruits are rubbed on cracked heels for quick healing. Seed paste is applied on pimples. |
| *Mimosa pudica* L. MBBT/0153 | Mimosaceae | Lajuki bon | Sensitive plant | Jan-Dec | Herb | Root-paste is applied on septic ulcer for quick healing. |
| *Moringa oleifera* Lamk. MBBT/0070 | Moringaceae | Sajina | Drumstick | Oct-May | Tree | Juice of the bark is applied on abscess. |
| *Neolamarckia cadamba* (Roxb.) Bosser MBBT/0190 | Rubiaceae | Kadam | ___ | Dec-May | Tree | Leaf-paste is applied on abscess. |
| *Ocimum americanum* L. MBBT/0006 | Lamiaceae | Kola tulosi, Kolia tulosi | Hoary basil | Jan-Dec | Undershrub | Crushed leaves are mixed with a pinch of salt and applied on the ringworm. |
| *Ocimum tenuiflorum* L. MBBT/0277 | Lamiaceae | Tulosi | Sacred basil | Jan-Dec | Undershrub | Same as above |
| *Oxalis corniculata* L. MBBT/0119 | Oxalidaceae | Tengeshi | Indian sorrel | Jun-Dec | Creeper | Leaf-juice is applied to cure scabies and root-paste is applied on eczema. |
| *Paederia scandens* (Lour.) Merr. MBBT/0048 | Rubiaceae | Bhedai-lota | ___ | Sept-Dec | Climber | Leaf-juice is applied on allergy. Leaf-juice with garlic is eaten to get relief from allergy. |
| *Phyllanthus emblica* L. MBBT/0161 | Euphorbiaceae | Amlokhi | ___ | Dec-April | Tree | Dried fruits with little amount of molasses are eaten to get relief from allergy. |

| | | | | | | |
|---|---|---|---|---|---|---|
| *Plumbago rosea* L. MBBT/0274 | Plumbaginaceae | Agiachita | Red lead wort | July-Feb | Shrub | Root-paste is applied to cure leucoderma. |
| *Polyalthia longifolia* (Sonn.) Thw. MBBT/0275 | Annonaceae | Debodaru | Mast tree | July-Oct | Tree | Crushed bark is made into a paste and applied on scabies. |
| *Ricinus communis* L. MBBT/0041 | Euphorbiaceae | Era | Castor | Jan-Dec | Shrub | Leaf-paste is used to cure carbuncles. |
| *Senna alata* (L.) Roxb. MBBT/0026 | Caesalpiniaceae | Khorpat | Ringworm bush | April-Oct | Shrub | Leaf-juice is applied on ringworm. |
| *Senna tora* (L.) Roxb. MBBT/0181 | Caesalpiniaceae | Dadigdiga, Bilokhoni | | July-Nov | Undershrub | Leaf-paste is applied on scabies and ringworm. |
| *Solanum myriacanthum* Dunal | Solanaceae | Kotahi bengena | ___ | May-Nov | Undershrub | Fruits are crushed and made into a paste and applied on abscess. |
| *Stephania japonica* (Thunb.) Miers. | Menispermaceae | Tubuki lota | ___ | April-Dec | Climber | Leaf-paste is applied on septic ulcer for quick healing. |
| *Terminalia bellirica* (Gaertn.) Roxb. MBBT/0247 | Combretaceae | Bhumora | Beleric myrobalan | Aug-Dec | Tree | Crushed fruit bark is applied on septic ulcer. The bark of the fruit is removed and the remain- ing part is crus-hed into a paste and applied on sore éyes. |
| *Tinospora glabra* (Burm. f.) Merr. | Menispermaceae | Shaguni lota | ___ | Jan-May | Climber | The juice of the leaves with little honey is taken to cure leprosy. |
| *Typhonium trilobatum* (L.) Schott MBBT/0117 | Araceae | Chomahu, Choma-kachu | ___ | April-May | Climber | Latex is applied on abscess and pimples. |
| *Urena lobata* L. MBBT/0031 | Malvaceae | Bor-sonborial | Aramina, Cadilla | July-Dec | Shrub | Leaf-paste is applied on septic ulcer for quick healing. |
| *Vigna mungo* (L.) Hepper. | Fabaceae | Mati mah | Black gram | Oct-Dec | Herb | Seeds are crushed into a paste and applied on scabies and ringworm. |

| | | | | | | |
|---|---|---|---|---|---|---|
| *Vitex negundo* L. MBBT/0016 | Verbenaceae | Pochotia | Chinese chaste tree | April-Aug | Tree | Leaf-paste is applied on abscess & allergy. Root-juice is prescribed for septic ulcer. |
| *Zanthoxylum oxyphyllum* Edgew. MBBT/0233 | Rutaceae | Mezenga | — | Mar-July | Shrub | Young leaves are eaten as vegetable to cure leucoderma. |
| *Zingiber officinale* Rosc. MBBT/0254 | Zingiberaceae | Ada | Ginger | April-Jun | Herb | The juice of the rhizome with little amount of molasses is eaten to get relief from allergy. |

## CONCLUSION

The information provided in the paper is limited and there is a scope to initiate further ethnobotanical study among the communities to gather information as far as possible. The medicated claims incorporated in the study need to be evaluated through phyto and pharmacochemical investigations to discover their potentiality as drugs. There is an urgent need to explore and document the ethnomedicinal plants used by the tribal & other communities of Assam before such valuable knowledge vanishes.

## ACKNOWLEDGEMENTS

The first author is thankful to the UGC, New Delhi for granting teacher fellowship for doing Ph.D. research under the Faculty Improvement Programme. The authors also acknowledge the help provided by the local people for carrying out the exploration activities.

## REFERENCES

BORA, P.J. & Y. KUMAR 2003. *Floristic Diversity of Assam.* Daya Publishing House, New Delhi.

CHOPRA, R.N., S.L. NAYAR & I.C. CHOPRA 1956. *Glossary of Indian Medicinal Plants.* Council of Scientific and Industrial Research, New Delhi.

DUTTA, A C. 1983. *Some common weeds of Tea States in North-East India.* Tea Research Association, Tocklai Experimental Station, Jorhat, Assam.

JAIN, S.K. 1991. Dictionary of Indian folk medicine & ethnobotany. Deep Publications, New Delhi.

KANJILAL, *et al.* 1940. *Flora of Assam,* Vol. 1-5. Government of Assam Publication.

KIRTIKAR, K.R. & B.D. BASU 1935. *Indian Medicinal Plants,* Vol. I-IV. Lalit Mohan, Allahabad, India.

PURKAYASTHA, J., S.C. NATH & M. ISLAM 2005. Ethnobotany of medicinal plants from Dibru-Saikhowa Biosphere Reserve of North-East India. *Fitoterapia* 76 : 121-127.

WARRIER, P.K., V.P.K. NAMBIAR & C. RAMANKUTTY 1995. *Indian Medicinal Plants – a Compendium of 500 species,* Vol. 1-5. Orient Longman, Madras.

# TRADITIONAL MEDICINAL PLANTS OF COLD DESERT LADAKH – USED IN TREATMENT OF JAUNDICE

BASANT BALLABH, O.P. CHAURASIA & Z. AHMED
*Field Research Laboratory, Defence Research & Development Organisation (DRDO), Leh-Ladakh, Jammu & Kashmir, India*

**ABSTRACT**

The Ladakh region comes under cold deserts of India and it has very thin population about 2-3 person per sq km area. The major Ladakhi population belongs to the Boto (the Budhists) tribal community, however, some minor tribal groups such as Bakarwal, Balti, Dards, Changpa etc are also found. The tribal communities always rely on traditional medical system and medical practitioners of Ladakh are still able to provide cure and relief to more than 60 per cent of population.

This paper deals with 17 medicinal plants of Ladakh belonging to 12 families of angiosperms which are being used in the treatment of jaundice by the local Amchis. Due to modernization and easily available facilities in the Government and private hospitals, the fast disappearing traditional knowledge on medicinal plants has an urgent need of documentation and preservation for future.

## INTRODUCTION

Ladakh is the northern most part of India in the State of Jammu and Kashmir, covering more than 70,000 sq km geographical area of the State, lying between 32° 15'-36° N latitude and 75°15'-80°15' E longitude. Being the highest inhabited plateau in the World, it is known for its inaccessible remoteness and cold climatic conditions. At a first glimpse, the entire region looks barren and vegetationless, but more than 700 plant species have been recorded so far (Chaurasia & Singh, 1996-2001; Chaurasia *et al.*, 2000).

Ladakh is also called the 'cold desert' and known for its harsh climatic conditions, extreme low temperature (-30°C to -70°C at various locations), very low rain and snow fall and rare floral and faunal wealth. About 16 per cent total World landmass is similar to Ladakh which falls in interior of Asia, North America and Indian trans-Himalayan zone. Himalayan regions are well known for rich wealth of medicinal and other valuable plants since ancient times, including the Ladakh Himalayas.

The traditional system of medicine in Ladakh, popularly called as Amchi system of medicine, is principally based on Tibetian system of medicine. The Amchi system is nothing but a skillful use of herbal plants, minerals, salts and animal products in their remedies. Due to harsh climatic conditions and remoteness of the region, the tribal communities still depend on traditional system of medicine. Above 200 Amchis (the herbal practitioners) are still looking after the health care of more than 60 per cent of tribal population in Ladakh.

## METHODS AND MATERIALS

To study the use of medicinal plants among tribal communities of the region an ethno-medico-botanical survey has been carried out at regular intervals since 1998 at various locations and far-flung villages of Ladakh during various seasons of the year.

The ethnobotanical information was gathered from shepherds, aged citizens, village heads, the Amchis & other knowledgeable people. The informations were also cross checked by raising quarries among tribal populations.

The collected specimens were identifie with the help of various floras (Murti, 2001; Kachroo *et al.,* 1977; herbaria of BSD, DD & J & K. A set of voucher herbarium specimens was deposited in the herbaria of Field Research Laboratory for future record.

## RESULTS

The use of medicinal plants among tribal communities is multifarious, but this investigation covers only 17 species of cold desert Ladakh utilized in the treatment of jaundice. The important medicinal plants, viz. *Aquilegia fragrans, Capparis spinosa, Datura stramonium, Euphrasia officinalis* and *Picrorhiza kurrooa* etc are commonly used by local herbal practitioners in the remedies of jaundice. Sometimes, they also involve spiritual and magical practices in their method of remedy. The main plant parts used in preparations are fresh or dried roots, stems, leaves, bark and whole plant. Plant material is used in various ways, but powder or round tablets, extract or decoction of fresh or dried plant parts are common. The common mode of administration is oral in case of internal use of medicine. The determination of dose is also based on the basis of age group, health of patient and body weight.

The jaundice is not a frequent disease in this region, but sometimes seen among children and youngsters. The traditional system of medicine claims the full proof remedy of jaundice, and as an evidence some cases of successful cure were observed during the investigation. The uses of certain medicinal plants belonging to this region have already been recorded in the literature (Anonymous, 1990; Bhattacharyya, 1991; Jain, 1991, 1994; Kachroo *et al.*, 1977; Navchoo & Buth, 1990; Vidyarthi, 1997; Singh, 1995; Uniyal & Issar, 1988).

The medicinal plants used in treatment of jaundice are arranged alphabetically with their botanical name, family, vernacular name (H-Hindi, E-English & L-Local names), collection number and utilizations.

***Aquilegia fragrans*** Benth. (Ranunculaceae)

*Vern.:* Fragrans Columbine (E); (*B.* 017)

The mature seeds are collected in September-October and crushed into powder. The powder is either mixed with other herbs and minerals and made into round tablets or given directly two teaspoonful with boiled water thrice a day for a period of two weeks to recover from jaundice during starting days.

***Artemisia annua*** L. (Asteraceae)

*Vern.:* Artemisia (E); (*B.* 103)

The leaves and flowers contain a drug named *'Artemisinin'*. The leaves and flowers are mixed with other herbs and mineral salts and made into round tablets, then two or three tablets are used with boiled water against remedy of jaundice.

***Berberis lyceum*** Royle. (Berberidaceae)

*Vern.:* Berberry (E), Daruhaldi (H); (*B.* 200)

An extract prepared in water from root, stem and branches called *'Ruset'*, is given twice a day to control jaundice upto 7-8 days.

***Capparis spinosa*** L. var. ***himalayensis*** (Jafri) Jacobs. (Capparaceae)

*Vern.:* Ber (H), Kaper bush (E), Kabra (L); (*B.* 589)

The plant extract is given twice or thrice a day upto 2-3 weeks against treatment of viral hepatitis, jaundice and cirrhosis of liver. An extract of the plant is one of the constituents of the Ayurvedic preparation 'Liv 52'.

***Coriandrum sativum*** L. (Apiaceae)

*Vern.:* Dhaniya (H), Coriander (E), Oshu (L); (*B.* 339)

An extract of leaves and seeds is prepared in water and used thrice a day to treat jaundice upto two weeks.

***Datura stramonium*** L. (Solanaceae)

*Vern.:* Datura (H), Thornapple (E); (*B.* 700)

A decoction prepared in water from leaves is given twice a day in small doses to recover from jaundice.

***Daucus carota*** L. (Apiaceae)

*Vern.:* Gajar (H), Carrot (E), Sarakhturam (L); (*B.* 390)

Carrot juice is popular remedy of jaundice in traditional system of medicine. It is given three-four times a day. Carrots are also eaten frequently during sickness.

***Echinops cornigerus*** DC. (Asteraceae)

*Vern.:* Globe Thistle (E), Eczema (L); (*B.* 435)

An extract of whole plant prepared in fresh water is used to wash yellowness of eyes due to jaundice.

***Euphrasia officinalis*** L. (Scrophulariaceae)

*Vern.:* Eyebright (E), Kangchuk (L); (*B.* 511)

One teaspoonful extract of whole plant is given twice a day to treat jaundice among children. The whole plant in combination with other herbs, minerals and rocksalt is grind into powder and prepared into round tablets. Three tablets are taken thrice a day with boiled water against jaundice for 7-8 days.

***Picrorhiza kurrooa*** Royle ex Benth. (Scrophulariaceae)

*Vern.:* Kutki (H), Hanglang (L); (*B.* 244)

The roots are good source of bitter tonic which is given twice a day in small doses against jaundice for a week period.

***Raphanus sativus*** L. (Brassicaceae)

*Vern.:* Muli (H), Radish (E), Labook (L); (*B.* 343)

The roots are eaten either cooked or raw as a remedy of jaundice. The radish juice is also used in jaundice and liver problems.

***Rubia cordifolia*** L. (Rubiaceae)

*Vern.:* Manjith (H), Indian Madder (E), Btsod (L); (*B.* 369)

The roots are made into a decoction or dry powder. One teaspoonful decoction or powder is recommended twice a day in treatment of jaundice.

***Scorzonera virgata*** DC. (Asteraceae)

*Vern.:* Rtsa-mkhris or Sanmaho or Tharnu (L); (*B.* 426)

The decoction of whole plant is given twice a day in small doses against jaundice upto 5-7 days.

***Sonchus oleraceus*** L. (Asteraceae)

*Vern.:* Milk Thistle (E), Khala (L); (*B.* 468)

The leaves are boiled and added with little salt and given to jaundice patient two-three times a day.

***Spinacea oleracea*** L. (Chenopodiaceae)

*Vern.:* Palak (H), Spinach (E), Vilayati palak (L); (*B.* 474)

The leaves are boiled in water and added with little salt and given to the patient suffering from jaundice. The seeds are also used against jaundice.

***Urtica dioica*** L. (Urticaceae)

*Vern.:* Bichhu-booti (H), Stinging Nettle (E); (*B.* 574)

The whole plant is washed, boiled, again washed then mixed with little salt and given to the patient two-three times a day to recover from jaundice.

***Vitis vinifera*** L. (Vitaceae)

*Vern.:* Angoor (H), Grape (E), Gun (L); (*B.* 565)

The ripe fruits of local grapes are eaten directly against jaundice. The juice of local grapes is also found useful against jaundice.

## DISCUSSION

The medicinal plants of cold desert are considered a symbol of purity and potential quality due to harsh and cold conditions. The jaundice is not very common due to low growth of microorganism in water for a long period because of severe cold conditions prevailing in Ladakh, however, some cases have been found among children and youngsters who are generally cured by the traditional medical practitioners. A total of 17 species belonging to 12 families of angiosperms are found to be used in treatment of jaundice by the Amchis of Ladakh. Out of these 17 species, roots, leaves and whole plant, of 5 species each, seeds of two species and flowers of *Artemisia annua* are used in the treatment.

The Amchis have a general tendancy to transfer the traditional knowledge to their coming generations. The entire medical system depends upon wild plant species and only few species are grown in the gardens near houses. Some species, however, are carried from other parts of the country. The continuous use and over-exploitation for various other purposes of these natural resources have created a serious threat of extinction for many wild species. Certain efforts are being carried out by research institutions and State Government departments for conservation of these valuable resources for future utilization through establishment of medicinal park, herbal garden and *ex-situ* and *in-situ* conservation practices during past few years. Still, there is need of scientific exploitation, collection, cultivation and sustainable utilization. This will not only conserve the valuable natural resources but also create awareness among tribal population for the sustainable utilization of rare and endangered cold desert species.

## ACKNOWLEDGEMENTS

Authors are highly thankful to Defence Research & Development Organisation (DRDO) for providing support and facilities at different stages of this investigation. We are also thankful to the people and Amchis of Ladakh for sharing their valuable knowledge with us for preparing this paper.

## REFERENCES

ANONYMOUS 1990. *Ethnobiology in India : A status Report*. Ministry of Environment & Forests, Govt. of India, New Delhi.

BALLABH, B. 2002. *Ethnobotany of Boto tribe of Ladakh Himalaya*. Unpublished, KUC, Nainital.

BHATTACHARYYA, A. 1991. Ethnobotanical observation in the Ladakh region of Northern J&K State, India. *Econ. Bot.* 45 (3) : 305-308.

CHAURASIA, O. P. & B. SINGH 1996-2001. *Cold desert plants*, (I-V). Field Research Laboratory, Leh, Ladakh.

——————, B.B. BHATT & COL. D. V. SINGH 2000. Ethnobotany of Ladakh Himalaya. In Jina, P.S. (Ed.) *Past Ladakh : Past & Present.* Gyan Publishing House, New Delhi.

JAIN, S.K. 1991. Dictionary of Indian folk-medicine and ethnobotany. Deep Publications, New Delhi.

—————— 1994. *Medicinal plants.* National Book Trust, New Delhi.

KACHROO, P., B.L. SAPRU & U. DHAR 1977. *Flora of Ladakh*. Bishen Singh Mahendra Pal Singh, Dehra Dun.

MURTHI, S.K. 2001. *Flora of cold deserts of western Himalaya,* Vol. I. (Monocotyledons). BSI, Kolkata.

NAVCHOO, I.A. & G.M. BUTH 1990. Ethnobot - any of Ladakh, India : bevarages, narcotics, food. *Econ. Bot.* 44 (3) : 318-321.

SINGH, B., O.P. CHAURASIA & K.L. JADAV 1995. An ethnobotanical study of Indus valley (Ladakh). *J. Econ. Taxon. Bot.* Addl Ser. 12 : 92-101.

SINGH, N.P., D.K. SINGH & B.P. UNIYAL 2002. *Flora of Jammu & Kashmir,* Vol. I. BSI, Kolkata.

UNIYAL, M.P. & R.K. ISSAR 1988. Utility of hitherto unknown herbal drugs traditionally used in Ladakh & possible alternative medicine. *Bull. Medico-Ethnobot. Res.* 9(3-4) : 96-105.

VIDYARTHI, OM PRAKASH 1997. *Wild & cultivated plants of Jammu, Kashmir & Ladakh.* Directorate of Social Forestry Project, Jammu & Kashmir Govt.

# ETHNOBOTANY OF SOME PLANTS OF DANG REGION, RAJASTHAN

LEENA SHARMA, K.P. SHARMA* & SITARAM KHANDELWAL

*Department of Botany, Govt. P.G. College, Karauli (Raj.) India*
**Professor of Botany, University of Rajasthan, Jaipur, India*

## ABSTRACT

The Dang region is distributed in three districts of Rajasthan. This region is inhabited mainly be tribals and traditional communities. Main tribal group is Meena and traditional communities are Gurjar, Mali, Kumhar, Kahaar, Mongia etc. These people are mostly settled and live on hillocks in small villages surrounded by dense forests and thereby, they have close relationship with natural resources. They use these plants for their daily requirements like food, fibre, medicine and others.

About fifty plants are described in this paper which are widely used by them for various purposes.

## INTRODUCTION

Sawai Madhopur, Karauli and Dholpur districts constitute the Dang region of Rajasthan. In Sawai Madhopur, Khandar Tehsil, in Dholpur Bari and Sarmathura Tehsils and in Karauli Mandrayal, Sapotra and Karauli Tehsils comprise the Dang region. The word Dang means undulating and hilly area covered with vegetation. According to Hindi Shabd Sagar the meaning of Dang is Jungle. Physiographically this region comprises of Vindhyan ranges and Chambal basin.

The Dang region has large concentration of tribal and traditional communities. The main tribe is Meena & other traditional communities are Gurjar, Mali, Chamar, Jogi, Kumhar, Kahaar, Mongia etc.

From the traditional point of view, the origin of Meena tribe is related to the Fish God. According to Matsya, Puran Manu is the predecessor of Meenas. The main occupation of Meenas is agriculture. Gurjars are a large community scattered throughout in this region. They are mainly cattle breeders and milk traders. They have their abodes generally near forest areas or in an unoccupied portion of a village. Mali are considered as have originated from a dirt of Lord Shiva. The main occupation of Malis is agriculture. The word Chamar is derived from Sanskrit word charm kar. Their main profession is processing of skin (leather). Besides this, they take away dead animals and peel off skin. They beat drum also.

The villages of above tribal and traditional communities are known as Gam, Pal, Pura or Dhani and they live together in cluster of houses made by them. They dwell in plant made huts and use locally available plants for their daily needs i.e. food, medicine, construction etc.

Vegetation of this region is mixed deciduous type. The main plant species are *Anogeissus pendula, Diospyros melanoxylon, Albizia lebbeck, Butea monosperma, Holoptelia integrifolia, Lannea coromandelica* etc. Land in the east of Aravallis is fertile and characterized by Hadoti plateau, Deccan plateau and Vindhyan scrapland.

## PREVIOUS WORKS

Ethnobotanical work in different parts of Rajasthan during recent years was started in late seventies. Joshi (1987, 1995) worked on Aravallis, Shekhawat (1986) and Sebastian (1985) in desert region, Singh & Pandey (1980) done a little work on medicinal plants of Eastern Rajasthan, Khandelwal (1998) worked on Bhils of Rajasthan. However, none of them have made efforts to work in this region. Ethnobotanically this area is not explored, except some references are available in Singh & Pandey (1998) contribution.

## METHODOLOGY

The uses of plants presented in this paper are a result of number of surveys conducted in different areas of Dang region. Some of the roadside localities were covered by vehicle while to reach in interior areas authors travelled on foot. Various uses of plants in tribal and traditional communities were recorded and discussed with a group and individually. The plants were collected and herbarium sheets prepared. The species are arranged alphabetically by botanical names, followed by family, local names and their uses.

## ENUMERATION

**Abrus precatorius** L. (Fabaceae)

*Local name :* Chirmi, Ratti, Chirmoti.

The leaves are chewed to cure mouth sores and cough. The seeds are used as weight by goldsmiths.

**Acacia catechu** (L.f.) Willd (Mimosaceae)

*Local name :* Kattha, Khair.

The twigs are used as tooth-brush. The bark is chewed orally to cure mouth sores, cough and for reddening of lips. The gum eaten raw to cure joint pain and backache and general weakness in female. The gum is used to made laddoos and eaten as cooling agent.

**Acacia nilotica** (L.) Willd. ex Del. subsp. **indica** (Benth.) Brenan (Mimosaceae)

*Local name :* Babul, Bambul, Desi babul, Bamboor.

The tender twigs are used as tooth-brush. Leaves and stem bark are chewed orally with salt to cure mouth sores. The gum eaten raw as cooling agent. The decoction of stem bark is used by ladies for washing the organ in leucorrhoea and gargled to cure mouth sores and cough. The gum is used to make laddoos and eaten to increase sexual potentiality. Fresh exuding gum is eaten by children with joy. The pods are cooked as vegetable and pickled. Stem bark is used to prepare liquor. The bark kept in an earthen pot with jaggary and kept below the earth surface for 6-7 days. Then pot is taken out and distilled and used as liquor which is locally called *"Desi Sharab."* The wood is used for making agricultural implements (*hal, hor*) huts, household articles, toys, carts, an handle of swords.

**Achyranthes aspera** L. (Amaranthaceae)

*Local name :* Onga, Andhijhara.

The root is used as tooth-brush. Decoction of root is taken to cure cold and cough. The whole plant is crushed into a paste and applied locally against snake-bite, scorpion-sting and mad dog bite. The root is pounded, mixed with honey and applied locally to eyes to cure styes. Seeds powder is mixed with oil and massaged on diseased skin. The leaves are crushed with jaggary in 3 : 1 ratio, made into 5-6 pills and taken orally to cure thorn injury. The root powder is taken against leucorrhoea and typhoid. The root is tied around neck to cure jaundice. Seeds are boiled with milk and sugar to prepare *"Kheer"*, and eaten as tonic. The wood is used in sacred pyre.

**Aegle marmelos** (L.) Corr. (Rutaceae)

*Local name :* Bel, Bel patra, Bil.

The fruit pulp is mixed with curd and taken for curing chronic dysentery, constipation and dyspepsia. The leaves are chewed for relief in diarrhoea. The pulp of fruits is given orally to cattle due to cooling properties in summers. The powder of root is mixed with sugarcandy and given orally against protrusion of uterus. The leaves are offered to Lord Shiva. The stem bark is mixed with buttermilk and cooked as vegetable. The wood is used for making pestles of oil expellers, cart wheels, frame of cots, hut making etc.

**Albizia lebbeck** (L.) Benth. (Mimosaceae)

*Local name :* Saras, Siras, Siris.

The decoction of stem bark is used to gargle to cure weak and spongy gums and pharyngitis. Seeds are crushed into a paste and applied locally on neck against tumour and swellings. The decoction of leaves is taken in night blindness.

**Aloe vera** (L.) Burm. f. (Liliaceae)

*Local name :* Ghee ganwar, Gwar patha.

The pulp is used for making laddoos and eaten daily in winters to cure rheumatism and leucorrhoea. The pulp is smeared locally to relief in burn. Leaves are peeled off and filled with turmeric powder and tied over boils and pimples to cure.

**Anogeissus pendula** Edgew. (Combretaceae)

*Local name :* Dhok, Dhau.

The seeds are crushed and given orally to cattle to increase milk quantity. The strong and durable wood is used for agricultural implements, supporting pillars in huts and for household articles.

**Asparagus racemosus** Willd. (Liliaceae)

*Local name :* Satawari, Safed musali.

The root is powdered and taken with milk as lactagogue. Men use root powder with milk as a tonic against impotency. The root powder is given to buffaloes as galactagogue.

**Azadirachta indica** A. Juss. (Meliaceae)

*Local name* : Neem.

The tender twigs are used as tooth-brush. Leaves juice is dropped in eyes in conjunctivitis and taken orally to purify the blood and by ladies as contraceptive. The fresh leaves are chewed against snake-bite. The stem bark is rubbed on a stone and smeared on boils and pimples. Leaves are crushed into a paste and applied locally against prickly heat. The leaves boiled with leaves of *Ocimum tenuifolium* and mustard oil and applied on scalp daily against baldness. The leaves are used as an insecticide and put with the grains and cloths to protect them from insects. The fruits are crushed into paste and smeared on scalp to kill lice and ticks. The Bhopa, Shaane and Gothiya use the twigs in Jhada rituals. The wood is used for furniture, cart wheels, beam, gate of huts, pole, legs of cots, toys, drums and agricultural implements. In the freshly pierced lobes of ears and nose, the petiole or rachis of leaves is introduced in the hole to prevent the aperture from closing.

**Balanites aegyptiaca** (L.) Del. (Balanitaceae)

*Local name :* Hingota, Hingot.

The tender twigs are used as toothbrush. The fruit is warmed, pulp is mixed in water and gargled to cure spongy gums. The roasted kernel eaten to get rid from cough. Mesocarp and kernel taken orally against stomach disorders. Roasted kernel is rubbed on the stone, mixed with cow's ghee and this paste is applied locally on boils and pimples. The pulp of fruits is used as a detergent and for washing the hairs. The timber is used for making household articles, agricultural implements and in hut construction.

**Boerhavia diffusa** L. (Nyctaginaceae)

*Local name :* Santhi, Santho.

The root is crushed into a paste, mixed with water and sugar and given orally against vomiting and stomachache. The root and leaves are cooked as vegetable.

**Boswellia serrata** Roxb. ex Colebr. (Burseraceae)

*Local name* : Salar.

To cure dysentery the decoction of stem bark, honey and sugar are taken orally. The powder of stem bark is mixed with honey and taken to cure bronchitis. The gum is burnt to make kajal applied in eyes. The aromatic gum and resin are used in sacred pyre.

**Butea monosperma** (Lam.) Taub. (Fabaceae)

*Local name :* Dhak, Palash, Cheela, Chhola.

The leaves are boiled in water and tied locally in arthritis. The seeds are rubbed on a stone in lemon juice and smeared over ringworm. One or two seeds are given orally against intestinal worms. The leaves are used for making of saucers and bowls (*Pattal and Dauna*). The wood is used in marriage ceremony for making *toran*; so tribals and traditional communities do not cut this plant for fire and conserve it. The bast fiber is obtain from the roots.

**Calotropis procera** (Ait) Ait. f. subsp. **hamiltonii** (Wight) Ali (Asclepiadaceae)

*Local name :* Aak, Ankda, Akdo, Ankota.

The latex is dropped in the nostrils to get relief from scorpion-sting pain. The root is crushed into a paste and applied locally on fractured bone. Mature yellow leaves are warmed, squeezed and dropped in earache. Leaves smeared with oil, warmed and tied locally on boils and pimples and over abdomen to get relief from pain. The flowers are eaten with jaggary to cure cough and asthma. The flowers are mixed with jaggary and made into pills, used as appetizer. The leaves are boiled in oil and this oil is used for massage as pain-killer. The root or twig is warmed and used as tooth-brush in molarache. Latex is smeared locally over thorn injury. Grains of pearl millet are soaked in the latex for few days, dried, powdered, mixed with ghee and applied locally on eczema to cure. The ash of leaves is sprayed over agricultural crops to keep away insects. Fumes of leaves are used as mosquito repellent.

**Capparis decidua** (Forssk.) Edgew. (Capparaceae)

*Local name :* Kair, Karil, Ker, Tainty.

The juice of tender twigs is dropped in earache. The juvenile leaves are chewed for relief in toothache and diarrhoea. The leaves are tied locally on boils and swellings. The ash of root is mixed with turmeric, alum and tied over fractured bone. The unripe fruits are pickled and useful in diarrhoea and dysentery. The children are very fond of mature ripe fruits, locally called *Penchu*. A wooden peg is inserted in the stem of sterile plant of papaya to make fertile. This practice is done on Saturday.

**Cassia fistula** L. (Caesalpiniaceae)

*Local name :* Amaltas, Burr.

The decoction of stem bark is used to gargle to get relief in swelling of throat. The leaves juice is given orally to polio patient and massaged over affected organ. The fruit pulp is eaten to cure constipation and abdomenache. The dry pods are used as toy by children. The pod is kept under the pillow to keep away nightmares and kept in cot to keep away snakes.

**Chlorophytum tuberosum** (Roxb.) Baker (Liliaceae)

*Local name :* Dhauli musli, Safed musli.

The flowers are cooked as vegetable which is restorative. Roots are powdered and taken orally as tonic.

**Commiphora wightii** (Arn.) Bhandari (Burseraceae)

*Local name :* Gugal, Googal.

The leaves are cooked as vegetable and eaten to cure rheumatism. The stem yields a gum – resin which is used in fumigation, perfumery and in sacred pyre. The tribals collect stem bark and gum and sale in the market.

**Cordia dichotoma** Forst. f. (Ehretiaceae)

*Local name :* Lisoda, Lehsua.

The leaves are warmed and tied to get relief from throbbing pain. The ash of leaves is mixed with oil and applied as ointment on burn. The decoction of bark is taken orally to cure inflammation in urine. The unripe fruits are cooked as vegetable and pickled.

**Cuscuta reflexa** Roxb. (Cuscutaceae)

*Local name* : Amarbel.

The stem is boiled in water and tied slight warm locally to get rid from lumbago, sprain and throbbing pain. The stem is boiled in water and this water is used to take bath twice a day to cure paralysis. Plant juice is boiled with oil and massaged on scalp against baldness.

**Dalbergia sissoo** Roxb. (Fabaceae)

*Local name :* Shisham, Sisso.

The leaves are boiled in water and tied on foot to cure sprain and swellings. The oil of wood is used for healing of wounds and scabies. Decoction of leaves is taken to cure sexual diseases. 4-5 leaves are crushed into a paste and taken in the morning against menstruation cramps. Stem bark is chewed for reddening lips. The wood is used in making of plough, frame of cot and other agricultural implements.

**Datura innoxia** Mill. (Solanaceae)

*Local name :* Dhatura, Dhaturo.

The leaves are smeared with ghee, warmed and tied locally on boils and pimples. Seeds are given orally as anesthesia. Leaves are cooked with rice flour and taken orally to cure rheumatic pain and swellings. Dried leaves are powdered and smoked in chilam to avoid respiratory diseases. The root is boiled in milk and given orally with butter against insanity. Fresh root is crushed into a paste and applied locally against scorpion sting. The fruits are offered to Lord Shiva.

**Dendrocalamus strictus** (Roxb.) Nees (Poaceae)

The decoction of leaves is taken to get rid from cough (with blood). The leaves are given to cattle against retention of placenta. The wood is used in making of huts and agricultural implements. The stripes of culms are used to made baskets, mats etc.

**Euphorbia neriifolia** L. (Euphorbiaceae)

*Local name :* Dandathor, Thuar.

Latex is applied with cotton on the loose and aching tooth for its speedy removal. Latex is also used as styptic.

**Ficus benghalensis** L. (Moraceae)

*Local name :* Bad, Bargad, Bar.

The latex is smeared externally to cure pain of rheumatism and lumbago. 2-3 drops of latex are dropped in *batasha* and taken in epistaxis and to avoid nocturnal emission. Decoction of stem bark is taken against increase frequency of micturition. The leaves are warmed and tied over boils. Latex is smeared over ringworm and foot-sores. The leaves are used for making bowls and saucers which are used for taking food.

**Ficus racemosa** L. (Moraceae)

*Local name :* Gular, Goolar.

The latex is dropped on cuts and wounds to check bleeding and applied locally on cracked skin, foot-sores and in toothache. 2-3 drops of latex are taken orally with sugar (*batasha*) in diarrhoea. Stem bark is pounded and mixed in whey and taken orally in diarrhoea. 1-2 drops of latex is mixed with mother's milk and given orally to infants in dysentery and vomiting. Extract of fresh root and fruits is given orally to diabetic patient. Boiled fruits are tied over abdomen against any disorder. The fruits are given with fodder to cattle against retention of placenta.

**Ficus religiosa** L. (Moraceae)

*Local name :* Pipal, Peepad, Peepli.

Fruits, root and stem bark are boiled in milk, mixed with honey and sugar and taken orally to progenerate. Stem bark is burnt, dipped in water and this water is taken orally to avoid vomiting. The decoction of stem bark is given orally to cattle against exposure to disease.

**Gloriosa superba** L. (Liliaceae)

*Local name :* Lolokadi, Ladokari.

The paste of root is applied locally against scorpion sting. Leaves are crushed into a paste mixed with milk and given orally against pneumonia. It is believed that when the plant is brought to the home, it creates dispute in the family.

**Grewia hirsuta** Vahl (Tiliaceae)

*Local name :* Gangru, Gangren.

The root powder is taken orally against impotency in men. The root powder is made into a poultice and tied over boils for early healing. Root powder is taken with whey to relief in throbbing pain.

**Justicia adhotoda** L. (Acanthaceae)

*Local name :* Adusa, Safed basoda, Adsuta.

The decoction of leaves with seeds of *Piper nigrum* is taken for cough, cold, bronchitis, asthma and pneumonia. The leaves are smeared with ghee, warmed and tied over fractured bone. Flower juice is dropped in nostrils in epistaxis. Plant twigs are used to make brooms.

**Leptadenia pyrotechnica** (Retz.) Wt. & Arn. (Asclepiadaceae)

*Local name :* Kheep, Kheemp.

Extract of whole plant is warmed and massaged to cure dropsy. Tender twig is crushed, boiled and tied on thorn injury. The leaves are crushed into a paste and applied on thorn injury. The pods are cooked as vegetable and eaten to cure rheumatism and dropsy. Tender twigs are used for making basket, and fibre obtain from stem is used in hut construction.

**Madhuca longifolia** (Koen.) Macbr. (Sapotaceae)

*Local name :* Mahuwa, Mahu.

Stem bark is rubbed on the stone and smeared over swellings and scabies. Warmed up leaves are tied locally on breast against tumor. The leaves and fruits are tied on teats against exposure to disease. The oil of seeds is edible.

**Mangifera indica** L. (Anacardiaceae)

*Local name* : Aam, Keri.

Powder of stem bark is taken orally to cure epistaxis and piles. Kernels are roasted, powdered and taken in cough problem. The unripe fruits are roasted, pulp mixed with water, salt and sugar to make a drink which has cooling effect during summers and locally known as "*Keri Ki Chhachh*". The sap of plucked raw fruit is applied on scorpion sting.

**Mimusops elengi** L. (Sapotaceae)

*Local name :* Molsiri, Monsadi.

The decoction of stem bark is gargled to cure spongy gums, pyorrhoea and ulcerated throat and taken orally in fever. Twigs are used as tooth-brush.

**Moringa oleifera** Lam. (Moringaceae)

*Local name :* Sehajan, Sainjna.

The fresh root and stem bark are crushed and made into a paste which is applied on joints for relief in rheumatic pain, sprain, swellings and wounds. The leaves are boiled in water and tied locally to relief in pain from sprain, rheumatism, arthritis and abodomenache. The oil of seeds is massaged over joints in arthritis and rheumatic diseases. Leaves, flowers and pods are cooked as vegetable and eaten.

**Morus australis** Poir (Moraceae)

*Local name :* Shehtut, Atut.

The extract and decoction of stem bark is taken against fever and intestinal worms. The mature fruits are eaten raw. Tender twigs are used for making basket.

**Nerium oleander** L. (Apocynaceae)

*Local name :* Kaner, Kander.

Leaves juice is applied locally in skin diseases. The flowers are offered to deity.

**Ocimum basilicum** L. (Lamiaccac)

*Local name :* Marua, Marva.

Leaves juice is mixed with honey and taken to cure cough and cold. Plant juice is taken against snake bite after every 4 hrs. Leaves are crushed into a paste and smeared against scorpion sting. The twigs are pierced in cots to keep away mosquitoes.

**Ricinus communis** L. (Euphorbiaceae)

*Local name :* Arand, Andi, Andora.

The leaves are smeared with ghee or oil, warmed and tied locally on boils and to cure throbbing pain, sprain and swellings. 3-4 drops of seed oil is mixed with milk and taken as a purgative. To get light, a wire pierced into fruits like a chain and lighted.

**Salvadora oleoides** Decne. (Salvadoraceae)

*Local name* : Pilu.

The tender twig and root are used as tooth-brush. Seeds oil is used for massage to get relief from rheumatic pain. The leaves are chewed orally to cure cough. Powder of stem mixed with curd is taken orally to avoid nocturnal emission. Seed powder and leaves are used as insecticide and put with the grains to protect them from insects. Twigs are used to make basket, broom and stake.

**Senna tora** (L.) Roxb. (Caesalpiniaceae)

*Local name :* Pamar, Pamad.

The leaves and seeds are crushed into a paste and applied locally on boils and pimples. Seeds are crushed into a paste, mixed in water and given 1-2 spoons daily to infants to cure bronchitis. Fresh young leaves are eaten to cure throat sores and to increase eyesight. The leaves are cooked as vegetable and eaten daily against night blindness. Mature stem is used to make baskets, and dried whole plant is used as broom.

**Syzygium cumini** (L.) Skeels (Myrtaceae)

*Local name :* Jamun, Jaman.

Stem bark is boiled in water and taken orally to cure dysentery and abdomenache. Fresh stem bark extract is mixed with goat's milk and given to infants to cure diarrhoea. Leaves juice is taken against excess bleeding during menses.

**Tecomella undulata** (Sm.) Seem. (Bignoniaceae)

*Local name :* Roila, Rohida.

Stem is used for making pillars in hut construction. Wood is used for making cots and in construction of agricultural implements. Flowers are offered to the deity.

**Tephrosia purpurea** (L.) Pers. (Fabaceae)

*Local name* : Dhamasa.

The paste of leaves made with the leaves of *Cannabis sativa* (Bhang) is applied on bleeding and painful piles. Dried whole plant is used as a broom.

**Terminalia bellirica** (Gaertn) Roxb. (Combretaceae)

*Local name :* Baheda, Bahera, Guter.

The powder of epicarp of fruits is taken orally to cure cough and abdomenache. The fruits are eaten against enlarged spleen. Kernels are eaten by children.

**Trianthema portulacastrum** L. (Aizoaceae)

*Local name* : Vishkapra.

Leaves and stem are cooked as vegetable and eaten to cure pain and swellings of joints and indigestion. Whole plant is uprooted, cooked and eaten in cystitis. Plant juice is taken orally to cure jaundice.

**Tridax procumbens** L. (Asteraceae)

*Local name :* Kalo bhanvaro, Rukhadi, Peele Phoolki rukhadi.

The plant juice is dropped on cuts and wounds to check bleeding, and taken orally to cure dysentery and diarrhoea. Leaves are crushed into a paste and applied locally on boils, ringworm and scabies. Paste of leaves is mixed in water or whey candy and taken due to cooling properties in summers. Leaves juice is dropped over burnt skin due to cooling effect and applied locally on mouth-sores. Plant juice mixed with sugar is taken orally empty stomach against excessive bleeding during menses.

**Urginea indica** (Roxb.) Kunth (Liliaceae)

*Local name :* Koli Kanda, Kori Kanda.

The bulb is roasted in ash, crushed into a paste and applied on the boils. The paste of bulb is used for tighting the kite flying thread.

**Ziziphus nummularia** (Burm. f.) Wt. & Arn. (Rhamnaceae)

*Local name :* Jharber, Beri, Jhara, Jhad.

The leaves are crushed into a paste and applied against scorpion sting and skin diseases. The root is boiled in water and this water is given to woman after delivery and used to take bath due to antibiotic and antiseptic properties. Powdered gum is given orally to children to cure diarrhoea. Fresh root bark is crushed into a paste and used as ointment on boils and pimples. The root is boiled in water and this water is taken as lactagogue. Root bark is dried, grounded finely, mixed with wheat flour, made into laddoos and given after delivery to woman. The leaves juice is smeared on scalp daily to prevent falling of hairs. Ripe fruits are eaten raw. The leaves serve a good fodder for goats and sheeps, locally called *'Palo'*.

## ACKNOWLEDGEMENTS

The first author is thankful to Director, College Education, Govt. of Rajasthan, Jaipur for permission of research.

## REFERENCES

JOSHI, P. 1987. *Ethnobotany of Aravallis.* Ph.D. Thesis, University of Rajasthan, Jaipur.

——— 1995. Ethnobotany of the primitive tribes of Rajasthan. Printwell, Jaipur.

KHANDELWAL, S.R. 1998. *Ethnobotany of the Bhil tribe in Rajasthan.* Ph.D. Thesis, University of Rajasthan, Jaipur.

SEBASTIAN, M.K. 1985. *Ethnobotanical survey in some forest areas of Rajasthan.* Ph.D. Thesis, Jodhpur University, Jodhpur, India.

SHEKHAWAT, G.S. 1986. *Ethnobotanical survey of desert areas of Rajasthan.* Ph.D. Thesis, University of Jodhpur, Jodhpur, India.

SINGH, V. & R.P. PANDEY 1980. Medicinal Plant lore of the tribals of eastern Rajasthan, India. *J. Econ. Taxon. Bot.* 1 : 137-147.

——— & ——— 1998. *Ethnobotany of Rajasthan.* Scientific Publishers, Jodhpur, India.

# TRADITIONAL INVENTORY OF SOME POTENTIAL ECONOMIC PLANTS RELATED TO ANGAMI TRIBE IN NAGALAND

KUNTALA N. BARUA, INDRANI P. BORA & JASBIR SINGH

*Rain Forest Research Institute, Jorhat, Assam, India*

## ABSTRACT

Indigenous knowledge is the total achievements traditionally attained by the local people in its place. The folk culture and knowledge is still vital in Nagaland and the Angami tribe expressed itself and the sign of culture through it. The present study deals with the utility of plant resources practiced by Angami tribe and a total of 75 species belonging to 59 genera and 41 families are enumerated. These include 31 species of edible food, 19 species of natural dyes, 11 species of fodder, 6 species of timber, 4 species as fish poison, 2 species as fiber and 2 species as gum and oil yielding that grow in the forests or found as weed in the cultivated or open fields.

## INTRODUCTION

Tribals constitute a notable segment of Indian population representing 8% of the total population. Since time immemorial different tribal ethnic groups have practiced the art of living through forest ecosystem and screened in his own way the wild growing plants for their edible, medicinal and other domestic purposes. Arora & Pandey (1996) reported that about 1000 species are consumed by tribal inhabitants as food in India. Approximately 80% of the forest dwellers of Bihar, Orissa, Madhya Pradesh, West Bengal and Himachal Pradesh depend for 25-50% of annual food requirements on forests (Sinha & Lakra, 2005).

The tribal people harbour rich and diverse knowledge about the surrounding biodiversity and its importance. They are not only familiar with the plant species in their ecosystem but also understand ecological interaction of the various components of their resources. Therefore, cataloging and documentation of the indigenous knowledge is essential in a region (Maheshwari, 1994; Sahai, 2003). The indigenous societies of Nagaland discovered various uses of natural resources around their habitat based on close association with nature over ages; necessities, trial and long experiences are passed down through generations. Fourteen major tribes inhabit in Nagaland with their own folk cultural patterns, traditions and typical food practices. In the present study various indigenous plant resources used by ethnic group 'Angami' for their dietary pattern, dyeing agent and other miscellaneous uses have been documented. Use of some plants are somehow endemic to this region, particularly to the aboriginals.

Fig. 1. Sketch map of Nagaland.

The Angami is the oldest aboriginal tribe which subscribes to the *kheza-kenoma* legend belonging to Mongoloid race. It is spread over in Kohima and Dimapur districts of Nagaland. According to 1991 census, 97,433 persons in Nagaland are declared Angami as their mother tongue. Angami language, in which a vast printed literature has been produced, is based on the *Tengima* dialect written in the roman script language. The main source of livelihood of the Angami tribe is agriculture. Traditionally they have practiced slash and burn (Jhum) cultivation and additionally developed a sophisticated system of wet terrace rice cultivation. Other then agriculture, trade and business have always attracted individuals of Angami. The Angami trade in the past was mainly based on barter and exchange system.

Nagaland is a narrow stripe of mountainous territory and the land of enchanting beauty and endowed with rich natural resources. The area is lying between 25°60' and 27°40' N latitude and between longitudinal lines 93°20' and 95°15' E (Fig. 1). The climate of Nagaland ranges from subtropical to subtemperate with an annual rainfall of 2500 mm. The temperature varies from 0° to about 35°C, but occasionally extreme weather conditions also prevail in the areas. Relatively dry winter and wet summer are the major features of this region. Soil is derived from tertiary rocks belonging to Barail and Disang series made up of alluvial soils. pH value ranges between 4.8 to 6.5.

The native home of Angami tribe is bounded by Assam in west and north, by Manipur in the south, Wokha and Phek district in North east. It has very severe topography with full of hill ranges. The vegetation represents the characteristic features of Subtropical Moist Deciduous and Temperate Evergreen Highland type.

## MATERIALS AND METHODS

The study was conducted in Dimapur and Kohima districts of Nagaland. Data were collected during the period of 2003 to 2005 through a combination of technique of questionnaires, interview and informal discussion with tribal head and other local informer. Repeated enquires were made to authenticate the information. The collected information was crosschecked critically, analyzed and documented. Questionnaires may differ from place to place and new questions may be added according to the ability to the surveyor. Field trips were organized to the different tribal pockets, markets of tribal villages and forest area at regular intervals. Detail information regarding local name, purpose of use were recorded in field book. The collected specimens were processed and identified with the help of regional floras and voucher specimens were deposited in RFRI, Jorhat. The plants are arranged alphabetically by their botanical name, followed by family.

## RESULT AND DISCUSSION

Forest resources play a significant role in boosting up the economy of Angami tribe. Many indigenous plants which grow in Nagaland have got multiple uses. A year round survey among the Angami in Kohima and Dimapur district has revealed 33 species of edible food, 18 species of natural dyes, 10 species of fodder, 6 species of timber and 8 were recorded for miscellaneous uses.

### Wild edible plants

Wild food plants play a significant role in income and survival strategies besides nutrition for Angami dominated areas. These plants are the major components of biodiversity but many species are vanishing now-a-days due to deforestation. The present study recovered 9 numbers of leafy vegetable, 16 numbers of edible fruits and 5 numbers of tuber / roots / seed which form part of tribals diet. The plants either grow in the forests or found as weed in the cultivated or open fields. Majority of the species are commonly used, but some are consumed during scarcity of conventional food. In due course of time, the plants may get disappear due to heavy consumption. Commercial cultivation is, therefore, needed to have a renewable source of non conventional food to their dietary pattern.

Wild fruits make important contribution to the nutrition and subsistence of food of the rural people. Some fruits are eaten raw either ripe or unripe while others are taken after cooking as curries. Some of the fruits are prickled and some are made into chutney. During the crucial time the tribal people collect tubers to supplement their meager food available at home. The tubers and roots mostly belong to Araceae and Dioscoreaceae families (Table 1).

Table 1. Wild edible plant used by Angami tribe.

| **Scientific name** | **Common name (Angami name)** | **Family** | **Use pattern** |
|---|---|---|---|
| *Amaranthus spinosus* Linn. | - | Amaranthaceae | Leaves are consumed as vegetables. Whole plant is boiled and eaten. |
| *Amorphophallus* spp. | Taro | Araceae | Tuberous roots are used as food. |
| *Ananas comosus (*Linn.) Merr. | Castard apple (Alipiang) | Bromeliaceae | Cultivated on hill slopes for edible fruits. |
| *Artocarpus chaplasha* Roxb. | Wild Jackfruit (Chierie) | Moraceae | Fruit is eaten raw. |
| *Bauhinia purpurea* Linn. | - | Caesalpiniaceae | Leaves are consumed as vegetables. |
| *Calamus tenuis* Roxb. | Ratten | Arecaceae | Young stem is boiled and eaten. |
| *Chenopodium album* Linn. | - | Chenopodiaceae | Leaves are consumed as vegetables. |

| | | | |
|---|---|---|---|
| *Coix lacryma-jobi* Linn. | Menjang | Poaceae | Grains are used as food. |
| *Colocasia* spp. | Taro | Araceae | Tuberous root is used as food. |
| *Dendrocnide sinuata* (Bl.) Chew. | - | Urticaceae | Flowers are used in curries. |
| *Dioscorea bulbifera* Linn. | Yam | Dioscoreaceae | Tuberous root is used as food. |
| *Dioscorea pentaphylla* Linn. | Yam | Dioscoreaceae | Tuberous root is used as food. |
| *Diospyros melanoxylon* Roxb. | Kendu | Ebenaceae | Fruit is eaten raw. |
| *Ficus fulva* Bl. | Fig (Chiede) | Moraceae | Fruit is eaten raw and used in preparation of jam. Young leaves are eaten as vegetables. |
| *Ficus semicordata* Buch.-Ham. ex Sm.. | Fig (Thotsenno) | Moraceae | Fruit is edible and used in preparation of jam. |
| *Ipomoea aquatica* Forssk. | - | Convolvulaceae | Leaves are consumed as vegetables. |
| *Juglans regia* Linn. | Walnut (Pfhu) | Juglandaceae | Seeds are used for edible purpose. |
| *Litsaea cubeba* (Lour.) Pers. | Mejankori (Cie ) | Lauraceae | Fruits are eaten raw or mixed in pickle after roasting. |
| *Malus* spp. | Crap apple (Cipho) | Rosaceae | Fruits are used in preparation of squash, dry fruit is eaten. |
| *Manihot esculenta* Crantz. | Alicha | Euphorbiaceae | Tuber and young leaves are eaten as a vegetable. Tubers are also used in preparation of local wine. |
| *Passiflora edulis* Sims. | Passion fruit | Passifloraceae | It is cultivated for edible fruits. The juice extract is used for squash making. |
| *Plantago major* Linn. | - | Plantaginaceae | Young leaves are boiled and eaten. |
| *Rhus semialata* Murr. | Naga tenga (Zombou) | Anacardiaceae | Fruits are edible and used for preparation of prickle and cold drink. |
| *Rubus ellipticus* Sm. | Raspberry (Ruomva) | Rosaceae | Fruits edible and has a good flavour and taste. |
| *Senna tora* (Linn.) Roxb. | - | Caesalpiniaceae | Leaves are consumed as vegetables. |
| *Solanum anguivi* Lamk. | - | Solanaceae | Fried fruits are eaten. |
| *Spilanthes paniculata* Wall. ex DC. | - | Asteraceae | Young boiled leaves are eaten. |
| *Tamarindus indica* Linn. | Tamarin tree | Fabaceae | Fruits are used to prepáre jam. |
| *Trema orientalis* (L.) Bl. | Neetle tree (Thedie) | Urticaceae | Fruits are edible. |
| *Zanthoxylum acanthopodium* DC. | Changpet | Rutaceae | Plant is used as spice. Young leaves, shoots and seeds are eaten as vegetables. |
| *Ziziphus mauritiana* Lamk. | Palm | Rhamnaceae | Fruits are eaten raw and used for preparation of prickle. |

Table 2. Dye-yielding plants used by Angami tribe.

| Scientific name | Family | Use pattern |
|---|---|---|
| *Achyranthes aspera* Linn. | Amarant-haceae | Whole plant is dried, burnt to ash and is mixed with water. This dye is used as adhesive for various colours for brightening. |
| *Bauhinia purpurea* Linn. | Caesalpinia-ceae | A rose purple dye is obtained from the fresh flowers for dying and painting. |
| *Bixa orellana* Linn. | Bixaceae | The plant imparts red colour to the cloth, yarn and painting. Fruits are plucked and the seeds are soaked in the water and rubbed to obtain dye. Due to non toxicity it can be used for food stuff. |
| *Buddleja macrostachya* Benth. | Buddlejaceae | Leaves give yellow colour dye for dying cloths. |
| *Butea monosperma* (Lam.) Taub. | Fabaceae | Flowers give yellow dye to cotton clothes. |
| *Curcuma longa* Linn. | Zingiberaceae | The fresh rhizomes are crushed, allowed to soak in water to obtain yellowish colour. Cloths and threads are soaked overnight, add lime water or an alkali, a brick red dye is obtained. It is used in foods as edible dye. |
| *Cuscuta reflexa* Roxb. | Cuscutaceae | A yellow dye is prepared from the whole plant by crushing in cold water. It is best for dyeing cotton cloths and in paintings. |
| *Diospyros melanoxylon* Roxb. | Ebenaceae | A reddish dye is obtained from the fruits and used for dyeing fishing nets. |
| *Erythrina stricta* Roxb. | Fabaceae | The flowers are dried, powdered and added sufficient cool water to obtain coral red dye for dyeing paintings. |
| *Mallotus philippensis* (Lam.) Muell.-Arg. | Euphorbi-aceae | A crimson red dye is obtained from dried fruits by soaking in cool water. It is better for dyeing silk fabrics. The edible bright colour can also be used for dyeing food stuff. |
| *Michelia champaca* Linn. | Magnoliaceae | The flowers also yield a yellow dye for textiles. |
| *Persicaria chinensis* (Linn.) H. Gross. | Polygonaceae | Whole plant is crushed and soaked in water for 3 days to obtain dye. It is also used as an adhesive for brown and black colours. |
| *Phyllanthus emblica* Linn. | Euphorbiaceae | Crushed fruits and bark are soaked for 4 to 5 days in water. A reddish black dye is obtained. It is used as adhesive for dark colour dyes and for dyeing fishing nets. |
| *Punica granatum* Linn. | Punicaceae | After removal of seeds, pieces of fruit are soaked in cold water and the extract is used as an adhesive for black/brown colours. |
| *Solanum stramon-ifolium* N. Jacq. | Solanaceae | Fruits are crushed and soaked in water. Extract is concentrated by boiling and added alkalies. The dye is used for printing of design and dyeing cotton and silk fabrics. |
| *Strobilanthes decurrens* T. Anders. | Acanthaceae | Boiling of dry leaves gives blue, and black colour is obtained by boiling fresh leaves in water. |
| *Tectona grandis* Linn. | Verbenaceae | Leaves and bark are cut into pieces and soaked in water for 2-3 days. Yarn thread is dipped into dye for giving reddish colour. |

| | | |
|---|---|---|
| *Terminalia citrina* Roxb. | Combretaceae | A blackish dye is obtained from the bark and fruits by soaking or boiling in water. It is used for dyeing and painting. |
| *Wrightia tomentosa* Roem. & Schultes | Apocynaceae | Bark yields milky latex from which a yellow dye is obtained which is used for cotton fabrics. |

Table 3. Plants in miscellaneous uses by Angami tribe.

| Scientific name | Common name (Angami name) | Family | Use pattern |
|---|---|---|---|
| ***Timber*** | | | |
| *Acrocarpus fraxinifolius* Wight & Arn. | Mundani (Cietha) | Mimosaceae | Wood is used for timber, veneers and plywood. It is also suitable for packing cases of heavy machinery. |
| *Albizia lebbeck* (Linn.) Benth. | Black siris (Zoë) | Mimosaceae | Wood is used for high quality furniture, internal decoration, paneling, flooring, construction and carving. It is also used as joint post, placed in the soil as it withstands moisture and termite attack for long periods. |
| *Michelia champaca* Linn. | Titachap (Seire) | Magnoliaceae | Wood is used for posts, boards, veneers, furniture, decorative fittings and carriages. Flowers are used in the preparation of perfumes and hair oils. |
| *Morus macroura* Miq. | Bhola (Mau) | Moraceae | Wood is used for furniture, sports equipments, agricultural implements, ornamental paneling. |
| *Schima wallichii* (DC.) Korth. | Needle tree (Mecho) | Ternstroemiaceae | Wood is used for construction of buildings, planks, canoes, agricultural implements and poles. It is a good source of fuel. The bark is also mixed in feed and used to expel tapeworms in livestock. |
| *Wrightia arborea* (Denst.) Mabb. | Dudhi (Gasiase) | Apocynaceae | Wood is used in making cups, plates, combs and frames, and in carving. |
| ***Fiber*** | | | |
| *Sterculia villosa* Roxb. | Udal (Ketsho) | Sterculiaceae | Bark yields coarse, strong fiber used for cordage and rough bags. Fiber is specially valued for breast bands of elephants for dragging timber and for tying cattle. Wood is used for tea boxes and light packing cases. |
| *Trema orientalis* (Linn.) Bl. | Neetle tree (Thedie) | Urticaceae | The plant extract is used for tannins. Bark also yields fiber and used for ropes and coarse cloths. |
| ***Gum and oil*** | | | |
| *Anthogonium gracile* Wall. ex Lindl. | (Akongtong) | Orchidaceae | The pseudobulb of the plant is used as an adhesive gum. |
| *Litsea cubeba* (Lour.) Pers. | (Cie) | Lauraceae | Oil from the flowers and fruits is used in perfumes. It is also used as insect repellent. Seeds yield fat used as an illuminant. |

| | | | |
|---|---|---|---|
| ***Fodder and pasture grasses*** | | | |
| *Albizzia chinensis* (Osbeck) Merr. | - | Mimosaceae | Leaves used as fodder. |
| *Albizzia lebbeck* (Linn.) Benth. | - | Mimosaceae | Leaves used as fodder. |
| *Bauhinia variegata* Linn. | (Togo) | Caesalpiniaceae | Leaves used as fodder. |
| *Coix lacryma-jobi* Linn. | Menjang- | Poaceae | Plants used as fodder and pasture grasses. |
| *Cordia dichotoma* Forst. | - | Ehretiaceae | Leaves and twigs are used as fodder. |
| *Cynodon dactylon* (Linn.) Pers. | - | Poaceae | Plants used as fodder. |
| *Imperata cylindrica* (L.) Raeusch. | - | Poaceae | Plants used as fodder. |
| *Morus mucroura* Miq. | Bhola (Mau) | Moraceae | Young branches are used as fodder. |
| *Panicum repens* Linn. | - | Poaceae | Plants used as fodder. |
| *Trema orientalis* (Linn.) Bl. | Neetle tree (Thedie) | Urticaceae | Leaves are used as fodder. |
| *Saccharum spontaneum* Linn. | | Poaceae | Whole plant is used as fodder and pasture grasses. |
| ***Fish poison*** | | | |
| *Acacia pennata* (Linn.) Willd. | - | Mimosaceae | Leaves are used in fish poison. |
| *Aphanamixis poly-stachya* (Wall.) Parker | Akawa | Meliaceae | Ground leaves are applied for fish poison. |
| *Pueraria hirsuta* Schneider | Anetli | Fabaceae | The plants are crushed and used to kill fish. |
| *Zanthoxylum acanthopodium* DC. | Changpet | Rutaceae | Fruits are used in fish poison. |

## Dye yielding plants

Since time immemorial, the Angami ethnic group has practiced the art of dyeing cloths and other items. The natural dyes are obtained from plants for imparting different shades of colour to clothes, baskets, fishing nets, other decorative items and edible purposes also (Table 2). But, now a days the indigenous system of dyeing has been slowly replaced by the introduction of synthetic dyes. Large scale use of synthetic dyes results in some unfair effects on human life. Therefore, screening of biodegradable vegetable dyes is essential (Ghosh, 2003; Kumar & Rao, 2003; Akimpou *et al.,* 2005). Many plants have been identified as potentially rich in natural dye contents (Table 2).

From the above account it can be observed that the people still extensively use these natural resources for edible, dyeing, fodder and other miscellaneous purposes and have rich economic potentiality. This deserve much attention and scientific innovation is imperative for protection of plants associated with indigenous knowledge, as many of the plants are lesser known.

## ACKNOWLEDGEMENTS

Authors are thankful to the Director, RFRI, Jorhat for facilities and encouragement. They are also grateful to the G.B. Pant Institute of Himalayan Environment & Development, Koshi-Katarmal, Almora, Uttaranchal for the financial support to carry out the study.

## REFERENCES

AKIMPOU, G., K. RONGMEI & P.S. YADAV 2005. Traditional dye yielding plants of Manipur, North East India. *Ind. J. Trad. Knowledge* 4 (1) : 33-38.

ARORA, R.K. & ANGULA PANDEY 1996. *Wild edible plants of India.* National Bureau of Plant Genetic Resoureces, New Delhi.

GHOSH, A. 2003. Traditional vegetable dyes from central West Bengal. *J. Econ. Taxon. Bot.* 27 (4) : 825-826.

KUMAR, S.V. & P.S.N. RAO 2003. Dye yielding plants of Andaman and Nicobar Island. *J. Econ. Taxon. Bot.* 27 (4) : 827-838.

MAHESHWARI, J.K. 1994. *Interdisciplinary approaches in ethnobotany.* In Jain, S.K. (Ed.) *A manual of ethnobotany,* pp. 19-27. Scientific Publishers, Jodhpur.

SAHAI, S. 2003. Importance of indigenous knowledge. *Ind. J. Trad. Knowledge* 2 (1) : 11-14.

SINHA, R. & V. LAKRA 2005. Wild tribal food plants of Orissa. *Ind. J. Trad. Knowledge* 4 (3) : 246-252.

# STUDIES ON MEDICINAL PLANTS OF BURDWAN - III

GITA HOTWANI & AMBARISH MUKHERJEE

*Department of Botany, Burdwan University, Golapbag, Burdwan-713 104, West Bengal (India)*
*E-mail : ambarishmukherjee@rediffmail.com; hotwani@rediffmail.com*

## ABSTRACT

The present paper deals with the inventorization of 132 species of medicinal plants belonging to 58 families and 108 genera growing in different parts of the Burdwan district. This work is in conformity with the first and second phase wherein 230 such species alongwith their family and local name, plant part(s) and/or material used and medicinal property/uses have been brought into light. The work is still in progress.

## INTRODUCTION

Having realized the need to inventorize the most wanted patient-friendly medicinal plants, atleast on regional basis, a work was undertaken, of which the present communication is the third part. During field survey in different parts of the district, the practice of folk-medicine has been found as the most important data-source since it is capable of providing remedial material and information to modern medicine as well as to traditional systems like ayurvedic, unani, siddha etc.

Burdwan, between 22° 56′ and 23° 53′N latitude and between 86° 48′ and 88°25′E longitude, is one of the western districts of West Bengal. The district is naturally divisible into the alluvial delta land to the east and the rocky undulating land towards the west. The natural vegetation is of tropical dry deciduous type, dominated by Sal which is mostly confined to the western uplands of Asansol subdivisions. The eastern alluvial plain is mostly covered with rice fields. The plant diversity of the district includes many medicinal plants which need to be documented. In view of this the present work was undertaken.

## MATERIALS AND METHODS

During ethnomedicinal survey in Burdwan district, 132 species were picked up in the third phase (January to June, 2005). The specimens of each species were processed as voucher specimens for preservation following the standard herbarium techniques (Jain & Rao, 1977). The identification, nomenclature and systematic position were confirmed in consultation with authentic literature (Guha Bakshi, 1984; Panigrahi & Murti, 1989; Prain, 1903). Appropriate medical terms were judiciously used for precision (Ghosh, 1956). The concerned species are enumerated in alphabetical succession along with the respective family and local names, plant part(s) or material used and medicinal properties cum uses. The medicinal uses have been worked out on the basis of emperical (own experience) and authoritative approaches (see references).

## ENUMERATION

Table 1. An inventory of medicinal plants in Burdwan district with their uses.

| Scientific name with synonym (s) | Family | Local name | Part(s) or material used | *Medicinal Property/Uses* | Voucher specimen no. |
|---|---|---|---|---|---|
| **Abroma augusta* (L.) L.f. | Sterculiaceae | Ulatkambal | Root-bark | Emmenagogue, uterine tonic, in dysmenorrhoea | GH 285 |
| *Acacia suma* Buch.-Ham. | Mimosaceae | Saikanta | Bark | Astringent | GH 271 |
| **Acorus calamus* L. | Araceae | Bach | Rhizome | Emetic, stomachic, nerve tonic, in dyspepsia, colic, bronchitis, in purification of drinking water | GH 287 |
| *Ailanthus excelsa* Roxb. | Simaroubaceae | Mahaneem | Bark, leaves | Tonic, febrifuge, expectorant, antiseptic, astringent, in chronic bronchitis and asthma, diarrhoea, dysentery | GH 277 |
| *Alocacia macrorrhiza* (L.) G. Don | Araceae | Mankachu | Tubers, leaves | Styptic, astringent, in piles, anasarca, constipation | GH 242 |
| **Aloe vera* (Tourn. ex L.) Burm.f. | Liliaceae | Ghrita kumari | Whole plant, root | Stomachic, purgative, emmenagogue, anthelmintic, cathartic, in piles, rectal fissures, constipation, fevers, colic, menstrual suppression | GH 272 |
| *Amaranthus tricolor* L. | Amaranthaceae | Lal-sag | Whole herb, root | Diuretic, demulcent | GH 259 |
| *Ammannia baccifera* L. | Lythraceae | Dadmari | Leaves | Acrid, rubefacient, in blisters, rheumatic pains, fever, skin diseases. | GH 281 |

| | | | | | |
|---|---|---|---|---|---|
| *Amomum subulatum* Roxb. | Zingiberaceae | Bara-elachi | Seeds | Stomachic, stimulant, aphrodisiac, antidote, in neuralgia, gonorrhoea, eyelid inflammaton | GH 251 |
| *Annona squamosa* L. | Annonaceae | Ata | Root, leaves, fruits, seeds | Purgative, insecticidal, abortifacient, in conjuctivitis | GH 267 |
| **Areca catechu* L. | Arecaceae | Supari | Nut | Aphrodisiac, astringent, anthelmintic, nervine tonic, emmenagogue, in urinary disorders, snake-bite | GH 231 |
| *Artabotrys hexapetalus* (L.f.) Bhandari | Annonaceae | Katchampa | Leaves | Cholera | GH 288 |
| **Aristolochia indica* L. | Aristolochiaceae | Isharmul | Root, leaves | Tonic, stimulant, emmenagogue, emetic, in leucoderma, fever | GH 307 |
| **Bambusa arundinacea* (Retz.) Willd. | Poaceae | Bans | Leaves | Emmenagogue, in haemetemesis, coughs, colds | GH 342 |
| **Barleria prionitis* L. | Acanthaceae | Kantajati | Root-bark, leaves | In catarrhal affections, cough, anasarca, toothache, boils, glandular swellings | GH 298 |
| **Bauhinia variegata* L. | Caesalpiniaceae | Raktakanchan | Root, bark, buds | Alterative, tonic, astringent, antidote, in skin diseases, ulcers, scrofula, dysentery, piles, diarrhoea, worms, dyspepsia | GH 341 |
| **Borassus flabellifer* L. | Arecaceae | Tal | Whole plant, root, pulp | Cooling, restorative, diuretic, stimulant, antiphlegmatic, demulcent, in inflammatory affections, dropsy | GH 318 |
| **Calendula officinalis* L. | Asteraceae | Calendula | Whole plant, leaves, flowers | Astringent, vulnerary, styptic, stimulant, antiseptic, emmenagogue, resolvent, diaphoretic | GH 246 |
| *Callicarpa macrophylla* Vahl | Verbenaceae | Mathara | Root, leaves | Stomachic, in rheumatism | GH 350 |

| | | | | | |
|---|---|---|---|---|---|
| *Calophyllym inophyllum* L. | Clusiaceae | Sultan champa | Bark, leaves, gum, seeds | Astringent, emetic, purgative, in internal haemorrhage, skin diseases, rheumatism | GH 233 |
| **Calotropis gigantea* (L.) R. Br. | Asclepiadaceae | Akanda | Root-bark, leaves, latex | Emetic, expectorant, diaphoretic, in dysentery, elephantiasis, intermittent fevers | GH 280 |
| *Canna indica* L. | Cannaceae | Sarbajaya | Root, stalks | Diaphoretic, diuretic, demulcent, antidote, stimulant, in fevers, dropsy | GH 324 |
| **Canscora decussata* (Roxb.) Schult. & Schult. f. | Gentianaceae | Dankuni | Whole plant | Laxative, alterative, nerve tonic, in insanity, epilepsy, nervous debility | GH 343 |
| *Capparis zeylanica* L. | Capparaceae | Kalokera | Root-bark, leaves | Sedative, stomachic, cholagogue, counter irritant, in cholera, boils, swellings, piles | GH 297 |
| *Cassia javanica* L. | Caesalpiniaceae | | Root-bark, leaves, fruits, seeds | Laxative, cathartic, emetic, tonic, astringent, febrifuge, purgative | GH 243 |
| *Chamaecrista mimosoides* (L.) Greene | Caesalpiniaceae | | Roots | In spasm of stomach | GH 358 |
| *Chenopodium ambrosioides* L. | Chenopodiaceae | | Whole plant | Anthelmintic | GH 344 |
| **Chrysanthemum indicum* L. | Asteraceae | Chandra mallika | Whole plant | Stomachic, aperient, in gonorrhoea | GH 260 |
| **Cinnamomum tamala* (Spreng.) Nees & Eberm. | Lauraceae | Tejput | Bark, leaves | Stimulant, carminative, in rheumatism, colic, diarrhoea, scorpion-sting, gonorrhoea, in sore throat | GH 293 |
| **C. verum* Presl | Lauraceae | Dalchini | Bark | Astringent, stimulant, carminative, in nausea and vomiting | GH 319 |
| **Cissus quadrangularis* L. | Vitaceae | Horjora | Stem, leaves | Alterative, stomachic, in asthma, irregular menstruation, scurvy, digestive troubles | GH 323 |
| *Citrus grandis* (L.) Osbeck | Rutaceae | Mahanimbu | Leaves, fruits | Cardiotonic, nutritive, refrigerant, in epilepsy, cholera and convulsive cough | GH 316 |

| | | | | | |
|---|---|---|---|---|---|
| **Coleus amboinicus* Lour. & *C. aromaticus* Benth. | Lamiaceae | Patherchur | Leaves | Carminative, in colic, dyspepsia, urinary diseases, vaginal discharges | GH 328 |
| *Commelina maculata* Edgw. | Commelinaceae | Jatakanchura | Root | Laxative, refrigerant, antidote, in vertigo, fever, bilious affections | GH 244 |
| *Corchorus aestuans* L. | Tiliaceae | Titapat | Seeds | Stomachic, in pneumonia | GH 331 |
| *C. capsularis* L. | Tiliaceae | Narcha | Root, leaves, fruits | Demulcent, stomachic, laxative, carminative, stimulant, in dysentery, fever, dyspepsia | GH 290 |
| *C. olitorius* L. | Tiliaceae | Mithapat | Leaves | Tonic, febrifuge, demulcent, diuretic, in chronic cystitis, gonorrhoea, dysuria | GH 325 |
| **Coriandrum sativum* L. | Apiaceae | Dhane | Fruits, seeds | Stimulant, carminative, diuretic, tonic, stomachic, refrigerant, aphrodiasiac | GH 317 |
| **Crateva nurvala* Buch.-Ham. | Capparaceae | Barun | Root-bark, stem-bark, leaves | Demulcent, stomachic, laxative, diuretic, antipyretic, alterative, tonic | GH 329 |
| *Cucurbita maxima* Duch. ex Lam. | Cucurbitaceae | Saphuri komra | Seeds, oil, fruit pulp | Anthelmintic, diuretic, tonic, in burns, inflammations, boils | GH 309 |
| *C. pepo* L. | Cucurbitaceae | Sada kumra | Leaves, seeds | Anthelmintic, in burns | GH 289 |
| **Curcuma amada* Roxb. | Zingiberaceae | Amada | Rhizome | Carminative, stomachic, cooling, applied over contusions, sprains | GH 235 |
| **C. aromatica* Salisb. | Zingiberaceae | Bonhalud | Rhizome | Tonic, carminative | GH 236 |
| *C. caesia* Roxb. | Zingiberaceae | Kalahaldi | Rhizome | Stimulant, carminative | GH 359 |
| **C. longa* L. | Zingiberaceae | Haldi | Rhizome | Stimulant, tonic, carminative, blood-purifier, antiperiodic, alterative, anthelmintic, in skin care and skin diseases | GH 245 |
| *Cymbopogon flexuosus* (Nees ex Steud.) Wats. | Poaceae | Lemon grass | Leaves | Germicidal, in aroma therapy | GH 261 |
| *C. martinii* (Roxb.) Wats. | Poaceae | Palma-rosa | Inflorescence, flowers, fruits | In aroma therapy | GH 262 |

| | | | | | |
|---|---|---|---|---|---|
| *C. winteranus* Jowitt | Poaceae | Citronella | Leaves | Insect repellent, germicide | GH 234 |
| *Dioscorea alata* L. | Dioscoreaceae | Chupri-alu | Tubers | Anthelmintic, in leprosy, piles, gonorrhoea | GH 252 |
| *Ecbolium viride* (Forsk.) Alston | Acanthaceae | Udajati | Whole plant, root, leaves | In jaundice, menorrhagia, gout, rheumatism, dysuria | GH 345 |
| *Echinochloa crus-galli* (L.) Beauv. | Poaceae | Burashama | Whole plant | In haemorrhage, diseases of spleen | GH 322 |
| **Elettaria cardamomum* Maton | Zingiberaceae | Choti-elachi | Seeds | Stimulant, stomachic, carminative, diuretic | GH346 |
| **Emilia sonchifolia* (L.) DC. | Asteraceae | Sadhimodi | Whole plant, root, leaves | Febrifuge, in bowel complaints, eye inflammation, night blindness, sore ears, diarrhoea | GH 312 |
| **Erythrina variegata* L. | Fabaceae | Paltemadar | Bark, leaves | Astringent, febrifuge, diuretic, anthelmintic, galactagogue, emmenagogue, cathartic | GH 333 |
| **Eulophia dabia* (D.Don) Hochr. | Orchidaceae | Salibmisri | Rhizome | Tonic, aphrodisiac, in stomatitis, purulent cough, heart troubles | GH 269 |
| **Foeniculum vulgare* Mills. | Apiaceae | Pan-mouri | Root, leaves, seeds | Stimulant, stomachic, carminative, emmenagogue, diuretic, purgative | GH 254 |
| **Gloriosa superba* L. | Liliaceae | Bishalanguli | Root | Purgative, cholagogue, anthelmintic, in leprosy, parasitical affections of skin, piles, colic, gonorrhoea | GH 311 |
| **Glycyrrhiza glabra* L. | Fabaceae | Jashti-madhu | Root | Tonic, laxative, demulcent, emollient, in genito-urinary diseases, coughs, sore throat | GH 321 |
| **Gmelina arborea* L. | Verbenaceae | Gamari | Whole plant, leaves | Demulcent, in gonorrhoea, cough, foetid discharges, ulcers | GH 274 |
| **Gymnema sylvestre* (Retz.) R. Br. ex Schult. | Asclepiadaceae | Meshasingi, Gurmar | Root, leaves | Emetic, expectorant, in diabetes, glycosuria | GH 332 |
| *Hedyotis diffusa* Willd. | Rubiaceae | | Whole plant | Blood-purifier, in biliousness, fever, gonorrhoea | GH 348 |

| | | | | | |
|---|---|---|---|---|---|
| *Hyptis suaveolens* (L.) Poit. | Lamiaceae | | Whole plant | In parasitical skin diseases | GH 268 |
| *Ipomoea aquatica* Forsk. | Convulvulaceae | Kalmisak | Whole plant | Emetic, purgative, antidote, in nervous and general debility of females | GH 237 |
| *I. obscura* (L.) Ker-Gawl. | Convulvulaceae | | Leaves | In aphthous affections | GH 310 |
| *Ixora arborea* Roxb. ex Smith | Rubiaceae | Rangan | Root, fruit | Given to females when the urine is highly coloured | GH 253 |
| *I. coccinea* L. | Rubiaceae | Rangan | Flowers | In dysentery, dysmenorrhoea | GH 360 |
| *Jasminum sambac* (L.) Ait. | Oleaceae | Bel, Mogra | Whole plant, root, leaves, flowers | Cooling, emmenagogue, lactifuge, in insanity, weakness of sight and affections of the mouth, abscess, indolent ulcers | GH 286 |
| *Jatropha curcas* L. | Euphorbiaceae | Bagh bherenda | Whole plant, twigs, leaves, nuts, seeds | Purgative, lactagogue, rubefacient, in scabies, eczema, ringworm | GH 291 |
| *Kaempferia galanga* L. | Zingiberaceae | Chandu-mula, Ekangi | Tubers | Stimulant, expectorant, diuretic, carminative, in coughs, pectoral affections | GH 339 |
| *Kyllinga brevifolia* Rottb. | Cyperaceae | | Rhizome | In sore legs | GH 353 |
| *Lablab purpureus* (L.) Sweet | Fabaceae | Makham-sim | Seeds | Febrifuge, stomachic, antiseptic, aphrodisiac | GH 273 |
| **Lawsonia inermis* L. | Lythraceae | Mehendi | Bark, leaves, flowers, oil | Astringent, refrigerant, alterative, soporific, in jaundice, skin diseases, spermatorrhoea | GH 352 |
| *Leucas cephalotes* (Koen. ex Roth.) Spreng. | Lamiaceae | Barahalkusa | Whole plant, flowers | Stimulant, diaphoretic, insecticidal, in scabies, cold, cough | GH 354 |
| *Limnophila indica* (L.) Druce | Scrophulariaceae | Karpur gandha | Whole plant | Antiseptic, in elephantiasis, pestilent fevers, dysentery | GH 238 |
| *Lipocarpha sehacelata* (Vahl) Kunth | Cyperaceae | Nirbisi | Root | In fever, diabetes, pruritus | GH 336 |
| **Litsea glutinosa* (Lour.) Robinson | Lauraceae | Kukurchita | Bark, leaves, berries | Demulcent, astringent, aphrodisiac, anodyne, antidote, emollient, | GH 347 |

| | | | | | |
|---|---|---|---|---|---|
| | | | | antispasmodic, emetic, in bruises, rheumatism | |
| **Martynia annua* L. | Martyniaceae | Baghnokh | Leaves, fruits | Alexiteric, in epilepsy, sore throat, tuberculous glands, inflammations | GH 338 |
| *Melilotus indica* (L.) All. | Fabaceae | Bon-methi | Whole plant, seeds | Emollient, discutient, in bowel complaints, infantile diarrhoea, swellings | GH 256 |
| *Mentha arvensis* L. | Lamiaceae | Podina | Whole plant | Carminative, stomachic, antispasmodic, refrigerant, stimulant, emmenagogue, diuretic | GH 335 |
| **Mesua ferrea* L. | Myrsinaceae | Nagkesar | Bark, leaves, flowers, fruits | Astringent, stomachic, expectorant, sudorific, in cough, bleeding piles, burning of the feet, dysentery | GH 340 |
| *Michelia champaca* L. | Magnoliaceae | Champa | Root, root-bark, stem-bark, leaves, flowers, fruits, seeds | Stimulant, expectorant, astringent, carminative, cooling, purgative, antiseptic, in dyspepsia, nausea, gonorrhoea, colic, ophthalmia | GH 270 |
| *Mollugo pentaphylla* L. | Molluginaceae | Julpapra | Whole plant, leaves | Stomachic, aperient, antiseptic, emmenagogue, antiperiodic | GH 239 |
| **Momordica charantia* L. | Cucurbitaceae | Karela | Root, leaves, fruits | Purgative, emetic, anthelmintic, astringent, vermifuge, stomachic, in bilious affections, piles, leprosy, jaundice, haemorrhoids, burning sensations | GH 284 |
| **Musa paradisiaca* L. | Musaceae | Kela | Root, stem, flowers | Astringent, anthelmintic, in otalgia, haemoptysis | GH 255 |
| *Mytragyna parvifolia* (Roxb.) Korth. | Rubiaceae | Kelikadamba, Khelkadam | Bark, root | In fever, colic, muscular pains | GH 355 |
| *Naringi crenulata* (Roxb.) Nicolson | Rutaceae | Naringi | Root, leaves, fruits | Purgative, sudorific, in colic, cardalgia, epilepsy | GH 294 |
| **Nigella sativa* L. | Ranunculaceae | Kalojire | Seeds | Stimulant, diuretic, carminative, emmenagogue, galactagogue | GH 326 |

| | | | | | |
|---|---|---|---|---|---|
| *Nymphoides indicum* (L.) Kuntze | Menyanthaceae | | Whole plant | In fever, jaundice | GH 327 |
| *Ochna obtusata* DC. var. *pumila* (Buch.-Ham. ex G. Don) Kanis | Ochnaceae | | Root | Antidote, in menstrual complaints, consumption, asthma | GH 330 |
| **Ocimum basilicum* L. | Lamiaceae | Babuitulasi | Root, leaves, flowers, seeds | Carminative, diuretic, stimulant, demulcent, in gonorrhoea, dysentery, chronic diarrhoea | GH 301 |
| *O. gratissimum* L. | Lamiaceae | Ramtulsi | Whole plant, leaves, seeds | In seminal weakness, gonorrhoea, rheumatism, paralysis, headache | GH 351 |
| *Pandanus foetidus* Roxb. | Pandanaceae | Keya | Leaves, bracts | Stimulant, antiseptic, in leprosy, small-pox, syphilis, scabies, leucoderma, headache, rheumatism | GH 250 |
| *Paspalum scrobiculatum* L. | Poaceae | Kodoadhan | Whole plant | In scorpion-sting | GH 356 |
| *Pavetta indica* L. | Rubiaceae | Kukurachura | Root, leaves | Aperient, in visceral obstructions, haemorrhoidal pains | GH 303 |
| *Pentapetes phoenicea* L. | Sterculiaceae | Bandhuli, Dupurmani | Whole plant | Emollient, demulcent, in snake-bite | GH 266 |
| *Persicaria hydropiper* (L.) H. Gross. | Polygonaceae | Packur-mul | Whole plant, root, leaves | Stimulant, diuretic, emmenagogue, tonic, amenorrhoea | GH 362 |
| *P. orientale* L. | Polygonaceae | Panimarich | Whole plant | Tonic, vulnerary | GH 315 |
| *Phoenix sylvestris* (L.) Roxb. | Arecaceae | Khajur | Root, fruit | Tonic, restorative, in toothache | GH 361 |
| *Phyllanthus reticulatus* Poir. | Euphorbiaceae | Panjuli | Bark, leaves | Diuretic, cooling, alterative, attenuant, in bleeding gums, diarrhoea in infants | GH 248 |
| **Piper chaba* Hunter | Piperaceae | Choi | Fruit | Stimulant, carminative, in cough, cold, haemorrhoidal affections | GH 337 |
| **P. longum* L. | Piperaceae | Piplamul, Pipul | Root, fruit | Alterative, tonic, antidote, in bronchitis, cough, cold | GH 249 |
| **P. nigrum* L. | Piperaceae | Golmarich | Fruit | Stimulant, antiperiodic, alterative, rubefacient, tonic, in dyspepsia, flatulence, paraplegia | GH 300 |

| | | | | | |
|---|---|---|---|---|---|
| **Pistia stratiotes* L. | Araceae | Takapana | Whole plant, root, leaves | Demulcent, refrigerant, emollient, laxative, diuretic, in dysuria, skin diseases | GH 349 |
| *Pisum sativum* L. | Fabaceae | Mattar | Seeds | Emollient, resolvent, in dysentery | GH 240 |
| **Plumbago zeylanica* L. | Plumbaginaceae | Chita | Root, root-bark, milky-juice | Appetizer, sudorific, antiperiodic, in diarrhoea, dyspepsia, piles, anasarca, skin diseases | GH 334 |
| **Polianthes tuberosa* L. | Amaryllidaceae | Rajanigandha | Bulb, flowers | Diuretic, emetic, in gonorrhoea | GH 302 |
| *Polygonum plebeium* R. Br. | Polygonaceae | | Whole plant, root | In pneumonia, bowel complaints | GH 275 |
| *Pterospermum acerifolium* Willd. | Sterculiaceae | Kanak champa | Bark, leaves, flowers | Tonic, haemostatic, in small-pox | GH 247 |
| **Punica granatum* L. | Punicaceae | Dalimgachi | Root-bark, stem-bark, fruit, seeds, pulp | Astringent, anthelmintic, stomachic, cooling or refrigerant | GH 263 |
| *Quamoclit pennata* Bojer | Convulvulaceae | Tarulata | Whole plant, leaves | Cooling, in carbuncles, bleeding piles | GH 299 |
| *Rumex dentatus* L. | Polygonaceae | | Root | Astringent | GH 320 |
| *Rungia pectinata* (L.) Nees | Acanthaceae | | Root, leaves | Cooling, aperient, febrifuge, in small-pox, contusions | GH 295 |
| *Rungia repens* (L.) Nees | Acanthaceae | | Whole plant, leaves | In fevers, cough | GH 279 |
| **Senna alexandrina* Gars. ex Miller | Caesalpiniaceae | Sanna-makki | Leaves, fruits | Laxative, purgative | GH 357 |
| **Senna obtusifolia* (L.) Irwin & Barneby | Caesalpiniaceae | | Root, leaves, seeds | Purgative, in skin diseases, cough, colds, asthma, eye diseases, liver complaints | GH 232 |
| *Senna siamea* (Lam.) Irwin & Barneby | Caesalpiniaceae | | Root, leaves | Purgative, in intestinal worms, convulsions in children | GH 308 |
| **Sesamum indicum* L. | Pedaliaceae | Til | Leaves, seeds, oil | Tonic, emollient, diuretic, lactagogue, demulcent, in dysentery, piles, ulcers, in removal of facial spots | GH 314 |
| **Sida cordata* (Burm. f.) Borss. | Malvaceae | Junka | Leaves, flowers, fruits | In diarrhoea of pregnancy, cuts, wounds, burning sensations | GH 283 |

| | | | | | |
|---|---|---|---|---|---|
| **Solanum indicum* L. | Solanaceae | Byakura | Root, leaves, fruits | Carminative, expectorant, in asthma, cough, catarrhal affections, toothache, dysuria | GH 258 |
| *S. torvum* Swartz | Solanaceae | Titbaigun | Fruits | In enlargement of the spleen | GH 264 |
| *Spermacoce articularis* L.f. | Rubiaceae | | Root, seeds | Alterative, stimulant | GH 292 |
| *Spinacia oleracea* L. | Chenopodiaceae | Palang | Whole plant, leaves, seeds | Cooling, laxative, in febrile affections, inflammations of liver, lungs and bowels, jaundice, urinary calculi, difficult breathing | GH 304 |
| **Terminalia bellerica* (Gaertn.) Roxb. | Combretaceae | Bahera | Fruits | Astringent, tonic, laxative, antipyretic, purgative, in piles, dropsy, diarrhoea, leprosy, dyspepsia, biliousness | GH 276 |
| **T. chebula* (Gaertn.) Retz. | Combretaceae | Haritaki | Bark, fruits | Astringent, laxative, alterative, diuretic, cardiotonic | GH 313 |
| *Tiliacora acuminata* (Lam.) Miers | Menispermaceae | Tiliakoru | Root | In snake-bite | GH 306 |
| **Trachyspermum ammi* (L.) Sprague | Apiaceae | Jowan | Root, fruit | Diuretic, carminative, antiseptic, stomachic, tonic, stimulant, in diarrhoea, colic, flatulence, indigestion, dyspepsia | GH 265 |
| **Trewia nodiflora* L. | Euphorbiaceae | Pitali | Whole plant, root | In bile, phlegm, rheumatic affections | GH 296 |
| **Tylophora indica* (Burm. f.) Merr. | Asclepiadaceae | Antamul | Whole plant, leaves | Emetic, diaphoretic, expectorant | GH 305 |
| *Trianthema portulacastrum* L. | Aizoaceae | Sepunney, Swetpunarnava | Root, leaves | Cathartic, abortifacient, diuretic, in amenorrhoea, hypertension, oedema, dropsy | GH 257 |
| **Withania somnifera* (L.) Dunal | Solanaceae | Asvagandha | Root, leaves, fruits, seeds | Alterative, aphrodisiac, tonic, deobstruent, diuretic, in rheumatism, carbuncles, ulcers, prevents pulmonary tuberculosis | GH 241 |

| | | | | | |
|---|---|---|---|---|---|
| **Woodfordia fruticosa* (L.) Kuntze | Lythraceae | Dhai | Flowers | Astringent, in dysentery, menorrhagia, haemorrhoids | GH 282 |
| *Zeuxine strateumatica* (L.) Schlechter | Orchidaceae | Shwethuli | Tubers | Used as salep | GH 278 |

* Pharmaceutically used plants (Kapoor & Mitra, 1979)

## ACKNOWLEDGEMENT

The authors are grateful to the Head, Department of Botany, Burdwan University for providing all working facilities.

## REFERENCES

ANONYMOUS 1985. *The Wealth of India, Raw Materials.* Council of Scientific and Industrial Research, New Delhi.

BENNET, S.S.R. 1986. Name Changes in Flowering Plants of India and Adjacent Regions. Triseas Publishers, Dehra Dun.

CHATTERJEE, A. & S.C. PAKRASHI 1991-1997. *The Treatise on Indian Medicinal Plants,* Vol. I-V. Council of Scientific and Industrial Research, P.I.D., New Delhi.

CHOPRA, R.N., S.L. NAYAR & I.C. CHOPRA 1956. *Glossary of Indian Medicinal Plants.* Council of Scientific and Industrial Research, P.I.D., New Delhi.

GHOSE, J.N. 1956. *A Dictionary of Medicinal Terms.* M. Bhattacharya & Co. Pvt. Ltd., Calcutta.

GUHA BAKSHI, D.N. 1984. *Flora of Murshidabad District, West Bengal, India.* Scientific Publishers, Jodhpur, India.

JAIN, S.K. & R.R. RAO 1977. *A Handbook of Field and Herbarium Methods.* Today and Tomorrow's Printers and Publishers, New Delhi and Calcutta.

JOSHI, S.G. 2000. *Medicinal Plants.* Oxford and IBH Publishing Co. Pvt. Ltd., New Delhi.

KAPOOR, S.L. & R. MITRA 1979. *Herbal drugs in Indian Pharmaceutical Industry.* Economic Botany Information Service, National Botanical Research Institute, Lucknow.

PANIGRAHI, G. & S.K. MURTI 1989. *Flora of Bilaspur District (Madhya Pradesh),* Vol. I & II. B.S.I., Calcutta.

PRAIN, D. 1903. *Bengal Plants,* Vol. I & II. B.S.I., Calcutta (Rep. ed. 1963).

RAO, R.R. 1989. Methods and techniques in Ethnobotanical Study and Research: Some basic considerations. In Jain, S.K. (Ed.) *Methods and approaches in Ethnobotany,* pp. 13-23. Society of Ethnobotanists, Lucknow.

# PLANTS USED AS MOUTH FRESHNER IN SAURASHTRA REGION OF GUJARAT

B.A. JADEJA, N.K. ODEDRA & K.B. SOLANKI

*Botany Department, M. D. Science College, Porbandar – 360 575, Gujarat*

**ABSTRACT**

The present paper discusses the use of plants for oral hygiene (mouth freshner) among various tribal and rural communities of Saurashtra region. Various plant parts like seed, fruit, leaves, bark, rhizome, root etc are used as mouth freshner and are beneficial for curing various gum disorders. Total 54 plant species are included with their local name, family and plant part used.

## INTRODUCTION

The Saurashtra region is arid to marginally semiarid ecoclimatic zone forming the western peninsular part of Gujarat. The region lies between 20° 42′ and 23° 33′ N and 69° 4′ and 72° 18′ E. The central part is comprised of hilly and undulating terrain, which extends into plains towards seashore and the Gujarat main land. The region is surrounded by sea on three sides. The overall climate is typical monsoonal. Total number of rainy days are 30 and average annual rainfall ranges between 270 mm and 700 mm and the temperature ranges from 11.5°C to 45.6°C. Forests in this area are dry deciduous type.

The man is using plants in various ways since the existence of his life on the earth. He used it in many ways, including food, fodder, medicine, timber and other miscellaneous uses. Our old time literature contains a lot of information on different uses of plants. The use of plants or plant parts - as mouth freshner or 'Mukhwas', for cleaning teeth and mouth and to increase digestion is much prevalent among the tribals and rural people, particularly Maldhari, Maher, Charan, Koli, Karadia, Kharva, Kangasia etc, and also among common rural folk of Saurashtra region. The medicinal uses of plants of Gujarat region are reported by several workers after Thaker (1910, 1926), especially during last decade (Bhatt & Mitaliya, 1999; Bhatt *et al.*, 2000, 2001; Bhattacharya, 1996; Jadeja, 1999; Kakrani & Saluja, 2001;

Mitaliya *et al.,* 2001, 2003; Punjani, 1998; Punjani & Kumar, 2003 etc).

## MATERIAL AND METHODS

The present information on dental care and to clean mouth by plants was collected through repeated interviews with tribals, rural people and older women during 2003-2005. Collections of plants were made from different areas of Saurashtra region and identification was confirmed in the light of local names and specimens in hand with the help of literature on authors disposal (Shah, 1978; Bole & Pathak, 1988).

## ENUMERATION

Table 1. List of plants used as mouth freshner.

| ***Botanical Name*** | ***Family*** | ***Local Name*** | ***Part used*** |
|---|---|---|---|
| *Acacia catechu* (L.f.) Willd. | Mimosaceae | Katho | Bark |
| *Anacardium occidentale* L. | Anacardiaceae | Kaju | Seed |
| *Anethum graveolens* L. | Apiaceae | Sawa | Seed |
| *Arachis hypogaea* L. | Fabaceae | Magphali | Seed |
| *Areca catechu* L. | Arecaceae | Supari | Seed |
| *Buchanania cochinchinensis* (Lour.) Almeida | Anacardiaceae | Charoli | Seed |
| *Carica papaya* L. | Caricaceae | Papaya | Fruit |
| *Cinnamomum tamala* (Buch.-Ham.) Nees & Eberm. | Lauraceae | Tamal patra | Leaves |
| *Cinnamomum verum* Presl | Lauraceae | Taj | Bark |
| *Citrullus lanatus* (Thunb.) Mat. & Nakai | Cucurbitaceae | Tarbuch | Seed |
| *Citrus limon* (L.) Burm. f. | Rutaceae | Limbu | Flower |
| *Cocos nucifera* L. | Arecaceae | Narial | Dried flesh (Copra) |
| *Coriandrum sativum* L. | Apiaceae | Kothmir | Seed |
| *Crocus sativus* L. | Iridaceae | Kesar | Style |
| *Cucumis callosus* (Rottl.) Cogn. | Cucurbitaceae | Kothimdu | Seed |
| *Cucumis melo* L. var. *melo* | Cucurbitaceae | Sakkartetee | Seed |
| *Cucumis melo* L. var. *momordica* Duthie & Fuller | Cucurbitaceae | Chibhadu | Seed |
| *Cucumis sativus* L. | Cucurbitaceae | Kakdi | Seed |
| *Cucurbita maxima* Duch. ex Lam. | Cucurbitaceae | Kolu | Fruit |
| *Cuminum cyminum* L. | Apiaceae | Jeeru | Fruit |
| *Curcuma longa* L. | Zingiberaceae | Haldar | Rhizome |
| *Cymbopogon citratus* (DC.) Stapf | Poaceae | Lili cha | Leaves |
| *Elettaria cardamomum* Maton | Zingiberaceae | Alachi | Fruit |
| *Ferula assafoetida* L. | Apiaceae | Hing | Root |

| | | | |
|---|---|---|---|
| *Ficus carica* L. | Moraceae | Angir | Fruit |
| *Foeniculum vulgare* Mills. | Apiaceae | Variyali | Seed |
| *Glycyrrhiza glabra* L. | Fabaceae | Jethimadh | Root |
| *Juglans regia* L. | Juglandaceae | Akharot | Seed |
| *Lagenaria siceraria* (Molina) Standl. | Cucurbitaceae | Dudhi | Seed |
| *Mangifera indica* L. | Anacardiaceae | Ambo | Cotyledon |
| *Manilkara zapota* (L.) Royen | Sapotaceae | Chiku | Milky juice |
| *Mentha spicata* L. | Lamiaceae | Pipermint | Leaves |
| *Morus alba* L. | Moraceae | Shetur | Leaves |
| *Myristica fragrans* Houtt. | Myristicaceae | Jayphal | Aril |
| *Nicotiana tabacum* L. | Solanaceae | Tamaku | Leaf |
| *Pandanus fascicularis* Lam. | Pandanaceae | Kevdo | Leaves |
| *Papaver somniferum* L. | Papaveraceae | Khaskhas | Seed |
| *Phoenix sylvestris* (L.) Roxb. | Arecaceae | Khajur | Fruit |
| *Phyllanthus emblica* L. | Euphorbiaceae | Ambala | Fruit |
| *Piper betle* L. | Piperaceae | Nagarvel | Leaf |
| *Piper longum* L. | Piperaceae | Lindipiper | Fruit |
| *Piper nigrum* L. | Piperaceae | Mari | Fruit |
| *Pistacia vera* L. | Anacardiaceae | Pista | Fruit |
| *Punica granatum* L. | Punicaceae | Dadam | Seed |
| *Rosa damascena* Mill. | Rosaceae | Gulab | Calyx |
| *Saccharum officinarum* L. | Poaceae | Sherdi | Stem |
| *Sesamum orientale* L. | Pedaliaceae | Tal | Seed |
| *Syzygium aromaticum* (L.) Merr. & Perry | Myrtaceae | Laving | Flower bud |
| *Tamarindus indica* L. | Caesalpiniaceae | Ambali | Seed |
| *Terminalia catappa* L. | Combretaceae | Badam | Seed |
| *Terminalia chebula* (Gaertn.) Retz. | Combretaceae | Harde | Fruit |
| *Trachyspermum ammi* (L.) Sprangue | Apiaceae | Ajmo | Seed |
| *Vitis vinifera* L. | Vitaceae | Draksh | Fruit |
| *Zingiber officinale* Rosc. | Zingiberaceae | Adu | Rhizome |

## RESULTS AND DISCUSSION

Rural, tribal and backward classes of people of this region use 54 plant species, belonging to 47 genera and 30 families as mouth freshner. Apiaceae and Cucurbitaceae are the largest families whose 13 species are used. Seed of 20 species, fruit of 12 species, leaves of 7 species, roots of 2 species, bark of 2 species, rhizome of 2 species, stem of 1 species, flower of 1 species, flower bud of 1 species, cotyledons of 1 species, aril of 1 species, style of 1 species, calyx of 1 species, milky juice of 1 species, dried flesh of 1 species are used as mouth freshner.

## ACKNOWLEDGEMENTS

Authors are thankful to Dr. D.C. Bhatt, Department of Marine Science, Bhavnagar University, Bhavnagar for valuable advice. The authors also thank to Prof. S.V. Buch, D.K.V. College, Jamnagar for providing the information and help in identifying the plants during filed trips.

## REFERENCES

BHATT, D.C. & K.D. MITALIYA 1999. Ethnomedicinal plants of Victoria Park (Reserved forest) of Bhavnagar, Gujarat, India. *Ethnobotany* 11 : 81-84.

————, ———— & S.K. MEHTA 2000. Plant twigs as tooth brushes. *Adv. Plant Sci.* 13 (1) : 19-22.

————, M.A. NURANI, K.D. MITALIYA & U.S. BAXI 2001. Some ethnomedicinal plants of Shetrunjava hill of Palitana, Gujarat. *Ethnobotany* 11 : 22-25.

BHATTACHARYA, G. 1996. Medico-ethnobotanical value of Saurashtra weeds. *J. Econ. Taxon. Bot.* Addl. Ser. 12 : 166-168.

BOLE. P.V. & J.M. PATHAK 1988. *Flora of Saurashtra,* Vols. 2 & 3. Botanical Survey of India, Kolkata.

JADEJA, B.A. 1999. Plants used by tribe Rabari in Barda hills of Gujarat. *Ethnobotany* 11 : 42-46.

KAKRANI, H.N. & A.K. SALUJA 2001. Plants used in diuretic activity in traditional medicine in Kutch district, Gujarat. *J. Natural Remedies* 1 (2) : 121.

MITALIYA, K.D., D.C. BHATT, D.M. PATEL & P.N. JOSHI 2001. Medicinal value of some selected stem-bark used by tribals and rurals folk in Gujarat. *Adv. Plant Sci.* 14(1): 255-259.

————, ————, J.N. PATEL & S.K. PATEL 2003. Observations on plants used for joint disorders by rural and tribal people of Gujarat. *J. Econ. Taxon. Bot.* 27 (1) : 212-216.

PUNJANI, B.L. 1998. *An ethnobotanical study of tribal areas of district Sabarkantha (North Gujarat).* Ph.D. Thesis, North Gujarat University, Patan.

———— & VIMAL KUMAR 2003. Plants used in traditional phytotherapy for hair care by tribals in Sabarkantha district, Gujarat, India. *Indian J. Trad. Knowledge* 2 (1) : 74-78.

SHAH, G.L. 1978. *Flora of Gujarat State,* 2 Vols. Sardar Patel University, Vallabh Vidhyanagar (Gujarat).

THAKER, J.I. 1910. *Vanaspati Shastra (Flora of Bharda hills).* Pravin Pustak Bhandar, Rajkot.

———— 1926. *Plants of Kutch & their utility.* Kutch Darbar, Kutch.

# STATUS SURVEY OF ETHNOBOTANICAL RESOURCES OF SATNA DISTRICT, MADHYA PRADESH

S.N. DWIVEDI

*Department of Botany, Janata PG College, A.P.S. University, Rewa – 486 002, India*
*For Correspondence : 14/473, Near Samiti Office, Sanjay Nagar, Rewa- 486 002 (M.P.) India*
*E-mail: drsndwivedi@yahoo.co.in.*

## ABSTRACT

Ethnobotany includes the study of natural and traditional relationship and the interaction between man and his surrounding knowledge about plant wealth and search of new resources of herbal drugs, edible plants, including conservation. The importance of ethnobotanical knowledge is exemplified by the fact that more than eighty percent live hood are based directly or indirectly on the uses of plant resources and associated traditional knowledge.

The present investigation indicates that Satna district of Madhya Pradesh is rich in biodiversity and inhabitants mostly depend on herbal resources for their basic requirements. Ethnobotanical survey and fieldwork in the study area throw the light on their ethnic culture and tradition. The floristic diversity has been utilized as food, vegetable, medicine, fuel and dyes, and also used in several social, cultural and religious ceremonies.

The biodiversity as well as the traditional knowledge is getting eroded rapidly due to ecological imbalance, indiscriminate exploitation and dearth of the knowledge holders. Besides, erosion also occurs due to the infiltration of the modern developments into the indigenous knowledge system. Inhabitants conserve the biodiversity and threatened species by faith, tradition and religious aspects. The present work throws the light on basic problems of tribal and rural people and also provides adequate infrastructure for the establishment of forest based minor cottage industries

## INTRODUCTION

Ethnobotanical uses of plants have been known since time immemorial and plants were used since birth to death of the inhabitants. Biodiversity is the very basis of human survival and economic well being (Singh *et al.,* 1994) and constitute the resources upon which families, communities, nation and future generations depend (Dowdeswell, 1995). Further, with proper management, the ethnobotanical resources can be used sustainably ensuring their use for future generation (Mac Neely *et al.,* 1990). The ethnic people of our country have emotional and symbiotic relationship with biodiversity, which they have been protecting and conserving since ancient time (Ford, 1994). These can be seen in sacred groves, the tracts of virgin forests harbouring rich biodiversity protected by the local people based on the ground of indigenous culture, religious beliefs and taboos.

Our ancestors were mainly dependent upon the plants for their basic requirements. The uses of plants were discovered by the method of trial and error. Indeed, most of the economic uses of plants and all systems of Indian traditional medicine had their roots in the uses of plants as understood by ethnic people. However, some of their empiric knowledge had found place in ancient scriptures, archeological records and religious and historical writings. Many of these earlier discoveries survived only by the oral communication from generation to generation among the tribal and rural communities.

Harshberger (1895) suggested the term 'Ethnobotany, and used vernacular name of plants as suggested by ethnic peolpe. However, the main credit to document indigenous knowledge of ethnomedicine goes to Roxburgh (1832), who during floristic studies recorded medicinal herbs from South India. Later, Watt (1889-1896) studied the economic plants of India and in addition to much information, he mentioned the sacred and religious uses of plants. No doubt, his work is the first true reflection of ethnobotany and indigenous knowledge of India. Moreover, he mentioned the name of some British Physicians, who practiced with native plants for the treatment of their patients.

Studies on ethnobotany in India were pioneered by Jankli Ammal (1956) and later on natured by Jain (1963). Thereafter, a number of sporadic investigations were carried out by various workers. The entire work on ethnobotany in India was summarized by Maheshwari (1970), Binu & Pushpanghadan (1992) and Wahal (2005). Consequently, a graphic review of ethnobotany for last two decades and its status and relationship with other discipline have been narrated by Jain & Shrivastava ( 2001) and Jain (2002).

Several works on ethnobotanical aspects in relation to utilization and conservation have been conducted in many part of India (Given, 1996; Airi *et al.*, 2000; Ved *et al.,* 2001; Dhar, 2002; Biswas *et al.*, 2003; Srinivasamurthy *et al.*, 2003; Upadhyaya, 2003; Chakraborty & Pal, 2004; Dwivedi, 2004). Except a little work (Dwivedi & Singh, 1984; Dwivedi & Pandey, 1992; Dwivedi, 1999, 2003; Dwivedi *et al.,* 2005, 2006) no body has studied the ethnobotanical secrets of Satna district. It was, therefore, worth to record the ethnobotanical resources among the inhabitants of Satna district because of :

1. The area is rich in floristic diversity and mostly unexplored.
2. There is a natural harmony between the inhabitants and flora of the surrounding places.

3. There are many traditional, cultural and religious ceremonies in which they use a large number of plants.
4. Minor forest products and medicinal herbs are main source of some additional income of tribal and rural people.

## STUDY AREA

Satna district of Madhya Pradesh lies in the northern part of Indian subcontinent and is situated on Vindhyan Plateau. It forms a part of the central highland region of our country and inhabited by a number of tribal and rural people. Out of nearly 1870104 population of the district, SC and ST are 304217 and 268104, which constitute 16.27% and 12.96% of total population. The population density of the district is high (249 per sq km). Edaphically, the district contains a variety of soils that have favoured the development of flora and fauna. There are 8-development blocks, viz. Uchehara, Rampur Baghelan, Amarpatan, Ramnagar, Nagod, Sohawal, Maghagawan and Maihar, and a number of religious places e.g. Chitrakoot, Maihar, Ramban etc in the district, which are rich in biodiversity

## METHODOLOGY

During the course of present investigation following methods were adopted :

1. The present investigation is based on fieldwork of 24 remote places scattered over 8-development blocks of the study area during April 2004 to January 2006. The survey and methodology was adopted to study the vegetation and gather ethnobotanical informations from the inhabitants as per method suggested by Jain (1989).
2. During the field trips author met with the villagers of various age groups and interacted with them in detail following standard ethnobotanical manual (Jain & Goel, 1995).
3. As much information possible on the uses of plants was collected and recorded in field book.
4. To the best of possibility the plants were identified in the field itself for their botanical name along with local name as spelled by the inhabitants. However, for further confirmation, the herbarium specimens were collected and pressed as per methods suggested by Jain & Rao (1977).
5. Further informations were gathered on the plants used in different social, cultural and religious ceremonies.
6. Confirmation of specimens were made with the help of floristic literatures (Verma *et al.*, 1985; Oommachan & Shrivastava, 1996).

## ETHNOBOTANICAL RESOURCES

Satna district of Madhya Pradesh is a vast emporium of ethnobotanical resources. The information gathered from the different area is mentioned below :

### 1. WILD FOOD PLANTS

The wild plants have been a source of food and medicine from the dawn of human civilization. These plants are more economic and often more acceptable during the days of famine and scarcity. The tribal and rural people mostly use wild plants as food and vegetable. The natives in local village markets have also sold these plants. Some of these are rich in nutrients and also consumed by urban people. The study of wild edible plants is important not only to identify the potential source which could be utilized as alternative food, but also to select promising types for domestication. The author have collected and identified some important wild food plants (Table 1) from the different remote places of the study area.

## 2. MEDICINAL PLANTS

Medicinal plants have been available in human societies since time immemorial. Indeed, the uses of plants were discovered by ancient people by the method of trial and error. The systems of traditional medicine had their roots in the uses of plants by these people and survived only by the oral communications from generation to generation. Obviously, medicinal plants have been prized for their aromatic, flowering and drug yielding qualities. Their drug values lie in alkaloids, saponins, glycosides, tannins, phenols, alcohols, acids, oils etc.

### (i) Plants used to cure human diseases

Satna district of Madhya Pradesh has a vast emporium of medicinal plant wealth. The medicinal plants used to cure various human diseases have been collected from different study sites and mentioned in Table 2.1.

### (ii) Plants used to cure cattle diseases

Animals play a great role in the economy and welfare of people of our country. Their healthcare is, therefore, a matter of significant concern. Herbal drugs have always played a significant role in animal healthcare as alleviating human ailments. Considerable traditional knowledge about animal healthcare has descended over generation through oral folklore, and in the last few decades, ethnobotanist have done fieldwork and documented good part of this knowledge. Some important herbal drugs used to cure cattle diseases are given in Table 2.2.

## 3. TIMBER AND FUEL PLANTS

Forests are renewable resources for raw materials like timber, firewood and other economic and medicinal plants and have a major contribution in the economic development of our country. Moreover, they play a vital role in the life and economy of tribal and rural people. The natives collect and utilized many wild woody shrubs and trees for timber and fuel wood (Table 3). They also sell timber and fuel wood plants in the local market, which is an additional source for their live hood.

## 4. DYE YIELDING PLANTS

Since time immemorial, tribal and rural people have practiced the art of dyeing the clothes and other items by vegetable and natural dyes. Indeed, the weavers for dyeing different domestic articles such as clothes, bamboo baskets, flowerpots, ropes, threads, mats etc used the natural dyes. Moreover, vegetable dyes have been reported to be used by confectioners and weavers (Chandra Prakash Kala, 2002; Ashish Ghosh, 2003; Sampath Kumar & Rao, 2003). In order to get the desired colour and better results dyeing is done in yarn stage. In addition to dyeing the domestic articles and textiles, these have also been used in colouring and painting. During the course of present investigation author collected data about the dye yielding plants of the study area.

Table 1. Wild food plants used by the natives of Satna district of Madhya Pradesh.

| **Botanical Name** | **Local Name** | **Month of Collection** | **Parts Used** | **Utilization** |
|---|---|---|---|---|
| *Achyranthes aspera* L. | Chirchiri | Jul-Oct | Tender shoot | Leafy vegetable |
| *Aegle marmelos* (L.) Corr. | Bel | May-Jun | Ripe fruits | Pulp edible |
| *Amorphophallus bulbifer* (Roxb.) Bl. | Suran | Dec- Mar | Roasted/ boiled corm | Vegetable & prickle |
| *Annona squamosa* L. | Sitafal | Feb-Apr | Ripe fruits | Pulp edible |
| *Anthum graveolens* L. | Soya | Oct-Jan | Tender shoot | Leafy vegetable |
| *Bambusa vulgaris* Schard. | Bans | Jul- Nov | Tender shoot | Vegetable |
| *Bauhinia variegata* L. | Kachnar | Jan-Mar | Flowers | Vegetable |
| *Boerhavia diffusa* L. | Punarnaba | Aug-Oct | Leaves | Leafy vegetable |
| *Chenopodium album* L. | Bathua | Nov-Feb | Tender shoot | Leafy vegetable |
| *Coccinia grandis* (L.) Voigt | Kundru | Sep-Dec | Unripe fruits | Vegetable |
| *Commelina benghalensis* L. | Tankaua | Jul-Nov | Tender shoot | Leafy vegetable |
| *Dioscorea bulbifera* L. | Khanima | Oct-Dec | Boiled tubers | Edible/vegetable |
| *Diospyros melanoxylon* Roxb. | Tendu | Feb-Apr | Ripe fruits | Edible |
| *F. racemosa* L. | Umar | Mar-Apr | Ripe fruits | Edible |
| *F. religiosa* L. | Peepal | Mar-Apr | Ripe fruits | Edible |
| *Feronia limonia* (L.) Swingle | Kaitha | Oct-Feb | Fruit pulp | Prickle/chutany |
| *Ficus benghalensis* L. | Bargad | Mar-Apr | Ripe fruits | Edible |
| *Ipomoea turbinata* Lag. | Khotalaiya | Aug-Nov | Floral pedicle | Vegetable |
| *Lathyrus aphaca* L. | Jangali matar | Feb-Mar | Ripe seeds | Famine food |
| *Momordica dioica* Roxb. ex Willd. | Parora | Jul-Sep | Unripe fruits | Vegetable |
| *Nelumbo nucifera* Gaertn. | Kamal | Mar-May | Rhizome | Vegetable |
| *Phoenix sylvestris* (L.) Roxb. | Khajoor | May-Jun | Ripe fruits | Edible |
| *Pithecellobium dulce* (Roxb.) Benth. | Jangal jalebi | Apr-May | Ripe fruit pulp | Edible |
| *Portulaca oleracea* L. | Kulfa | Sep-Dec | Tender shoot | Leafy vegetable |
| *Senna tora* (L.) Roxb. | Chakura | Jul-Sep | Tender shoot | Leafy vegetable |
| *Solanum nigrum* L. | Makoya | Oct-Dec | Ripe fruits | Edible |
| *Vicia sativa* L. | Akari | Feb-Mar | Seeds | Famine food |
| *Vigna radiata* (L.) Wilczek | Banmoog | Oct-Nov | Seeds | Famine food |
| *Ziziphus nummularia* (Burm. f.) Wt. & Arn. | Jhar-ber | Nov-Dec | Ripe fruits | Edible |

Table 2.1. Medicinal plants used to cure the human diseases.

| Botanical Name | Local Name | Family | Parts Used | Diseases |
|---|---|---|---|---|
| 1 | 2 | 3 | 4 | 5 |
| **Abrus precatorious* L. | Ratti | Fabaceae | Root chewing | Cough, cold & URI |
| *Abutilon indicum* (L.) Sweet | Kanghi | Malvaceae | Seed powder | Purgative |
| *Acacia catechu* (L.f.) Willd. | Khair | Mimosaceae | Bark decoction | Diarrhoea |
| **A. sinuata* (Lour.) Merr. | Shikakai | Mimosaceae | Fruit pulp | Hair shampoo |
| *A. nilotica* (L.) Willd. ex Del. | Babool | Mimosaceae | Gum paste | Burn |
| *Achyranthes aspera* L. | Latjira | Amaranthaceae | Roasted seeds | Bronchitis |
| **Acorus calamus* L. | Bach | Araceae | Rhizome juice | Cough & asthma |
| ****Aegle marmelos* (L.) Corr. | Bel | Rutaceae | Fruit pulp | Gastric disorder |
| *Ageratum conyzoides* L. | Shahdei | Asteraceae | Leaf paste | Headache |
| *Allium sativum* L. | Lahsun | Liliaceae | Fresh cloves | URI |
| *Aloe vera* (L.) Burm.f. | Gheekumar | Liliaceae | Leaf pulp | Burn & migraine |
| *Amorphophallus bulbifer* (Roxb.) Bl. | Suran | Araceae | Roasted corm | Piles, digestive disorder |
| **Andrographis paniculata* (Burm. f.) Wall. ex Nees | Kalmegh | Acanthaceae | Plant decoction | Malaria, jaundice |
| *Argemone maxicana* L. | Ghamoya | Papaveraceae | Leaf sap | Ringworm |
| *Asparagus racemosus* Willd. | Satavar | Liliaceae | Root powder | Leucorrhoea, debility |
| ****Azadirachta indica* A. Juss. | Neem | Meliaceae | Leaf & Seed oil | Skin eruption |
| *Bauhinia purpurea* L. | Kachnar | Caesalpiniaceae | Bark gargle | Oral ulcer |
| *Biophytum sensitivum* (L.) DC. | Khattamitha | Oxalidaceae | Leaf paste | Burn |
| **Bixa orellana* L. | Sinduri | Bixaceae | Leaf poultice | Swelling |
| **Boerhavia diffusa* L. | Punarnaba | Nyctaginaceae | Root extract | Viral hepatitis |
| *Bombax ceiba* L. | Semal | Bombacaceae | Bark sap | Diarrhoea |
| **Buchnania cochinchinensis* (Lour.) Almeida | Chironji | Anacardiaceae | Bark paste | Skin eruption |
| *Butea monosperma* (Lamk.) Taub. | Palas | Fabaceae | Bark extract | Dysentery |
| *Caesalpinia crista* L. | Gatayan | Caesalpiniaceae | Leaf decoction | Malaria |
| *Calotropis procera* (Ait.) Ait.f. | Madar | Asclepiadaceae | Roasted flower | Asthma |
| *Cassia fistula* L. | Amaltas | Caesalpiniaceae | Fruit pulp | Gastric disorder |
| *Catharanthus roseus* (L.) G.Don | Sadabahar | Apocynaceae | Leaf & Flower | Diabetes |
| **Centella asiatica* (L.) Urban. | Brahmi | Apiaceae | Leaf extract | Brain tonic |
| **Chlorophytum tuberosum* (Roxb.) Baker | Safed musali | Liliaceae | Finger powder | Debility |
| **Cissus quadrangularis* L. | Harjoor | Vitaceae | Stem extract | Bone fracture |
| *Clitoria ternatea* L. | Aparajita | Fabaceae | Seed powder | Purgative |

| | | | | |
|---|---|---|---|---|
| ***Commiphora wightii* (Arn.) Bhandari | Guggul | Burseraceae | Resin | Asthma |
| *Convolvulus arvensis* L. | Shankha-puspi | Convolvulaceae | Plant extract | Brain tonic |
| ***Costus speciosus* (Koen.) Sm. | Keokand | Costaceae | Tuber juice | Amoebiosis |
| *Croton roxburghii* Balk. | Banrendi | Euphorbiaceae | Seed oil | Skin diseases |
| *Curcuma longa* L. | Haldi | Zingiberaceae | Rhizome powder | Respiratory & joint disorders |
| ***C. angustifolia* Roxb. | Tikhur | Zingiberaceae | Tuber powder | Leucorrhoea |
| *Cuscuta reflexa* Roxb. | Amarbel | Cuscutaceae | Plant paste | Bone fracture |
| *Cynodon dactylon* (L.) Pers. | Doob | Poaceae | Plant extract | Menstrual trouble |
| *Cyperus rotundus* L. | Nagarmotha | Cyperaceae | Bulb powder | Dysentery |
| *Datura metal* L. | Dhatura | Solanaceae | Leaf poultice | Inflammation |
| *Diplocyclos palmatus* (L.) Jeffry | Shivalingi | Cucurbitaceae | Fruit & seed<br>Seed | Colitis, malaria<br>Sterility |
| *Echinops echinatus* Roxb. | Untakatar | Asteraceae | Plant extract | Malaria |
| *Eclipta prostrata* (L.) L. | Ghamira | Asteraceae | Plant juice | Cut & wounds |
| *Enicostema axillare* (Lam.) Raynal | | Gentianaceae | Plant extract | Diabetes |
| *Euphorbia hirta* L. | Chhoti dudhi | Euphorbiaceae | Plant juice | Diarrhoea |
| *Evolvulus alsinoides* (L.) L. | Neelpuspi | Convolvulaceae | Plant extract | Dysentery |
| *Feronia limonica* (L.) Swingle | Kaitha | Rutaceae | Leaf paste | Boils |
| ****Ficus benghalensis* L. | Bargad | Moraceae | Bark extract | Malaria |
| ****F. racemosa* L. | Umar | Moraceae | Unripe fruit | Diabetes |
| ****F. religosa* L. | Peepal | Moraceae | Bark decoction | Gonorrhoea |
| *Fumaria indica* (Haussk.) Pug. | Pitpapara | Fumariaceae | Plant juice | Malaria |
| ***Gloriosa superba* L. | Kalihaari | Liliaceae | Tuber extract | Child birth |
| **Gymnema sylvestris* (Retz.) R. Br. ex Schult. | Gurmar | Asclepidiaceae | Leaf powder | Diabetes, snake bite |
| *Hedychium coronarium* Koen. | Gulwakawali | Zingiberaceae | Petal sap<br>Rhizome poultice | Conjunctives<br>Inflammation |
| **Helicteres isora* L. | Marorfali | Sterculiaceae | Fruit powder | Colic |
| *Heliotropium indicum* L. | Hanthisura | Boraginaceae | Plant poultice | Eczema |
| **Holarrhena pubescens* (Buch.-Ham.) Wall. ex G. Don | Kurichi | Apocynaceae | Bark juice | Diarrhoea, dysentery |
| *Hypoxis aurea* Lour. | Bankand | Hypoxidaceae | Tuber poultice | Swelling |
| *Indigofera linifolia* (L.f.) Retz. | Chhotineel | Fabaceae | Seed paste | Boils |
| **Ipomoea turbinata* Lag. | Khotaliya | Convolvulaceae | Pedicle<br>Seed | Appetizer<br>Purgative |
| *Jasminum humile* L. | Chameli | Oleaceae | Leaf chewing | Oral ulcer |

| | | | | |
|---|---|---|---|---|
| *Jatropha curcas* L. | Ratanjot | Euphorbiaceae | Seed oil | Skin eruption |
| *Justicia adhatoda* L. | Adusa | Acanthaceae | Leaf decoction | Cough |
| *Lawsonia inermis* L. | Mehandi | Lythraceae | Leaf poultice | Feet sore |
| *Leonotis nepatiifolia* (L.) Br. | Barchibuti | Lamiaceae | Leaf sap | Eczema |
| *Lepidium sativum* L. | Chansur | Brassicaceae | Seed poultice | Inflammation |
| *Leucas lavandulaefolia* Sm. | Guma | Lamiaceae | Plant decoction | Cough & malaria |
| *Lippia javanica* (Burm. f.) Spr. | Butoni | Verbenaceae | Plant sap | Toothache |
| *Madhuca longifolia* (Koen.) Macbr. | Mahua | Sapotaceae | Roasted flower | Asthma |
| *Martynia annua* L. | Bichhu | Martyniaceae | Fruit paste | Scorpion sting |
| *Melia azedarach* L. | Bakayan | Meliaceae | Leaf poultice | Boils |
| *Mentha arvensis* L | Pudina | Lamiaceae | Leaf juice | Gastric disorder |
| *Mimosa pudica* L. | Lajawanti | Mimosaceae | Leaf extract | Piles |
| *Mirabilis jalapa* L. | Gulabbas | Nyctaginaceae | Leaf poultice | Piles |
| *Momordica charantia* L. | Karela | Cucurbitaceae | Fruit powder | Diabetes |
| *Moringa oleifera* Lam. | Munaga | Moringaceae | Fruit pulp | Appetizer |
| *Mucuna pruriens* (L.) DC. | Kemanch | Fabaceae | Seed powder | Night dreams |
| *Nigella sativa* L. | Karayal | Ranunculaceae | Roasted seed | Constipation |
| *Nyctanthes arbor-tristis* L. | Harsingar | Nyctaginaceae | Leaf decoction | Joint diseases |
| ****Ocimum tenuiflorum* L. | Tulsi | Lamiaceae | Leaf extract | Cough & fever |
| *Orthosiphon pallidus* Benth. | Naridaman | Lamiaceae | Shoot sap | Earache |
| *Oxalis corniculata* L. | Tinpatia | Oxalidaceae | Leaf poultice | Swelling |
| *Pandanus fascicularis* Lamk. | Kewara | Pandanaceae | Flower powder | Prickly heat |
| *Pentanema indicum* (L.) Ling | Banjhauri | Asteraceae | Plant extract | Contraceptive |
| *Pergularia daemia* (Forsk.) Chiov. | Dudhi bel | Asclepidiaceae | Plant extract | Blood purifier |
| *Peristrophe paniculata* (Forsk.) Brumm. | Chhoti harjori | Acanthaceae | Plant plaster | Bone fracture |
| ****Phyllanthus emblica* L. | Amla | Euphorbiaceae | Fruit powder | Gastric disorder |
| *Phyllanthus fraternus* Webster | Bhuamla | Euphorbiaceae | Root extract | Viral hepatitis |
| *Physalis minima* L. | Desi rasbhari | Solanaceae | Leaf juice | Earache |
| *Plumbago zeylanica* L. | Chitrak | Plumbaginaceae | Plant sap | Skin diseases |
| *Polygala arvensis* Willd. | Nilkanta | Polygalaceae | Leaf poultice | Swelling |
| *Pongamia pinnata* (L.) Pierre | Karanj | Fabaceae | Seed oil | Skin eruption |
| *Portulaca oleracea* L. | Kulfa | Portulacaceae | Plant sap | Blood purifier |
| ****Prosopis juliflora* (Sw.) DC. | Shami | Mimosaceae | Bark gargle | Oral ulcer |
| ***Rauvolfia serpentina* (L.) Benth. ex Kurz | Sarpagandha | Apocynaceae | Root extract | Blood pressure |
| *Ricinus communis* L. | Aarand | Euphorbiaceae | Seed oil | Purgative |
| *Ruellia tuberosa* L. | Kand | Acanthaceae | Tuber poultice | Swelling |
| **Sapindus emarginatus* Vahl | Ritha | Sapindaceae | Fruit pulp | Hair shampoo |
| ****Saraca asoca* (Roxb.) de Wilde | Ashoka | Caesalpiniaceae | Bark decoction | Leucorrhoea |

| | | | | |
|---|---|---|---|---|
| *Salvia plebeia* R.Br. | Karondha | Lamiaceae | Leaf extract | Common cold |
| *Senna tora* (L.) Roxb. | Chakura | Caesalpiniaceae | Leaf juice | Ringworm |
| *Sida acuta* Burm. | Bala | Malvaceae | Plant decoction | Anaemia, weakness |
| *Solanum virginianum* L. | Kanteli | Solanaceae | Root extract | Cough |
| *S. nigrum* L. | Makoya | Solanaceae | Plant sap | Viral hepatitis |
| *Sphaeranthus indicus* L. | Mundi | Asteraceae | Leaf juice | Cut & injury |
| ****Spilanthes calva** DC. | Akarkara | Asteraceae | Plant extract | Toothache |
| *Sterculia urens* Roxb. | Kullu | Sterculiaceae | Gum | Amoebiosis |
| ***Strychnos nux-vomica* L. | Kuchila | Loganiaceae | Seed powder | Nervous disorder |
| *Syzygium cumini* (L.) Skeels | Jamun | Myrtaceae | Seed powder | Diabetes |
| *Tamarindus indica* L. | Imli | Caesalpiniaceae | Fruit pulp | Laxative |
| *Tephrosia purpurea* (L.) Pers. | Silpoka | Fabaceae | Root chewing | Tonsillitis |
| *Terminalia cuneata* Roth | Arjun | Combertaceae | Bark decoction | Blood pressure |
| **T. bellerica* (Gaertn.) Roxb. | Bahera | Combertaceae | Fruit powder | Gastric disorder |
| *T. chebula* (Gaertn.) Retz. | Harra | Combertaceae | Fruit powder | Gastric disorder |
| *Tinospora cordifolia* (L.) Miers ex Hk. f. & Th. | Giloya | Menisper-maceae | Stem powder | Debility |
| **Tribulus terrestris* L. | Gokharu | Zygophyllaceae | Fruit powder | Asthma & cough |
| *Tridax procumbens* L. | Ghawapatti | Asterceae | Plant sap | Piles |
| *Vanda tessellata* (Roxb.) Hook. ex G.Don | Harjori | Orchidaceae | Plant plaster | Bone fracture |
| *Vernonia albicans* DC. | Jhurjuri | Asteraceae | Leaf juice | Cut, injuries & wounds |
| *Vetiveria zizanioides* (L.) Nash | Khas | Poaceae | Root powder | Prickly heat |
| *Vitex negundo* L. | Nirgundi | Verbenaceae | Leaf decoction | Joint diseases |
| **Withania somnifera* (L.) Dunal | Ashwagandha | Solanaceae | Root powder | Debility |
| *Woodfordia fruticosa* (L.) Kuntze | Dhawai | Lythraceae | Flower powder | Smallpox, prickly heat |
| *Xanthium strumarium* L. | Latakani | Asteraceae | Seed fume | Piles |
| *Zingiber officinale* Rosc. | Adarakh | Zingiberaceae | Rhizome juice | Cough & URI |
| *Ziziphus mauritiana* Lamk. | Ber | Rhamnaceae | Ripe fruit | Laxative |

*Threatened species, ** Rare species, *** Sacred and Religious species

Table 2.2. Medicinal plants used to cure cattle diseases.

| **Botanical Name** | **Local Name** | **Family** | **Parts Used** | **Disease** |
|---|---|---|---|---|
| *Achyranthes aspera* L. | Latjira | Amaranthaceae | Plant sap | Safe delivery |
| *Acorus calamus* L. | Bach | Araceae | Rhizome paste | Septicemia |
| *Allium sativum* L. | Lahsun | Liliaceae | Fresh clove | Mastitis |

| | | | | |
|---|---|---|---|---|
| *Aloe vera* (L.) Burm.f. | Gheekumar | Liliaceae | Pulp paste | Miscarriage |
| *Annona squamoa* L. | Sitsfal | Annonaceae | Seed powder | Maggots |
| *Asparagus racemosus* Willd. | Satavar | Liliaceae | Root powder | Lactation |
| *Azadirachta indica* A. Juss. | Neem | Meliaceae | Leaf & Seed oil | FMD |
| *Bombax ceiba* L. | Semal | Bombacaceae | Bark extract | Indigestion |
| *Boerhavia difussa* L. | Punarnaba | Nyctaginaceae | Root powder | Blackquarter |
| *Calotropis procera* (Ait.) Ait.f. | Madar | Asclepiadaceae | Leaf poultice | Swelling |
| **Centella asiatica* (L.) Urban | Brahmi | Apiaceae | Plant extract | Peptic ulcer |
| *Centratherum anthelminticum* (L.) Kuntze | Kalajira | Asteraceae | Seed powder | Pyrexia |
| *Curcuma longa* L. | Haldi | Zingiberaceae | Rhizome powder | Cut, wound & scabies |
| *Eclipta prostrata* (L.) L. | Ghamira | Asteraceae | Plant juice | Fever & wounds |
| *Ficus religiosa* L. | Peepal | Moraceae | Bark decoction | FMD |
| *Lawsonia inermis* L. | Mehandi | Lythraceae | Leaf poultice | FMD |
| *Leucas cephalotes* (Koen. ex Roth) Spr. | Guma | Lamiaceae | Plant decoction | Cough & cold |
| *Moringa oleifera* Lam. | Munaga | Moringaceae | Leaf extract | Anthrax |
| *Nicotiana tabacum* L. | Tambakhu | Solanaceae | Leaf poultice | Ectoparasite |
| *Nyctanthes arbor-tristis* L. | Harsingar | Oleaceae | Leaf poultice | Hoof infection |
| *Persicaria glabra* (L.) Hara | Puror | Polygonaceae | Plant sap | Stop clay eating |
| *Ricinus communis* L. | Aarand | Euphorbiaceae | Leaf poultice | Inflammation |
| *Sorghum bicolor* (L.) Moench | Jwar | Poaceae | Fresh leaf | Expel worms |
| *Tagetes erecta* L. | Genda | Asteraceae | Leaf juice | Feet ulcer |
| *Terminalia chebula* (Gaertn.) Retz. | Hara | Combretaceae | Fruit powder | Constipation |
| *Trachyspermum ammi* (L.) Sprangue | Ajwain | Apiaceae | Seed powder | Gastric disorder |
| *Trigonella foenum-graecum* L. | Menthi | Fabaceae | Seed powder | URI |
| *Vitex negundo* L. | Nirgundi | Verbenaceae | Leaf poultice | Mastitis |
| *Withania somnifera* (L.) Dunal | Asahwagandha | Solanaceae | Root powder | Strength & vigour |
| *Zingiber officinale* Rosc. | Adarakh | Zingiberaceae | Rhizome juice | Cough & indigestion |

FMD = Foot and Mouth Disease, URI = Upper Respiratory tract Infection

Table 3. Timber and fuel wood plants.

| Botanical Name | Local Name | Family | Habit | Status |
|---|---|---|---|---|
| *Acacia nilotica* (L.) Willd. ex Del. ssp. *indica* (Benth.) Brenan | Babul | Mimosaceae | MT | Common |
| *Aegle marmelos* (L.) Corr. | Bel | Rutaceae | MT | Occasional |
| *Albizia lebbeck* (L.) Benth. | Siris | Mimosaceae | T | L. Common |
| *Anogeissus latifolia* (Roxb. ex DC.) Wall. ex Guill. & Perr. | Dhawa | Combretaceae | T | L. Common |

| | | | | |
|---|---|---|---|---|
| *Azadirachta indica* A. Juss. | Neem | Meliaceae | MT | Common |
| *Boswellia serrata* Roxb. ex Colebr. | Salai | Bursaraceae | T | Common |
| *Butea monosperma* (Lam.) Taub. | Palas | Fabaceae | MT | Common |
| *Cassia fistula* L. | Amaltas | Caesalpiniaceae | MT | Common |
| *Dandrocalamus strictus* (Roxb.) Nees | Bans | Poaceae | A | Common |
| *Diospyros melanoxylon* Roxb. | Tendu | Ebenaceae | MT | Frequent |
| *Feronia limonia* (L.) Swingle | Kaitha | Rutacaeae | T | Frequent |
| *Ficus racemosa* L. | Umar | Moraceae | MT | Occasional |
| *Haldina cordifolia* (Roxb.) Ridsd. | Haldu | Rubiaceae | T | Occasional |
| *Holarrhena pubescens* (Buch.-Ham.) Wall. ex G.Don | Baridudhi | Apocynaceae | ST | Threatened |
| *Madhuca longifolia* (Koen.) Macbr. | Mahua | Sapotaceae | T | Common |
| *Mangifera indica* L. | Aam | Anacardiaceae | T | Common |
| *Nyctanthes arbor-tristis* L. | Harsingar | Oleaceae | Sh | Frequent |
| *Piterocarpus marsupium* Roxb. | Bijahara | Fabaceae | T | Common |
| *Pithecellobium dulce* (Roxb.) Benth. | Jangal jalebi | Mimosaceae | ST | L. Common |
| *Pongamia pinnata* (L.) Pierre | Karanj | Fabaceae | ST | L. Common |
| *Schleichera oleosa* (Lour.) Oken. | Kusum | Sapindaceae | T | Occasional |
| *Sterculia urens* Roxb. | Kullu | Sterculiaceae | T | Rare |
| *Syzygium cumini* (L.) Skeels | Jamun | Myrtaceae | T | Common |
| *T. bellirica* (Gaertn.) Roxb. | Bahera | Combretaceae | T | Threatened |
| *T. chebula* (Gaertn.) Retz. | Harra | Combretaceae | T | Common |
| *Tamarindus indica* L. | Imali | Caesalpiniaceae | T | L. Common |
| *Tectona grandis* L.f. | Sagon | Verbenaceae | T | Common |
| *Terminalia cuneata* Roth | Kahawa | Combretaceae | MT | L. Common |
| *Vitex negundo* L. | Nirgundi | Verbenaceae | Sh | Frequent |
| *Ziziphus mauritiana* Lamk. | Ber | Rhamnaceae | Sh | Common |

Abbreviation used : T=Tree, MT=Middle Tree, ST=Small Tree, Sh=Shrub, L=Least, A=Arborescent

Table 4. Dye yielding plants.

| Botanical Name | Local Name | Family | Habit | Parts Used | Colour |
|---|---|---|---|---|---|
| *Acacia catechu* (L.f.) Willd. | Katha | Mimosaceae | MT | Heartwood | Red-black |
| *Basella rubra* L. | Poi | Basellaceae | GT | Ripe fruits | Deep purple |
| *Bauhinia purpurea* L. | Kachnar | Caesalpiniaceae | ST | Flowers | Purple |
| *Bixa orellana* L. | Sinduri | Bixaceae | ST | Seeds | Sinduri |
| *Clitoria ternatea* L. | Aprajita | Fabaceae | WT | Flowers | Blue |
| *Curcuma longa* L. | Haldi | Zingiberaceae | RH | Rhizome | Yellow |
| *Cuscuta reflexa* Roxb. | Amarbel | Cuscutaceae | PT | Shoot | Yellow |
| *Hibiscus rosa-sinensis* L. | Gurhal | Malvaceae | SH | Flowers | Red |

| | | | | | |
|---|---|---|---|---|---|
| *Lawsonia inermis* L. | Mehandi | Lythraceae | SH | Leaves | Dark brown |
| *Persicaria chinensis* (L.) Gross | Nadighee | Polygoniaceae | AH | Plant | *Dark |
| *Phyllanthus emblica* L. | Amla | Euphorbiaceae | ST | Fruits | *Dark |
| *Piper betle* L. | Pan | Piperaceae | AC | Leaves | Deep red |
| *Psidium guajava* L. | Amrood | Myrtaceae | ST | Ripe fruits | *Brown |
| *Tagetes erecta* L. | Genda | Asteraceae | AH | Flowers | Yellow |
| *Tectona grandis* L. f. | Saugon | Verbenaceae | MT | Bark | Reddish |

AH= Annual Herb, AC=Annual Climber, RH= Rhizomatous Herb, PT=Parasitic Twiner, SH= Shrub, ST= Small Tree, MT= Middle Tree, GT= Glabrous Twiner, WT= Woody Twiner

Table 5. Commerce and cottage industry based plants.

| Botanical Name | Common Name | Family | Parts Used | Economic Product |
|---|---|---|---|---|
| *Acacia nilotica* (L.) Willd. ex Del. ssp. *indica* (Benth.) Brenan | Babool | Mimosaceae | Gum | Gum |
| *Bambusa arundinacea* (Retz.) Willd. | Bans | Poaceae | Stem | Basketry |
| *Phanera integrifolia* (Roxb.) Benth. | Kachnar | Caesalpiniaceae | Leaves | Plate making |
| *Buchanania cochinchinensis* (Lour.) Almeida | Chirongi | Anacardiaceae | Seeds | Edible seeds |
| *Butea monosperma* (Lam.) Taub. | Palas | Fabaceae | Leaves<br>Shoot | Plate making<br>Lac |
| *Crotalaria juncea* L. | Aprajita | Fabaceae | Shoot | Cordage |
| *Cyperus pangorei* Rottb. | Sarpat | Cyperaceae | Leaves | Mat making |
| *Diospyros melanoxylon* Roxb. | Tendu | Ebenaceae | Leaves | Bidi making |
| *Guizotia abyssynica* (L.f.) Cass. | Ramtil | Asteraceae | Seeds | Oil extraction |
| *Madhuca longifolia* (Koen.) Macbr. | Mahua | Sapotaceae | Flowers<br>Seeds | Alcohol extraction<br>Oil |
| *Oryza sativa* L. | Dhan | Poaceae | Grains | Paddy/Rice |
| *Phoenix sylvestris* (L.) Roxb. | Khajur | Arecaceae | Leaves | Broom and mat |
| *Sterculia urens* Roxb. | Kullu | Sterculiaceae | Gum | Gum |
| *Tamarindus indica* L. | Imli | Caesalpiniaceae | Fruits | Fruits |
| *Terminalia cuneata* Roth | Kahawa | Combretaceae | Bark | Tannin and tea |
| *Vetiveria zizanioides* (L.) Nash | Khas | Poaceae | Roots<br>Leaves | Perfume<br>Broom |

## 5. COMMERCE AND COTTAGE INDUSTRY BASED PLANTS

Commerce and small-scale cottage industries are based on plants and minor forest products. Balanced exploitation of these can help the tribal and rural people to achieve economic prosperity and consequently a better standard of life. However, forest depletion has been rapid and the direct demand on forest has been mounting. Yet, the situation is not very bad, because the natives if left on their own to manage their affairs, have an inbuilt system of balancing their needs and of their environment and ecology. The inhabitants utilize a number of plants (Table 5) for commercial, economic and industrial purpose.

## THREAT STATUS AND CONSERVATIONAL STRATEGIES

Biodiversity of the study area is at stake for a variety of reasons. The plants of ethnomedicinal value are largely threatened (Table 2*) due to over-exploitation. Consequently, other plants have been subjected to direct pressure for a number of factors such as change in agriculture system, illegal trading, industrialization, urbanization process etc. These factors operating simultaneously, resulting the destruction of habitat and extinction of flora. Moreover, loss of genetic diversity, catastrophes and variation in individual birth and death also affect biodiversity degradation.

During the field trip author has found 16 plants, viz. *Abrus prectorius, Acacia sinuata, Acorus calamus, Andrographis paniculata, Bixa orellana, Boerhavia diffusa, Buchanania cochinchinensis, Ipomoea turbinata, Centella asiatica, Cissus quadrangularis, Chlorophytum tuberosum, Gymnema sylvestris, Helicteres isora, Holarrhena pubescens, Withania somnifera, Terminalia bellirica*, which are scarce in the area. Some of these may be common in other part of India, but, to preserve the diversity, this scarcity needs attention. Likewise, author tried to have some idea of rare plants in the area and has come across 9 plants i.e. *Commiphora wightii, Curcuma angustifolia, Costus speciosus, Gloriosa superba, Rauvolfia serpentina, Sterculia urens, Tribulus terrestris,* which find mention in recent literature like Red Data Book and other publications on rare and threatened plants of India.

Tribal and ethnic people have been actively engaged in conservation of threatened flora since time immemorial. They conserve and protect the surrounding species by myth, taboos, social, cultural, traditional and religious aspects. Furthermore, the Forest Department with the participation of local people is doing conservation of widely separated species by raising stone wall fencings. By this process the naturally generated plant populations would be increased. In addition to this, *in-situ* conservation of biodiversity within the protected areas, viz. sanctuaries, parks, natural reserves etc, provide intact protection exclusively. The *ex-situ* conservation through the establishment of ethnobotanical gardens, kitchen gardens, drug farms etc are sustainable to the local conditions ranging from traditional agro-ecosystems to modern agro-forestry technology. Cultivation of medicinal plants is inversely linked to prevalence of easy and cheap collection from the wild, leak of regulation in trade, concerning of the profits from the wild collection by a vast network of traders and middleman and absence of industries interested in providing buy-back guarantees to growers. It is a best alternative conservational strategy of economically and widely grown species and also equally economic as compared to other crops. Likewise, tissue culture techniques are suitable for commercial protection and utilization of rare and vanishing species. Sustainable pre and post harvesting methods are absolutely essential to balance eco-existence of utilization and conservation.

## EMERGING IDEAS AND CONCLUSION

India is known for its valuable heritage of herbal resources. The ethnic people living in remote areas are still depending to great extent on biodiversity. The present study reveals a very rich tradition of utilization of plants by the natives. Though the study could not be done very extensively, yet the informations gathered are encouraging. The plants play a vital role in the society of tribal and rural people (Table 1-5). They are depending on bioresources for their basic needs. These people are very knowledgeable about the uses of plants in their day-to-day life. Meanwhile, they use the plants in a different, effective and novel manner. Scientific investigations on ethnic practices, especially the use of herbal medicines, would open new horizon of knowledge.

The informations gathered from the present study may be helpful in formulating conservational strategies for threatened species, which is vital for future ethnobotanical researches. This aspect of the investigation should also be given its due importance and the research work of ethnobotanically important plants must be continued so that these species can be conserved before they are lost for ever.

To boost the rural economy with flora, modern technological knowledge is prerequisite. Based on the above observations and discussions following suggestions can be made as recommendations :

1. A systematic survey and documentation of floristic diversity occurring in the division/ State.
2. Mapping of identified areas where identified plants are available.
3. Identification of plants used by the natives and their documentation for safeguarding the Intellectual Property Right (IPR).
4. Proper emphasis should be given to identify the threatened, endangered, vulnerable and rare flora. Consequently, conservational strategies like *ex-situ* and *in-situ* should be taken up with the development and utilization of modern technology and the participation of local people.
5. Initiation of steps for cultivation of economically important aromatic and medicinal plants which are locally available or by introduction of some selected ones having great demand with proper agro-technology should be taken, as modern pharmaceutical industries requires a large quantity of plant materials.
6. Imparting training in drying, packing, storage and analytical quality control. This may be helpful in the establishment of forest based minor scale cottage industries.
7. Essential training to develop manpower as skilled woodcrafters for collection of minor forest products from wild sources.
8. Providing infrastructural facilities for trade and marketing, which is very much lacking in the State.

9. The ethnobotanical knowledge and culture of ethnic people should be protected from modernization, urbanization and the careless use of bioresources.
10. Research and development work, especially in the field of clinical and pharmacological evaluation of traditional drugs should be encouraged by making available plant materials. Moreover, the marker compounds of the drug should be identified and patented towards eventual protection of IPR and marketing the drug.

## ACKNOWLEDGEMENTS

The author wishes to express his sincere gratitude to University Grants Commission for providing the financial assistance. Thanks are also due to the natives of the study area for lucid discussion and completing the questionnaire.

## REFERENCES

AIRI, S., R.S. RAWAL, U. DAR & A.N. PUROHIT 2000. Assessment of availability and habitat preference of Jatamasi – A critically endangered plant of West Himalaya. *Curr. Sci.* 7 (10) : 1467.

AHISH GHOSH 2003. Traditional vegetable dyes from Central West Bengal. *J. Econ. Taxon. Bot.* 27 (4) : 825.

BINU NAIYAR, T.S.R. & P. PUSHPANGHADAN 1992. An outline of ethnobotanical research in India. *J. Econ. Taxon. Bot.* (Addl. Ser.) 10 : 405-428.

BISWAS, S., S. S. JAIN & M. PAL 2003. Research needs and priorities for conservation of Indian medicinal flora. *Ind. For.* 129 (2) : 179.

CHAKRABORTY, M. K. & D. C. PAL 2004. Plant conservation through tribal deities in some sacred groves of Purnia district, West Bengal. *Ethnobotany* 16 : 99-102.

CHANDRA PRAKASH KALA 2002. Indigenous knowledge of Bhotiya tribe community on wood dyeing and its present status in the Garwal Himalaya. *Curr. Sci.* 83 (7) : 814.

DHAR, U. 2000. Conservation implication of plant endemism in high altitude Himalayas. *Curr. Sci.* 82 (2) : 141.

DOWDESWEL, ELIZABETH 1995. *Global Biodiversity Assessment.* UNEP, CUP, UK.

DWIVEDI, S.N. 1999. Traditional health care among the tribals of Rewa district of Madhya Pradesh. *J. Econ. Taxon. Bot.* 23 (2) : 315-320.

——————— 2003. Ethnobotanical studies and conservational strategies of wild and natural resources of Rewa district of Madhya Pradesh. *Ibid.* 27 (1) : 233-244.

——————— 2004. Herbal remedies among the tribals of Sidhi district of Madhya Pradesh. *Ibid.* 28 (3) : 675-687.

——————— & ARCHANA PANDEY 1992. Ethnobotanical studes on wild and indigenous species of Vindhyan Plateau. I. Herbaceous Flora. *J. Econ. Taxon. Bot* (Addl. Ser.) 10 : 143-150.

——————— & H. SINGH 1984. Ethnobotany of Kols of Rewa Division, Madhya Pradesh. *Proc. Nat. Sem. on Envt.* EPCO. II : 34-44.

———————, DWIVEDI, SANGEETA & P.C. PATEL 2005. Herbal remedies used by the rural people in the treatment of joint diseases.*Ethnobotany* (Silver Jubilee Volume) 17 (1 & 2) : 193-196.

———————, ——————— & ——————— 2006. Medicinal plants used by the tribal and rural people of Satna district, Madhya Pradesh for the treatment of gastrointestinal diseases and disorders. *Nat. Prod. Rad.* (In Press).

FORD, R. I. 1994. *The Nature and Status of Ethnobotany.* Anthrop. Pap. 67, 2nd ed Mus Anthrop Univ, Michigan Ann. Arbor.

GIVEN, A. T. 1996. *How plants become extinct? Principal and practices of plant conservation,* pp. 15-17. Timber Press, Portland, Oregon.

HARSHBERGER, J.W. 1895. Some New Ideas : The plants cultivated by aboriginal people and how used in primitive commerce. *Philadelphia Evening Telegraph* 64 (134) : 2.

JANKI AMMAL 1956. Introduction to the subsistence economy of India. In William, L.T. Jr (ed.) *Mans Role in Changing the Face of the Earth,* pp. 324-335. University of Chicago Press, Chicago.

JAIN, S. K. 1963. Studies in Indian Ethnobotany – Plants used in medicine by tribals of Madhya Pradesh. *Bull. Reg. Res. Lab. Jammu* 1 : 126-128.

——— 1989. *Method and Approaches in Ethnobotany.* Society of Ethnobotanists, Lucknow.

——— 2002. *Bibliography in indian Ethnobotany.* Scientific Publishers, Jodhpur.

——— & A.K. GOEL 1995. Proforma for field work. In Jain, S.K. (ed.) *A Manual of Ethnobotany,* pp. 142-159. Scientific Publishers, Jodhpur.

——— &. R.R. RAO 1977. *A Hand Book of Field and Herbarium.* Today & Tomorrow, New Delhi.

——— & S. SHRIVASTAVA 2001. Indian ethnobotanical literature in last two decades- A graphic review and further direction. *Ethnobotany* 13 (1) : 1-8

MAHESHWARI, J.K. 1970. New vistas in ethnobotany. *J. Econ. Taxon. Bot.* (Add. Ser.) : 1-11.

MAC NEELY, J., K.R. MILLER, W.V. REID, R.A. MITTERMEIR & T.B. WEMER 1990. *Conserving the words biological diversity.* Global Biodiversity, IUCN, Gland, Switzerland.

OOMMACHAN, M. & J.L. SHRIVASTAVA 1996. *Flora of Jabalpur.* Scientific Publishers, Jodhpur.

ROXBURGH, W. 1832. *Flora Indica.* Today & Tomorrow Publications, New Delhi.

SAMPTH KUMAR, V. & P.S.N. RAO 2003. Dye yielding plants of Andaman & Nicobar Islands. *J. Econ. Taxon. Bot.* 27 (4) : 827.

SINGH, J.S., A.S. RAGHUBANSHI & C.K. VARSHNEY 1994. Integrated biodiversity research in India. *Curr. Sci.* 66 (7) : 109.

SRINIVASAMURTHYY, T.S., K. MOHAN, V. PRABHAKARAN, S.N. JADHAV, E. SATISH, K. RAVIKUMAR & G. UTRASH 2003. Medicinal plant conservation and sustainable use through forest gene bank. *Indian For.* 129 (1) : 179.

UPADHYAYA, R. 2003. Ethnobotanical observation on some rare and endangered herbs of Singarauli in Madhya Pradesh. *J. Econ. Taxon. Bot.* 27 (4) : 989-992.

VED, D.K., C.L. PARTHIMA, NACY MORTON & S. DARSHAN 2001. Conservation of Indian medicinal plant diversity through a novel approach of establishing a network of *in situ* gene banks. In Uma Shanker, R., K.N. Ganeshaiah & K.S. Bawa (Eds.) *Forest Genetic Resources : Status, Threats and Conservation Strategies*, pp. 183. Oxford & IBH, New Delhi.

VERMA, D.M., P.C. PANT & M.I. HANFI 1985. *Flora of Raipur, Durg and Rajnandgaon.* Botanical Survey of India, Calcutta.

WATT, G. 1889-1896. *A Dictionary of the Economic Products of India.* Periodical Experts, Delhi (Reprinted 1972).

WAHAL DEEPAK 2005. Selected references of ethnobotany. *Ethnobotany* 17 (1 & 2) : 206-209.

# A PRELIMINARY STUDY OF WILD VEGETABLES OF BONGAIGAON DISTRICT OF ASSAM

ANANTA BORAH & PARTHA PRATIM SARKAR*

*Deptt. of Botany, Birjhora Mahavidyalaya, Bongaigaon - 783 380, Assam, India*
**Forest colony, Aie Valley Forest Div., Bongaigaon- 783 380, Assam, India*

## ABSTRACT

Bongaigaon district of Assam has been inhabited by different ethnic groups, viz. Koch-Rajbangshi, Bodo, Garo, Rabha etc, the traditional food of which is comprised of different wild plants and plant parts. Most of such plants are not documented and analyzed. Investigation has been carried out to know the species, parts and mode of use of wild vegetables in the Bongaigaon district and adjacent areas. The present preliminary report enumerates fifty six species of plants that have been collected from their wild habitat and are used as vegetable traditionally in different forms by different ethnic groups in the area. The botanical names along with vernacular ones, habit and habitat, use as vegetable, traditional medicinal utilization and seasonal variation of such plants are recorded. It has also been revealed that due to change in socio-economic condition, a lot of information regarding wild vegetables has been lost and some of such plants are on verge of rarity.

## INTRODUCTION

Growing interest in the search of new economically viable plants has influenced many investigators to devote attention to the study of wild edible plants. Knowledge of potential food plants and utilization of these in genetic improvement of food and medicinal plant resources as well as acceptance of new food plants is the prime importance of such type of study. Documentation of wild edible plants is an important step towards their conservation. Investigations have been carried out in different parts of India on wild edibles, i.e. Western Himalaya (Nagi & Gaur,1994), North-East hills (Deb, 1975; Biswas & Bhuyan, 1983; Arora, 1981), Western Ghats (Uthaiah, 1994), Maharashtra (Vartak, 1959), Jammu & Kashmir (Sharma & Gupta, 1994), Bihar (Sing & Sing, 1994) etc. Many floras also account for some information on uses of wild plants in different ways. Kanjilal *et al.* (1934-40) mentioned uses of some plants as wild edibles while describing plants in Flora of Assam. Still the use and knowledge of wild edibles is intact among different ethnic groups and the uses of wild plants as food may vary in different tribes and among same tribe inhabited in different areas. In Assam, despite of its rich biodiversity and ethnic inhabitants, the knowledge of wild vegetables is yet fragmentary. Some of the published works and unpublished Ph. D. thesis also dealt with about the wild vegetables of different parts of Assam (Sarmah, 1989; Gogoi, 1981; Boruah, 1992; Brahma, 1992; Sing, 1993; Sarkar, 1993, Chowdhury *et al.,* 2005), but the district Bongaigaon and its periphery remain uncovered except occasionally mentioned.

The study area, Bongaigaon district (90°21′33″ E to 90°54′32″ E and 26°41′56″ N to 26°10′27″ N) of Assam (89°40′44″ E to 95°53′03″ E and 28°01′54″ N to 24°04′21″ N) has been inhabited by different ethnic groups, viz. Koch- Rajbangshi, Bodo, Garo, Rabha etc. The traditional food of these people is comprised of different wild plants and plant parts. Floristically, the area is not fully explored and only a little has been known in this regard (Borah, 2003; Agarwala & Borah, 2001). The natives use to collect wild plants from their surroundings for various purposes and some of them sale such plants in the local market to earn for their daily requirements. Most of such plants are not biochemically analyzed and documented. Due to change in demographic profile and socio-economic condition of the area, some information regarding wild vegetables has been lost. In addition, contraction of forest areas as well as destruction of habitat due to natural and anthropogenic factors, some of such plants are on verge of rarity. Investigation has been carried out to know the flora and species, parts and mode of use of wild vegetables in the Bongaigaon district and adjacent areas since 2001. The aim of present paper is to document wild edible plants, especially as vegetable, by different ethnic groups in the Bongaigaon and adjacent areas.

## MATERIALS AND METHODS

Frequent field trips among the natives of different localities and market areas in the study area have been taken during different seasons. People from different communities, i.e. Koch-Rajbangshi, Bodo, Garo and Rabha and plant sellers in the markets were interacted. Plants used as vegetable by these people are not in practice of cultivation. The information gathered from the villagers was compared with existing ethnobotanical literature from that locality and enumerated accordingly. The habitat of the plants, season of use as vegetable and traditional medicinal uses are also mentioned. Families are arranged as per Bentham and Hooker's (1862-83) system of classification with slight modifications. The botanical names within each family are arranged alphabetically. Vernacular names followed by language within parentheses are given as far as possible. The plants which are not mentioned as wild vegetable in the chapter "Plants of economic importance" in the "Assam's Flora (present status of vascular plants)" by Chowdhury *et al.* (2005) are marked with an asterisk (*).

## ENUMERATION

*Abbreviaitions :* **Habit :** H = Herb, Us = Undershrub, Sh = Shrub, T = Tree, ST = Small tree, Cl = Climber, Rh. H = Rhizomatous herb, Co. H = Cormous herb, Tu. H = Tuberous herb, Tu. Cl = Tuberous climber; **Habitat :** Aq = Aquatic, Trs = Terrestrial, ML = Marshland; **Vernacular name :** As. = Assamese, Bo. = Bodo; **Season :** Spr.= Spring, Sum. = Summer, Aut. = Autumn, Win. = Winter.

| Family, Species & Vernacular name | Habit & Habitat | Parts & mode of use | Season of use |
|---|---|---|---|
| WOODSIACEAE | | | |
| *Diplezium esculentum* (Retz.) Sw.<br>Dhekiya (As.), Denkhiya (Bo.) | Fern<br>Trs | Tender circinate fronds are eaten cooked. | Spr.- early Aut. |
| DILLENIACEAE | | | |
| *Dillenia indica* L.<br>Ou-tenga (As.), Thaigir (Bo.) | T<br>Trs | Fleshy calyx is eaten cooked with pulses, pickles are also made. Ripe ones are considered as medicinal for jaundice patients. | Spr. |
| NYMPHAEACEAE | | | |
| *Euryale ferox* Salisb.<br>Mokhona (As.) | H<br>Aq | Fruits eaten fresh and in curry. | Sum.- Aut. |
| *Nymphaea nouchali* Burm. f.<br>Boga bhet (As.), Thoblo (Bo.) | H<br>Aq | Flowers and leaf petioles are eaten fried. Seeds are eaten fresh and dried. | -do- |
| *N. rubra* Roxb.ex Endrews*<br>Ronga bhet (As.) | H<br>Aq | -do- | -do- |
| NELUMBONACEAE | | | -do- |
| *Nelumbo nucifera* Gaertn.<br>Podum (As.), Thoblo (Bo.) | H<br>Aq | Flowers are cooked as vegetable. | |
| CARYOPHYLLACEAE | | | |
| *Drymaria diandra* Bl.*<br>Lai-jabori (As.), Jabsri (Bo.) | H<br>Trs | Tender leaves and shoots are used as vegetable. | Aut.- Spr. |
| *Stellaria media* (L.) Vill.<br>Morolia sak, Thun-thuni (As, Bo.) | H<br>Trs | -do- | -do- |
| PORTULACACEAE | | | |
| *Portulaca oleracea* L.<br>Malbhog-sak (As.), Hangsuramai (Bo.) | H<br>Trs | Tender shoots with leaves are eaten cooked. | -do- |
| OXALIDACEAE | | | |
| *Oxalis corniculata* L.<br>Tengesi sak (As.), Singri-gakhwi (Bo.) | H<br>Trs | Tender shoots and leaves are eaten cooked; considered as medicinal for dysentery and high blood pressure. | Throughout the year |

| | | | |
|---|---|---|---|
| *Oxalis debilis* H. B. & K. var. *corymbosa* (DC.) Lourt. Bor-tengesisak (As.) | H Trs | Same as above. | -do- |
| PAPILIONACEAE | | | |
| *Sesbania grandiflora* (L.) Poir.* Bok phul (As.) | Sh Trs | Flowers are eaten fried. | Aut.-Win. |
| TRAPACEAE | | | |
| *Trapa natans* L. var. *quadrispinosa* (Roxb.) Makino* Pani singori (As.), Singreb (Bo.) | H Aq | Leaves and fruits are used as vegetable; fruits are also taken raw. | Sum.- Aut. |
| CUCURBITACEAE | | | |
| *Coccinea grandis* (L.) Voigt | Cl Trs | Fruits are used as vegetable. | -do- |
| APIACEAE | | | |
| *Centella asiatica* (L.) Urban* Korola, Tatoka tita (As.) | H Trs | Leaves and shoots are used as vegetable; also as medicine for cut and wounds, stomachache etc. | Throughout the year |
| *Eryngium foetidum* L.* Kanta-dhaniya, Man dhaniya (As.), Gongar-dundia (Bo.) | H Trs | Leaves are used for flavouring curry, sometimes as vegetable, and as salads. | -do- |
| *Hydrocotyle sibthorpioides* Lamk.* Saru-manimuni (As.) | H Trs | Same as *Centella asiatica* | -do- |
| RUBIACEAE | | | |
| *Paederia foetida* L. Paduri-lewa, Bhedai lota (As.), Khipi bandung (Bo.) | Cl Trs | Leaves are taken as vegetable, as medicine for stomachache and as anti-gas; sometimes as vegetable cakes. | -do- |
| ASTERACEAE | | | |
| *Eclipta prostrata* (L.) L.* Keheraj, Alanchi (As.) | H Trs/ ML | Tender leafy shoots are used as vegetable. | -do- |
| *Enhydra fluctuans* Lour. Helanchi sak (As., Bo.) | H Trs/ ML | Tender leafy shoots are used as vegetable. | -do- |
| *Sonchus wightianus* DC.* Usumai-pagla (Bo.) | H Trs | Tender leafy shoots are used as vegetable by Bodo people. | -do- |
| *Spilanthes paniculata* Wall. ex DC.* Mahabhringraj (As.), Usumai(Bo.) | H Trs/ ML | Tender leafy shoots, sometimes with flowers, are eaten cooked as vegetable and as medicine for body ache, fever etc. Flowers are used as medicine for toothache and to treat mouth ulcer. | -do- |
| *Xanthium strumarium* L. Agora, Agra (As.) | H Trs | Seedlings are cooked mixing with other vegetables. | Spr. |
| HYDROPHYLLACEAE | | | |
| *Hydrolea zeylanica* (L.) Vahl* Leheti (As.) | H Trs/ ML | Tender leafy shoots are taken as vegetable. | Sum.- Aut. |

*Alternanthera philoxeroides*

*Amaranthus viridis*

*Amorphophalus paeoniifolius*

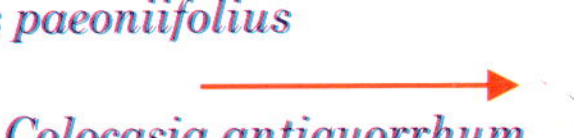

*Colocasia antiquorrhum*

Corm of *Amorphophalus campanulatus*

Different herbs in the market

**Plate 1**

*Diplezium esculentum*

Fruits of *Solanum myriacanthum*

*Houttuynia cordata*

*Ipomoea aquatica*

*Nymphaea nouchali* and *Ipomoea aquatica*

*Paederia foetida* and *Diplezium esculentum*

**Plate 2**

| | | | |
|---|---|---|---|
| CONVOLVULACEAE | | | |
| *Ipomoea aquatica* Forssk.<br>Kolmou (As.) | Cl<br>Aq | Tender leafy shoots and fruits are taken as vegetable. | -do- |
| SOLANACEAE | | | |
| *Solanum nigrum* L.<br>Koisi, Laskosi (As.) | H<br>Trs | Leaves in fewer amounts are mixed with other vegetables and taken cooked during *Sat Bihu.* | Spr. |
| *S. anguivi* Lamk.*<br>Bilkuli tita (As.) | Us<br>Trs | Fruits are eaten fried as vegetable and also as medicine against worm. | Win. |
| *S. viarum* Dunal<br>Bhekuri, Bilkuli tita (As.) | Us<br>Trs | Fruits are eaten fried as vegetable and also as medicine against worm. | Win. |
| BIGNONIACEAE | | | |
| *Oroxylum indicum* (L.) Vent.<br>Bhatghila, Dingdinga (As.)<br>Kharong khandai (Bo.) | T<br>Trs | Flowers are eaten cooked. | Spr- Sum. |
| ACANTHACEAE | | | |
| *Justicia adhatoda* L.*<br>Baska tita, Boga bahak (As.)<br>Barsikha (Bo.) | Sh<br>Trs | Flowers are eaten fried.<br>Leaves as medicine for cough and fever. | Through out the year |
| VERBENACEAE | | | |
| *Clerodendrum indicum* (L.) O. Ktze.*<br>Akalabrikha (As.), Akalbir (Bo.) | Sh<br>Trs | Flowers are taken as vegetable by Bodos. | Win. |
| *Pygmaeopremna herbacea* (Roxb.) Moldenke*<br>Matiphesua (As.) | Us<br>Trs | Leaves are eaten fried as vegetable; also used as medicine for jaundice. | Sum. |
| LAMIACEAE | | | |
| *Leucas aspera* (Willd.) Link*<br>Doron, Kansisa (As.) | H<br>Trs | Tender leafy shoots are eaten cooked; also medicinal for jaundice, sinusitis, indigestion, snake-bite etc. | Aut.- Win. |
| AMARANTHACEAE | | | |
| *Alternanthera philoxeroides* (Mart.) Griseb.<br>Menmeni (As.) | H<br>ML | Tender leafy shoots are eaten cooked. | Sum.- Aut. |
| *A. sessilis* (L.) R. Br. ex DC.<br>Matikaduri (As.) | H<br>Trs/ ML | Tender leafy shoots are eaten cooked. | -do- |
| *Amaranthus spinosus* L.<br>Kanta-khutura (As.) | H<br>Trs | Tender leafy shoots are eaten cooked. | Aut.- Spr. |
| *A. viridis* L.<br>Khuduna, Khutura sak As.) | H<br>Trs | Tender leafy shoots are eaten cooked. | -do- |
| *Celosia argentea* L.<br>Nakitenga (As.), Thathelpa maigong (Bo.) | H<br>Trs | Tender leafy shoots are eaten cooked. | -do- |

| | | | |
|---|---|---|---|
| *Deeringia amaranthoides* (Lamk.) Merr.<br>Methok-thoka (As.) | H<br>Trs | Tender leafy shoots are eaten cooked. | -do- |
| CHENOPODIACEAE | | | |
| *Chenopodium album* L.<br>Bathua, Jilmil sak (As.) | H<br>Trs | Tender leafy shoots are eaten cooked. Also cultivated. | -do- |
| *C. ambrosioides* L.<br>Bathua, Jilmil sak (As.) | H<br>Trs | Tender leafy shoots are eaten cooked. Also cultivated. | -do- |
| POLYGONACEAE | | | |
| *Rumex maritimus* L.*<br>Singri-gidirjat (Bo.) | H<br>Trs/ ML | Tender leafy shoots are eaten cooked. | -do- |
| EUPHORBIACEAE | | | |
| *Euphorbia hirta* L.<br>Samlaothi (Bo.) | H<br>Trs | Tender leafy shoots are eaten cooked. | Throughout the year |
| DIOSCOREACEAE | | | |
| *Dioscorea alata* L.<br>Kath alu (As.), Thaganda (Bo.) | Tu. Cl<br>Trs | Tubers are eaten cooked | Win. |
| *D. bulbifera* L.<br>Kath alu (As.), Thaganda (Bo.) | Tu. Cl<br>Trs | -do- | Win. |
| *D. esculenta* (Lour.) Burkill<br>Kath alu (As.), Thaganda (Bo.) | Tu. Cl<br>Trs | -do- | Win. |
| *D. pentaphylla* L.<br>Panch-patia alu (As.), Tha gusm (Bo.) | Tu. Cl<br>Trs | -do- | Win. |
| ZINGIBERACEAE | | | |
| *Alpinia nigra* (Gaertn.) Burtt.*<br>Tora (As.), Tharai (Bo.) | Rh. H<br>ML | Tender shoots are used as vegetable by Bodo and others. | Spr. |
| PONTEDERIACEAE | | | |
| *Monochoria hastata* (L.) Solms<br>Jonaki phul (As.), Ajnai (Bo.) | H<br>Aq/ ML | Flowers are eaten cooked. | Sum.- Aut. |
| *M. vaginalis* (Burm.f.) C. Presl<br>Jonaki phul (As.) | H<br>Aq/ ML | Flowers are eaten cooked. | -do- |
| ARACEAE | | | |
| *Alocasia macrorrhiza* (L.) G. Don<br>Man-kachu (As.), Man-thasu (Bo.) | Tu. H<br>Trs | Tender leaves, leaf petioles and rhizomes are eaten cooked with chili or lemon. | Aut., Spr. |
| *A. odora* (Roxb.) Koch & Bouche<br>Dudh kachu (As. & Bodo) | Tu. H<br>Trs | Leaf petioles and rhizomes are eaten cooked with chili or lemon. | -do- |
| *Amorphophallus paeoniifolius* (Dennst.) Nicolson.<br>Ol-kachu (As.) | Co. H<br>Trs | Rhizomes are eaten cooked often with chili or lemon. | -do- |

| | | | |
|---|---|---|---|
| *Colocasia esculenta* (L.) Schott* Kala kachu (As.), Thasu (Bo.) | Tu. H Trs | Tender leaves and rhizomes are eaten cooked with chili or lemon. | -do- |
| *Lasia spinosa* (L.) Thw.* Chengmora (As.), Sibru (Bo.) | H ML | Tender stem tips are used as vegetable. | Spr.- Sum. |
| *Typhonium trilobatum* (L.) Schott Sam-gach, Sam- kachu (As.) | Tu. H Trs | Leaves, petioles and inflorescences are eaten as vegetable. | Aut.- Spr. |

## ANALYSIS AND DISCUSSION

### Total Number of species recorded : 56

Pteridophytes: 01

Angiosperms: Dicotyledons: 42;

Monocotyledons: 13

### Utilization of plant parts as vegetable:

Leaves: 32 spp.; Stem and tubers : 07 spp.; Flowers and floral parts : 10 spp.; Leaf petioles : 05 spp.; Fruits : 05 spp.

The investigated region serves as the home for numerous food plants. Possibility is that some of them may have not been recorded in this preliminary study. Detailed study on such plants may result in more knowledge on wild edibles, some of which may be accepted as new potential food in future by other communities. Biochemical analysis of such plants may open a new door to the commercialization of such plants because they may be the source of many important minerals, organic compounds, alkaloids, vitamins etc. Several of wild edible plant species could also be investigated for other economic uses, as source of fibre, dyes, edible oil, medicines etc.

## ACKNOWLEDGEMENT

The authors are thankful to all the persons who cooperated in this study in various ways.

## REFERENCES

AGARWALA, N.K. & A. BORAH 2001. On the biodiversity of the Bhairab hills of Bongaigaon district of Assam, part I Flora. *Journal Economic Taxonomic Botany* 25 (2) :

ARORA, R. K. 1981. Native food plants of North Eastern Tribals. In Jain, S.K. (Ed.) *Glimpses of Indian Ethonobotany,* pp. 91-106. New Delhi.

BENTHAM, G. & J.D. HOOKER 1862-83. *Genera Plantarum,* Vols. 1-3. London.

BISWAS, S. & T.C. BHUYAN 1983. On the identity of some food plants of Garo hills, Meghalaya. *Indian J. Forest.* 6 : 208-213.

BORAH, A. 2003. Flora of Kakoijana (proposed) WildLife Sanctuary. *Biodiversity of Eastern Himalayan Protected Areas,* pp. 65-71. Dept. of Botany, Handique Girls' College, Guahati.

BORUAH, I.C. 1992. Systematic studies of the Angiosperms of Kamrup district, Assam, 3 Vols. Ph.D. Thesis, G.U.

BRAHMA, B.K. 1992. A study on the ethnobotany of Bodos of Kokrajhar district, Assam. Ph. D. Thesis, G.U.

CHOWDHURY, S., ASHIS KR. NATH, ANANTA BORAH, PARTHA PRATIM DAS & UTPAL PHUKAN 2005. *Assam's Flora (Present Status of Vascular Plants).* Assam Science, Technology and Environment Council, Silpukhuri, Guwahati, Assam.

DEB, D.B. 1975. Economic plants of Tripura State- edible plants: food grains, vegetables and fruits. *Indian Forester* 101:399.

GOGOI, P. 1981. *Flora of Golaghat and Neighbouring area,* 2 Vols. Ph.D. Thesis, G.U.

KANJILAL *et al.* 1934-40. *Flora of Assam,* Vol I-IV. Govt. of Assam. Calcutta.

NAGI, K. S. & R. D. GAUR 1994. Principal wild food plants of Western Himalaya, Uttar Pradesh, India. *Higher plants of Indian subcontinent.* 3: 1-78. Bishen Singh Mahendra Pal Singh, Dehra Dun (India).

SARKAR, S. 1993. Studies on the herbaceous plants of Karbi-Anglong district of Assam with reference to their taxonomy and economic utilization. Ph.D. Thesis, G.U.

SARMAH, J. N. 1989. *Flora of Sibsagar district, Assam.* Ph.D. Thesis, G.U.

SING, R. 1993. Systematic studies on the dicotyledonous plants of Lakhimpur district, Assam. Ph.D. Thesis, G.U

SHARMA, C.K. & B.K. GUPTA 1994. Note on some ethnobotanical plants of Billawar and Tehsils of Kathua district (J&K). *Higher plants of Indian subcontinent* 3 : 233-241. Bishen Singh Mahendra Pal Singh, Dehra Dun (India).

SING, L.B. & C.L. SING 1994. An ethnobotanical study of Deoghar district (Bihar). *Higher plants of Indian subcontinent* 3 : 273-279. Bishen Singh Mahendra Pal Singh, Dehra Dun (India).

UTHAIAH, B.C. 1994. Wild edible fruits of Western Ghats - a survey. *Higher plants of Indian subcontinent* 3 : 87-98. Bishen Singh Mahendra Pal Singh, Dehra Dun (India).

VARTAK, V.D. 1959. Some edible wild plants from hilly region of Pune district, Bombay State. *J. Bombay Nat. Hist. Soc.* 58 (1) : 7-25.